Intercalation in
Layered Materials

NATO ASI Series

Advanced Science Institutes Series

*A series presenting the results of activities sponsored by the NATO Science Committee,
which aims at the dissemination of advanced scientific and technological knowledge,
with a view to strengthening links between scientific communities.*

The series is published by an international board of publishers in conjunction with the
NATO Scientific Affairs Division

A	**Life Sciences**	Plenum Publishing Corporation
B	**Physics**	New York and London
C	**Mathematical and Physical Sciences**	D. Reidel Publishing Company Dordrecht, Boston, and Lancaster
D	**Behavioral and Social Sciences**	Martinus Nijhoff Publishers
E	**Engineering and Materials Sciences**	The Hague, Boston, Dordrecht, and Lancaster
F	**Computer and Systems Sciences**	Springer-Verlag
G	**Ecological Sciences**	Berlin, Heidelberg, New York, London,
H	**Cell Biology**	Paris, and Tokyo

Recent Volumes in this Series

Series B: Physics

Intercalation in Layered Materials

Edited by

M. S. Dresselhaus

Massachusetts Institute of Technology
Cambridge, Massachusetts

Springer Science+Business Media, LLC

Proceedings of the 10th Course of the Erice Summer School on
Intercalation in Layered Materials,
held July 5–15, 1986,
in Erice, Trapani, Sicily, Italy

Library of Congress Cataloging in Publication Data

Intercalation in layered materials.

(NATO ASI series. Series B, Physics; v. 148)
"Published in cooperation with NATO Scientific Affairs Division."
Proceedings of the 10th course of the Erice summer school, International
School of Materials Science and Technology, Erice, Sicily, July 5–15, 1986.
Bibliography: p.
Includes index.
1. Clathrate compounds—Congresses. 2. Layer structure (Solids)—Congress-
es. I. Dresselhaus, M. S. II. North Atlantic Treaty Organization. Scientific Affairs
Division. III. International School of Materials Science and Technology (1986:
Erice, Sicily) IV. Series.
QD474.I564   1986   530.4'1   86-30680
ISBN 978-1-4757-5558-9   ISBN 978-1-4757-5556-5 (eBook)
DOI 10.1007/978-1-4757-5556-5

© 1986 Springer Science+Business Media New York
Originally published by Plenum Press, New York in 1986

International School of Materials Science and Technology
INTERCALATION IN LAYERED MATERIALS

Summer School at Erice
10th Course
Erice, Trapani, Sicily, July 5-15, 1986

Sponsored by
NATO
US Army European Research Office
Italian Ministry of Education
Italian Ministry of Science and Technological Research
Sicilian Regional Government
US National Science Foundation
European Physical Society

Principal School Support:
NATO
European Physical Society
Provincial Government of Sicily
US Army Research Office
US National Science Foundation

Industrial Support:
General Motors Research Laboratory
Intercal

PREFACE

This volume is prepared from lecture notes for the course "Intercalation in Layered Materials" which was held at the Ettore Majorana Centre for Scientific Culture at Erice, Sicily in July, 1986, as part of the International School of Materials Science and Technology. The course itself consisted of formal tutorial lectures, workshops, and informal discussions. Lecture notes were prepared for the formal lectures, and short summaries of many of the workshop presentations were prepared. This volume is based on these lecture notes and research summaries. The material is addressed to advanced graduate students and postdoctoral researchers and assumes a background in basic solid state physics.

The goals of this volume on Intercalation in Layered Materials include an introduction to the field for potential new participants, an in–depth and broad exposure for students and young investigators already working in the field, a basis for cross–fertilization between workers on various layered host materials and with various intercalants, and an elaboration of the complementarity of intercalated layered materials with deliberately structured superlattices.

The physics of intercalation into layered materials is presented in this volume in a series of chapters by workers active in this field. Emphasis is given to the complementarity of intercalated materials and deliberately structured superlattices formed using molecular beam epitaxy, organometallic chemical vapor deposition, sputtering and other techniques. Common aspects of intercalation into graphite, layered chalcogenides, clays, and layered ionic hosts are found throughout the volume. An in–depth study of the structure and electronic, lattice, magnetic and superconducting properties of graphite intercalation compounds is presented with particular emphasis given to the unique aspects associated with staging. The specific topics included synthesis; kinetics; staging; electronic, phonon and optical properties; charge transfer; transport properties; structures and phase transitions; Raman and infrared spectroscopy; magnetic resonance; magnetism; superconductivity and graphite fibers. The chapters on the transition metal dichalcogenides and MPS_3 layered materials emphasize and contrast many of the same structure–property relations as are considered in the chapters on graphite intercalation compounds, though other phenomena such as charge density waves are also highlighted. In a similar vein, the presentation on the similarities and contrasting properties of GICs and CICs (clay intercalation compounds) is of particular interest as are also the briefer reports on the intercalated oxides and InSe layered materials. By juxtaposing tutorial material with short research papers on intercalated graphite, transition metal dichalcogenides, MPS_3 compounds, layered oxides, clays and other hosts, some degree of cross–fertilization is achieved between these related fields.

From an intercomparison between intercalation in various host materials, an interesting complementarity emerges. The transition metal dichalcogenides are remarkable in the variety of metal species and polytype structures that can be synthesized as host materials and subsequently intercalated. The graphite intercalation compounds are remarkable for their atomically abrupt interfaces and high degree of in–plane structural order and c–axis staging order. The deliberately structured superlattices are remarkable for their flexibility in terms of chemical constituents and periodicity of the superlattice. Thus study of specific physical problems (such as low dimensional magnetism or super-conductivity) benefits from the availability of complementary physical systems enabling the exploration of different physical limits in each of these systems.

Many new, important and stimulating physical phenomena were presented for the first time in the workshops at Erice, some of importance to low dimensional physics, some bearing on localization in two dimensions, some of significance to an illucidation of structure–property relations, and some of relevance to possible future applications of these materials. In the course of the lectures, workshops and discussions, many of the fundamental and unanswered problems in intercalated layered materials were identified for and by the participants with many suggestions given for Ph.D. thesis topics. It is likely that the Course on Intercalation in Layered Materials will have a significant impact on future work in this field. It is hoped that the preparation of a book containing pedagogic and review articles on the major topics covered at the School will extend its impact to a more general audience in Materials Research.

The environment and facilities in Erice, Sicily were ideal for stimulating interactions between the participants from 10 countries in a relaxed, enjoyable setting. For most of the participants, the School provided a unique opportunity for professional and personal growth. The living accommodations and meals were superb, as was the dedication of the support staff and the quality of the various social and cultural events, planned and sponsored by Ettore Majorana Center for Scientific Culture in Erice. The unique, historical heritage of Erice and its surroundings added to the richness of the experience by the participants.

M.S. Dresselhaus
September 15, 1986

ACKNOWLEDGMENTS

It is a pleasure to acknowledge with gratitude the financial support for the Course on Intercalation in Layered Materials by NATO, the European Physical Society, the U.S. Army Office of Research, and the Ettore Majorana Centre for Scientific Culture in Erice.

We owe special thanks to the director of the International School of Materials Science and Technology, Professor M. Balkanski who worked tirelessly in organizing the course. Special thanks are also extended to the directors, Dr. A. Zichichi and Dr. A. Gabrieli, and the staff at the Ettore Majorana Centre for Scientific Culture at Erice who worked many long hours to make the course a success.

We also wish to thank our colleagues in the field for many helpful suggestions on the organization of the Course. In addition, we would like to thank all the lecturers, particularly to my co–director Professor J–P. Issi, the workshop chairmen and the participants for making the program so stimulating and successful. Special thanks are extended by all to Dr. G. Dresselhaus who shared equally with M.S. Dresselhaus in the organization of the course and in the preparation of this volume. Finally we wish to thank Ms. P. Cormier and Mr. Eliot Dresselhaus for help with the production of this volume.

CONTENTS

PART I SUPERLATTICES

PART II INTERCALATION COMPOUNDS IN TRANSITION METAL DICHALCOGENIDES, CLAYS AND OTHER HOSTS

PART III   GRAPHITE INTERCALATION COMPOUNDS

SUPERLATTICES AND INTERCALATION COMPOUNDS

M.S. Dresselhaus

Massachusetts Institute of Technology
Cambridge, MA 02139, USA

INTRODUCTION

A crystallographic superlattice denotes a structure with additional periodicity arising from a large unit cell containing multiples of the primitive unit cells of the constituents of the superlattice. Typically, the periodicity is smaller than the electron mean free path, so that the various layers of the superlattice are electronically coupled to each other. In this volume we will be concerned with superlattices associated with layered materials where the superlattice periodicity occurs in one–dimension (1D) perpendicular to the layer planes (the z–direction). In many interesting cases, electronic transport is largely confined to the basal planes (x–y planes) normal to the z–direction, giving rise to two–dimensional behavior, as for example the 2D electron gas found in semiconductor heterostructure quantum wells. Furthermore, for commensurate intercalation compounds described below, 2D superlattices are frequently found in the layer planes, in addition to the 1D superlattices in the z–direction. By patterning in the layer planes, 1D superlattices in the xy plane can also be observed in semiconductors heterostructure superlattices.

Two important and complementary classes of layered superlattice materials are intercalation compounds and deliberately structured materials (not found in nature). We first present a discussion of complementary features of graphite intercalation compounds (GICs) and deliberately structured materials. This is followed by a more detailed summary of the structure and properties of the deliberately structured materials. Deliberately structured superlattices are typically prepared by molecular beam epitaxy (MBE) or by metal organic chemical vapor deposition (MOCVD), through a controlled layer by layer deposition of the constituent species under high vacuum conditions and computer control.[1]

A schematic representation of a deliberately structured heterostructure superlattice is shown in Fig. 1 where d is the superlattice periodicity, comprising a distance d_1, of material M_1, and d_2 of material M_2. Deliberately structured superlattices can be formed with large variety of choices for materials M_1 and M_2, including:

1. Two semiconductors with band offsets of the same sign (Type I–GaAs/Al$_x$Ga$_{1-x}$As) and of different sign (Type II–InAs/GaSb);

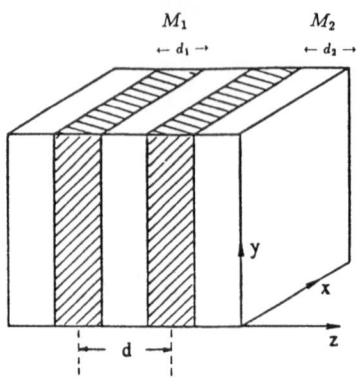

Figure 1: A heterojunction superlattice of periodicity d. Each superlattice unit cell consists of a thickness d_1 of material M_1 and d_2 of material M_2.

2. The same semiconductor (e.g., GaAs) with alternating n- and p-type regions ("nipi" superlattice);

3. Metal/semiconductor or magnetic metal/semiconductor or superconductor/semiconductor sequences;

4. Two different metals or superconductor/normal metal or magnetic/non–magnetic sequences of materials;

5. Two different amorphous materials.

In the typical semiconductor superlattice sample (thickness $\sim 1\mu$m), the periodic unit of distance $d = d_1 + d_2$ is repeated many times (e.g., 100 times). Each period typically varies between a few layers and many layers (10Å to 500Å). Superlattices of all types are today the subject of an extremely active international research field, both from the point of view of fundamental physics and for practical applications.

INTERCALATION COMPOUNDS AS SUPERLATTICES

Intercalation compounds are formed by the insertion of atomic or molecular layers of a guest chemical species between layers in a host material. This is illustrated in Fig. 2 for potassium–intercalated graphite. The intercalation process occurs in highly anisotropic layered structures where the *intraplanar* binding forces are large in comparison with the *interplanar* binding forces. Thus graphite intercalation compounds (GICs) form a limiting case of superlattices where one of the constituents of Fig. 2 is a single molecular or atomic layer with an atomically abrupt interface to the other constituent. Examples of host materials for intercalation compounds are graphite, transition metal dichalcogenides, some silicates and metal chlorides, clays, polymers and gels. Intercalation provides a means for controlled variation of many physical properties of the host material over wide ranges. Of the various types of intercalation compounds, the graphite compounds are of particular physical interest because of their high degree of structural ordering and potential for practical applications. At present, several hundred different chemical species are known to intercalate into graphite. By variation of the intercalant species and concentration, a large number of compounds with different properties can be prepared.[2,3,4]

Graphite intercalation compounds exhibit a high degree of ordering.[5] The most striking type of ordering is the staging phenomenon, which is defined by a periodic arrangement of n graphite layers between sequential intercalate layers, where n is the stage index (see Fig. 3). In practice it is possible to prepare single–staged materials with only small (1 to 5%) admixtures of secondary–staged regions. A number of other host materials also exhibit the staging phenomenon, but usually to a lesser degree, consistent with their lower bonding anisotropy.[6,7]

The simplest method for stage characterization of GICs is x–ray diffraction based on (00ℓ) reflections (see Fig. 4). Such measurements show that well–staged materials can be prepared up to stage $n \sim 10$, indicating a repeat distance I_c between consecutive intercalate layers of $I_c \sim 40$ Å. The physical basis for the staging phenomenon is the strong interatomic intercalant–intercalant binding relative to the intercalant–graphite binding, thereby favoring a close–packed in–plane intercalant arrangement. The introduction of each intercalate layer adds a substantial strain energy as the crystal expands to accommodate the intercalate layer, thereby favoring the insertion of a minimum number of intercalate layers, consistent with a given average intercalate concentration. Thus for a given intercalate concentration, the minimal energy state corresponds to a close packed in–plane intercalate arrangement with the largest possible separation between a minimum number of intercalant layers. This combination of conditions results in the staging phenomenon and superlattice formation.

Analysis of (00ℓ) x–ray diffractograms such as in Fig. 4 by means of Bragg's law

$$\ell\lambda = 2I_c\sin\theta_\ell \qquad (1)$$

shows that the c–axis repeat distance I_c for GICs obeys the simplest possible stage dependence

$$I_c = d_s + (n-1)c_o \qquad (2)$$

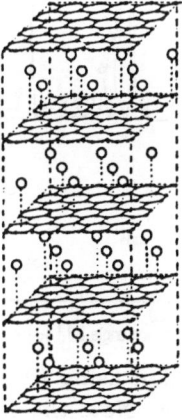

Figure 2: Model for C_8K showing the stacking of graphite layers and of potassium layers (networks of spheres). The graphitic and intercalate layers are arranged in an A α A β A γ A δ stacking sequence, where A refers to the graphitic layers and the Greek letters to the intercalate layers.[2]

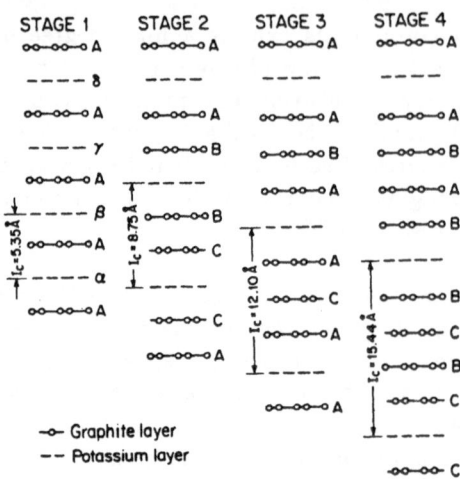

Figure 3: Schematic diagram illustrating the staging phenomenon in K–GICs for stages $1 \geq n \geq 4$. The potassium layers are indicated by dashed lines and the graphite layers by solid lines connecting open circles, and indicating schematically a projection of the carbon atom positions. The ABAB graphite layer stacking for stages $n \geq 2$ is maintained between intercalate layers. However, a rhombohedral stacking arrangement appears across intercalate layers. For each stage, the distance I_c between adjacent intercalate layers is indicated. For first stage C_8K, the unit cell includes intercalate layers with the stacking sequence α, β, γ, δ (see Fig. 2).[2,5]

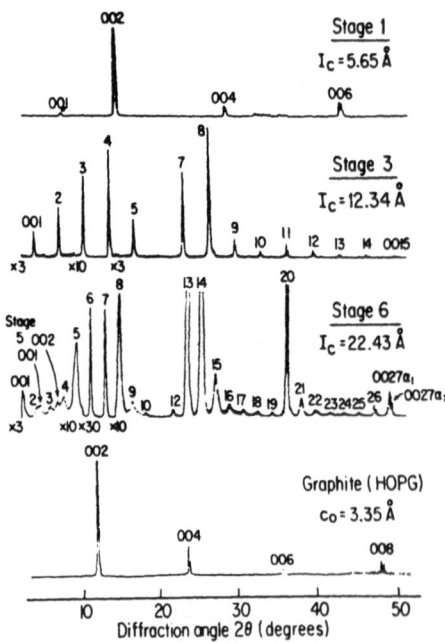

Figure 4: X-ray stage characterization using (00ℓ) diffractograms for stages 1, 3 and 6 Rb–GICs and for pristine graphite. Note the correlation in the diffraction angle 2θ of the high intensity reflections for the high stage intercalation compounds with the occurrence of (00ℓ) reflections in graphite.[8] Since many GICs are unstable in air, the x-ray diffractograms are often taken through protective glass ampoules using Mo Kα radiation.

Table 1: Intercalate sandwich thickness d_s for several common GICs. The repeat distance I_c can be calculated from Eq. 2 using the data in this table.

Donor d_s(Å)	Li	K	Rb	Cs	Ba	Eu	KHg	CsBi	$KH_{0.8}$
	3.706	5.35	5.65	5.94	5.28	4.87	10.22	11.48	12.08
Acceptor d_s(Å)	Br_2	AsF_5	SbF_5	$AlCl_3$	$SbCl_5$	HNO_3	$FeCl_3$	$CoCl_2$	H_2SO_4
	7.04	8.15	8.46	9.54	9.42	7.84	9.37	9.50	9.86

where d_s is the distance between two graphitic layers between which the intercalant layer is sandwiched, n is the stage index and c_o (3.35 Å) is the interlayer separation in pristine graphite. The distance d_s is usually observed to be stage independent on the scale of ~ 0.01 Å, though several examples of a larger stage dependence of d_s have been reported, particularly in the ternary GICs. The relative intensities of the peaks provide information on the in–plane intercalant density.[8] In the case of multiple layer intercalants, the peak intensities also provide information on the layer separation within the intercalate sandwich and between the intercalant and adjacent graphitic layers. For example, the air stable $SbCl_5$–GICs contain a three layer intercalate sandwich structure consisting of an Sb layer surrounded by two Cl layers which interface with the graphite bounding layers. A table of values for d_s for a number of intercalants is given in Table 1, from which we can in general obtain the c–axis repeat distance I_c using Eq. (2).

The analysis of the integrated intensities of the (00ℓ) lines is related to the structure factor $F_{00\ell}$ by $I_{00\ell} \propto |F_{00\ell}|^2$ where the structure factor is given by

$$F_{00\ell} = \sum_j f_j n_j \exp(2\pi i \ell z_j) \tag{3}$$

in which n_j is the number of atoms per unit area in layer j, and z_j denotes the zth coordinate of the jth layer, with and the atomic scattering of the jth atom given by f_j. For a simple atomic intercalate, the structure factor is given by

$$|F_{00\ell}| = \left| \xi f_C \frac{\sin\pi\epsilon\ell}{\sin\pi\epsilon} + (-1)^\ell \sum_X f_X n_X \exp(-2\pi i \epsilon_X \ell) \right| \tag{4}$$

where f_C and f_X are atomic scattering factors for carbon on graphitic layers and the intercalant on layer X, while $\epsilon = c_0/I_c$, and $\epsilon_X = z_X/I_c$, where z_X is the distance of an intercalate layer relative to the center of the intercalate sandwich, and $1/\xi$ is the intercalate density.

This formula assumes perfect layer periodicity. Once a model is introduced to relate the phases to the structure factors, the scattering density projection onto the c–axis is obtained from the Fourier summation

$$\rho(z) = \frac{1}{c} \sum_\ell F_{00\ell} \exp(-2\pi i \ell z). \tag{5}$$

The distances d_1 and d_2 for the deliberately structured superlattices can be obtained from (00ℓ) x-ray diffractograms by a similar analysis.[9]

5

STRAINED LAYER SUPERLATTICES AND COMMENSURATE STRUCTURES

In synthesizing semiconductor superlattices, a great deal of attention has been devoted to minimizing lattice strain by lattice matching of the two semiconductors (i.e., by choosing materials with in–plane unit cells of the same size, $\Delta a/a < 0.1\%$) and by choosing materials M_1 and M_2 with similar thermal expansion coefficients, so that lattice strains would not be introduced upon cooling the samples from the temperature of fabrication to the temperature of use. Lattice matching minimizes "misfit dislocations" and other imperfections, and also reduces surface states and charge accumulation at the interfaces which adversely affect transport properties and device performance. For this reason, a great deal of work has been done using the $GaAs/Al_xGa_{1-x}As$ system which is lattice matched for a wide range of x values. When the two constituents are properly lattice matched, atomically sharp interfaces can be achieved as shown in Fig. 5a) for the $Si/NiSi_2$ interface using high resolution transmission electron microscopy.[10]

However, the restriction of lattice matching severely restricts the number of different semiconductors that can be used for heterojunctions, quantum well structures and superlattices. One method of circumventing this restriction is the introduction of quarternary compounds such as $Ga_xIn_{1-x}As_yP_{1-y}$ and where x and y are appropriately chosen to achieve lattice matching and control of band gap. While quarternary compounds provide the desired flexibility in adjusting the lattice constant and band gap, their use introduces considerable complications into synthesis.

An alternate approach to the lattice matching problem is provided by strained layer superlattices (see Fig. 6) which exploit the concept that if the semiconducting layers are sufficiently thin, the mismatch will be taken up by both of the materials provided that the strain energy per unit area is less than that needed to form dislocations.

For strained layer superlattices that are thin, the lattice mismatch $\Delta a/a$ can be several percent (e.g., $\Delta a/a \sim 1.8\%$ for $GaAs/GaAs_{0.5}P_{0.5}$) if the layer thicknesses are on the order of 250Å or less. The fundamental idea is that the epitaxial growth of a very thin layer onto a lattice mismatched substrate introduces strains which force the lattice constant of the epitaxial layer parallel to the interface to conform to that of the substrate.[11] Thus the superlattice will consist of alternating materials successively in tension and compression as shown in Fig. 6. The superlattice is grown on a lattice matched graded $Al_xGaAs_yP_{1-y}$ substrate, where the composition of the surface layer is chosen to have an in–plane lattice constant equal to the average lattice constant of the mismatched layers of the superlattice. Strained layer superlattices allow growth of high quality superlattices from lattice mismatched semiconductors, allowing a great deal of flexibility in choice of semiconductors, and therefore in tailoring electrical and optical properties. In practice, strained layer superlattices have been prepared with no misfit dislocations at the interface, using for example $GaAs_yP_{1-y}/GaP$ for the superlattice materials. The graded layer is formed by deposition of $GaAs_xP_{1-x}$ on the GaP substrate, where x is varied in the graded layer in accordance with the above discussion. The tension and compression in the superlattice causes the lattice constants $a_\perp^{(i)}$ normal to the layers to be less than the average lattice constant ($a_\perp^{(i)} < a_0^{(s)}$) for semiconductor (1) and greater than average ($a_\perp^{(i)} > a_0^{(s)}$) for semiconductor (2). The in–plane lattice matching requires $a_0^{(s)} = a_\parallel^{(1)} = a_\parallel^{(2)}$, where the $\parallel$ subscript denotes the in–plane lattice

6

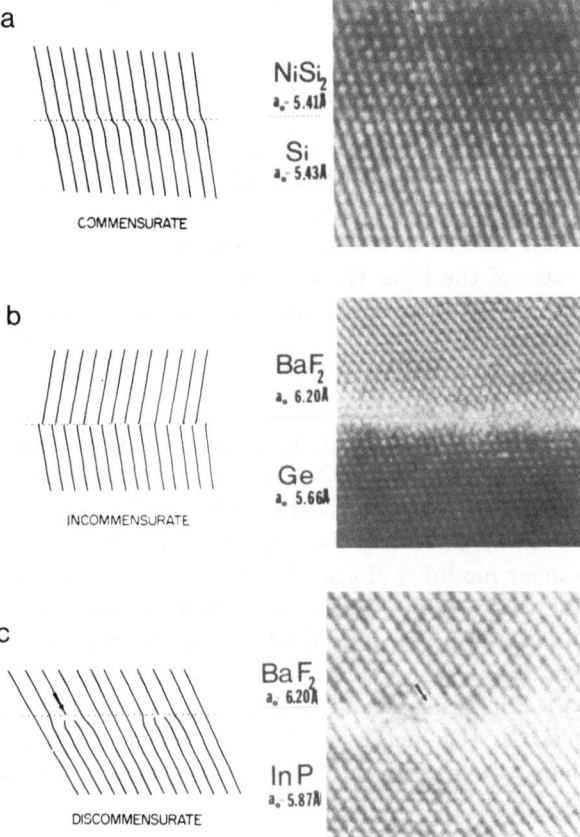

Figure 5: Several types of "misfit" dislocations at a heterojunction interface: (a) a commensurate interface with no dislocations when $a_0^{(1)} \approx a_0^{(2)}$. (b) An incommensurate interface where the lattice constants are so different that no interface accommodation occurs. (c) A discommensurate interface where the lattice mismatch is a few % and accommodation occurs, giving rise to misfit dislocations.[10]

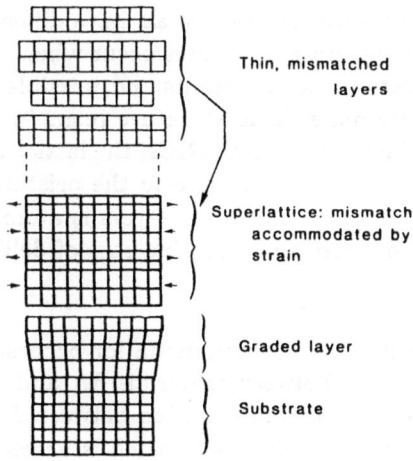

Figure 6: A schematic representation of a strained layer superlattice.[11]

constant, and $a_0^{(s)}$, the in–plane lattice constant of the strained layers, is given by

$$a_0^{(s)} = a_1\left[1 + \frac{fG_2d_2}{G_1d_1 + G_2d_2}\right] = a_2\left[1 - \frac{fG_1d_1}{G_1d_1 + G_2d_2}\right]. \qquad (6)$$

Here a_1 and a_2 are the unstrained lattice constants of semiconductors 1 and 2, G_i and d_i are the shear moduli and layer thicknesses of semiconductors i ($i = 1,2$) and f is the lattice mismatch ($\Delta a/a$) of the unstrained semiconductors. Thus the ratio of the components of the strain parallel to the interfaces for the two types of layers is inversely proportional to the ratio of the layer thicknesses, so that the thinner of the two types of layers experiences the greater layer strain. The lattice constant $a_0^{(s)}$ is measured by x–ray diffraction techniques.[11,12]

Strained layer superlattices also can be found in metallic superlattices where the two metals have the same crystal structure and similar lattices constants, such as BCC Nb ($a_1 = 3.3003$ Å) alternating with BCC Ta ($a_2 = 3.3024$ Å).[13] Furthermore, commensurate GICs,[5] where the intercalate layers are in registry with the adjacent graphite bounding layers (see Fig. 3), provide a limiting case for the strained layer superlattice. Because of the large shear modulus of graphite relative to the intercalant, $a_0^{(s)}$ is almost entirely determined by the in–plane lattice constant of graphite $a_{gr}^0 = 2.46$ Å, while the lattice constant of the intercalant will be modified to accommodate the commensurate structure of the intercalation compound. In addition to lattice strain considerations, charge transfer plays a dominant role in determining the in–plane lattice constants of GICs.[5,14] For donor GICs, the increased electron carrier concentration results in an increase in the in–plane lattice constant, while the opposite is found for acceptor compounds.

For small thicknesses d_1 and d_2, relatively large lattice mismatches can be accommodated in the strained layer superlattice. When the mismatch or layer thicknesses become too large to support lattice accommodation, incommensurate structures with a high density of dislocations are formed, as shown in Fig. 5(b). For such interfaces, a significant amount of interdiffusion occurs, so that the interfaces become smeared out over several atomic layers in the z–direction. Such incommensurate structures commonly occur in metal superlattices where the two metals have the same crystal structure but a large lattice mismatch, or where the two metals have different crystal structures.[17,18] In the case of GICs, incommensurate structures occur when there is a large lattice mismatch between the intercalate lattice constants and possible commensurate structures compatible with the graphite honeycomb structure (e.g., $(\sqrt{3} \times \sqrt{3})R30°$, $(2 \times 2)R0°$, $(\sqrt{7} \times \sqrt{7})R19.1°$ etc). For incommensurate GICs, the lattice constant of the intercalant usually remains essentially unchanged relative to the pristine parent material and the interface between the graphite and intercalant layers is atomically abrupt. Thus intercalation compounds form an interesting, limiting case of deliberately structured superlattices.

For intermediate cases where some accommodation occurs, a discommensurate structure is achieved at the interface between materials M_1 and M_2 as shown in Fig. 5(c). For discommensurate structures, accommodation between M_1 and M_2 is achieved over a commensurate domain, and the lattice mismatch is accommodated by the insertion of an extra lattice plane of the small lattice constant material in the domain walls. As the commensurate domains become smaller, larger lattice mismatches can be accommodated. When the insertion of the extra lattice planes is periodic, a striped domain phase is achieved.[15] Discommensuration phenomena are frequently observed in alkali metal GICs.[5] For Br_2–GICs, the domain walls are periodic in the 7–fold direction of the intercalate unit cell, forming a striped domain phase above the commensurate-incommensurate transition temperature ($T_c = 342.2$ K for a stage 4 Br_2–GIC).[15]

Accommodation at the interface between materials M_1 and M_2 can give rise to novel metastable structures. For example, the lattice constants and crystal structure of the intercalate layer in an intercalation compound can be different from the pristine bulk material, especially for commensurate GICs. An example of a metastable structure in a deliberately structured metallic multilayer, is the Nb/Zr superlattice where it is possible to grow a BCC phase of Zr on a BCC Nb substrate,[16] in contrast to the hexagonal structure found in bulk crystalline Zr. The study of novel metastable metallic phases is one exciting research area that will be greatly enriched by the wider availability of metallic superlattices and multilayers. One important complication of the metallic superlattices has been the lack of a sharp interface between M_1 and M_2 (see Fig. 1), due to extensive interdiffusion and chemical reaction in the solid state.[17,18] Sample preparation at lower temperature and under higher vacuum conditions (MBE) significantly decreases this problem.[19]

BOUND ELECTRONIC STATES IN SEMICONDUCTOR SUPERLATTICES

Since the intercalate layer (or sandwich) in a GIC is a single atomic or molecular layer with an abrupt interface to the layers of the host material, the electronic structure for a GIC is a limiting case of the electronic structure of a deliberately structured superlattice. In particular, for a semiconductor heterojunction superlattice with sufficiently large thicknesses d_1 and d_2, the electronic structure near the Fermi level will be dominated by bound states in the quantum well. But as the thickness of the quantum well decreases, the number of bound states decreases until a critical thickness is reached, below which no bound states can form. The electronic structure is then dominated by nearly free electron states in zone folded energy bands, similar to those characteristic of GICs.

Because of the different band gaps in the two semiconductors of a deliberately structured semiconductor heterojunction superlattice, potential wells and barriers are formed, with barrier heights or band offsets in the conduction and valence bands given by $\Delta \mathcal{E}_c$ and $\Delta \mathcal{E}_v$, respectively (see Fig. 7). In principle, these band offsets are determined by matching the Fermi levels for the two semiconductors. In most of the early work, the band offsets were approximated by empirical rules, and it is only recently that sophisticated calculations of band offsets are becoming available.[20]

From the diagram in Fig. 7, we see that the type I semiconductor heterojunction superlattice consists of an array of potential wells. In the limit where the width of the potential well contains only a small number of crystallographic unit cells ($L_z < 100$ Å), the number of bound states associated with the well is a small number. The simplest case to consider is an infinitely deep rectangular potential well. In this case, a particle of mass m^* in a well of width L_z in the z direction satisfies the free particle Schrödinger equation

$$-\frac{\hbar^2}{2m^*}\frac{d^2\psi}{dz^2} = \mathcal{E}\,\psi \tag{7}$$

with eigenvalues

$$\mathcal{E}_n = \frac{\hbar^2}{2m^*}\left(\frac{n\pi}{L_z}\right)^2 = \left(\frac{\hbar^2\pi^2}{2m^*L_z^2}\right)n^2 \tag{8}$$

and eigenfunctions

$$\psi_n = A\sin(n\pi z/L_z) \tag{9}$$

where n = 1, 2, 3... The phase of the wave functions of Eq. (7) is arbitrary, except that the wave functions must vanish at the walls of the quantum wells ($z = 0$ and $z = L_z$).

Although Schrödinger's equation (Eq. 7) is satisfied by both sine and cosine functions, it is only the sine functions with arguments ($n\pi z/L_z$) that vanish at the two sides of the infinite potential well. We note that the energy levels are not equally spaced, but have energies $\mathcal{E}_n \sim n^2$, though the spacings $\mathcal{E}_{n+1} - \mathcal{E}_n$ are proportional to $(2n + 1)$. We also note that $\mathcal{E}_n \sim L_z^{-2}$, so that as L_z becomes large, the levels become very closely spaced as expected. However when L_z decreases, the number of states in the quantum well decreases, so that for a well depth E_d (the band offsets of Fig. 7), the critical width L_z^c below which there will be no bound states is

$$L_z^c = \frac{\hbar\pi}{(2m^*E_d)^{\frac{1}{2}}}. \tag{10}$$

An estimate for L_z^c found by taking $m^* = 0.1m_0$ and $E_d = 0.1$ eV is $L_z^c = 61$ Å. For $L_z < L_z^c$, there will be no bound state. The closer level spacing of the valence

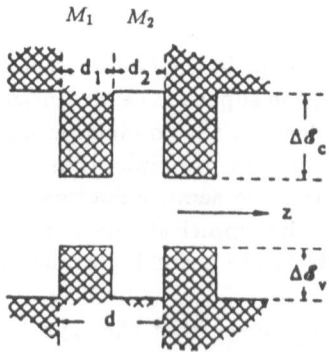

Figure 7: A schematic diagram of the superlattice heterojunction energy band gaps. Each superlattice unit cell consists of a thickness d_1 of material M_1 and d_2 of material M_2. Because of the different band gaps, a periodic array of potential wells and potential barriers are formed.

band bound states in Fig. 8 reflects the heavier masses in the valence band. Since the states in the potential well are quantized, the structures in Figs. 7 are called quantum well structures. If the potential energy of the well V_0 is not infinite but finite, the wave functions within the well are similar to those given in Eq. 9, but will have decaying exponentials on either side of the potential well walls.

When the electron has an energy greater than the band offset, its eigenfunction corresponds to a continuum state $exp(ik_z z)$. The eigenvalues and eigenfunctions can be found by appropriately matching boundary conditions in the narrow gap and wide gap regions of the semiconductor superlattice. The effect of the finite size of the well is most pronounced near the top of the well. For the finite potential well, the wave functions decay exponentially

$$\psi = \psi_0 e^{-\beta z} \tag{11}$$

so that tunneling through the potential well becomes possible. The probability P that the electron tunnels through the potential barrier of thickness z_0 is given by

$$P = exp\left\{ -2 \int_0^{z_0} \beta(z)dz \right\} = exp\left\{ -2\left(\frac{2m^*}{\hbar^2}\right)^{\frac{1}{2}} (V_0 - E)^{\frac{1}{2}} z_0 \right\} \tag{12}$$

where V_0 is the barrier height, E is the electron energy and $\beta^2 = (2m^*/\hbar^2)(V(z) - E)$. As z_0 increases, the probability of tunneling decreases exponentially. Semiconductor superlattices are fabricated with barrier widths such that significant tunneling occurs and electronic superlattices can be realized. When the electron energy corresponds to a bound state of the quantum wells, the tunneling probability is greatly increased. This resonance effect is utilized in the experimental determination of bound states in semiconductor superlattices.[1,21]

With the sophisticated computer control available using state of the art molecular beam epitaxy systems, it is now possible to produce quantum wells with specified poten-

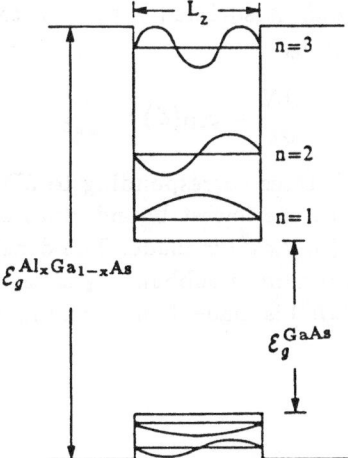

Figure 8: The eigenfunctions of an infinitely deep potential well shown schematically in two finite wells. The upper well applies to electrons and the lower one to hole bound states. This configuration is appropriate to GaAs / $Al_x Ga_{1-x}As$, a type I heterojunction semiconductor superlattice.

tial profiles $V(z)$ for semiconductor heterojunction superlattices. Potential wells with non–rectangular profiles also occur in the fabrication of various types of superlattices (e.g., by modulation doping, as discussed below).[1]

In cases where the potential well has an arbitrary shape, solution for the bound state energy levels is found by the WKB (Wentzel–Kramers–Brillouin) approximation.[22] According to this approximation, the energy levels satisfy the Bohr–Sommerfeld quantization condition

$$\int_{z_1}^{z_2} p_z dz = \hbar\pi(r + c_1 + c_2)$$

(13)

where $p_z = (2m^*[E - V])^{\frac{1}{2}}$, the quantum number $r = 0, 1, 2 \ldots$, while c_1 and c_2 are the phases which depend on the form of $V(z)$ at the turning points z_1 and z_2 where $V = E$. If the potential has a sharp discontinuity at a turning point, then $c = 1/2$, but if V depends linearly on z at the turning point then $c = 1/4$. Common applications of the WKB method are made to quantum wells that are approximated by symmetric and asymmetric triangular potential wells and harmonic oscillator potential wells.[22]

The thin films used for the fabrication of quantum well structures are very thin in the z–direction but have macroscopic size in the perpendicular x–y plane. The motion of electrons in the x and y directions is similar to that in the corresponding bulk solid which can be treated by the conventional one–electron approximation and the Effective Mass Theorem. Thus the potential can be written as a sum of a periodic term $V(x, y)$ and the quantum well term $V(z)$. The electron energies thus have superimposed on the bound state energies, the periodic solutions obtained from the eigenvalues of the 2D periodic potential, so that

$$\mathcal{E} = \mathcal{E}_n + \frac{\hbar^2(k_x^2 + k_y^2)}{2m^*}$$

(14)

in which the quantized energies $\mathcal{E}_n$ are given by the previous discussion of the bound states. A plot of the energy levels is given in Fig. 9. At $(k_x, k_y) = (0,0)$ the energy is precisely the quantum well energy $\mathcal{E}_n$ for all n. The band of energies associated with each state n is called a *subband*. Associated with each two–dimensional subband is a constant density of states given by

$$\frac{\partial N}{\partial \mathcal{E}} = g_{2D}(\mathcal{E}) = \frac{m^*}{\pi\hbar^2}.$$

(15)

If we now plot the density of states corresponding to 3D motion in a 1D rectangular well, we have $g_{2D}(\mathcal{E}) = 0$ until the lowest bound state energy $\mathcal{E}_1$ is reached, when a step function contribution of $(m^*/\pi\hbar^2)$ is made. The density of states $g_{2D}(\mathcal{E})$ will then remain constant until the minimum of subband $\mathcal{E}_2$ is reached when an additional step function contribution of $(m^*/\pi\hbar^2)$ is made, hence yielding the staircase density of states shown in Fig. 10.

ZONE FOLDING EFFECTS

When the thickness of the small gap semiconductor is too small $(L_z < L_z^c)$ for the localization of a bound state within the quantum wells, then the carriers will occupy continuum states where the periodic potential of the superlattice modulates the crystal periodic potential within each of the layers M_1 and M_2. The superlattice introduces a periodicity of *larger real* space dimensions and hence *smaller reciprocal* space dimensions. In the reduced Brillouin zone for the superlattice (the *folded zone*), a band gap of magnitude $2V_{G'_n}$ will open up at each zone boundary $\pm n\pi/d$ for the superlattice, where $|\vec{G}'_n| = 2n\pi/d$ and $n = 1, 2, \ldots$ (see Fig. 11).

Translation of each subband by the appropriate reciprocal lattice vector $G'_n = 2\pi n/d$ results in the $E(\vec{k})$ diagram for the folded zone shown in Fig. 11 for a superlattice where $d = 6a$. We note that the bandwidths of the zone folded bands are fractions of electron volts and the resulting $E(\vec{k})$ relations describe the propagation of electrons along the superlattice. Zone folding allows low lying energy states with negative masses to occur. In fact, the desire to create such states led Esaki and coworkers to pioneer the fabrication of semiconductor superlattice structures in the 1960's.[1,23]

The corresponding zone folding effects (see Fig. 11) also occur for the phonon dispersion relations $\omega(\vec{k})$, giving rise to new $\vec{k} \sim 0$ phonons in the folded zone. These new $\vec{k} \sim 0$ phonons can often be excited by Raman and infrared spectroscopy. In addition, the superlattice layering causes the z direction to have a different symmetry relative to the x–y plane. Hence a cubic material prepared in a superlattice structure will no longer have the full symmetry of the cubic groups. The lowered tetragonal symmetry will cause additional phonon modes to become Raman and infrared active.[24,25]

Zone folding also occurs in intercalation compounds. Here zone folding along the c–axis arises from the 1D superlattice associated with staging. In the layer planes, zone folding also occurs for commensurate intercalation compounds where the intercalate lat-

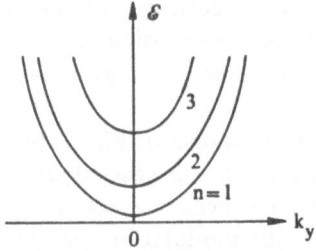

Figure 9: Simple picture of subbands associated with bound states. The 2D intralayer mass m^* is assumed to be isotropic.

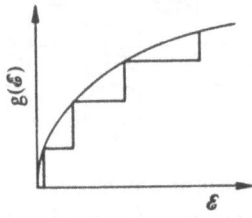

Figure 10: Two dimensional (staircase) and three dimensional (solid parabola) density of states for 2D rectangular quantum well structures and simple 3D bands.

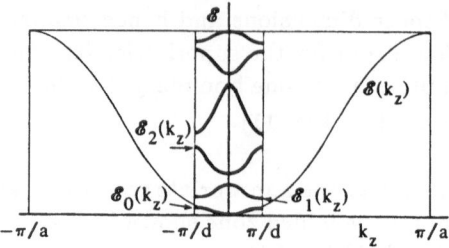

Figure 11: Zone folding of the electronic energy bands of a superlattice with lattice constant d, where $d = 6a$, and a is the lattice constant of the primitive cell.

tice constant is a multiple of that of the host material. The c–axis zone folding is similar to that discussed above for the deliberately structured superlattices. The in–plane zone folding also maps $k \neq 0$ points into the zone center, giving rise to new interband transitions for the electronic properties[26] and new phonon modes in the Raman spectra.[27] A number of zone folding phenomena have been observed in the electronic and lattice properties of GICs.[2,4,26,27]

CHARGE TRANSFER AND MODULATION DOPING

To discuss the question of charge transfer in semiconducting superlattice structures, consider the quantum well structures shown in Fig. 12 appropriate for a type I (e.g., GaAs/Al$_x$Ga$_{1-x}$As) superlattice. In the undoped case (a), we assume as a first approximation that the system is uncharged. In the case of uniform electron doping (b), the electrons on the donor sites in the Al$_x$Ga$_{1-x}$As layers become ionized as the electrons find lower lying states below the band edge of the GaAs. Thus the GaAs becomes negatively charged and the Al$_x$Ga$_{1-x}$As becomes positively charged, so that band bending at the heterojunction occurs due to the resulting electric field. The transfer of electrons from the Al$_x$Ga$_{1-x}$As to the GaAs causes the Fermi level to rise, as shown in the figure.

The use of modulation doping is shown in Fig. 12(c), where only the Al$_x$Ga$_{1-x}$As is doped and the GaAs is not doped.[28] In this case, there are no donor states in the GaAs layers, but only electrons that have been transferred from the doped Al$_x$Ga$_{1-x}$As as shown in the figure. In practice, the modulation doping is carried out using the molec-

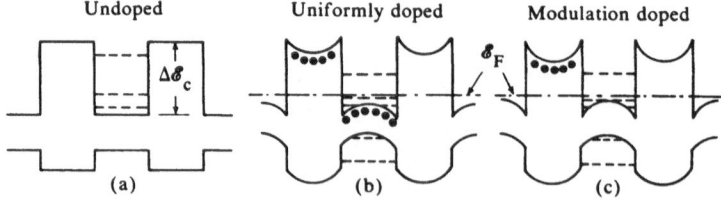

Figure 12: A semiconductor heterojunction superlattice with various methods of doping: (a) undoped, (b) uniformly doped, and (c) modulation doped. The black dots denote donor states and the dashed lines denote electron bound quantum well states. Doping causes the Fermi level to rise above the band gap.[28]

ular beam epitaxy technique, by opening the Sn shutter only when the $Al_xGa_{1-x}As$ is being deposited. Thus the donor defects are only present in the $Al_xGa_{1-x}As$ while the carriers are in the GaAs region, which is spatially separated from the donor defects. By this mechanism, very high mobilities can be achieved. Charge transfer in graphite intercalation compounds also leads to high mobility carriers in the graphite layers, since the carriers are confined to the high mobility layers while the ionized impurities are in the intercalate layers.

Further tailoring the superlattice structure by insertion of undoped $Al_xGa_{1-x}As$ layers between the doped $Al_xGa_{1-x}As$ layers and the GaAs layers, yields even higher mobility materials (see Fig. 13). The greatest increase in mobility is found at low temperatures where scattering from ionized impurities is the dominant mechanism for reducing the mobility in uniformly doped material. Improvements in compositional tailoring of the superlattice unit cells have led to more than two orders of magnitude increase in the in–plane mobility within the GaAs region. Low temperature mobility values up to 2×10^6 cm^2/Vsec have been measured, close to the theoretical limit.[29] The high in–plane mobility of this GaAs material has been exploited in such devices as modulation–doped field effect transistors which have high switching speeds and low power consumption.[30] The main effort thus far has been directed toward achieving high mobility n–type GaAs material. In addition, modulated doping has recently been applied successfully to p–type GaAs. This is an important technological advance for making possible ultra–low power logic based on n– and p–channel transistors with high transconductance.

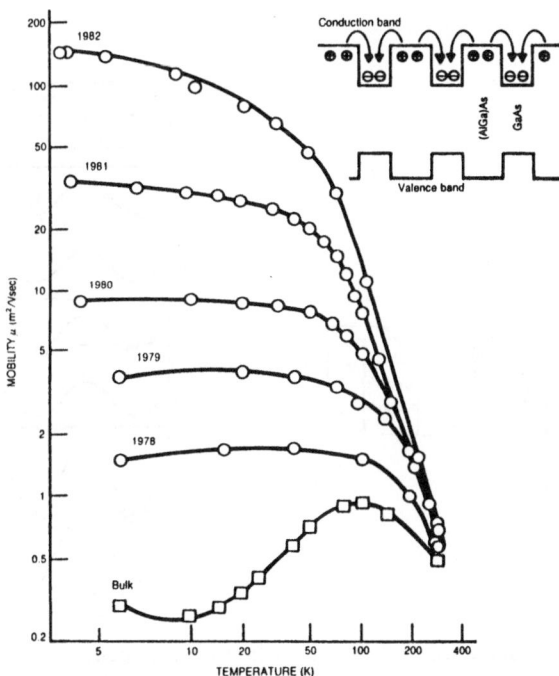

Figure 13: The mobility of electrons in GaAs prepared from modulation doped $Al_xGa_{1-x}As$/GaAs heterostructures. The inset at the top of the figure shows a schematic representation of the band edges, and the charge transfer mechanism. The donor states are denoted by $\oplus$ and the conduction electron states by $\ominus$.[10]

One especially interesting type of semiconductor superlattice is a periodic sequence of n–doped and p–doped layers (see inset to Fig. 14) of thicknesses d_n and d_p of the same semiconducting material such as GaAs[31] where $d = d_n + d_p$. These structures are called "nipi" superlattices and can be prepared from a variety of semiconductors where both n–type and p–type material can be synthesized. Structurally, "nipi" superlattices consist of an assembly of alternating p–n and n–p junctions. "Nipi" structures have many advantages for device applications since there are no mismatches in lattice constant or band offsets at the p–n and n–p junctions. The space charge induced periodic modulation of the energy bands leads to a confinement of electrons and holes in alternate layers (indirect gap in real space). This effective spatial separation of charge leads to very long recombination lifetimes for the excess carriers so that large deviations of electron and hole concentrations from their equilibrium values are possible. These long recombination lifetimes also lead to a wide tunability of the effective bandgap.

If there is no charge transfer between the n and p regions, the bandgap is independent of z and the charge accumulation in the n and p regions is as shown in Fig. 14a. If charge transfer is allowed, then the electrons created by ionization of the donor impurities in the n–type material are transferred to the p–type regions and become attached to the acceptor sites, creating a region with a density of N_A^- negatively charged acceptors.

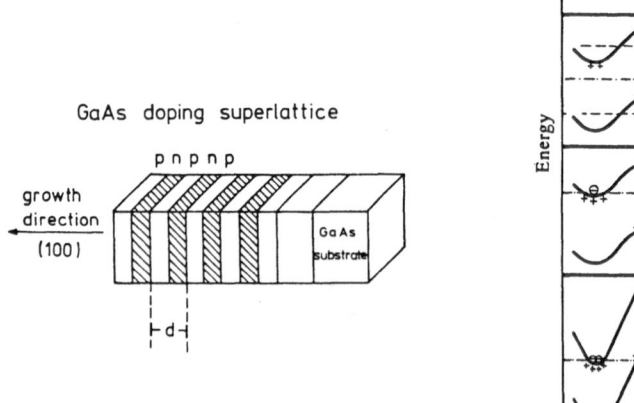

Figure 14: Various aspects of "nipi" superlattice structures. The inset on the left denotes a periodic doping multilayer semiconducting "nipi" structure. (a) The periodic donor and acceptor doping of a semiconductor using a "Maxwell demon" to prevent ionization of the impurity states. (b) Modulated bands are produced when the donors and acceptors ionize and come to thermal equilibrium. (c) A "nipi" structure with an excess of donors. (d) A semimetal "nipi" structure.[31]

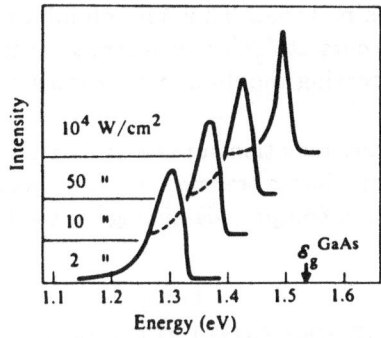

Figure 15: The photoluminescence spectra of a GaAs "nipi" superlattice at 4°K showing the tuning of the effective bandgap by variation of the light intensity to generate electron–hole pairs. The watts/cm² refer to the intensity of the incident light,[32] and the bandgap of bulk GaAs is also indicated.

Likewise the holes from the p–type region attach themselves to the donor sites, creating a region with a density of $N_D{}^+$ charged donors. Due to the periodic arrangement of the ionized impurities, a periodic space charge is created, which in turn gives rise to a periodic electric field (i.e. an additional periodic potential) of periodicity $d = d_n + d_p$. For the case of a superlattice compensated so that the total charge in the n and p regions is equal, we have the condition

$$d_n N_D = d_p N_A. \tag{16}$$

Fig. 14b shows the spatially modulated periodic potential associated with a compensated "nipi" structure while Fig. 14c shows a system with an excess of n–type carriers, such that the Fermi level is in the conduction band. The doping levels in "nipi" structures are usually heavy ($\sim 10^{18}$/cm³) so that impurity bands are formed. In that case the electrons in the conduction band can be itinerant. For very heavy doping, the Fermi level crosses both electron and hole pockets (see Fig. 14d) and a semimetal results.

One unique feature of the "nipi" structures is their wide tunability (see Fig. 15). This can be understood from the following argument. If an electron–hole pair is generated by light, thermal excitation or other means, long lived electrons and holes will be formed near the conduction and valence band extrema because the electrons and holes are spatially separated and do not readily recombine. To recombine, an electron must tunnel through a potential barrier of height $2V_0$ and width d_n. As a consequence, large deviations from equilibrium for electron and hole concentrations can be maintained for long times, thereby making possible the study of semiconductor systems in highly non-equilibrium phases. Because of these long lifetimes, electrons are more likely to fall into donor states and neutralize them than to recombine with holes. Likewise the holes are more likely to neutralize the acceptor $N_A{}^-$ states. As the donor and acceptor states become neutralized, the potential V_0 decreases and the effective bandgap $\mathcal{E}_g^{eff}$ increases, where $\mathcal{E}_g^{eff}$ defined as the energy spacing between the lowest conduction band subband extremum $\mathcal{E}_{c,0}$ and the uppermost valence subband extremum $\mathcal{E}_{v,0}$ becomes

$$\mathcal{E}_g^{eff} = \mathcal{E}_g^0 - 2V_0 + \mathcal{E}_{c,0} + |\mathcal{E}_{v,o}|, \tag{17}$$

where $\mathcal{E}_g^0$ is the bandgap of the bulk material and $\mathcal{E}_{c,0}$ and $\mathcal{E}_{v,0}$ are measured relative to the band extrema in the bulk material. Thus photoluminescence (recombination radiation) for "nipi" structures occurs at $\mathcal{E}_g^{eff}$, which increases toward the bulk bandgap for the semiconductor upon increasing the illumination intensity (see Fig. 15).

Although M_1 and M_2 in intercalation superlattices are different species (thus differing from "nipi" superlattices), charge transfer causes opposite signs in the constituents M_1 and M_2 of the intercalation compounds analogous to the "nipi" superlattice structures.

TYPE II SEMICONDUCTOR HETEROSTRUCTURE SUPERLATTICES

Semimetallic superlattices also can be formed from semiconductor heterojunction superlattices, such as InAs/GaSb where the band offsets are such that the conduction band edge of one of the semiconductors (InAs) lies below the upper valence band edge of the other semiconductor (GaSb).[23] Figure 16 shows that for the $In_{1-x}Ga_xAs$ / $GaSb_{1-y}As_y$ system in the composition range $0 < x < 0.3$, the conduction band edge $\mathcal{E}_{c_1}$ of $In_{1-x}Ga_xAs$ lies below the valence band $\mathcal{E}_{v_2}$ for $GaSb_{1-y}As_y$, so that *semimetallic behavior* can in principle be achieved, even though the constituents InAs and GaSb are both *semiconductors*. The band gap is read from Fig. 16 as $\mathcal{E}_g = \mathcal{E}_{c_1} - \mathcal{E}_{v_2}$ where the band edge energies are all measured with respect to a common vacuum level. By varying both x and y it is possible to adjust the bandgap and at the same time to achieve lattice matching in the superlattice. For convenience, the lattice constants for $In_{1-x}Ga_xAs$ and $GaSb_{1-y}As_y$ are plotted in Fig. 16 *vs.* x and y respectively in the dashed curves indicated by a_1 and a_2. When a superlattice is now made from the constituents $In_{1-x}Ga_xAs$ and $GaSb_{1-y}As_y$, band states form in the quantum wells and the effective bandgap is given by $\mathcal{E}_{gap}^{eff} = \mathcal{E}_{1e} - \mathcal{E}_{1h}$ where again all energies are referred to a common vacuum level, and $\mathcal{E}_{1e}$ and $\mathcal{E}_{1h}$ are the ground state levels in the electron and hole quantum wells, respectively.

In Fig. 17 (right hand panel), the familiar band diagram for the $GaAs/Al_xGa_{1-x}As$ superlattice, a type I heterogeneous semiconductor superlattice, is shown.[23] In this fig-

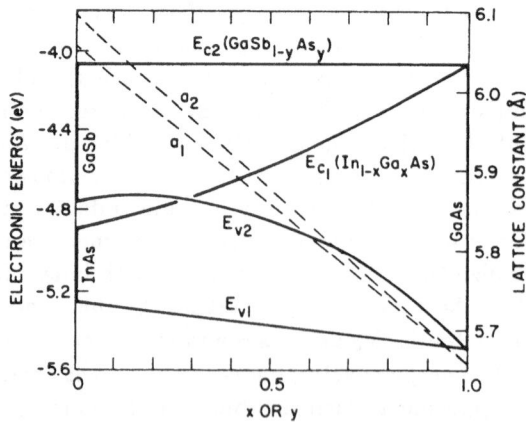

Figure 16: Band edge energies with respect to the vacuum level, and lattice constants versus alloy compositions in the ternaries $In_{1-x}Ga_xAs$ and $GaSb_{1-y}As_y$.

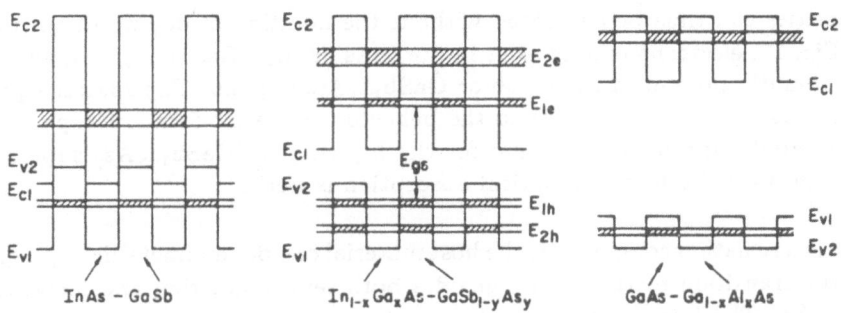

Figure 17: Schematic energy diagrams for the three indicated superlattice systems. Indicated are the band edges of the host semiconductors and the subbands for both electrons and holes. Shaded areas in the subbands show the spatial regions where the carriers are concentrated. The extremum of each bound state is assigned a width.[23]

ure, all electron and hole bound quantum states are assigned a width, consistent with fluctuations in the d–spacing of the superlattice. The hole ground state pertains to that for the heavy holes. The central panel of Fig. 17 shows a schematic of the valence and conduction bands for a type II heterogeneous semiconductor superlattice formed from $In_{1-x}Ga_xAs/GaSb_{1-y}As_y$, for x and y values such that the resulting superlattice is a semiconductor (i.e., there exists a band gap between the valence and conduction states). What is different for the type II and type I cases is the fact that the band offsets have the same sign for the central panel (type II) and opposite signs for the right hand panel (type I). Thus for $GaAs/Al_xGa_{1-x}As$, the bound states for electrons and holes occur in the same spatial region, while for the case of $In_{1-z}Ga_zAs/GaSb_{1-y}As_y$, the electron and hole bound states occur in different spatial regions. Now when x is decreased below $x = 0.3$ (see Fig. 16), then $\mathcal{E}_{v_2}$ lies higher than $\mathcal{E}_{c_1}$ for the band edges and semimetallic behavior can occur, as shown in the left panel of Fig. 17. In this panel, the width of the quantum well is sufficiently small so that the lowest bound state for the conduction band lies above that for the corresponding state in the valence band, so that *semiconducting* behavior results. Now, without changing the composition x and y, but only by increasing the well width L_z (i.e., d_1 and d_2 for the superlattice), the ground state energies approach the band edge levels, and at some critical values of d_1 and d_2, the $\mathcal{E}_{1e}$ level will cross the $\mathcal{E}_{1h}$ level. Thus for $\mathcal{E}_{1h} > \mathcal{E}_{1e}$, *semimetallic* behavior will result. Clearly as x decreases, semimetallic behavior can occur for smaller lattice constants d_1 and d_2. This is illustrated in Fig. 18 where the carrier density from Hall measurements is plotted *vs.* the layer thickness d_1 for InAs in an InAs/GaSb heterostructure.[23] Because of the much higher mobility for the electrons than holes in InAs/GaSb heterostructures, the thickness of the GaSb layers is not so important in the determination of the mobility of the heterostructure; this effect has been verified experimentally. The results in Fig. 18 show a rapid increase in the carrier concentration for the layer thickness $d_{InAs} \sim 150$Å, corresponding to the onset of the semiconductor–semimetal transition. The decrease in carrier concentration at larger layer thicknesses is explained by the decreasing importance of the band bending effects at the interface.

For pure InAs/GaSb superlattices, the semiconductor–semimetal transition occurs at a critical thickness of $d = 170$ Å. It is indeed remarkable that a semimetal can be synthesized by forming a heterostructure of two direct gap semiconductors. Large carrier densities can thus be generated without the addition of doping impurities. As shown in Fig. 17, electrons mainly exist in the InAs or $In_{1-x}Ga_xAs$ regions, while holes are predominantly present in the GaSb or $GaSb_{1-y}As_y$ regions. The spatial separation of the electrons and holes is similar to the behavior in the modulation doped and the modulated "nipi" superlattices, so that for the $In_{1-x}Ga_xAs/GaSb_{1-y}As_y$ superlattices, the carrier lifetimes are long and optical absorption is weak.

For the intercalation compounds, the host material can be semimetallic (e.g., graphite), while for the transition metal dichalcogenides both semiconducting and metallic hosts are possible.[6,33] The intercalation process greatly increases the carrier concentration, giving rise to metallic behavior for all hosts. For both intercalation compounds and heterogeneous semiconductor superlattices, the in–plane conductivity σ_a is much larger than σ_c along the c–direction.

METALLIC SUPERLATTICES

Metallic superlattices today offer many opportunities and challenges for the study of new physics and new materials science. Some of the new physics involves the novel behavior in thin film (anisotropic) magnetism[34] and superconductivity.[35] Because of the long coherence distances in superconductors, a detailed study of 2D superconductivity is now within reach with deliberately structured metallic superlattices.[17] On the other hand, to achieve 2D behavior in magnetic materials, it is probably necessary to pre-pare superlattices of magnetic monolayers, separated by much thicker non–magnetic spacer layers. This limit is more easily achieved with intercalation compounds.[34] Since intercalation compounds are generally metallic systems, the electronic structures of the deliberately structured metallic multilayers and superlattices have more in common with

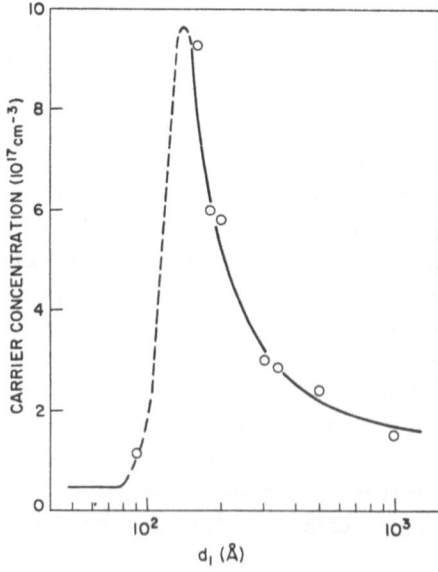

Figure 18: Carrier concentration from Hall measurements at 4.2 K *vs.* layer thickness of InAs for different InAs/GaSb heterostructures. The dashed region represents the semiconductor–semimetal transition.[23]

the intercalation compounds than the heterogeneous semiconductor superlattices. On the other hand, from a structural point of view, we will see below that the intercalation compounds have more in common with the heterogeneous semiconductor superlattices discussed above.

From a materials science point of view, the simplest metallic superlattices are formed from components with the same crystal structure and with matched lattice constants. Because of the severe restrictions on the lattice constants for physically interesting metallic superlattice systems, this is not a common case. A more common case is the one where the two metal constituents have the same crystal structure but different lattice constants (e.g., Cu/Ni), in which case the materials preparation for thin layers is conceptually similar to the case of a strained layer superlattice. However, even for constituents with the same crystal structure, the problem of solid state alloying and interdiffusion at the interface is a much more serious problem for the metallic superlattices than for the semiconducting superlattices.[36]

On the other hand, many systems of interest in the study of magnetism and superconductivity involve constituents with *different* crystal structures (e.g., BCC/FCC or FCC/hexagonal). The first successful metallic superlattice formed from constituents with different crystal structures was by Schuller in 1980.[37] Metallic superlattices have become an increasingly active field since that time. Though much progress has already been made, it would be safe to say that the field of metallic superlattices is still in its infancy.

The two most commonly used synthesis methods for metallic superlattices are MBE and sputtering. Magnetron sputtering guns permit relatively rapid growth (~ 50Å/sec) using a constant pressure of argon at a constant voltage and a sapphire substrate.[37] Multilayers with periodicities d between 10 Å and 100 Å can be grown. Recent work using MBE (at 4×10^{-11} torr) yields films with a much lower concentration of defects and much sharper interfaces.[19]

Of the various characterization techniques for metallic superlattices, x–ray diffraction has been the most informative.[36] Evidence for the formation of metallic superlattices from constituents of different crystal structures is given in Figs. 19 where the x–ray diffraction patterns for a superlattice of Nb(BCC) and Cu(FCC) are shown.[37] The x–ray results show that if the layer thickness is too small ($d_1 \sim 5$Å), no well defined superlattice is formed. Instead, a large distribution of layer thicknesses relative to the interface width is found. However, for larger layer thicknesses, well defined (00ℓ) x–ray diffraction peaks are found corresponding to the growth of distinct (111) FCC Cu layers and (110) BCC Nb layers in a quasi–strained layer–like structure. In this structure, the lattice constants for Nb (3.29Å) and Cu (3.61Å), are within 0.2% of their values in the pure materials. Qualitative agreement of the observed patterns with modulated diffraction theory is obtained. The calculations are sensitive to the nature of the Nb/Cu interface, whether it is commensurate or discommensurate, and whether significant interdiffusion occurs.

The growth of metallic superlattices has not yet reached a level of control so that the in–plane structure of one Nb layer is correlated with that of neighboring Nb layers and similarly for the Cu. In general, the coherence length in the layers is longer than that perpendicular to the layers. X–ray studies yield a c–axis coherence length of ~ 300Å, which is much larger than typical superlattice constants d. Diffuse x–ray scattering determines the average grain size and thus gives valuable information on the overall structural quality of the superlattice.[19]

The Nb/Cu metallic superlattice differs significantly from the semiconductor strained layer superlattices discussed above, insofar as there is no lattice matching of the in–plane lattice constants across the interface, and the in–plane structural coherence is much lower in each of the metal layers M_1 and M_2 as compared with the strained layer superlattice. From a structural point of view, the intercalation compounds are much closer to the semiconductor superlattices, with a direct analogy occurring between the commensurate intercalation compounds and the semiconductor strained layer superlattices. Although a metallic superlattice like Cu/Nb is conceptually related to a binary donor GIC, the two types of superlattices have fundamental structural differences due to the lack of a sharp definition of the interface in the Cu/Nb system.

It has been shown that the modulation direction in a metallic superlattice could be controlled by the choice of substrate.[38] A common substrate for metal superlattices is the $(01\bar{1}2)$ face of sapphire. The temperature of the substrate is critical, as shown for studies on Nb/Ta grown on sapphire $(01\bar{1}2)$. The symmetry of the substrate is important, insofar as epitaxial films grow with planes of symmetry similar to that of the substrate to minimize interfacial strains. For the case of Nb/Ta superlattices grown on a sapphire $(01\bar{1}2)$ surface (Al_2O_3), it is found that the (110) grains of the Nb grow parallel to $(01\bar{1}2)$ grains of the sapphire, but the (001) grains of the Ta grow at an angle

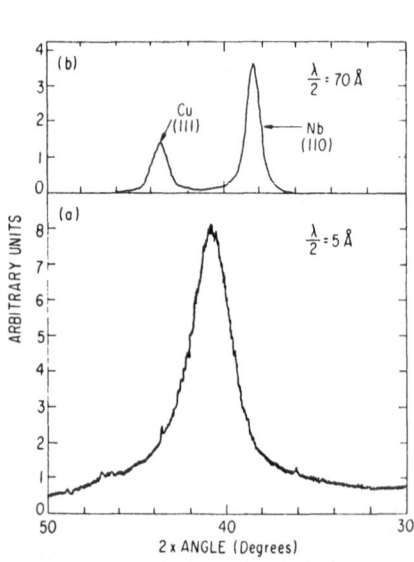
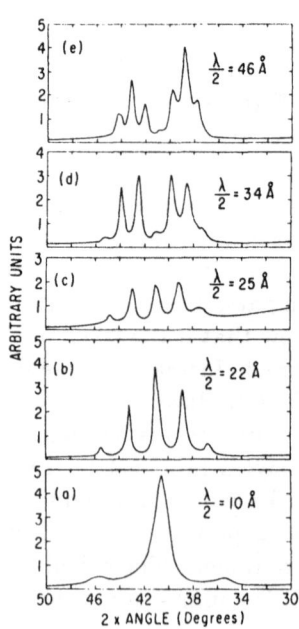

Figure 19: Evolution of the experimental (00ℓ) x–ray diffraction patterns as a function of layer thickness $\lambda/2$ for a metallic superlattice formed from FCC Cu (111) and BCC Nb (110).[37] Here $\lambda/2 = d_1 = d_2$.

of 2.6° with respect to the (01$\bar{1}$2) sapphire grains. We note that both Nb (3.3003Å) and Ta (3.3024Å) are BCC crystals with similar lattice constants. Despite the fairly close lattice match, the interfaces are 10 − 12Å thick.[38]

Another interesting metallic superlattice is that of Nb/Zr where Zr is normally a hexagonal metal. When forming a superlattice, the Zr forms a BCC structure, though the interface is thicker than for the Nb/Ta.[16] Perhaps the interdiffusion of Zr and Nb at the interface stabilizes the BCC structure in the Zr layer. The formation and study of metastable metallic phases (such as BCC Zr) are of fundamental interest.

It was found that a niobium buffer region on a sapphire (Al_2O_3) substrate prevents interaction between the rare earth metal and the substrate at the growth temperature (600–800°C).[19] This finding has been of significant importance in the synthesis of rare earth multilayers. Close packed (110) planes of BCC single crystal Nb can be grown on either (11$\bar{2}$0) or (0001) hexagonal substrates.[38] A Y/Gd metallic superlattice grown on the (110) Nb buffer shows the c−axes of the hexagonal Y and Gd to be aligned along the growth direction. Narrow rocking curves parallel and perpendicular to the film growth direction imply a high degree of crystallinity in the films. Further evidence for crystallinity in the layer planes is the 60° periodicity of the x−ray pattern upon rotation of the scattering vector in the basal plane.[19] The similarity of the lattice constants of Y (3.65 Å) and Gd (3.64 Å) is favorable for lattice matching at the interface. The Y/Gd superlattice is also of interest for its magnetic properties.[34]

The preparation of crystalline metal superlattices with structural ordering both parallel and perpendicular to the layer planes and with coherence between superlattice unit cells would open up new possibilities in the study of electronic transport and Fermi surfaces for these quasi−two−dimensional systems. Limiting cases of interest are cases where the superlattice constants d_1 and d_2 are either large or small compared to the mean free path for carriers.

The behavior of a metal/insulator or metal/semiconductor superlattice is of particular interest. For example, for the Nb/Ge superlattice[17] the in−plane conductance per square $G_\square$ is seen to decrease roughly proportionally to the thickness of the metal layer[17] (see Fig. 20), which is associated with boundary scattering. However for metal thicknesses $\leq$ 30Å, the conductivity per square becomes much smaller, due to interlayer diffusion of the Ge and Nb to form a low conductivity alloy. Significant studies of 2D superconductivity have been carried out on this system.[17,34]

Not only do superlattices offer the opportunity for new physics relating to novel structural arrangements, but also exciting new research opportunities are expected to arise from the possibility of preparing materials with novel doping profiles, compositional grading, metastable constituents, and metastable interface structures and compositions. Rapid progress in these areas of metals physics await further developments in the materials science of metallic superlattices. Early studies on metal superlattices have emphasized superconductivity,[17] though significant work on magnetic superlattices has also been carried out. These topics are discussed elsewhere in this volume.[34,35]

Intercalation causes large changes in electron concentration because of the transfer of charge between the intercalant and graphite. By transferring carriers from regions of low mobility (the intercalant) to regions of high mobility (graphite), large increases in the electrical conductivity can be achieved, in analogy to the charge transfer process in modulation doped semiconductor superlattices. The charge transfer process in donor GICs can be readily understood on the basis of Fig. 21. Suppose for example that the intercalant species is the alkali metal potassium. Each potassium atom has one valence electron. If this electron is totally transferred to the graphite layers, the intercalate layer becomes positively charged, thereby attracting electrons in the graphite layers. Thus most of the transferred charge resides in the graphite bounding layers adjacent to the intercalate. Because the in–plane intercalate concentration in a stage 1 compound C_8K is 1/8 of the carbon concentration in a graphite layer, one can achieve as much as a three orders of magnitude increase in the carrier concentration in K–GICs compared with pristine graphite. Thus, intercalation results in major changes in the transport properties.[14,39] Of the various types of layers within the unit cell (see Fig. 21), the conductivity of the graphite bounding layers (adjacent to the intercalant) is dominant because of the high carrier density in these layers relative to the graphite interior layers and because of the much higher carrier mobility in the graphite bounding layers relative to the intercalate layer. Though smaller, the contribution of the graphite interior layers can in some cases be significant.[2]

It is believed[2,4] that the alkali metal intercalants donate essentially all their free carriers to the graphite layers for the higher stage compounds (n ≥ 2), though the donated electrons remain predominantly in the graphite bounding layers which then screen the

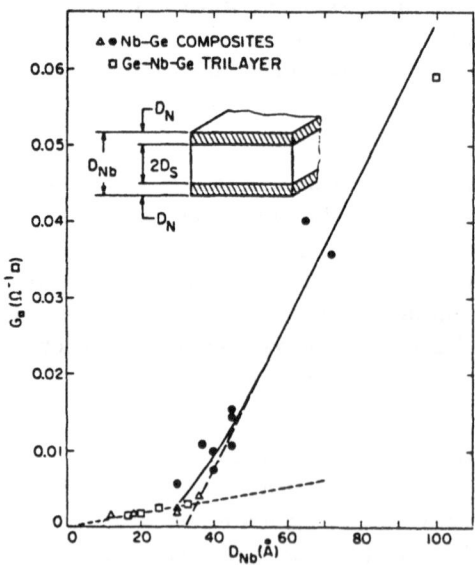

Figure 20: Low–temperature conductivity per square vs. niobium layer thickness. The data for thicker layers show a distinct offset of ∼ 30 Å from the origin. The conductivity data are explained assuming that the Nb layers D_{Nb} are composed of a central region $2D_S$ of well–ordered material of 9.8 $\mu\Omega$ cm resistivity on either side of which is a total of $D_N = 26$Å of disordered material of 114 $\mu\Omega$ cm resistivity (see inset).[17]

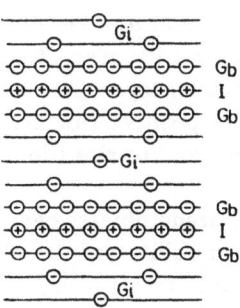

Figure 21: Schematic diagram of the electronic charge distribution in the graphite layers of a stage 5 donor compound. I, G_b, and G_i refer, respectively, to intercalate, graphite bounding, and graphite interior layers.

positively charged intercalate layers. A rapid decrease in carrier density occurs with distance into the graphite interior layers. The screening is calculated explicitly through the charge distribution given by self–consistent band calculations.[26]

The charge transfer in donor compounds is usually one or two electrons (for mono-valent or divalent species) per intercalate unit but sometimes the charge transfer is incomplete. But generally the transfer for donor GICs is greater than that for accep-tor compounds. For the acceptor compounds, the intercalate layer becomes negatively charged by extracting electrons predominantly from the graphite bounding layers, in contrast to the situation shown in Fig. 21 for the donor compounds. Thus a high hole concentration is found in the graphite bounding layers, and the hole concentration de-creases rapidly with distance into the graphite interior layers. The electronic charge distribution in the graphite and intercalate layers has been calculated self–consistently for a number of GICs.[26]

Intercalation generally results in major changes in the transport properties of the host material.[39] For example, the in–plane electrical conductivity σ_a of GICs typically increases from that for graphite by an order of magnitude or more, as shown for several alkali metal GICs in Fig. 22. The magnitude of the increase in σ_a depends on the intercalate species and its concentration. Of the various types of layers within the unit cell (see Fig. 23), the conductivity of the graphite bounding layers is dominant because of the high carrier density in these layers relative to the graphite interior layers and because of the much higher carrier mobility in the graphite bounding layers relative to the intercalate layer. Though smaller, the contribution of the graphite interior layers can in some cases be significant. We can account for the observed stage dependence of σ_a by considering the conductance of the total sample as a sum of conductances of the individual layer types, weighted according to their relative thicknesses within the unit cell. The conductance can then be approximated by

$$I_c(\sigma_a/\sigma_a^0) = d_i(\sigma_i/\sigma_a^0) + 2c_0(\sigma_{gb}/\sigma_a^0) + (n-2)c_0(\sigma_{gi}/\sigma_a^0) \quad n \geq 2 \tag{18}$$

and for stage 1

$$I_c(\sigma_a/\sigma_a^0) = d_i(\sigma_i/\sigma_a^0) + c_0(\sigma_{gb}/\sigma_a^0), \tag{19}$$

where n is the stage index, c_0 is the separation between adjacent graphite layers, I_c is the GIC repeat distance along the c–axis, d_i is the intercalate layer or sandwich thickness and $d_s = d_i + c_0$. Also σ_a, σ_a^0, σ_i, σ_{gb}, and σ_{gi} are respectively the in-plane conductivities for the intercalation compound, pristine graphite, the intercalate layer, the graphite bounding layer and the graphite interior layer. Hall effect measurements show that the sign of the dominant carrier can be positive (for acceptors) or negative (for donors). Thus intercalation causes the Fermi level to rise (donors) or to fall (acceptors).

For many GICs one carrier type predominates over the others and qualitative results for the carrier density and carrier mobility can be obtained approximately from Hall effect and magnetoresistance measurements through the simple expressions[2]

$$N_H = \frac{1}{R_H ec} \tag{20}$$

and

$$\langle \mu_H \rangle = \left(\frac{\Delta \rho}{\rho_0 H^2} \right)^2. \tag{21}$$

If two carriers contribute significantly then the resistivity, Hall effect and magnetoresistance are given by

$$\sigma = (1/\rho) = (p\mu_p + n\mu_n)e, \tag{22}$$

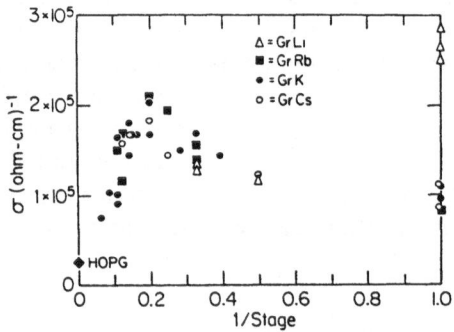

Figure 22: Dependence on reciprocal stage of the in-plane conductivity of several graphite–alkali metal compounds. The legend gives the symbols for the various alkali metal compounds.[40]

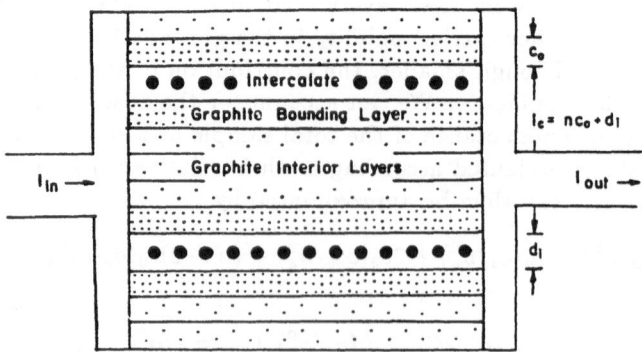

Figure 23: Additive conductance model for graphite intercalation compounds.[2]

$$R_H = \frac{1}{ec} \frac{p\mu_p^2 - n\mu_n^2}{(p\mu_p + n\mu_n)^2},$$ (23)

and

$$\frac{\Delta\rho}{\rho_0 H^2} = \frac{p}{n}\mu_n\mu_p \left(\frac{1 + (\mu_n/\mu_p)}{(p/n) + (\mu_n/\mu_p)} \right)^2$$ (24)

in which the two carrier types are denoted by the subscripts p (positive charge) and n (negative charge), while p and n denote carrier densities. Suitable generalizations can be made to two p or two n carrier types. Magnetoresistance and Hall measurements are difficult to make with accuracy because of the high electrical anisotropy of GICs.[39] The interesting behavior of c–axis transport is discussed elsewhere in this volume.[41]

As the temperature T is lowered, the in–plane electrical conductivity σ_a for GICs increases. Thus, at low temperature, both donor and acceptor GICs normally satisfy the condition $\omega_c\tau \gg 1$ (where ω_c is the cyclotron frequency and τ is the scattering time), so that electrons can complete at least one electron orbit before scattering. Thus quantum oscillatory phenomena can be observed readily in both donor and acceptor GICs at low temperatures. Quantum oscillatory phenomena are also used to study the constant energy surfaces of semiconducting superlattices in deliberately structured materials.[1]

There are many analogies between intercalation compounds and deliberately structured superlattices. Some of these analogies will be pursued explicitly in discussions of magnetic[34] and superconducting[35] GICs in relation to their magnetic and superconducting properties.

ACKNOWLEDGMENTS

The author thanks Dr. G. Dresselhaus and Prof. P.C. Eklund for many important discussions. This work was supported by AFOSR contract #F49620–83–C–0011.

REFERENCES

1. "Synthetic Modulated Structures", edited by L.L. Chang and B.C. Giessen, (Academic Press, Orlando, Florida, 1985); "Two–Dimensional systems, Heterostructures and Superlattices", edited by G. Bauer, F. Kuchar and H. Heinrich, (Springer Verlag, Berlin, 1984), Solid State Sciences **53**.

2. M.S. Dresselhaus and G. Dresselhaus, *Adv. in Phys.* **30**, 139 (1981).

3. P.C. Eklund, (this volume) p. 163.

4. S.A. Solin, *Adv. Chem. Phys.* **49**, 455 (1982).

5. R. Moret, (this volume) p. 185.

6. W.Y. Liang, (this volume) p. 31.

7. S.A. Solin, (this volume) p. 145.

8. S.Y. Leung, C. Underhill, G. Dresselhaus and M.S. Dresselhaus, *Solid State Commun.* **33**, 285 (1980); S.Y. Leung, C. Underhill, T. Krapchev, G. Dresselhaus, M.S. Dresselhaus, and B.J. Wuensch, *Phys. Rev.* **B24**, 3505 (1981).

9. W.P. Lowe, T.W. Barbee, Jr., T.H. Geballe, and D.B. McWhan, *Phys. Rev.* **B24**, 6193 (1981).

10. V. Narayanamurthi, *Physics Today*, October (1984), p. 24. The micrograph of Fig. 5 by J.M. Gibson, appears in this reference.

11. R.M. Biefeld, G.C. Osbourn, P.L. Gourley and I.J. Fritz, *J. Electr. Mat.* **12**, 903 (1983).

12. A. Segmueller and A.E. Blakeslee, *J. Appl. Cryst.* **6**, 19 (1973).

13. S.M. Durbin, J.E. Cunningham, M.E. Mochel, and C.P. Flynn, *J. Phys.* **F11**, L223 (1981).

14. P.C. Eklund, (this volume) p. 337.

15. A. Erbil, A.R. Kortan, R.J. Birgeneau and M.S. Dresselhaus, *Phys. Rev.* **B28**, 6329 (1983); S.G.J Mochrie, A.R. Kortan, R.J. Birgeneau and P.M. Horn, *Phys. Rev. Lett.* **53**, 985 (1984).

16. W.P. Lowe and T.H. Geballe, *Phys. Rev.* **B29**, 4961 (1984).

17. S. Ruggiero, T.W. Barbee, Jr., and M.R. Beasley, *Phys. Rev.* **B26**, 4894 (1982).

18. C.M. Falco and I.K. Schuller, in "Synthetic Modulated Structures", edited by L.L. Chang and B.C. Giessen, (Academic Press, Orlando, Florida, 1985) p. 339.

19. J. Kwo, D.B. McWhan, M. Hong, E.M. Gyorgy, L.C. Feldman and J.E. Cunningham, "Layered Structures, Epitaxy and Interfaces", *Proceedings of the Materials Research Society*, Vol. 37, edited by J.M. Gibson and L.R. Dawson, p. 509 (1985).

20. J.N. Schulman and T.C. McGill, in "Synthetic Modulated Structures", edited by L.L. Chang and B.C. Giessen, (Academic Press, Orlando, Florida, 1985) p. 77.

21. L. Esaki and L.L. Chang, *Phys. Rev. Lett.* **33**, 495 (1974).

22. W. Zawadski, in "Two–Dimensional systems, Heterostructures and Superlattices", edited by G. Bauer, F. Kuchar and H. Heinrich, (Springer Verlag, Berlin, 1984), Solid State Sciences **53**, p. 2.

23. L.L. Chang and L. Esaki, *Surface Science* **98**, 70 (1980).

24. C. Colvard, R. Merlin, M.V. Klein and A.C. Gossard, *Phys. Rev. Lett.* **45**, 298 (1980).

25. C. Colvard, T.A. Grant, M.V. Klein, R. Merlin, R. Fischer, H. Morkoc and A.C. Gossard, *Phys. Rev.* **B31**, 2080 (1985).

26. C. Rigaux, (this volume) p. 235.

27. P.C. Eklund, M.H. Yang, and G.L. Doll, (this volume) p. 257; P.C. Eklund, (this volume) p. 323.

28. R. Dingle, H.L. Störmer, A.C. Gossard and W. Wiegmann, *Appl. Phys. Lett.* **33**, 665 (1978).

29. H.L. Störmer, *Surf. Sci.* **132**, 519 (1983).

30. F.C. Capasso, W.T. Tsang and G. Williams, *IEEE Trans. on Electron Devices*, **ED–30**, 381 (1983).

31. G.H. Döhler, H. Künzel, D. Olego, K. Ploog, P. Ruden, H.J. Stolz and G. Abstreiter, *Phys. Rev. Lett.* **47**, 864 (1981).

32. H. Künzel, G.H. Döhler, P. Ruden and K. Ploog, *Appl. Phys. Lett.* **41**, 852 (1982).

33. R. Brec, (this volume) p. 75.

34. G. Dresselhaus and M.S. Dresselhaus, (this volume) p. 407.

35. G. Dresselhaus and A. Chaiken, (this volume) p. 387.

36. D.B. McWhan, in "Layered Structures, Epitaxy and Interfaces", *Proceedings of the Materials Research Society*, Vol. 37, edited by J.M. Gibson and L.R. Dawson, p. 493 (1985); and D.B. McWhan, "Synthetic Modulated Structures", edited by L.L. Chang and B.C. Giessen, (Academic Press, Orlando, Florida, 1985) p. 43.

37. I.K. Schuller, *Phys. Rev. Lett.* **44**, 1597 (1980).

38. S.M. Durbin, J.E. Cunningham and R.P. Flynn, *J. Phys. F.* **12**, L75 (1982).

39. J.–P. Issi, (this volume) p. 347.

40. E. McRae, D. Billaud, J.–F. Marêché and A. Hérold, *Physica* **B99**, 489 (1980).

41. E. McRae, J.–F. Marêché and A. Hérold, (this volume) p. 377.

ELECTRONIC PROPERTIES OF TRANSITION METAL DICHALCOGENIDES AND THEIR INTERCALATION COMPLEXES

W.Y. Liang

Cavendish Laboratory
Madingley Road
Cambridge   CB3 OHE, U.K.

1. INTRODUCTION

The transition metal dichalcogenides and their intercalate complexes belong to a large class of the so-called two-dimensional solids in which there is a good deal of interest at present. They are called two dimensional because they are formed in layered structures. Atoms within a layer are bound by strong covalent or ionic forces while individual layers are held together by relatively much weaker forces. The latter are frequently referred to as "van der Waals" type of interactions, though some contributions from covalent and ionic interaction are also possible, particularly in the intercalate complexes.

Typical layer structures would include solids such as graphite, mica (and related materials), gallium chalcogenides, a number of the metallic dihalides, and the β aluminas, but we will concentrate mainly on the layer type transition metal dichalcogenides, of which the best known example is of course molybdenum disulphide, MoS_2, which can be cleaved very readily. We must also remember that we are actually dealing with three dimensional solids having varying degrees of anisotropy in their physical properties. These anisotropic physical properties can be studied in many ways including band structure calculations, electronic and phonon excitations using plane polarized radiation, Raman spectroscopy, X-ray measurements of lattice parameters under hydrostatic pressure, electron and ion transport, superconductivity, and many others. We shall return to some of these aspects in more detail later. However we may note here that on the basis of force constants determined from phonon studies (Bagnall et al., 1980; Pereira and Liang, 1985), we can write down a general hierarchy of the degree of anisotropy for some of these materials, and such a series would be graphite, MoS_2, $MoSe_2$, SnS_2, HfS_2, $HfSe_2$, $SnSe_2$, CdI_2 in decreasing order of anisotropy. Metallic layer compounds are not included because there is insufficient data to deduce the anisotropy ratios of their force constants.

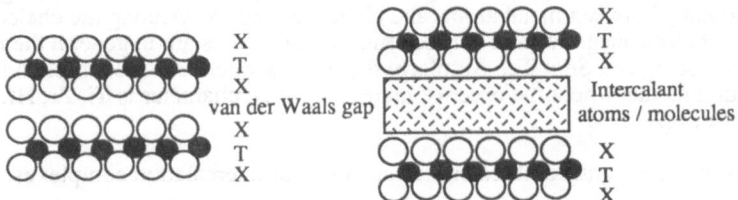

Figure 1. Schematic representation of (a) transition metal dichalcogenide TX_2 sandwiches and the interlayer van der Waals gap, (b) an intercalation complex with intercalant species inserted between layers of TX_2, where T=transition metal and X=S, Se or Te.

The weak interlayer forces offer the possibility of introducing foreign atoms or molecules between the layers and this is the process of *intercalation*. In this process intercalant species are inserted into the van der Waals gap, and in most cases they occupy the interstitial sites, and the process is generally accompanied by charge transfer between the intercalant species and the host layer. Figure 1(a) shows the van der Waals gap between two layers of transition metal dichalcogenide sandwiches, and Fig.1(b) shows a typical intercalation complex. It is believed that the tendency for this charge transfer process is the driving force for the intercalation reaction. The transition metal dichalcogenides are only known to form intercalation complexes with electron donor species, so the process here is of electron transfer from intercalant to the host material. Furthermore, since the intercalant species are to be found outside the sandwich layers of the host crystal and the local bonding within the sandwich is strong, a useful first approximation to describe the changes in electronic properties of the host material is the "rigid-band model" in which the only change to the host material electronic structure is the increased filling of the conduction band. The lowest-lying unoccupied energy levels in the host layers are the transition metal d bands, which are either empty or partially filled, and the increase in d band filling can be used to "fine tune" the electronic properties of the host material in a controllable way. It is this particular aspect of the science of intercalation that we shall explore in detail here. Control of the degree of band filling in this way is a feature unique to low-dimensional structures of this type, and provides an extra variable of enormous value. It is thus possible to achieve semiconductor to metal and metal to semiconductor transitions with intercalation. Changes in band filling also allow us to investigate the Fermi-surface driven structural instabilities demonstrated by the metallic layer structure materials, to be reviewed in later sections. In addition we shall cover order-disorder transitions of the intercalant layers, and of the magnetic moments on the intercalants (for the 3d transition metals), electron localisation and hopping conductivity, fast ion transport, and other phenomena.

Almost all transition metals from groups IV, V, VI, VII and VIII of the periodic table form layered structure dichalcogenides. However only the disulphides and diselenides of groups IV, V, and VI, with the exception of VS_2 and chromium compounds which are difficult or impossible to grow, have been studied in any detail, as shown in Table 1. Only certain transition metals of the group VIII form layer structure dichalcogenides. These are the tellurides of Ni, Pd, and Pt, as well as PtS_2 and $PtSe_2$. We will discuss the effects on band structures of increasing the number of d electrons later. The metal dihalides also form layer structures, but they are usually hygroscopic and are more ionic. Although interesting in their own right, they will not be discussed here.

Table 1. The transition metal elements, T, which readily form layered dichalcogenide crystals, TX_2, where X=S, Se.

Group IV	V	VI
Ti	V	[Cr]
Zr	Nb	Mo
Hf	Ta	W

In addition to the hierarchy of the degree of anisotropy, there is also a varying degree of ionicity in the bonding between metal atoms and chalcogen atoms. Among the chalcogens, sulphur is the most electronegative and consequently the sulphides are in general the most ionic with the sequence being S > Se > Te. Similarly there is a sequence, for example with the group IV transition metals where the order of increasing ionic character is Ti, Zr, Hf. In this case Hf is the most electropositive.

It is convenient to consider three broad categories of intercalation complexes. These are:

(a) *Alkali metal atoms* such as Li, Na, K, Rb. We include also in this group Ag and poly-valent atoms such as Ca, In, Sn, Pb, etc.

(b) *Organic molecules* which are mainly nitrogen-containing molecules such as amines, amides, pyridine, ammonia, hydrazine, etc. Although ammonia (NH_3) and hydrazine (N_2H_4) are not organic molecules, their intercalation is very similar to the organic molecules and are included here.

(c)   *The "3d" transition metal atoms* such as V, Cr, Mn, Fe, Co, Ni. In fact atoms of all the elements of Table 2 below have been successfully intercalated under headings (a) and (c).

Table 2. Elements which have been successfully intercalated into transition metal dichalcogenides

Li												
Na												
K	Ca											
Rb	Sr	Ti	V	Cr	Mn	Fe	Co	Ni	Cu	Zn	Al	Si
Cs	Ba						Rh		Ag	Cd	Ga	Ge
										Hg	In	Sn
											Tl	Pb

Two main effects of intercalation can be identified as being of interest. These are:

(a)   Increase of layer separation. This is most conveniently achieved by intercalation of organic molecules, particularly the long chain amines and related systems. For example, Gamble and Geballe (1976) and their colleagues found that in a solid such as 2H-TaS$_2$ intercalated with n-octadecylamine, the layers can be separated by a bilayer of octadecylamine molecules with a dimension approaching 60Å where the thickness of the single layer is ~ 6Å. Since the interactions between the layers are now considerably weakened, this makes it possible to study the physical properties of nearly individual layers, such as superconductivity, electrical conductivity, etc.

(b)   Charge transfer from the intercalated molecule to the host lattice. This can radically modify the topology of the Fermi surface of metal compounds, and in some cases to the extent of inducing semiconductor-metal phase transitions. The most effective electron donors are of course the alkali metal atoms which can give up to one electron for every formula unit of the host compound. In this case, as in the case of intercalation by the first row transition metal elements, the increase in layer separation need not be large. Under these conditions the relatively weak layer interactions may be replaced by strong Coulomb (alkali) or covalent (transition metal) interactions and the solid becomes more nearly "three-dimensional".

We note that all the intercalants listed above act as electron donors or Lewis bases. No one has yet succeeded in intercalating with electron acceptors, as is often done with graphite, polyacetylene and other systems, and it is possible to explain this simply by the presence of the Coulomb repulsions between an ionised acceptor and the neighbouring chalcogen layers carrying an effective negative charge. This is probably greater than any electronic energy to be gained by intercalation. Complexes such as $P_{0.2}VS_2$ have been synthesised by Brec and Ouvrard (1983), but the P is not acting here as an acceptor, but forms PS_3 groups in which there is P-P bonding. More recent band structure studies (Guo and Liang, 1985) have indicated that the group VIII metal ditellurides are nearly fully covalent with negligible effective charge on the tellurium atoms. These tellurides may therefore be intercalated by acceptors. As suggested by Ghorayeb et al. (1984) the dimetal chalcogenides such as Hf_2S, Nb_2S and Tl_2S may also be readily intercalated by acceptors since the intercalant species should now enter between two layers of metal atoms.

The properties of the intercalation complexes will be discussed in turn according to the category of intercalant species above. However, before entering into the detailed description of the intercalate complexes, we must first review the structure and electronic properties of the host materials, including the theoretical treatment of the Fermi-surface driven structural distortions observed in one and two dimensional metals, if we are to understand any changes in behaviour that may take place. There are several themes common to the first two classes of intercalation complexes, and these are the occurrence of Fermi-surface driven Charge Density Wave-Periodic Lattice Distortions (CDW-PLD) and superconductivity. The 3d transition metal intercalation complexes are of interest for their magnetic properties and are treated as a self-contained topic.

2. PROPERTIES OF THE HOST MATERIALS

2.1 Structure and Polytypism

As shown in Fig.1, the layer of transition metal dichalcogenide crystals TX_2 is made up of a sheet of metal atoms, T, sandwiched between two sheets of chalcogen atoms, X. The arrangements of atoms within a sheet and the stacking of the atomic sheets are now described. Each sheet consists of atoms in a hexagonally close-packed network as shown in Fig.2(a) which also gives the three inequivalent close-packed positions A, B and C. Two types of coordination structure are possible, and either one or both can form the basic units of the crystal. In the trigonal prismatic coordination, the sandwich of X-T-X follows the AbA sequence, Fig.2(b), and in the octahedral coordination, the AbC sequence, Fig.2(c). I have used capital letters to denote the chalcogen positions and lower case letters for the metal atoms. In an ideal structure the distance between chalcogen atoms across the sandwich is the same as that in the plane, which is also equal to the **a** and **b** lattice constants as well as the distance between metal atoms (in plane). However, varying degrees of distortion are frequently observed. These distortions have a profound influence on the electronic structures, which in the first place have been determined by the two coordination types, as discussed in the next section. In addition to the two *intra*-sandwich coordination arrangements, there are many variations for *inter*-sandwich packings, and a large number of polytypes are known. The six most common polytypes are shown in Fig.3. The nomenclature usually followed to identify these polytypes is to follow the number of sandwiches required by obtaining a unit cell perpendicular to the plane of the layers by the overall symmetry (trigonal, hexagonal or rhombohedral) of the structure. Lower case subscripts are sometimes further required to distinguish polytypes otherwise similarly labelled (e.g. $2H_a$ and $2H_b$). Thus the simplest polytype with octahedral coordination, labelled 1T, has a repeat perpendicular to the layers of one sandwich, and trigonal symmetry; and the simplest trigonal prismatic coordination polytypes have two sandwich repeats, and are designated $2H_a$ and $2H_b$. The $2H_a$ structure is adopted by metallic group V materials such as NbS_2, and stacks the metal atoms directly above each other along the **c** axis, whilst in the $2H_b$ structure, adopted by semiconducting group VI materials such as MoS_2, the metal atoms are staggered. Octahedral interstitial sites between sandwiches are also shown in Fig.3 in parentheses. These are vacant sites into which intercalate atoms or molecules may enter. Note that in an ideal stacking, the

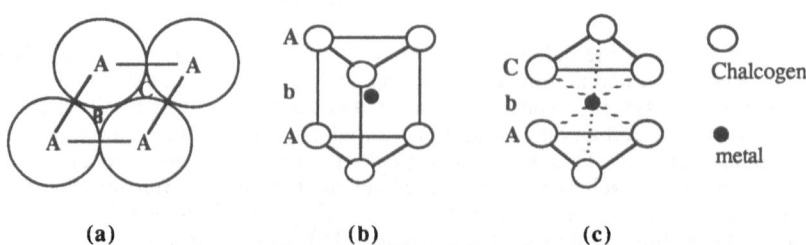

(a)   **(b)**   **(c)**

Figure 2. The octahedral and trigonal prismatic structures of transition metal dichalcogenides, TX_2. In (a), A, B and C represent the three inequivalent positions for close-packed stacking, and in (b) and (c), the capital letters designate chalcogen positions and the lower case letters the metal positions.

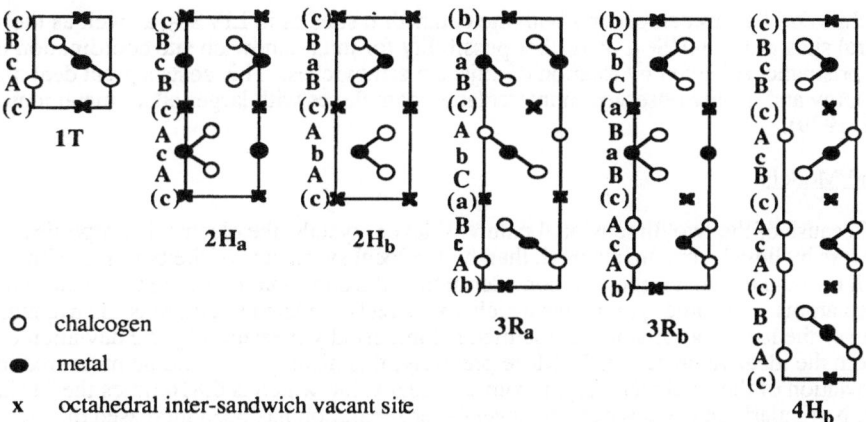

Figure 3. Unit cells of simple polytypes in the $(11\overline{2}0)$ projection.

chalcogen-chalcogen distance across two sandwich layers is again the same as that within a sandwich layer. In this case the c/a ratio is $\sqrt{(8/3)}$=1.633 or certain integral multiples of it.

The crystallographic axes for the T and H structures are those for standard hexagonal systems with $\mathbf{a}(x,y)$ parallel and $\mathbf{c}(z)$ perpendicular to the layers. Typical values are 3-4Å for the a parameter and 6Å for the sandwich repeat distance along the $\mathbf{c}$ axis. The 3R structures can have either trigonal prismatic or octahedral coordination. The primitive cell for the 3R structure is rhombohedral, with a $\mathbf{c}$ axis not perpendicular to the layer plane. The three layer repeat unit cell conventionally shown is therefore non-primitive. There are several mixed-coordination polytypes in which there is alternation of trigonal prismatic and octahedral coordination sandwiches; these include $4H_b$ shown in Fig.3. Stabilisation of these mixed coordination phases requires an interaction between different coordination layers, and it is considered that there is charge transfer from the octahedral to the trigonal prismatic coordination layer (Friend, Beal and Yoffe, 1977). Numerous other polytypes also exist and a more complete discussion can be found, for example, in the reviews by Hulliger (1976) and Marseglia (1983).

In addition to the octahedral interstitial sites, there are also tetrahedrally coordinated inter-sandwich sites which lie immediately below and above the chalcogen atoms, and are displaced from the symmetric plane between sandwiches. These are shown as the "a" and "b" sites in Fig.4. Octahedral interstitial sites are usually favoured in intercalation of group 1A alkali metals, Li, Na, etc., and the 3d transition metals, and the maximum uptake will give a limiting stoichiometry of MTX_2. Tetrahedral site occupation is found in intercalation complexes formed with silver and copper, and hydrazine; in this case the limiting stoichiometry is M_2TX_2 which is

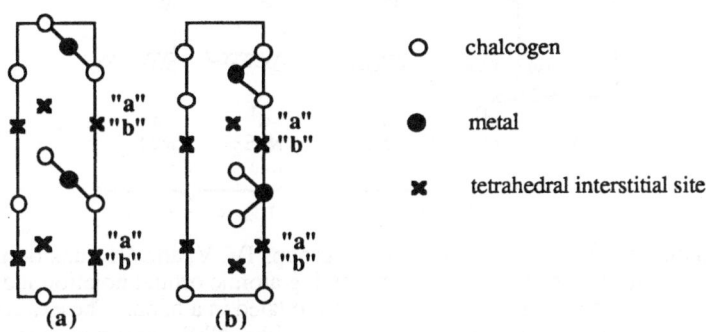

Figure 4. Tetrahedral inter-sandwich sites ✖ in (a) 1T, and (b) $2H_b$ structure.

rarely achieved. Li ions can occupy both the octahedral sites as in $LiVSe_2$, as well as the tetrahedral sites as in Li_2VSe_2. A further possibility for inter-sandwich site coordination is trigonal prismatic, achieved by rotation of adjacent sandwiches. This costs a great deal more lattice energy and is only observed in intercalation complexes with larger alkali metal ions (Rouxel, 1979).

2.2 Band Models

Because of the two dimensional nature of layer crystals, the electronic properties are affected less by how layers are stacked, than by the local symmetry of the bonding within a sandwich layer. The most important consideration is the coordination of the metal atom by chalcogen atoms. The other parameter which also affects the band structure is the trigonal distortion of the local coordination of the metal atom, usually measured by the deviation of the c/a ratios from the ideal value of 1.633. More precisely, this distortion should be measured in terms of the deviation of the sandwich height from the ideal value which is 0.816 times the "a" lattice constant, particularly in cases where the layer-layer separation has been increased due to intercalation. However, this information is not readily avalable, and we need to use c/a ratios as an indicator of such distortions. These ideas provide some clues to why one structure is preferred over another structure as electron filling increases, and why the c/a ratios for the group VIII dichalcogenides, for example, have values much smaller than 1.633. The schematic density of states diagrams for groups IV, V, and VI transition metal dichalcogenides are shown in Fig.5. These band diagrams were determined originally by energy loss experiments (Bell and Liang, 1976), but have since been modified to take into account results of recent optical measurements, photoemission, particularly angle resolved photoemission, and numerous band structure calculations. The effects of increasing electron filling on c/a ratios and hence on crystal structures and band structures are also shown.

There are two main energy terms to be considered in relation to the stability of a given coordination structure. These are the lattice energy term on the one hand which is based on ionicity and the Coulomb interactions between ions (the Madelung energy), and the electronic band energy term on the other, which depends on the energy position of the filled bands, particularly that of the occupied d_{z^2} band. The so-called d_{z^2} band is the lowest lying d sub-band of the d band manifold. As shown in Fig.5, the d_{z^2} band can either lie just below and overlap

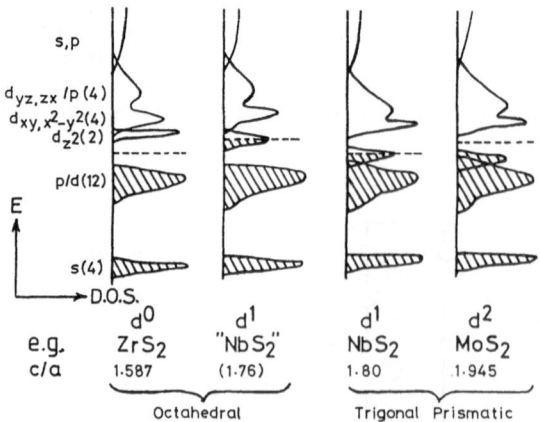

Figure 5. Schematic density of states diagrams for groups IV, V, and VI transition metal dichalcogenides. The numbers in parentheses following atomic orbital notation indicate the number of states per formula unit that can be accommodated in a band. The examples compound are taken from metals of the same row of the periodic table. NbS_2 does not crystallise in octahedral form and the second diagram represents an extrapolated hypothetical band diagram.

slightly with the remaining d conduction bands in the octahedral structure, or be detached from the remaining d conduction band by up to 1.5 eV as in the trigonal prismatic structure. If we assume the same charges are to be found on the ions, it can be shown (Shen and Liang, 1983; Johnstone and Scholl, 1984) that the ideal octahedral coordination symmetry has lower Madelung energy than the ideal trigonal prismatic coordination by about 20%. For this reason, the octahedral coordination structure is preferred when electronic energy is unimportant. The group IV metal dichalcogenides, having 16 valence electrons per formula unit, which fill the mainly chalcogen based p and s valence bands, are without any electrons in the metal d orbital based conduction bands (Fig.5). These materials are generally semiconductors with band gaps ranging from about 2 eV, such as in ZrS_2 and HfS_2, to 0.17 eV in TiS_2, and semimetals such as $TiSe_2$ and $TiTe_2$. The group IV metal dichalcogenides therefore naturally adopt the 1T octahedral coordination structure. With one or more electrons filling the d_{z^2} band as to be found in the group V and VI metal dichalcogenides, the crystals may trade off the lower Madelung energy of the octahedral coordination for the gain in electron band energy by lowering the d_{z^2} band relative to the other bands, which can be achieved by adopting the trigonal prismatic coordination.

The group VI metal dichalcogenides such as MoS_2 have two more electrons than the group IV, and the trigonal prismatic coordination is clearly advantageous for the energy of electrons. Thus all group VI compounds adopt trigonal prismatic coordination and they are semiconductors having a band gap of about 1.5 eV. The exceptions are β-WTe_2 and β-$MoTe_2$ (α-$MoTe_2$ is trigonal prismatic) which have distorted octahedral coordination and are metallic. It appears that the smaller gap between the top of the Te "p" band and the metal "d" band due to a higher energy tellurium p atomic level (cf. S and Se) has meant that the relatively small gain in electronic energy in lowering the d_{z^2} band is just insufficient to offset the cost in higher Madelung energy if the trigonal prismatic structure is adopted.

With only one electron more than group IV, the group V metal dichalcogenides such as TaS_2 are metals and both coordination types are found to be stable. In fact the energies that stabilise the two structures are finely balanced with only a small energy difference between them. It should be noted that a change in coordination must also be accompanied by a readjustment in the charges on the ions, so that the actual energy difference between the two structures is not as large as has been implied so far. This suggests that the trigonal prismatic coordinated structure is more covalent and evidence for this can be found in the study of infrared phonon spectroscopy (Lucovsky and White, 1973). The 1T materials exhibit strong Fermi-surface driven charge-density-wave induced periodic lattice distortions which begin forming above room temperature, and their room temperature conductivity is in the region of Mott's criterion for minimum conductivity for metallic conduction (Fazekas and Tosatti, 1979). The 2H structures, on the other hand, show rather better metallic properties than the 1T materials; the CDW distortions occur below room temperature and they are all superconductors. We will return to the topic of charge density waves in later sections.

The relationship between electron filling and c/a ratios, or trigonal distortion, and the energy of the system can be extended to account for the very low c/a values, ~1.3, found in group VIII metal compounds. There are, for example, 22 valence electrons in $NiTe_2$ which will fill the next lower d band and the crystal structure distorts in such a way as to result in a change-over between the position of the d_{z^2} band (which can take two electrons) and that of the d_{xy,x^2-y^2} band (which can take four electrons). A more detailed account of the effects of trigonal distortion on energy and crystal structure can be found in a separate review (Liang, 1984). Another way of looking at the shift in energy of the d_{z^2} band in relation to its occupation is to regard this as the splitting of energy bands due to Jahn-Teller distortions to lower the energy of the system.

Using linear muffin-tin orbitals and the atomic sphere approximation and including exchange-correlation corrections, Guo has calculated the self-consistent band structures of a number of layered compounds (Guo and Liang, 1985; 1986 and unpublished work). From these band structures the density of states curves can be calculated and Fig.6 shows a series of such curves for 1T-HfS_2, 1T-$PtSe_2$ and the antiprism structure 2H$_a$-Hf_2S. For HfS_2 in Fig.6(a), it can be seen that the Hf (d) orbitals make up most of the empty conduction bands, and S (s and p) orbitals most of the valence bands, and there is a relatively small amount of hybridisation between the two kinds of atoms. Although a considerable portion of the d band is occupied in the case of Hf_2S in Fig.6(c), there is still relatively little overlap in energy between the Hf and the

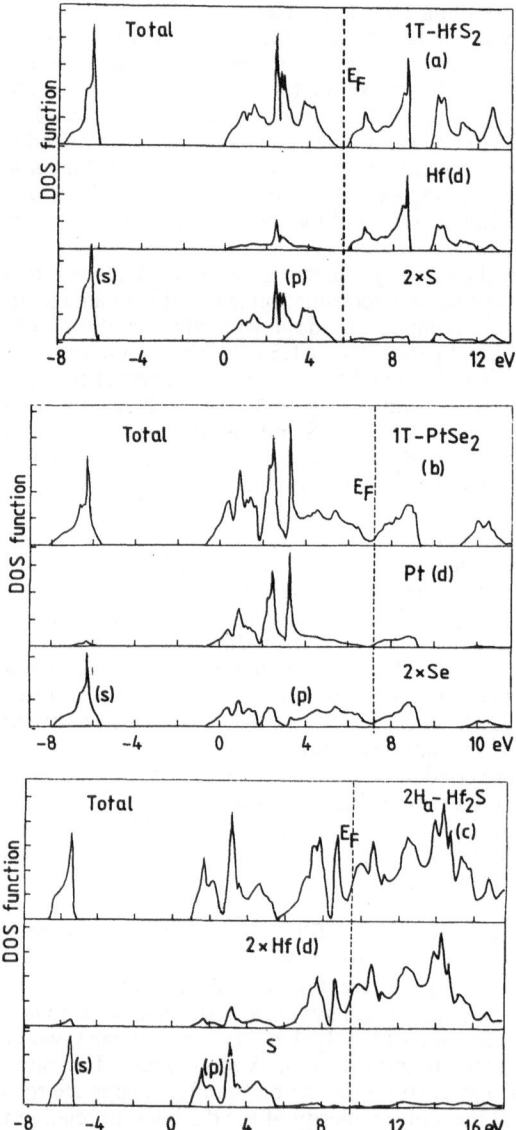

Figure 6. Density of states (DOS) curves from self-consistent band structure calculations using a linear muffin tin orbital method and an atomic sphere approximation, (a) $1T$-HfS_2, (b) $1T$-$PtSe_2$, and (c) $2H_a$-Hf_2S. For each compound, the top curve is the total DOS which includes contributions from all atomic orbitals used in the calculations, the second and the third curves represent only the contributions from metal d and chalcogen s, p orbitals respectively.

S bands. The resultant band structure clearly indicates that we have a d band metal. For $PtSe_2$, which is a semimetal, in Fig.6(b), the situation is quite different and the Pt (d) orbitals as well as the Se (s and p) orbitals are seen to contribute equally to the conduction and valence bands at both sides of the Fermi level. We expect $PtSe_2$ to be a covalent compound with the chalcogen layers carrying negligible net charge, and this (and other group VIII compounds) may be the only transition metal dichalcogenides which can be intercalated by electron acceptors. It should be noted that the self-consistent band structures and the density of states curves of Fig.6 give band gaps which are consistently smaller than those measured experimentally.

2.3 Rigid Band Approximation

The rigid band model is a useful first approximation for describing the changes in electronic properties of the host material after intercalation. How good such an approximation is will depend on a number of factors. The model assumes first that there is no mixing between the electron wave functions of the intercalant species and those of the host layers, otherwise new hybridised states can form which will modify the host electronic band structure. However, substantial mixings can take place if allowed by symmetry and if the two wavefunctions have similar energies. This is most likely to be the case with transition metals as intercalants, with the result that, $Fe_{1/4}TaS_2$ and $Cr_{1/3}NbSe_2$ for example, become more "three dimensional", the band structures are more dispersive along the c direction and the crystals are difficult to cleave. Next the integrity of the intra-sandwich structure is assumed. This assumption rarely holds in practice since the charge transfer which accompanies intercalation would normally be expected to change the ratio between sandwich height and the "a" lattice constant. As can be seen in Fig.5 that c/a ratios vary more or less continuously as the number of valence electrons increases from left to right, culminating in the change in coordination symmetry. It is proposed here that the position of the d_{z^2} band does not change suddenly from that corresponding to the octahedral structure on the left to that corresponding to the trigonal prismatic structure on the right with increased d_{z^2} band filling, but that it gradually shifts to lower energies following the increasing c/a ratios. Therefore, when an octahedrally coordinated material is intercalated and electrons are transferred to its empty or half full d_{z^2} band, there will be the tendency to increase the c/a ratio, with the consequence of lowering the d_{z^2} band energy. On the other hand, when a trigonal prismatic material with a full d_{z^2} band such as MoS_2 is intercalated, electrons enter the next higher d band and a lower c/a ratio with a change to the octahedral structure is possible. The arguments are the same as those given to explain the structure of the group VIII metal dichalcogenides, and this change in structure has been observed in $Fe_{1/2}MoS_2$ and $Na_{1/2}MoS_2$ (Py and Haering, 1983). We must therefore be aware of the limitations of the rigid band model. It is probably appropriate when states within a given band alone are considered, such as the situation of changing the Fermi

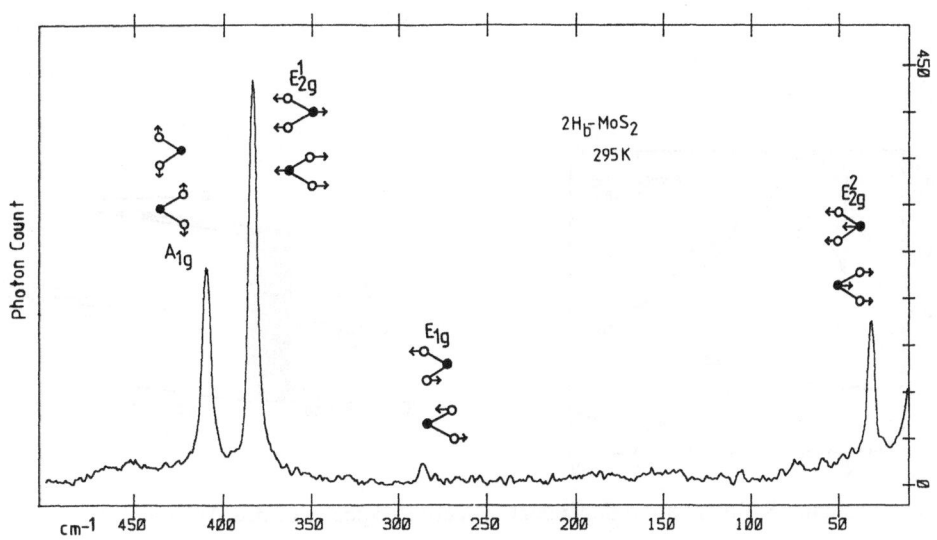

Figure 7. Room temperature Raman spectrum of $2H_b$-MoS_2.

surface in the d_{z^2} band following intercalation with a highly electro-positive metal. Another case when the rigid band model may be applied is in the group V compounds having trigonal prismatic structure, in which case the d_{z^2} band is already in its low energy position and has yet to be completely filled. The intercalant species should in all cases be highly electro-positive such as Na, K, or Cs, or strong Lewis base organic molecules.

2.4 Anisotropic Properties

Layered compounds exhibit anisotropic properties in a variety of ways. For example, mechanical properties are determined by the average interacting forces and are perhaps the most representative when comparing anisotropy. Optical properties are a measure of dielectric response, and transport properties are affected mostly by the electrons at the Fermi level. By fitting phonon frequencies to the linear-chain model, Bagnall et al. (1980) have calculated the force constants for a number of layered solids, for compressional as well as shear strains. For $2H_b$-MoS_2 the intra- and inter-layer shear force constants are 152 Nm^{-1} and 2.67 Nm^{-1} respectively, giving a ratio of 57. Next to graphite, $2H_b$-MoS_2 is perhaps the most anisotropic. The Raman spectrum, shown in Fig.7, very clearly demonstrates this aspect of anisotropy. The interlayer shear mode, $E_{2g}{}^2$, is at 33.5 cm^{-1}, while the intralayer shear modes, E_{1g} and $E_{2g}{}^1$, are at 297 cm^{-1} and 382.7 cm^{-1} respectively. Figure 8 shows the reflectivity spectra of

$3R_b$-WS_2, obtained using polarized light. The two spectra demonstrate not only the different selection rules for optical transitions with respect to electrical polarisation, but also the underlying anisotropy in the dielectric functions for fields in the layer as compared with fields perpendicular to the layer, which at low frequencies have a ratio of about 7.5. From the similarity in their band structures and the sulphur bounding layers of the van der Waals gap, we expect $2H_b$-MoS_2 and $3R_b$-WS_2 to be similarly anisotropic in both the force constants and the dielectric functions.

Conductivity measurements with current flow parallel and perpendicular to the layers often give rise to unexpected results. There are, to begin with, considerable inherent experimental difficulties associated with measuring conductivity perpendicular to the layers. We are concerned here, however, with theoretical expectations and parameters which may affect the outcome of measurements. Consider for example $1T$-TaS_2 and $1T$-$TaSe_2$. In their undistorted forms, their conductivity ratios are about 500 and 20 (Hambourger and Di Salvo, 1980), and yet in almost all other respects these materials are similar. To understand this wide difference in their conductivity ratios we must consider their band structures and Fermi surfaces in some detail. Figure 9 shows the band structures near the Fermi level and the Fermi surface cross sections for $1T$-TaS_2 and $1T$-$TaSe_2$, based on calculations by Wooley and Wexler (1977). The band

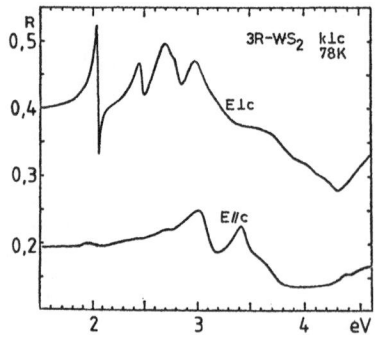
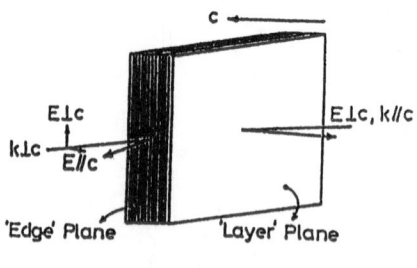

Figure 8. Polarisation dependent reflectivity spectra of $3R_b$-WS_2 at 78K.

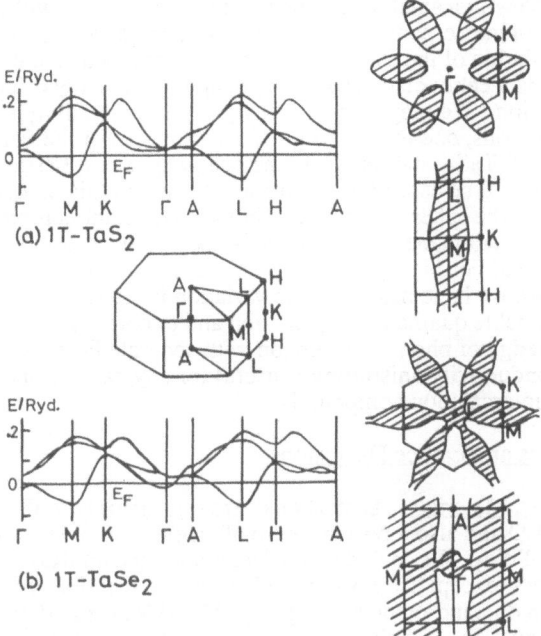

Figure 9. Band structures and Fermi surface cross-sections for (a) 1T-TaS$_2$ and (b) 1T-TaSe$_2$ (after Wooley and Wexler, 1977). The undulations of the Fermi surface cross sections have been exaggerated for the purpose of illustration.

structures are generally similar as would be expected, but for the small dip at Γ below the Fermi level in 1T-TaSe$_2$. This small difference has manifested itself in the topology of the Fermi surface as shown on the right of the figure. In the ΓMK plane of the Brillouin zone, the six ellipsoidal cross sections appear to be joined at Γ in 1T-TaSe$_2$, and in planes containing the AΓA axis there is now a small but closed Fermi surface with more or less spherical shape. We argue that it is the presence of this extra Fermi surface, which closes in the $\mathbf{c}$ direction, that is responsible for the low conductivity ratio in 1T-TaSe$_2$.

We can write the conductivity formula (see also Issi in this volume) as in (1). This is particularly useful when the Fermi surface is strongly anisotropic as is usually found in layer materials.

$$\sigma_n = e^2 <\tau\, v_F> . \, S_n \,/(2\,\pi^2\, h) \tag{1}$$

where $<\tau\, v_F>$ is the product of the mean free time of collision for the carriers and the Fermi velocity, averaged over the Fermi surface, S_n is the projection (an area in $\mathbf{k}$-space) of the Fermi surface on a plane normal to the applied field, and σ_n is the conductivity tensor along the field direction. The term containing τ becomes the mean free length in the classical limit with a directional independent τ, and S_n contains information on both the carrier density and the effective mass tensor component in the relevant field direction. τ is, in general, a function of direction, as for example, when τ is due to phonons, then the wave vectors and the density of states spectra of the phonons must be taken into account. The term $<\tau\, v_F>$ is in fact rather difficult to estimate but it would seem reasonable to assume that it varies with direction by no more than a factor of 10, and that the variation is similar between materials such as 1T-TaS$_2$ and 1T-TaSe$_2$. The effects on σ_n due to S_n on the other hand can be much greater. We note that a cylindrical Fermi surface with a small undulation of its cross section will have a small value of S_n (of a thin annulus) on a plane perpendicular to the c-axis, and a low conductivity in the c direction. A weak undulation of the cylindrical Fermi surface means of course that the energy band at the Fermi level is weakly dispersive along c, and has a heavy effective mass tensor component in that direction. We may therefore analyse the conductivity in 1T-TaSe$_2$ as being made up of two contributions, one from the cylindrical portions of the Fermi surface which is similar to that found in 1T-TaS$_2$, and one from the nearly spherical Fermi surface centred at Γ. The second contribution will be essentially isotropic. Based on this model, we may deduce that the cross sectional area of the sphere is about 25 times the total annular projection of the cylindrical Fermi surface on a plane perpendicular to the c-axis, and about 1/20 of the total projection of the cylindrical Fermi surface on a plane containing the c-axis. These estimates must be an upper limit because we have neglected the variation in $<\tau v_F>$ with direction. Nevertheless they give at least a reasonable qualitative explanation and to obtain a more quantitative picture we need to combine knowledge of phonon dispersion with accurate Fermi surface calculations. This method of analysing conductivity anisotropy is useful for any two- or one-dimensional system including graphite and intercalation compounds.

2.5 Electronic Properties and Lattice Distortions

2.5.1 Undistorted Materials. Distortions here refer mainly to those caused by the Fermi surface driven CDW-PLD and also a possible case of an electron-hole coupled or ferroelectric type phase transition in TiSe$_2$. We include therefore among the undistorted materials all the semiconducting members (except TiSe$_2$) and NbS$_2$ from the metallic group V compounds. From the foregoing discussion on band models, the group IV and VI metal dichalcogenides are expected to be semiconductors, with the exception again of TiSe$_2$ and many tellurides, where overlap between valence and conduction bands gives rise to semimetallic properties.

The lower bandgap semiconductors, such as ZrSe$_2$ with a gap of ~ 1.2eV, and TiS$_2$ with a gap of ~ 0.17eV can only be grown with a small excess of metal present, and this results in partial filling of the d conduction band and "extrinsic" semiconducting properties. The metal excess occupies intersandwich sites, and the crystals can be regarded as self-intercalates. Crystals of ZrSe$_2$ investigated by Klipstein et al. (1984) showed electron concentrations of ~ 5 x 10^{19}cm^{-3} but TiS$_2$ which has been much more extensively investigated, shows a range from 1.4 x 10^{20}cm^{-3} up to 3 x 10^{21}cm^{-3} for crystals with ~ 5% excess titanium. These extrinsic

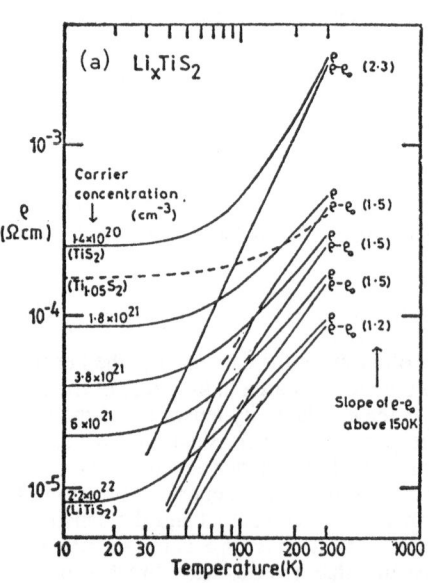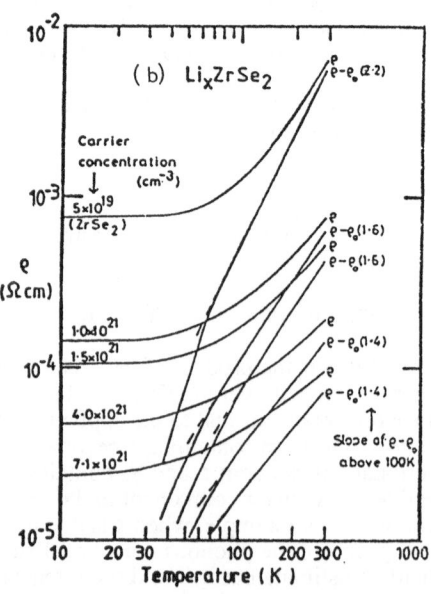

Figure 10. Logarithm of the resistivity, ρ, and of the temperature dependent part of the resistivity, ρ–ρ$_0$, plotted against the logarithm of the temperature for samples of (a) Li$_x$TiS$_2$ and (b) Li$_x$ZrSe$_2$.

electrons occupy states in small regions centred on the Brillouin zone boundary faces along LML (see Fig.9). Because of the mass anisotropy of these electron states, the electrons may be regarded as being confined to move in the two dimensional layers only. Furthermore, since the charged impurities are lodged in the intersandwich sites, the situation here is not unlike that of semiconductor heterojunctions with modulation doping. The main difference is of course in the thickness of the junction layers and in the fact that we are dealing with d electrons which are heavy and have low mobility. Nevertheless, these excess electrons offer an opportunity to study the temperature dependence of intervalley scattering and the effect this has on the non-linear temperature dependent resistivity as shown in Fig.10 (Klipstein et al., 1984). Figure 10 also contains the temperature dependent resistivity curves of Li intercalated samples, to which we will return later. The carrier concentrations indicated on the left of each curve have been determined from the temperature independent Hall coefficients, and the numbers in parentheses on the right give the slopes of $\log(\rho - \rho_0)$ versus $\log T$ above 150K in the case of TiS_2 and above 100K in $ZrSe_2$, where ρ_0 is the temperature independent residual resistivity. Writing resistivity ρ in the form

$$\rho = \rho_0 + AT^n \tag{2}$$

we find the exponent n to be 2.3 and 2.2 for the highest purity TiS_2 and $ZrSe_2$ samples respectively over a wide temperature range (30K < T < 300K for TiS_2) and n falls to just over one with samples containing a higher concentration of carriers. This superlinear temperature dependence of the resistivity has been modelled (Klipstein et al., 1981) by a combination of a linear dependent intra-valley scattering by acoustic phonons and an exponentially dependent inter-valley scattering by much higher energy zone boundary longitudinal acoustic phonons. As the valley states are filled by increased carrier concentration coming from impurities or intercalation, the distinction between the phonons involved in inter- and intra-valley scattering is removed, and "normal" linear temperature dependent metallic behaviour is restored.

As mentioned already, $TiSe_2$ and many group IV metal tellurides such as $TiTe_2$, and $HfTe_2$, are semimetals with overlap between the top of the chalcogen p valence band and the bottom of the transition metal d conduction band. Strong evidence for this is the pressure-dependence of conductivity and Hall coefficient in both $TiSe_2$ (Friend et al., 1982) and $HfTe_2$ (Klipstein et al., 1986), and angle-resolved photoemission measurements, which indicate a band-overlap of about 100meV for $TiSe_2$ (Anderson et al., 1985) and 600meV for $TiTe_2$ (de Boer et al., 1984). Photoemission measurements confirm directly the band models discussed earlier, and that the top of the valence band is at the zone centre, and the bottom of the conduction band at the Brillouin zone faces, for both $TiSe_2$ and $TiTe_2$ (there are no reports of measurements on $HfTe_2$). The transport properties of these materials are determined by both the electrons and the holes present, and a more complicated behaviour in the temperature dependence of the Hall coefficient, thermopower, etc. than in the extrinsic semiconductors discussed above is observed. There is no evidence for a structural distortion in either $TiTe_2$ or $HfTe_2$, as in all semiconductor group IV compounds, but $TiSe_2$ undergoes a structural distortion below 202K which will be discussed in the next section.

2H-NbS_2 is the only group V metal dichalcogenide which does not show structural distortions. All other compounds from the group exhibit charge density wave induced lattice distortions at low temperatures related to the shape of their Fermi surfaces. The trigonal prismatic structure group VI metal dichalcogenides such as MoS_2 and WS_2 are all semiconductors with band gaps about 1.5eV wide and are stable at all temperatures.

2.5.2 <u>Charge Density Wave induced Periodic Lattice Distortions (CDW-PLD)</u> For a metal having Fermi wavevector k_F, the period of the density fluctuations of electrons near the Fermi surface is π/k_F. Under certain conditions, the potential set up by these charges is sufficient to displace the atoms of the lattice so that the system can be driven to a lower energy state by undergoing a periodic lattice distortion (PLD). The possibility of such a phase transition was first conceived by Peierls in 1955 who proposed that a one-dimensional metal was inherently unstable, being able to lower its band energy by distorting so that an energy gap, 2Δ, was opened at the Fermi level; the one-dimensional metal underwent a phase transition and became an insulator with its new unit cell given by π/k_F. This is illustrated in Fig.11 for the case of a half-filled one-dimensional band, i.e. $k_F = \pi/2a$ (Fig.11(c)) and the density of electrons at the Fermi level fluctuates with a period = 2a (Fig.11(a)), where a is the regular interatomic distance.

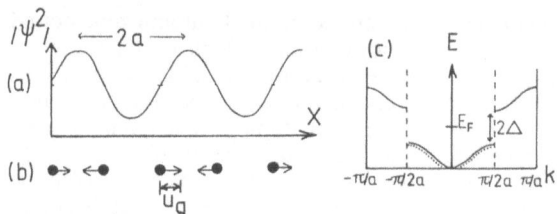

Figure 11. Peierls' distortion for a one-dimensional metal with a half-full band. (a) represents the density, $|\psi^2|$, of electron wavefunctions at the Fermi level varying with period = 2a, (b) indicates the displacements of atoms locking into a low energy phase with respect to the electron density fluctuations, and (c) shows the resultant band structure of the distorted lattice with a band gap 2Δ opened at $k_F = \pi/2a$.

If u_q represents the displacement of the atoms following the distortion, and $q = 2 k_F$, then

$$\Delta = 2 v \sin (\pi u_q/a) \qquad (3)$$

where v is the deformation potential and here represents the strength of the electron-phonon interactions. For small u_q, then Δ is proportional to u_q. The integrated band energy of the occupied states, as a result of opening a band gap of 2Δ at $\pm k_F$, may be estimated to be

$$\Delta E_{band} \sim -\Delta^2 N(E_F) \qquad (4)$$

where $N(E_F)$ is the density of states at the Fermi level.

We must also consider two further terms which are the consequence of the periodic lattice distortion. One is the strain or elastic energy

$$\Delta E_{elastic} = +M\omega^2 u_q^2/2 \qquad (5)$$

where M is the ion mass and ω is the mean phonon frequency. Another term is the Coulomb repulsion energy which increases as a result of localising electrons. This term is more difficult to estimate but must be of the form

$$\Delta E_{Coulomb} \sim +(e \Delta N(E_F))^2 / (8\pi\varepsilon_o R) \qquad (6)$$

where R represents some effective distance between the electrons involved. R may typically be, say, five to ten times the CDW period (2a), since we are interested in evaluating the difference in Coulomb energy before and after the formation of CDW's. It should be pointed out that while ΔE_{band} is linearly proportional to $N(E_F)$, the Coulomb repulsion term is proportional to the square of $N(E_F)$. This makes the $\Delta E_{Coulomb}$ term very important for materials with a narrow conduction band, since a narrow band implies a high density of states. It is a simple exercise to carry out further estimates using the nearly free electron approximation in one dimension by substituting $N(E_F)$ by $16m^*a/h^2$.

We are content, however, merely to note that the linear metal will be driven by the CDW to lattice distortions only if the gain in electron band energy can overcome the increase in the elastic and Coulomb energies, i.e.

$$\Delta E_{band} + \Delta E_{elastic} + \Delta E_{Coulomb} < 0, \qquad (7)$$

and that to first order, each of these energy terms is proportional to the square of u_q. This means that periodic distortions are not inevitable and that we need to look for higher order terms (e.g. fourth power in u_q from the expansion of Δ^2) in order to obtain a minimum in the total energy of the system. The strength of the CDW-PLD will of course be determined by the size of this energy minimum. It should also be noted that as is usually the case when both linear and quadratic terms (of $N(E_F)$) are present, the linear term dominates at low values, so that $\Delta E_{Coulomb}$ does not become prominent until $N(E_F)$ is quite large such as the case in VSe_2 to be discussed later. At finite temperatures, within the mean-field theory, excitation of electrons across the Peierls gap will reduce the size of the gap to zero at the Peierls transition temperature, T_P, which is related linearly to Δ.

The classic form of the band energy ΔE_{band} evaluated in one dimension for a nearly free electron band is a logarithmic expression involving the Fermi energy E_F of the undistorted form,

$$\Delta E_{band} = -(\Delta^2 / E_F) \ln (E_F/\Delta) \tag{4'}$$

This is equivalent to (4) when expanded in a polynomial form. Although (4') can lead to the BCS type of expression for the Peierls gap which can be made to compare with the super-conducting gap, there is no special merit in this since it neglects the Coulomb repulsion energy and has assumed a nearly free electron band. (4) on the other hand, emphasises the importance of the density of states, $N(E_F)$, at the Fermi level on the nesting energy. There are many materials which have been synthesised and investigated during the past decade, with chain or stack structures, which do show extremely anisotropic quasi one-dimensional conduction band structures. These include organic charge transfer salts such as TTF-TCNQ, linear chain platinum salts such as KCP, and several transition metal trichalcogenides, including $NbSe_3$. The formation of low temperature superstructures together with a loss of metallic properties in these materials is well-established (Friend and Jerome, 1979; Monceau, 1985). For materials with a two- or three-dimensional band structure the system is less susceptible to a Fermi-surface driven distortion of the kind discussed above. We can however generalise our analysis of the response of the electron system to such a distortion. The necessary "one-dimensional" characteristic that is exhibited by the Fermi surface of metals likely to distort in this way is that it should be possible, by uniform displacement of one section of the Fermi surface by a wavevector q, to superimpose it onto another section. This feature is termed Fermi surface "nesting" and is illustrated in Fig.12. The "nesting" characteristics of the Fermi surface can be seen by evaluating the non-interacting electron gas susceptibility, χ_q, defined as

$$\chi_q = \Sigma (n_{k+q} - n_k) / (\varepsilon_k - \varepsilon_{k+q}) \tag{8}$$

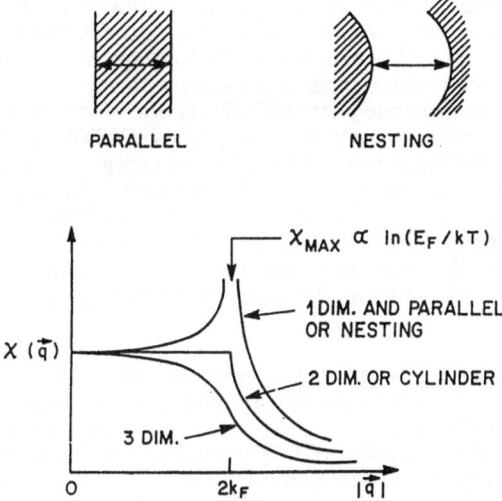

Figure 12. Nesting and free-electron gas susceptibilities for one, two and three dimensional metals.

where n_k is the Fermi occupation factor for state $\mathbf{k}$ with energy ε_k. The variation of χ_q with $\mathbf{q}$ for simple one-dimensional, two-dimensional (cylindrical) and three-dimensional (spherical) free-electron Fermi surfaces is shown in Fig.12. For strong "nesting", χ_q function should have a maximum for a specific $\mathbf{q}$, and it is clear that for the one-dimensional case, χ_q diverges at $\mathbf{q} = 2k_F$ as $\ln(E_F/kT)$, but that, although there is a sharp change in slope at $2k_F$ for the cylindrical Fermi surface, no peak is obtained. Note that these calculations of χ_q have not taken the effects of Coulomb interactions (6) into account.

Many of the metallic layer structure transition metal dichalcogenides do show structural distortions which are related to the Fermi surface as discussed in the following section, and it might thus be expected that these materials exhibit significant nesting over at least a fraction of the Fermi surface. However calculations based on band structures for these materials show only very weak peaking of χ_q in the vicinity of the observed PLD wavevector (Myron et al., 1977; Doran et al. 1978). To obtain a strong peak in χ_q there needs to be a very precise matching of sections of the Fermi surface and it is not certain if the level of accuracy in the calculations is sufficient for this purpose. Doran et al. (1978) nevertheless argued that the presence of periodic lattice distortions in these materials must be attributed to strong electron-phonon coupling instead of Fermi surface nesting. The two-dimensional nature of the bonding in these layer materials is an important factor in reducing the elastic strain energy, $\Delta E_{elastic}$. The relatively weak inter-sandwich force constants allow the strain energy associated with intralayer distortions which modulate the sandwich thickness to be low, and in this respect the distortions occurring in layer structure materials should be compared with the surface reconstructions commonly observed on clean metal surfaces such as W or Mo.

2.5.3 <u>Incommensurate and Commensurate Superlattices</u>. The Fermi-surface controlled charge density wave (CDW) coupled periodic lattice distortion (PLD) is determined by the wavevector spanning the "nested" portions of the Fermi surface. Although calculations indicate weak "nesting" in the layer materials, its contribution to superlattice formation is clearly significant, and a direct relationship exists between the size of the Fermi-surface and the wavevector of the PLD-CDW, as will be discussed next.

The CDW may be commensurate with the crystal lattice forming a superlattice with wavelength

$$\lambda_{superlattice} = (n/m)\,a \tag{9}$$

where n, m are integers ($n \geq 2m$) and a is the ion-ion spacing in the undistorted lattice. More generally the CDW is incommensurate with the crystal lattice in which case it has no preferred phase relation to the lattice. The commensurate superlattice can adjust its phase with respect to the lattice to minimise potential energy, and the strength of this "lock-in" energy is also a strong function of the superlattice period, favouring smaller periods. CDW-PLD's with a large lock-in energy will produce large displacements of atoms. In consequence, a number of the layer-structure materials show a sequence of phase transitions with varying temperature. Cooling from high temperatures, an incommensurate phase (ICDW) is often formed initially, followed by a transition, or series of transitions, to commensurate phases (CCDW) as the amplitude of the distortion builds up at lower temperatures and the lock-in energy becomes stronger. These transitions have been beautifully captured by the electron diffraction experiments taken over a range of temperatures on $1T\text{-}TaS_2$ by Williams et al. (1974) and reproduced in Fig.13. This series of transitions has been extensively modelled by Landau theory, initially by McMillan (1975), and there is considerable complexity possibly as a result of interactions between distortions in the three trigonal symmetry directions in the plane of the layers, and between CDW's across the layers.

The incommensurate CDW is an unusual phenomenon and has many unfamiliar properties. For the simple sinusoidal distortion amplitude so far considered, the CDW has no preferred phase with respect to the lattice. This phase invariance allows the possibility of a frictionless or lightly-pinned sliding of the CDW with respect to the lattice. This possibility was first noted by Frohlich (1955) as a current-carrying process, and there is excellent experimental evidence in the non-linear transport properties of linear chain conductors such as $NbSe_3$ for this "sliding mode" conductivity (Monceau, 1985). The invariance of the incommensurate distortion energy with respect to the phasing with the lattice may be lost if the distortion no longer takes a

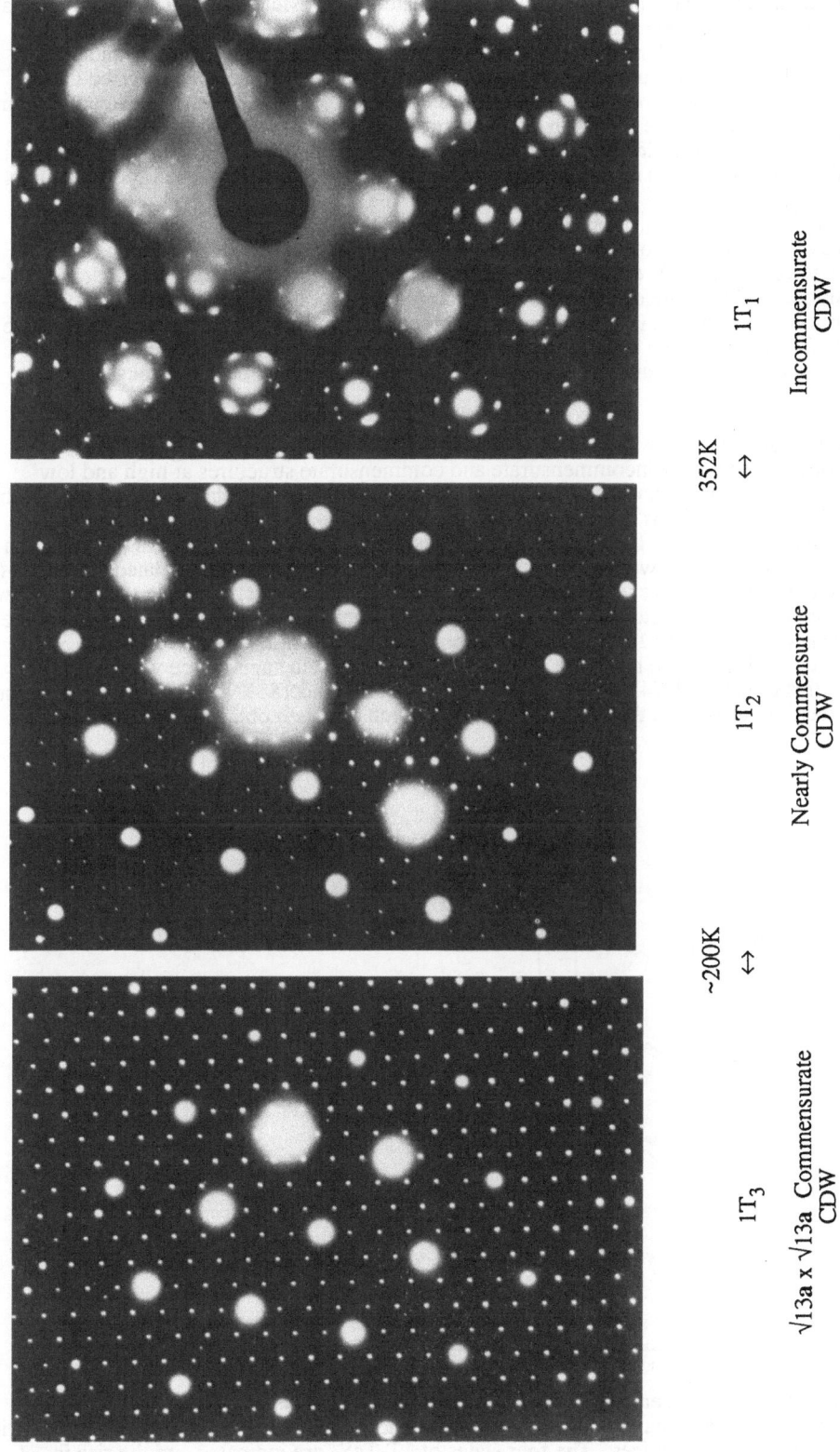

1T₃
√13a x √13a Commensurate
CDW

~200K
↔

1T₂
Nearly Commensurate
CDW

352K
↔

1T₁
Incommensurate
CDW

Figure 13. Electron diffraction in 1T-TaS₂ (after Williams et al., 1974). The high intensity spots in 1T₃ represent the undistorted lattice, while the weaker spots are due to the CDW-PLD superlattice.

simple sinusoidal form. This possibility was first considered by McMillan (1976) who realised that when close to commensurability, an incommensurate superlattice can still arrange to take advantage of the preferred phasing to the lattice if it arranges to be mostly in-phase and to accommodate the incommensurability in regions of rapid phase-slip. McMillan termed these regions discommensurations but they are also loosely called solitons in one-dimensional systems. Discommensurate superlattice phases in the layer structure materials can in some cases be observed directly by dark-field electron microscopy, and the studies of $2H_a$-$TaSe_2$ by Fung et al. (1981) and Chen et al. (1982) reveal a wealth of complexity and detail in these superstructures.

2.5.4 <u>CDW-PLD Transitions in Group V Metal Dichalcogenides</u>. The temperature dependence of the resistivity of the octahedral coordination 1T and trigonal prismatic 2H poly-types of some of the group V compounds is shown in Fig.14. The anomalies are due to CDW-PLD transitions, and they are clearly much stronger in the 1T polytypes. Since these are quasi-two-dimensional structures it is not expected that all of the Fermi surface should be destroyed by CDW-PLD formation, and so these materials should remain metallic at all temperatures. The behaviour of the two coordination structures is very different and it is best to discuss them separately.

(i) <u>1T polytypes</u>. The two 1T polytypes of tantalum, TaS_2 and $TaSe_2$, both exhibit strong distortions, with both incommensurate and commensurate structures at high and low temperatures. Their complex CDW-PLD's at various temperatures have been observed by both electron diffraction (Williams et al., 1974; Wilson et al., 1974, 1975) and transport measurements. On cooling 1T-TaS_2, the lattice transforms at 543K (Bayliss et al., 1984) to an incommensurate CDW state with a triple distortion along the three symmetry-related directions (at 120° to each other) with wavevector $0.283a^*$ ($\lambda = \sqrt{13}a$). This is the $1T_1$ phase. At 352K there is a second transition to a "nearly-commensurate" or "discommensurate" phase, $1T_2$, where the superlattice vector rotates by $11^\circ6'$. The distortion finally locks in to a commen- surate $3a + b$ superlattice ($\lambda = \sqrt{13}a$) phase below 183K, the $1T_3$ phase, and the superlattice vector rotates once more, this time by $13^\circ54'$. The two transitions at 352K and 183K are clearly visible in the resistivity data in Fig.14, and the transition at 543K which has been obtained by calorimetric

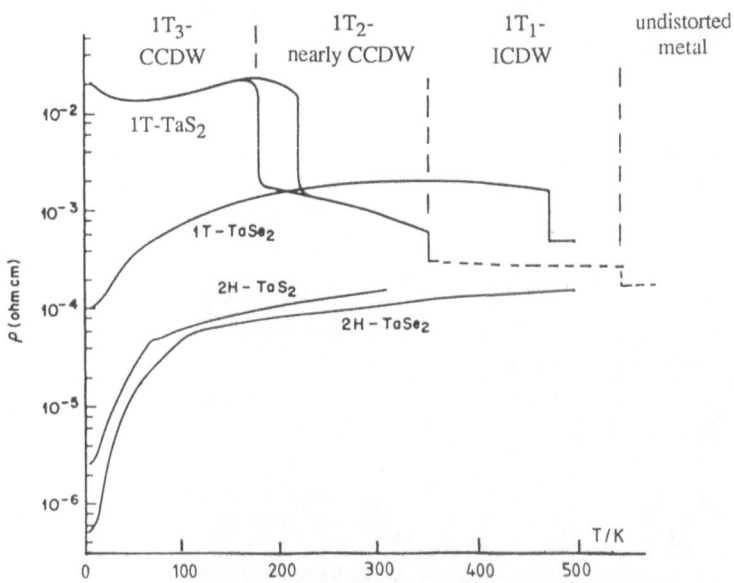

Figure 14. Temperature dependence of the in-plane resistivity of several group V metal dichalcogenides. Sudden change in slopes or in absolute values are associated with the formation of CDW-PLD's and superlattices. The four states of 1T-TaS_2 are indicated and the rise in resistivity at low temperatures is associated with localisation effects.

measurements is added here by the dotted curve. The strong hysteresis evident at the 183K transition actually arises because the transition from the low temperature $1T_3$ phase on warming is to a different discommensurate phase triclinic symmetry which returns to the $1T_2$ phase at 283K (Bayliss et al., 1984).

The "nesting" of the Fermi surface in $1T$-TaS_2 is considered to occur across the individual electron pockets and over large sections of the Fermi surface as shown in Fig.15(a). The clearest evidence that the periodicity of the CDW-PLD superlattice is related to the size of the Fermi surface comes from the electron diffraction study by Wilson et al. (1974) of the alloy system $1T$-$M_xTa_{1-x}S_2$. When M=Ti which has one fewer electron than Ta, the size of the superlattice wavevector is observed to scale as $(1-x)^{1/2}$ over nearly the entire range $0 < x < 1$ in the same way as the linear dimensions of the two-dimensional Fermi surface are expected to change. On the other hand, when M=V or Nb from the same group, the superlattice wavevector remains unchanged with x. This is of course what we would expect if the d band filling remains at the same level.

The CDW-PLD is particularly strong in $1T$-TaS_2. Indeed from X-ray diffraction experiments, Brouwer and Jellinek (1976, 1980) estimated Ta atom displacements parallel to the layer of approximately 0.25Å in the commensurate ($1T_3$)phase. The displacement pattern is

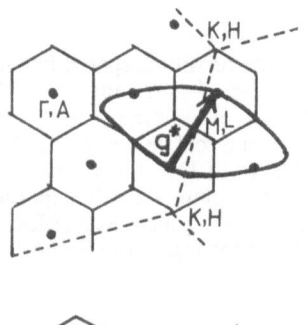

(a) $1T$- TaS_2, cross-section of one of the six Fermi surfaces along LML axes of the Brillouin zone. Very good "nesting" is obtained with $\mathbf{g^*} = \mathbf{a^*}/\sqrt{13}$.

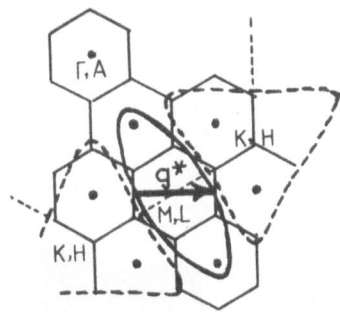

(b) $1T$- VSe_2, cross-sections of Fermi surfaces of both electron pockets - solid curve taken from the ALH plane, and hole pockets-dashed curves taken from the

ΓMK plane. The electron surfaces are providing a reasonably good "nesting", but hole surfaces do not nest. Tilting the superlattice vector in the **c** direction with $\mathbf{g^*} = \mathbf{a^*}/4 + \mathbf{c^*}/3$ can improve nesting.

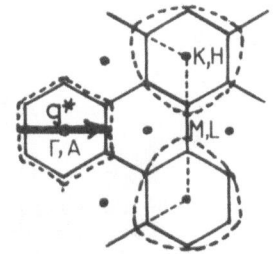

(c) $2H$-$NbSe_2$. At high temperatures, Fermi surfaces consist of hole surfaces

with one cylinder along $A\Gamma A$ and six cylinders along HKH axes of the Brillouin zone. Any nesting occurs only at the centre pocket. The Fermi surfaces become electron pockets after reconstruction, with $\mathbf{g^*} = \mathbf{a^*}/3$.

Figure 15. High temperature (undistorted) Fermi surface cross sections of several group V metal dichalcogenides. The reciprocal lattice of the normal structure and that of the superlattice structure are also shown. The Brillouin zone notations (refer to Fig. 9) are for the undistorted lattice.

shown in Fig.16(a) and we see the formation of 13 atom clusters or "stars". The star consists of two rings of six atoms with a 13th atom in the middle. The twelve atoms in the outer and inner rings are in different field environments, and this gives rise to a very large Ta 4f level splitting, of 0.6eV, seen in XPS measurements (Hughes and Pollak, 1976) which are interpreted by Wertheim et al. (1975, 1976) to indicate a CDW amplitude of as much as one electron. Recent high-resolution angle-resolved photoemission experiments performed by Smith et al. (1985) show directly the appearance of multiple d-band dispersion as a consequence of Brillouin zone reconstruction. Smith et al. (1985) have also carried out simple LCAO calculations for the Ta d band reconstruction using the structural data of Brouwer and Jellinek (1976, 1980), and the resultant density of states, N(E), is shown in Fig.16(c). The flat, two-dimensional N(E) in the undistorted state is broken into several sub-bands separated by typically 0.2eV. The lower two can each hold 6 electrons of the √13**a** x √13**a** superlattice cell, so the 13th electron must go into the bottom of the next band. The formation of these sub-bands thus removes 12 out of 13 electrons which were available for transport in the undistorted form at high temperatures. The fraction of 12/13 is in fact theoretically the largest fraction of electrons that can be removed by a √13x√13 reconstruction. This explains jumps in the resistivity of two orders of magnitude between 400K and 180K. Furthermore, the occupied width of the partially filled sub-band is only 50meV, and it is thus susceptible, in the presence of disorder, to a Mott-Anderson transition. This accounts for the upturn in resistivity at low temperatures seen in Fig.14.

1T-TaSe$_2$ shows similar behaviour to 1T-TaS$_2$, with the same low-temperature commensurate superlattice, but the transition from this phase at 470K is directly to the incommensurate phase.

1T-VSe$_2$ is the only vanadium dichalcogenide that can be readily prepared in a nearly stoichiometric form. It is an unusual material in a number of ways. Its c/a ratio of 1.829 is the highest known for 1T octahedrally coordinated materials. It cleaves extremely easily and when intercalated tends to decrease the c/a ratio. We will deal with its intercalate complexes later. In contrast to the 1T tantalum materials, 1T-VSe$_2$ forms weak CDW-PLD's at low temperatures. It shows a transition from the undistorted state to a commensurate in-plane **4a** x **4a** superlattice at 112K, although, curiously, it chooses an incommensurate repeat along the **c** axis (Moncton et al., 1979; Tsutsumi, 1982). The behaviour of the resistivity through the transition is similar to that

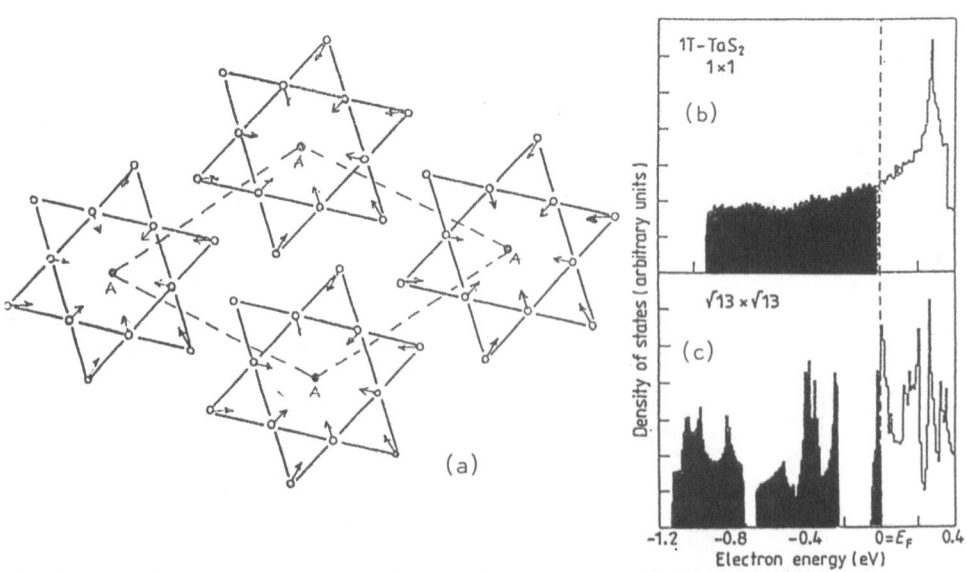

Figure 16. CDW-PLD in 1T-TaS$_2$, (a) displacement (~0.25Å) of Ta atoms in the superlattice state, (b) density of states for the normal (undistorted) state, and (c) density of states for the √13**a** x √13**a** superlattice state.

seen for the 2H polytypes in Fig.14. The anomalous behaviour of 1T-VSe$_2$ is related to the size of the vanadium atom compared with that of the selenium atom. As the **a** lattice parameter is determined largely by the larger of the two ions, the V atoms are separated by distances almost too large for its d orbitals to have sufficient overlap to form bands. The result is that very narrow d bands are formed, and Hughes et al. (1978) and Johnson et al. (1986) report an occupied d bandwidth of only 0.3eV from photoemission measurements, in contrast to the corresponding values of 1.5eV and 1.15eV reported by Traum et al. (1974) for 1T-TaS$_2$ and 1T-TaSe$_2$ respectively. As discussed earlier, the narrow d band and the high density of states at the Fermi level enhances the effects of Coulomb repulsion, which opposes the formation of a CDW-PLD. Support for such a view comes from high pressure experiments. Hydrostatic pressure which usually depresses CDW formation, raises the transition temperature in VSe$_2$ (Friend et al. 1978), shown in Fig.17, and this is considered to arise through a pressure-induced broadening of the vanadium d band. In addition to the significant Coulomb repulsion, the "nesting" of the Fermi surface is also considered to be poor in 1T-VSe$_2$ as shown in Fig.15(b). It appears that good "nesting" can only be obtained with a distortion vector $\mathbf{g}^* = \mathbf{a}^*/4 + \mathbf{c}^*/3$. We shall return to the consequences of narrow d conduction bands in relation to CDW-PLD formation when we discuss the behaviour of intercalated group IV dichalcogenides, and to the magnetic properties in later sections.

(ii) <u>2H polytypes</u>. CDW-PLD formation in the 2H polytypes is weaker than in the 1T polytypes, and the characteristic commensurate superlattice is 3a x 3a. Three 2H polytypes exhibit CDW-PLD formation, TaS$_2$, TaSe$_2$, and NbSe$_2$. The transition temperatures to the superlattice states are normally found to be well below room temperature, as can be seen from the resistivity anomalies in Fig.14. The largest fraction of electrons that can be removed in a 3x3 superlattice is 8/9 of the original lattice, and it is clear from the change in resistivity near the anomalies that a much smaller fraction of the carriers are actually removed by the CDW-PLD. This is confirmed also from the change in Pauli susceptibility (Wilson et al., 1975). As an example, cross-sections of the Fermi surface of NbSe$_2$ are shown in Fig.15(c). It can be seen that there is only a rather incomplete nesting of the Fermi surface, mostly from the section centred at $\Gamma(A)$ of the Brillouin zone. 2H-TaSe$_2$ which has been thoroughly investigated because the sequence of transitions from undistorted to first incommensurate, then to commensurate superstructures, is particularly interesting. Several different incommensurate structures are seen, and lines of discommensurations in the nearly-commensurate phases can be imaged directly in electron microscopy (Fung et al., 1981; Chen et al., 1982). Structural measurements of the distorted lattice have been performed with neutrons (Moncton et al., 1977) and X-rays (Brouwer

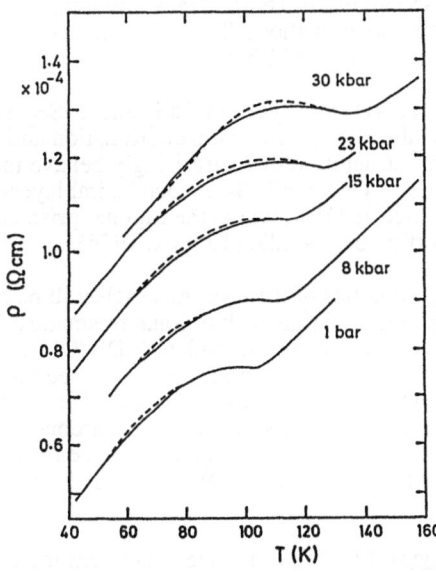

Figure 17. Resistivity versus temperature curves of 1T-VSe$_2$ under various hydrostatic pressures (Friend et al., 1978). The CDW transition, which coincides with the minimum in resistivity, clearly shifts to higher temperatures with increasing pressure.

Table 3. CDW-PLD superlattices and transition temperatures for the layer compounds.

		Commensurate Superlattice	% Incommensurate at ~ T_o	T_o (K)	T_d (K)	Reference
Octahedral	$1T\text{-}TaS_2$	$\sqrt{13}a \times \sqrt{13}a$	+3.8	543	352 200	Bayliss et al., 1984
	$1T\text{-}TaSe_2$	$\sqrt{13}a \times \sqrt{13}a$	2.6	~ 600	473	Wilson et al., 1975
	$4H_b\text{-}TaS_2$	$\sqrt{13}a \times \sqrt{13}a$	-5.2	~ 500	315	Wilson et al.,1975; Tatlock, 1976
	$4H_b\text{-}TaSe_2$	$\sqrt{13}a \times \sqrt{13}a$	-4.3	~ 550	410	Di Salvo et al., 1976b
	$1T\text{-}VSe_2$	$4a \times 4a$	2 ~ 6	110	~ 75	Williams, 1976
Trigonal Prismatic	$2H_a\text{-}NbSe_2$	$3a \times 3a$	-2.5	33	-	Moncton et al., 1975 1977
	$2H_a\text{-}TaS_2$	$3a \times 3a$	small	75	75?	Tidman et al., 1974
	$2H_a\text{-}TaSe_2$	$3a \times 3a$	-2.5	122	90	Moncton et al., 1975 1977
	$4H_b\text{-}TaS_2$	$3a \times 3a$	+7	22	-	Tatlock, 1976
	$4H_b\text{-}TaSe_2$	$3a \times 3a$	+4	75	-	Di Salvo et al., 1976b

T_o is the onset transition temperature from the normal state to the incommensurate CDW-PLD, and T_d is the commensurate-incommensurate superlattice transition temperature. The transitions in the two types of layer in the $4H_b$ polytypes have been grouped according to the coordination of the layer involved in the transition:

and Jellinek, 1980). Atomic displacements are estimated at ~ 0.1Å and 0.05Å for Ta and Se atoms respectively by Moncton et al. (1977), and Brouwer and Jellinek (1980) consider that the phasing of the three symmetry-related distortions creates 7-atom clusters in contrast to the 13-atom clusters in $1T\text{-}TaS_2$ shown in Fig.16(a). Transition temperatures from undistorted to incommensurate CDW-PLDs (onset transitions, T_o) and from incommensurate to commensurate (lock-in transitions, T_d) are summarised in Table 3.

(iii) 4H_b polytypes. The $4H_b$ polytypes of TaS_2 and $TaSe_2$ have been prepared, and studied. In many respects, the alternating octahedral coordination and trigonal prismatic coordination sandwiches in this structure rather surprisingly behave independently, and this is true of their CDW-PLD structure. Thus for TaS_2 the octahedral layers show a lock-in to the expected $\sqrt{13}a \times \sqrt{13}a$ superlattice at 315K, whilst the trigonal prismatic layers set up an incommensurate $3a \times 3a$ distortion below 22K (Tatlock, 1976).

The superconducting properties of the layer materials will be discussed later in relation to intercalation, but the unintercalated materials with trigonal prismatic coordination generally show superconducting behaviour. This is true even though a PLD-CDW distortion is present, and there has been interest in establishing the inter-relationship between the two phenomena. The most convenient experimental variable to employ is high pressure, since pressure generally suppresses the CDW-PLD distortion. It is found that the superconducting transition temperature is depressed by the presence of a CDW-PLD, and that T_c can be restored to its "undistorted" value at pressures high enough to drive out the CDW-PLD. This has been achieved in $2H\text{-}NbSe_2$ and $4H_b\text{-}TaS_2$.

2.5.5 Phase Transition in 1T-TiSe_2. $1T\text{-}TiSe_2$ is a semimetal with the d like conduction band just overlapping the p like valence band. It forms at low temperatures a $2a \times 2a \times 2c$ superstructure (Wilson and Mahajan, 1976; Di Salvo et al., 1976). The transition is second-order, and is most easily identified with a maximum in resistivity at 165K as seen in Fig.18(a).

However, structural measurements indicate that superlattice diffraction spots are formed at 202K, coinciding with the temperature at which the derivative of the resistivity has a peak. The superstructure that develops below the transition temperature couples the p band hole pocket at Γ with the d band electron pockets at the zone boundary, L, and many of the carriers present above the transition are removed. Indeed it is probable that the only remaining carriers, which are n-type as measured from the Hall coefficient, are those present as a result of excess, intersandwich titanium, as discussed above for the semiconducting group IV dichalcogenides. The removal of electrons and holes as a result of the p-d hybridisation that is possible following the appearance of the superlattice is indicated schemically in Fig.18(b). This will in general result in a lowering of the occupied band states, in a manner analogous to the Peierls distortion. The rise in resistivity towards low temperatures immediately after the transition is associated with this loss of carriers (as a second order phase transition, the process continues to 0K but most of the effects are felt just below the transition temperature). However, mobility improves at low temperatures and once the number of carriers is relatively stabilised, the resistivity falls as in metallic systems.

Like 1T-VSe$_2$, there is also a considerable difference between the size of the ions of Ti and Se. This provides little resistance elastically to the displacement of Ti ions within the "cage" of Se atoms. A number of models has been proposed for the driving force of the phase transition in 1T-TiSe$_2$. Di Salvo et al. (1976) consider that the coupling forces between electrons (L) and holes (Γ) are important. However, since the numbers of band states affected are very small (electron and hole concentrations above 202K $\sim 2 \times 10^{20}$cm^{-3}), it is not evident that this mechanism is sufficiently strong to be wholly responsible. White and Lucovsky (1977) who regard the distortion as a phonon-driven antiferroelectric transition, point out that the infra-red effective charge for TiSe$_2$ is very large, and comparable to that found in materials such as PbTe and SnTe, which do show such distortions. In their model, therefore, the electron-hole coupling that is achieved in TiSe$_2$ is accidental. A further mechanism proposed is the band Jahn-Teller model by Hughes (1977). In this the displacement of the Ti and Se atoms in the distorted structure from octahedral towards trigonal prismatic coordination (Di Salvo et al., 1976) results in a lowering of the bottom of the d band. This qualitative description has been supported by recent calculations (Yoshida and Motizuki, 1980; Motizuki et al., 1981; Suzuki et al., 1984,1986). It is interesting to note that the large difference between the sizes of the two types of ions has provided many examples, such as the tellurides of Nb, Ta, W and Mo (β type), in

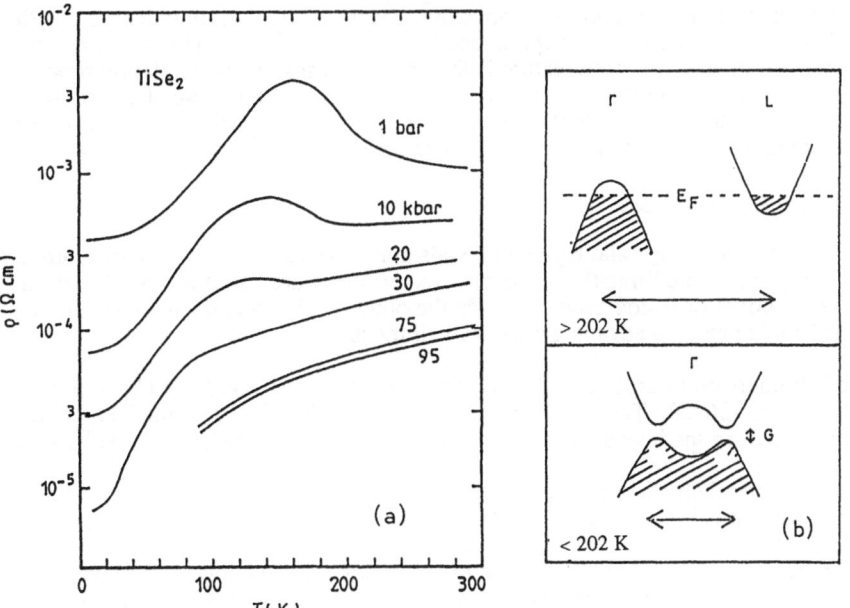

Figure 18. (a) Effects of temperature and pressure on the resistivity in 1T-TiSe$_2$, (b) The proposed electron-hole coupled model for the phase transition; the lower diagram represents the hybridised bands in the folded Brillouin zone with a finite gap between the filled and empty bands.

which a phase transition is evident (Wilson and Yoffe, 1969), and which may be driven by one of a number of the possible mechanisms mentioned above as well as by magnetic interactions.

In addition to resistivity data at ambient pressure, Fig.18(a) also shows the temperature dependence of the resistivity at elevated pressures. The structural phase transition is slowly depressed, but its influence on the resistivity anomaly is more rapidly reduced. Pressure increases the p-d band-overlap, at a rate of ~ 5meV/kbar, so the electron and hole concentrations are increased. Up to about 10kbar the low temperature resistivity is little altered, indicating that the distortion can still remove all the electrons and holes present that arise through band overlap. At higher pressures, however, the low temperature resistivity is much lower, and we can deduce that an increasing fraction of the electrons and holes present are unaffected by the distortion.

3. INTERCALATION WITH ALKALI METALS

There have been very extensive studies of intercalation compounds of transition metal dichalcogenides with alkali metals such as Li, Na, K, Rb, Cs and with other simple metals such as Ag and Cu. In this short review, it is not possible to cover many of the interesting aspects of this system and more detailed treatments can be found in a series of reviews by Subba Rao and Shafer, Rouxel, and Beal, collected in Volume 6 of "Physics and Chemistry of Materials with Layered Structures" edited by F.A. Levy (1979), and by Brec (1986) in this volume. Electrode applications of Na and Li have been reviewed by Johnson and Worrell (1982). There are also excellent reviews on graphite intercalation compounds by Dresselhaus and Dresselhaus (1981), Dresselhaus et al. (1983), and by Clarke and Uher (1984), and others. We concentrate instead, on those aspects which relate to the electronic properties and band structures discussed in the previous section. Furthermore we will limit our discussions mostly to Li intercalation compounds since these have been the most thoroughly investigated both for their structures and electronics properties.

Many features of intercalation with alkali metals and with other simple metals have made this system interesting for investigation. Because of the highly electropositive nature of alkali metals, complete ionisation may in a first approximation be assumed. The concentration of intercalant atoms may be varied over a wide range up to the saturation ratio of 1:1 or even higher, and once intercalated the ions appear to move in the van der Waals gap with ease. With certain concentrations commensurate with superlattice formation, the ions can become ordered at low temperatures. Another interesting feature is that in most cases, the intercalation process can be easily reversed and the "pure" host compounds are recovered. It is these characteristics which have given this system the important application in secondary batteries. First we review a number of the more widely used methods for preparing the intercalate complexes.

3.1 Methods for intercalation

All methods of intercalating alkali metals must take care to eliminate moisture and oxygen. This poses some limitations and most experiments are therefore carried out in carefully prepared "dry" boxes or inside a vacuum. On the other hand, Ag and Cu are stable in air and offer considerable practical advantages over alkali metals.

(i) Immersion of crystals in metal-ammonia solutions, e.g. Na/NH_3, at low temperatures (e.g. -36°C). The disadvantage with this method is that some NH_3 molecules are also inserted between the layers and this makes any quantitative analysis of results very difficult.

(ii) Electrochemical method or electrolysis, with for example the host dichalcogenide compound acting as cathode, an alkali metal salt in a non-aqueous solvent as electrolyte, and the alkali metal acting as anode. This is illustrated in Fig.19 for a Li / TiS_2 cell. Non-aqueous solvents are required if water molecules are to be excluded from between the layers. The Li /TiS_2 electrochemical cell has been widely studied since its characteristics are thought to be particularly suitable for high energy density secondary batteries. A suitable non-aqueous solvent is propylene carbonate, and lithium perchlorate is commonly used as the electrolyte. Other electrolytes make use of amorphous B_2S_3 or B_2O_3 with SiO_2 or organic polymers as bases and the lithium salt preferred is $LiCF_3SO_3H$. The intercalation reaction $Li + TiS_2 \rightarrow LiTiS_2$ is energetically favoured, giving an open-cell voltage which varies between 2.4 V (x=0) and 1.8 V

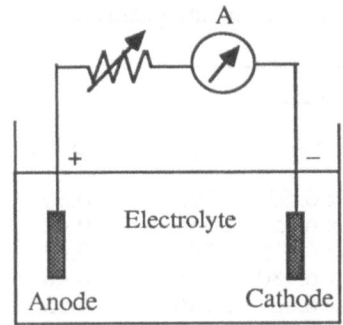

Anode: Li / Li$_3$Al

Cathode: TiS$_2$

Electrolyte: LiClO$_4$ in organic solvents /B$_2$O$_3$ - SiO$_2$ glass
or polyethylene oxide polymer

Figure 19. Schematic diagram for an electrochemical cell for intercalation of Li into TiS$_2$. A = current meter, and the rate of intercalation may be controlled by changing the resistance in the circuit. The same arrangement can also be used for Ag intercalation.

($x=1$), where x is the atomic fraction of Li as in Li$_x$TiS$_2$. The fall in EMF is in part due to the increase in the Fermi energy in the titanium d band; that it falls by only 0.6V is a consequence of the high value of the density of states in this d band. With electrochemical methods both the rate and the total amount of intercalate can be precisely controlled, and de-intercalation can be achieved simply by biasing the cell with a field opposing the cell's EMF.

Intercalation with silver is particularly easy since aqueous solutions of e.g. silver nitrate can be used and the intercalate complexes are stable in air. Ag ions have a rather high diffusion coefficient of about 10^{-7} cm^2/s at room temperature (Bonino et al., 1980; Folinsbee et al., 1981), and this helps to achieve equilibration rapidly even when working with single crystals. The cell potentials are much smaller than those found for Li, typically about 100mV (Folinsbee et al., 1981; Gerhards et al., 1984), and a positive biasing is sometime needed to increase the rate of intercalation. The limiting Ag uptake is also less than that found for alkali metals, giving for example, Ag$_{0.7}$TaS$_2$ and Ag$_{0.4}$TiS$_2$ (Scholz and Frindt, 1980) at saturation.

(iii) Immersion in butyl-lithium (n-C$_4$H$_9$Li) in the liquid state or in n-octane solutions (Dines, 1975; Murphy et al., 1976). Butyl-lithium is a mild lithiating reagent, and is particularly well matched with the transition metal dichalcogenides, being sufficiently active to attain a limiting composition of LiTX$_2$, but not so reactive as to cause further irreversible reactions. This method gives particularly clean products as lithium alone is allowed to enter between the layers. The rate is controllable by changing the strength of the solutions (hexane is usually used as the solvent) but this method does not provide information on the stoichiometry of the intercalate complexes.

Other methods such as employing molten salts mixed with alkali metals have also been used, but these are clearly rather difficult to work with.

3.2 Structural Changes

As would be expected, the greatest change produced by intercalation is in the expansion of the c lattice parameter of the host compounds. To a first approximation this expansion can be attributed to an increase in inter-layer distance which is needed in order to accommodate the intercalant atoms or molecules. While this is true in graphite compounds, the assumption is rather questionable in transition metal dichalcogenides where the sandwich height can change after intercalation since electrons are transferred and populate the d$_{z^2}$ band (in group IV and V metal compounds). The consequence of this increased electron filling is to increase the trigonal distortions in group IV and V metal compounds, with an increase in c/a ratios, but to decrease the trigonal distortion and c/a ratios in group VI compounds, as already discussed in the previous section, based on energetic arguments. We examine below this thesis using Li intercalation as an example.

Table 4. Lattice parameters of pure and Li intercalated compounds (after Whittingham and Gamble, 1975).

	Initial			Products			
	a/Å	c/Å	c/a	a/Å	c/Å	c/a	Δc/Å
TiS_2	3.407	1 x 5.696	1.672	3.455	1 x 6.195	1.793	0.50
ZrS_2	3.665	1 x 5.835	1.592	3.604	3 x 6.25	1.734	0.42
HfS_2	3.635	1 x 5.856	1.611	3.56	3 x 6.375	1.791	0.52
1T- $TiSe_2$	3.535	1 x 6.004	1.698	3.644	1 x 6.480	1.788	0.48
$ZrSe_2$	3.771	1 x 6.129	1.631	3.73	1 x 6.66	1.786	0.53
$HfSe_2$	3.742	1 x 6.160	1.646	3.715	1 x 6.642	1.788	0.48
NbS_2*	3.324	2 x 5.98	1.799	3.354	2 x 6.45	1.923	0.47
2H-TaS_2	3.340	2 x 6.04	1.808	3.340	2 x 6.475	1.939	0.44
$NbSe_2$	3.45	2 x 6.27	1.817	3.496	2 x 6.772	1.937	0.50
$TaSe_2$	3.463	2 x 6.348	1.847	3.477	? x 6.817	1.961	0.47
MoS_2	3.16	2 x 6.15	1.946	-	? x 6.40	-	0.25
2H-WS_2	3.16	2 x 6.18	1.956	-	-	-	-
$MoSe_2$	3.30	2 x 6.50	1.970	-	-	-	-
WSe_2	3.30	2 x 6.50	1.970	-	-	-	-
1T- VSe_2	3.35	1 x 6.10	1.821	3.584	1 x 6.356	1.773	0.26

* Dahn et al., 1986.

The small size of the Li^+ ions makes Li intercalate complexes a particularly suitable system for studying the effects of increased electron filling. The radius of Li^+ ions is only ~0.6Å which is somewhat smaller than the radius of the interstitial octahedral hole of about ~0.7Å, whether the bounding layers are sulphur or selenium. Apart from the slight expansion due to Coulomb repulsion, we should therefore expect the Li^+ ions to fill the octahedral hole comfortably with relatively little expansion in the c direction. Table 4 lists the lattice parameters of some pure and Li intercalated layer compounds (after Whittingham and Gamble, 1975) for stoichiometric 1:1 complexes with the Li sitting in octahedral sites between the layers. There are two points to note.

(i) In all group IV and V compounds except 1T-VSe_2, there is an increase in c of about 0.49Å (±0.03Å). This is considerably greater than the expansion found in Li intercalated graphite, where the separation between the layers bounding Li atoms increases by 0.36Å. A possible explanation for this is that in transition metal dichalcogenides the sandwich height also expands upon intercalation with Li, and there are two contributions to the c lattice expansion. Since the octahedral holes in transition metal dichalcogenides should provide at least the same amount of room for the intercalant ions as in graphite, it seems reasonable to assume that the expansion in the van der Waals gap in transition metal dichalcogenides due to Li^+ ions is no more than 0.36Å, and the change in sandwich height is therefore at least 0.13Å when one electron is donated to each formula unit of the host compound. Comparison between graphite and transition metal dichalcogenides intercalated with other alkali metals such as K and Rb is more difficult, not only because these are larger ions which would dominate the c lattice expansion, but also they have not been as systematically studied as the Li intercalates.

Of the group VI compounds, only MoS_2 has been studied in any detail. We find nevertheless that here a Δc of 0.25Å is less than the expansion found in graphite. The difference is again accounted for by a change in the sandwich height, this time by a contraction. With the d_{z^2} band already full (Fig.5) and intercalation increasing the d band population from d^2 to d^3, the band structure energy can be lowered by interchanging the order of d_{z^2} and d_{xy,x^2-y^2} bands. This can be achieved by a negative trigonal distortion, i.e. a decrease in c/a ratios (Liang, 1984). Indeed transformation from the trigonal prismatic coordination to octahedral coordination (which has a lower c/a value) for the Mo atom in $Fe_{0.5}MoS_2$ and in $Na_{0.5}MoS_2$ has been reported by Py and Haering (1983).

(ii) If we consider the c/a ratios for a group as a whole, we can also see some interesting trends. We find that after intercalation, the 1T group IV products have c/a ratios in the region of 1.79 which we compare with 1.751 for $1T\text{-}TaS_2$ and 1.804 for $1T\text{-}TaSe_2$. At the same time, the c/a ratios for the 2H group V products are in the region of 1.94 which is similar to those of the 2H group VI compounds before intercalation. In each case, the comparison is made between isoelectronic materials. This shows that the number of electrons populating the bands is the dominant effect in determining the crystal structure, particularly with regard to the intra-sandwich coordination. Further studies on the intra-sandwich structure with varying degrees of intercalation using, say, EXAFS techniques, should provide useful information.

We note also in Table 4, a different kind of structural reorganisation after intercalation. For Li_xZrS_2 a phase change takes place from the 1T to the 3R structure at x=0.25 (see McKinnon et al., 1984). The reason for this is not clear, but Coulomb interactions between the electrons in the now partially occupied d_{z^2} orbitals may be responsible for displacing the Zr atoms so that they stagger between layers in order to lower the repulsive energy. Unfortunately an opposing argument, namely the bonding between these electrons is used to explain the different 2H structures adopted by the group V and group VI metal compounds (Fig.3). Interpolytypic transitions which involve changing the coordination structure of the transition metal or of the intercalant atom (Rouxel, 1979) though energetically favourable are likely to be very slow under ordinary conditions since the energy required to surmount the potential barrier for such a change is usually very large.

3.3 Ordering of Intercalant atoms

With certain concentrations of intercalant atoms, ordering at low temperatures into superlattice structures of the intercalate complexes has been observed. As shown in Figs.3 and 4, there are well-defined inter-sandwich interstitial sites for occupation by intercalant ions. The diffusion of these ions is normally by a thermally activated process. Therefore at some well-defined temperatures, we may expect order-disorder transitions to take place. These transitions can also be achieved by varying pressure or composition. The fact that many such transitions have been observed below room temperature indicates that the diffusion coefficients for alkali metal and Ag ions to move between layers are high. For Ag^+ in TiS_2, for example, the coefficient is 10^{-13} cm^2/s at 200°C and $< 10^{-15}$ cm^2/s at room temperature. The diffusivity for Li^+ is even higher, being 5×10^{-9} cm^2/s in Li_xTiS_2, for 0<x<1 at room temperature, and $\sim 3 \times 10^{-8}$ and $\sim 10^{-7}$ cm^2/s at 30 and 70°C respectively in Li_xTaS_2 for 0.2<x<0.8 (Basu and Worrell, 1979). Direct structural evidence for ordered intercalant superlattices in lithium complexes has been difficult to obtain, mainly because of the difficulty in achieving the stringent oxygen free conditions at all times. However, Hibma (1980) has shown by diffuse x-ray scattering experiments that superlattices at x = 1/4 and x = 1/3 exist in Li_xTiS_2, and Hallak and Lee (1983) confirm the existence of a $\sqrt{3}a \times \sqrt{3}a$ superlattice at x = 1/3. Thompson (1978) pointed out that features in the discharging curves of cell EMF versus concentration are associated with superlattice ordering. From the electro- chemical titration in Li_xTiS_2, he concluded that superlattice ordering exists at concentrations of x equals to 1/3, 1/9, 1/4, 5/7, 4/5 and 6/7. However, n.m.r. measurements suggest that perhaps not all these features are associated with superlattices (Chianelli et al., 1978).

The Ag intercalation complexes also exhibit several examples of intercalant ion ordering, but the situation here is much more complicated since Ag ions are known to occupy both the octahedral as well as the tetrahedral sites, giving rise to complex diffraction patterns and domain formation. The advantage with the Ag complexes is of course that they are stable in air, and both the intercalation process and the subsequent study of the products can be carried out under ordinary laboratory conditions. Many careful investigations of Ag intercalate complexes have been carried out by the Simon Fraser group (see for example, Scholz et al., 1982, 1982a). For Ag_xTiS_2, Ag ions occupy the octahedral sites just as the Li ions, and at x ~ 0.34, a $\sqrt{3}a \times \sqrt{3}a$ superlattice has been found just below room temperature (Leonelli et al., 1980, Suter et al., 1982, Gerhards et al., 1984). The phase transition is second order and has been described by Suter et al. as an example of continuous melting in three dimensions.

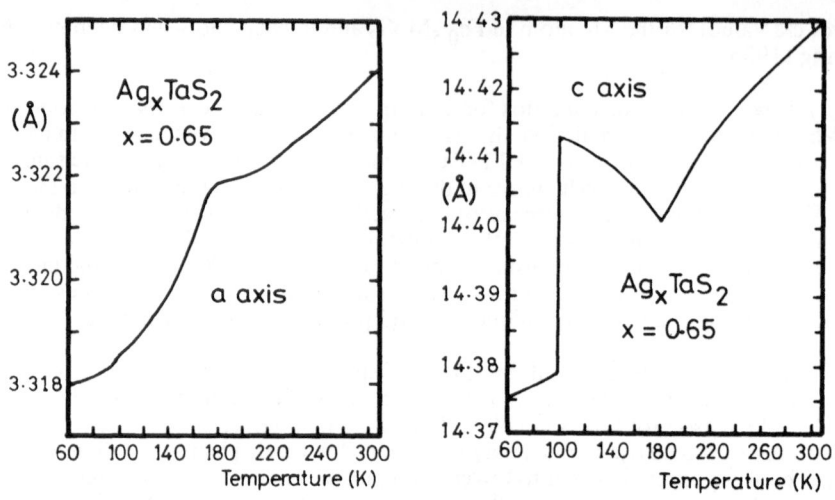

Figure 20. Lattice parameters for $2H_b$- $Ag_{0.65}TaS_2$ as a function of temperature.

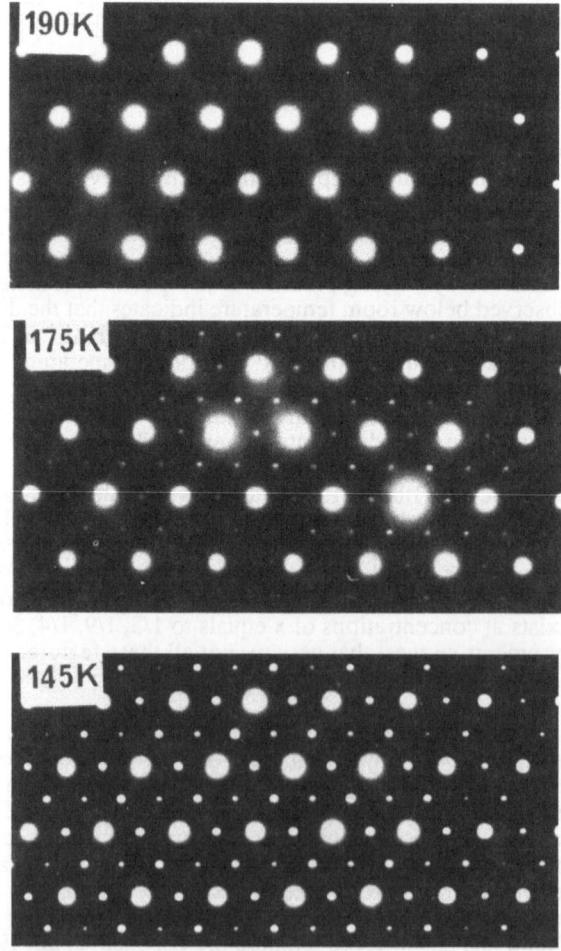

Figure 21. Electron diffraction in $Ag_{0.65}TaS_2$. In-plane $2a \times 2a$ superlattice is observed only below 190K. It grows in strength on cooling from 175K to 145K. The full structure is complicated which is revealed on tilting the specimen.

Ag_xTaS_2, provides examples for the occupation of tetrahedral sites. Boebinger et al. (1983) carried out transport and structural studies of the fully intercalated complex $Ag_{0.65}TaS_2$, and found that 30 to 33% of Ag occupies each of the two tetrahedral sites and an uncertain amount, but in any case less than 7%, occupies the octahedral sites. The stacking between layers adopts the $2H_b$ form similar to MoS_2 (Fig.3). Their transport measurements show evidence for phase transitions at 96K, 140K and 190K, replacing the CDW transition at 76K. Interestingly, lattice parameters determined by X-ray as a function of temperature show singularities only at 96K and 190K, as shown in Fig.20. Scholz et al. (1982a) and Tatlock et al. (1984) have reported electron diffraction studies which are in broad agreement. Electron diffraction taken at three temperatures is shown in Fig.21. Although they show quite clearly that ordering of the intercalant is absent at high temperatures but exists in a **2a x 2a** superlattice at low temperatures, the actual structure is considerably more complicated than is perhaps suggested by these pictures (see for example, Tatlock et al., 1984).

We have seen that CDW's and now ordering of intercalant ions can lead to the formation of superlattices. In the former case, a CDW induces atoms to displace from their regular high symmetry positions, and the potentials set up by such displacements cause a gap to be opened at the reciprocal lattice vector equal to $2k_F$, which also defines the new superlattice cell. Ordering of intercalant ions, on the other hand, is a thermodynamic effect with electrostatic interactions as the driving force. The atoms of the host lattice need not, in the first instance, displace to produce the superstructure. In M_xTX_2 complexes, therefore, both mechanisms are possible, though at a given superlattice transition, there is the likelihood that one of these primary forces may dominate. We remember that a CDW driven superlattice transition is usually accompanied by a marked change in the density of states at the Fermi level. Transport measurements, including magnetic susceptibility will be strongly influenced by such a phase change. To a first approximation, superlattice formation by ordering of intercalant ions should produce no gap at the new Brillouin zone boundary, so that at least any gap produced in such a process should be small. There is, therefore, the interesting distinction that can be made between these two primary mechanisms, namely intercalant ion ordering that involves merely zone-folding (in reciprocal lattice space) without a gap at the new zone boundary, and CDW-PLD's that creates gaps with loss of carriers at the Fermi level as well as zone-folding. Though both mechanisms are usually accompanied by changes in lattice constants, again to a first approximation, intercalant ion ordering should produce a smaller change in this respect than a CDW-PLD.

We can now examine the superlattice formation in intercalate complexes $Ag_{0.65}TaS_2$ and $Ag_{0.33}TiS_2$. For $Ag_{0.65}TaS_2$, we suggest that both transitions at 190K and 140K are due to Ag ions, probably making first an in-plane ordering then an out-of-plane ordering at the lower temperature. Only the transition at 96K is due to CDW's with a first order loss of carrier density. The energy involved in the out-of-plane ordering is probably too small to perturb the lattice constants. From the small discontinuity at 96K in the resistivity curve (Boebinger et al., 1983), we conclude that "nesting" is not strong, though the size of superlattice vector of 2a is in accord with the enlarged electron Fermi surface following the charge transfer to the conduction band of

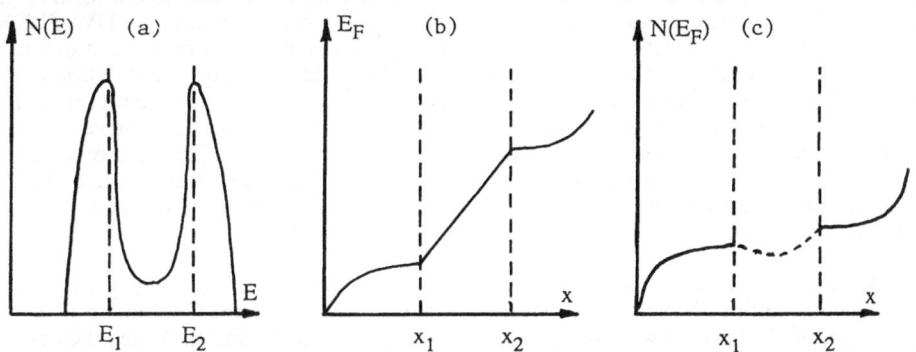

Figure 22. (a) Density of states function $N(E)$ near a CDW gap E_1E_2; (b) Effects of a CDW gap on the variation of E_F and cell EMF V with intercalant concentration x, $x=x_1$ corresponds to the concentration at which CDW induced superstructure is formed; (c) Effects of a CDW gap on the variation of $N(E_F)$ and χ_p with x.

the pure compound. It should be interesting to investigate the displacements of Ta and S atoms at the three transitions in order to determine if a real qualitative difference exists between them. Raman spectroscopy can also be used for such an investigation since the scattering cross sections of the folded zone boundary modes will be a strong function of atomic displacements.

$Ag_{0.33}TiS_2$ probably provides the best example of intercalant ion ordering without CDW. A second order phase transition is observed just below room temperature, with a Ag ion ordering onto a $\sqrt{3}a \times \sqrt{3}a$ superlattice. However, no anomaly is observed in resistivity or magnetic susceptibility associated with this transition, suggesting that the density of states at the Fermi level remains intact and no CDW gap exists. The Hall coefficient, which gives an electron concentration of $\sim 7 \times 10^{21}$ cm^{-3} at room temperature, in agreement with the value obtained by assuming complete ionisation of the Ag atoms, shows anomalous behaviour at low temperatures. This is of course just what would be expected as a result of zone folding following superlattice formation.

It is perhaps worth exploring further the relationships between CDW gaps, intercalant concentration, cell EMF and (Pauli) susceptibility in the intercalate complexes M_xTX_2. We already know that the drop in cell EMF, $V_{max}-V$, during an electrochemical discharge is mostly due to the rise in the Fermi energy E_F as a result of continuous electron transfer to the host material. The high density of states at the Fermi level, $N(E_F)$, normally ensures that the cell EMF drops very gradually with increasing x. Assuming a superlattice is formed at the concentration x $= x_1$ with a CDW gap at E_1E_2, a large fraction of $N(E_F)$ is removed from the gap region (Fig.22(a)). Consequently, for $x_1 < x < x_2$, both E_F and $V_{max}-V$ rise rapidly, and a mixture of the super- structure appropriate to $x = x_1$ and disordered states exist. A completely disordered state is restored for $x > x_2$ when E_F returns to the high density of states region and the cell EMF is stabilised. This is schematically shown in Fig.22(b). The value of x just prior to a steep rise in the $V_{max}-V$ curve thus corresponds to a CDW induced superlattice formation (Thompson, 1978).

The appropriate relationship between the Pauli paramagnetic susceptibility, χ_p, and x is more difficult to obtain. In general, χ_p follows the variation in $N(E_F)$, and in the absence of a CDW gap, $N(E_F)$ reaches a plateau with increasing x more rapidly than E_F. Depending on the fraction of $N(E_F)$ removed due to the CDW gap with respect to the rest of the Fermi surface, $N(E_F)$ may dip or stay more or less constant, and rise again only when x is somewhat greater than x_2. This is shown in Fig.22(c). Bernard et al. (1985) obtained χ_p as a function of x in Li_xTiS_2 and found a distinct plateau between $x = 1/9$ and $x = 1/3$. Although it is clear that a superlattice with a CDW is present, it is actually quite difficult in this case to pinpoint the exact value of x when this is formed.

3.4 Absence of CDW-PLD in Li intercalated group IV compounds

Lithium intercalated group IV complexes including $LiZrS_2$ and $LiZrSe_2$ have been widely studied, for example, by Berthier et al. (1981,1984), Onuki et al. (1980, 1982), and Klipstein et al. (1984). These complexes are iso-electronic with the 1T octahedral form of group V metal dichalcogenides such as $1T$-TaS_2 and $1T$-$TaSe_2$, in which very strong CDW-PLD's have been observed as discussed earlier. Although the group IV Li complexes do show metallic conduction and the resistivity curves for a series of Li_xTiS_2 and Li_xZrSe_2 as a function x and temperature have already been shown in Fig.10, they show no evidence for the presence of strong CDW's. A very weak slope discontinuity in the resistivity temperature curve whose position depends strongly on the concentration has been observed in Li_xZrS_2 and reported by Klipstein et al. (1984). The absence of strong CDW-PLD's in this case may be understood in terms of the energy expression (7) and our knowledge of the band structure of the un-intercalated compounds, as follows.

We have shown in Section 2.5.2 that for electrons to form charge density waves and induce a periodic lattice distortion, the system will have to overcome the Coulomb repulsion due to localisation of electrons as well as elastic strain energy due to distortion. When the conduction band is narrow, there is a very high density of states at the Fermi level which will enhance the Coulomb term. There is a good deal of evidence, coming from electron energy loss and optical measurements, photoemission, electronic heat capacity and Pauli paramagnetic susceptibility, to indicate that the conduction band to which electrons from Li in $LiZrS_2$ and $LiZrSe_2$ are transferred is a narrow band no more than 1 eV wide. This is the d_{z^2} band whose separation of

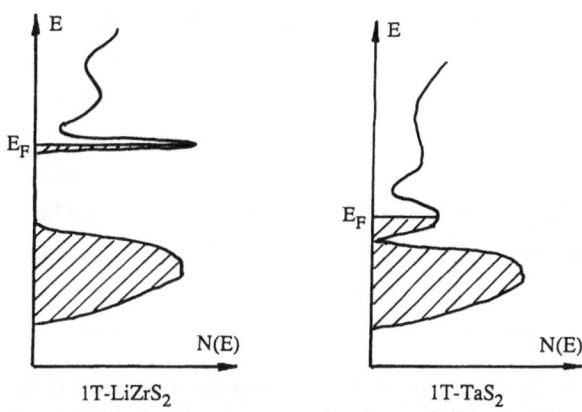

Figure 23. Density of states schemes for 1T-LiZrS$_2$ and 1T-TaS$_2$. The separation between the d$_z^2$ band and the main p-like valence band in 1T-LiZrS$_2$ may be smaller than indicated.

about 2 eV from the p like valence band prevents hybridisation and helps to produce the narrowing. By comparison, the d$_z^2$ band in 1T-TaS$_2$ and 1T-TaSe$_2$ is probably close to touching or overlapping the p like valence band, and p-d hybridisation is a dominant factor in making this a broader band, which is about 2.5 to 3 eV wide (see discussions on 1T-VSe$_2$). Of course the increased electron filling in the Li intercalated group IV compounds will lower and hence widen the d$_z^2$ band, but this is thought not to be sufficient to create the condition for the formation of CDW's. The different band structures are illustrated with the density of states schemes in Fig.23.

3.5 Special case of Li intercalation of 1T-VSe$_2$

We have already encountered the many examples of the unusual behaviour of 1T-VSe$_2$ such as the increase of CDW phase transition temperature under pressure, a large **a** axis expansion coupled with a small **c** axis expansion and a reduction in c/a ratio after intercalation. This behaviour can not be readily explained in the frame work of the band structures discussed so far. Instead we have to consider the strong interactions between local moments. The exchange energy which has so far been treated as a perturbation to the overall electron energy scheme plays a non-negligible role in 1T-VSe$_2$. Localisation of d orbitals is determined by the relative size of the transition metal ion radius to the distance between ions, and by the degree of ionicity or the strength of the ligand fields surrounding the transition metal ions. The tendency to localisation is found in the transition metal oxides (see Delmas in this volume) and in the first row transition metal compounds. Ionicity decreases from oxides, through sulphides, selenides, to tellurides, and the characteristics of the first row transition metals are their tightly bound d orbitals (without inner d orbitals there is little screening of the nuclear potentials). Thus 1T-VSe$_2$ (also 1T-CrSe$_2$) sits on the boundary, being the material with the narrowest d$_z^2$ band just before the d orbitals become localised. Many of the arguments given in this Section would also apply to the chromium dichalcogenides and vanadium disulphide, except that in these materials the d orbitals have become localised and the binary compounds are no longer stable. Lattice stability can only be achieved through Coulomb forces with the introduction of alkali or Ag or Cu ions into the intersandwich sites (Delmas in this volume). We have for example stable compounds of MCrS$_2$ where M=Li, Na, K, Ag and Cu, and MVS$_2$ where M=Li and Na, in which chromium is present as Cr^{3+}(d^3) and vanadium as V^{3+}(d^2). Both Cr and V atoms in these complexes have local moments consistent with their valencies, and become ordered at low temperatures, usually around 40-50K. The most interesting material from the point of view of band formation or orbital localisation, nevertheless, is still 1T-VSe$_2$. This can be seen from the variation of lattice constants with Li intercalation.

Table 5. Lattice constants of 1T-VSe$_2$ and its Li intercalated complexes (after Wiegers, 1980).

		a/Å	c/Å	c/a
1T-VSe$_2$	(d^1)	3.35	6.12	1.827
1T-LiVSe$_2$	(d^2)	3.58	6.36	1.773
1T-Li$_2$VSe$_2$	(d^3)	4.02	6.44 (300K)	1.605
		3.99	6.43 (4K)	1.612

As we have already mentioned, photoemission measurements (Johnson et al., 1986) give the width of the occupied part of the d$_{z^2}$ band in 1T-VSe$_2$ to be only ~0.3 eV. This is so narrow that the use of a one-electron band description must begin to be questionable. With the number of d electrons per V atom increased to 2 and 3 in the intercalated compounds, the possible local moments also increase, and the d electrons will gain energy by dismantling the former d band into localised states, in which they interact via an attractive exchange interaction. The process of localising d states is achieved in Li$_2$VSe$_2$ where local moments are detected at the V sites, and the system is antiferromagnetic at low temperatures. The transition from a band to localised states is accompanied by crystallographic changes, in which the sandwich is dramatically flattened with a decreased c/a ratio which favours the lattice Coulomb energy (the Madelung energy). Indeed Shen (1985) has shown by calculating the Coulomb energy involved between the Na^{+1}, V^{-1+2x} and Se^{-x} ions that lower c/a ratios would give the preferred lower energy configuration (where x is an arbitrary variable determined by the ionicity within the VSe$_2$ sandwich). It is not surprising, therefore, since the dominant stabilising energy in the Li intercalated VSe$_2$ system is electrostatic energy between ions and not electron band energy, that the model based on trigonal distortion with electron filling discussed in Section 2 does not apply here. It should be noted that while Li$^+$ ions occupy octahedral inter-sandwich sites in LiVSe$_2$, they occupy tetrahedral sites in Li$_2$VSe$_2$ (Tigehelaar et al., 1982). It would be most interesting to carry out an investigation on d orbital localisation (the Mott-Hubbard transition) by varying the Li concentration in the intercalate complexes.

It is worth mentioning also that the complexes AgCrS$_2$ and AgCrSe$_2$ are good ionic conductors at high temperatures. Hibma (1980) quotes an ionic conductivity of 0.3Ω^{-1} cm^{-1} for Ag in AgCrS$_2$ above 400K.

4 INTERCALATION WITH ORGANIC MOLECULES

4.1 Method of Preparation and Structure

Nearly all nitrogen-containing organic molecules, ammonia and hydrazine (N$_2$H$_4$) will intercalate with the transition metal dichalcogenides. In general the metallic dichalcogenides such as TaS$_2$ and TiSe$_2$ will intercalate more easily, but with the group VI semiconductors such as MoS$_2$ it is more difficult to form organic intercalate complexes. The simplest way to intercalate with these molecules is to immerse the crystal in the liquid or vapour at a suitable temperature or vapour pressure. When vapour is used, the amount taken up by the crystal can be monitored by a sensitive microbalance or fine quartz fibre spring appropriately calibrated. Figure 24 shows an interesting example of an intercalation-deintercalation cycle of 1T-TaS$_2$ by hydrazine, N$_2$H$_4$. Hydrazine is a particularly suitable species to use for intercalation since it has a convenient vapour pressure near room temperature (the range 6-12 torr is usually covered) and intercalation can be achieved by exposure of the host material to the hydrazine vapour, without the need for a solvent, and deintercalation is achieved by heating at a reduced pressure. Just as Li was used as the prime example for the alkali metal intercalation, we use N$_2$H$_4$ intercalation as representative here since a great deal of detailed investigation has been made on this system. Two intercalation complex structures for the 2/3 and 4/3 stoichiometry are formed as shown in Fig.24. X-ray measurements (Acrivos, 1979) indicate that they are of 3R rhombohedral symmetry, with a unit cell repeat in the c-axis direction of 29.79Å and 38.04Å respectively. For the 4/3 stoichiometry,

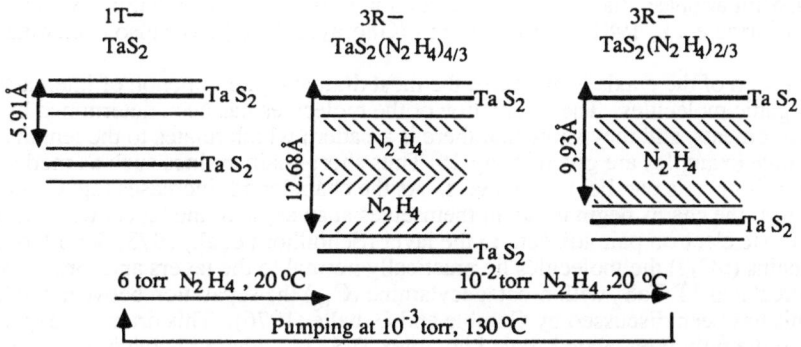

Figure 24. Intercalation processes and stoichiometry of 1T-TaS$_2$ by hydrazine, N$_2$H$_4$.

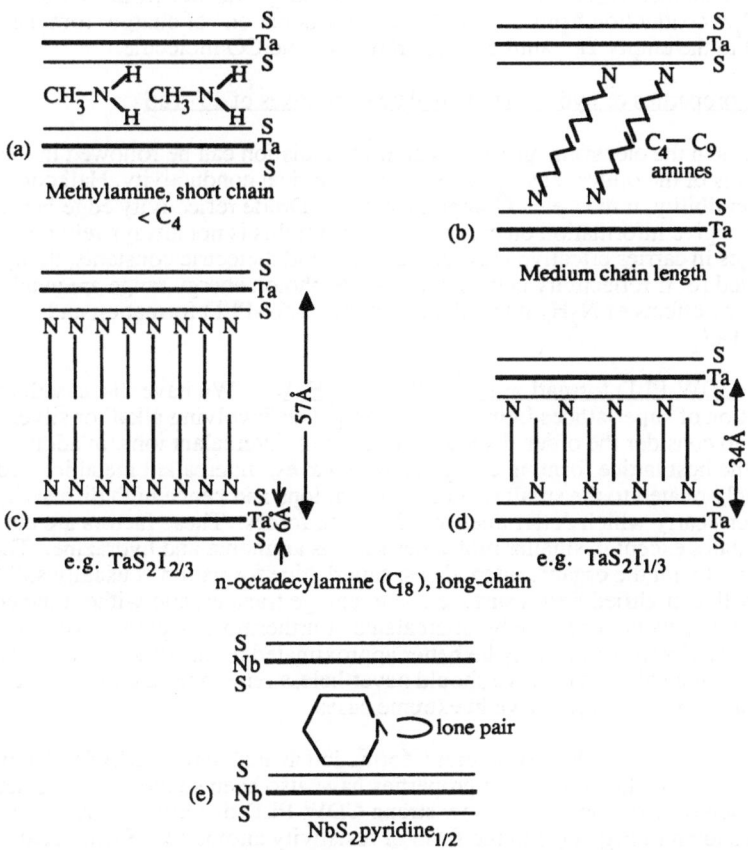

Figure 25. Some examples of the structures of intercalate complexes with organic molecules.

Acrivos (1979) shows that hydrazine forms a double layer within the van der Waals gap and the molecules occupy the tetrahedral sites, while with the 2/3 stoichiometry, only half as many sites are occupied. Similar phase diagrams for intercalation with hydrazine of $TiSe_2$ (Sarma et al., 1982), VSe_2 (Sarma et al., 1981), TiS_2 and ZrS_2 (Ghorayeb 1985) have also been obtained.

Expansion of the c axis is of course the most dramatic phenomenon of intercalation by long-chain organic molecules. The orientation of the molecules has been determined by X-ray or neutron measurements, and they show that there is a pattern which relates to the lengths of the molecules. Some examples are given in Fig.25. Thus short-chain amines such as methylamine stack with their long axis parallel to the layers. As the chain length increases (eg. $C_4 - C_9$ aliphatic amines) the chains begin to orient themselves at an angle to the layers with the nitrogen atom with its lone electron pair adjacent to the layer (Schollhorn et al., 1973; Schollhorn, 1980). For longer chains ($>C_{16}$) the molecules lie practically normal to the layers and form bilayers. For fully intercalated 1T-TaS_2 with n-octadecylamine (C_{18}) the separation between the layers is ~ 57Å and this has been discussed by Gamble and Geballe (1976). This dramatic expansion corresponds to the fully intercalated stage I complex $TaS_2.I_{2/3}$ where I is the intercalate molecule. Such stage I complexes can be partly deintercalated to give the complex $TaS_2.I_{1/3}$. We also show in Fig.25(e) one of the rather surprising orientations for small molecules such as pyridine (Riekel et al., 1976), in which the planes of the molecule are perpendicular to the layers with the lone pair on the nitrogen atom pointing in a direction parallel to the layers (Wada et al., 1978). This gives a distance of 4.4Å from the nitrogen atom to the nearest sulphur atom, a distance too large for appreciable orbital overlap. There has also been a suggestion by Dines (1978) that back bonding can occur from the host lattice to the intercalate molecule. This kind of behaviour has of course been worked out in detail during adsorption studies of simple molecules such as CO on to the surface of clean metals such as Ni, where there is charge transfer from a carbon 5σ lone pair to the d-like (d_{z^2}) conduction band of nickel and "back donation" of charge from the d_{xy} filled band (orbitals) to the empty $2\pi^*$ antibonding orbitals on the CO molecule.

4.2 Electronic properties of hydrazine intercalate complexes of 1T-TaS_2

Changes in the electronic properties after intercalation can be followed in the usual way by measurements of the optical absorption spectra, electrical conductivity, Hall constant, magnetic susceptibility, n.m.r., etc. Comparison of the Drude reflectivity edge before and after intercalation can give information on charge transfer, but this is not always reliable because of possible changes in carrier effective mass and background dielectric constants, though the latter can be accounted for if reflectivity curves over a wide photon energy range are available. We discuss below the effects of N_2H_4 intercalation on the CDW-PLD formation and localisation of carriers in 1T-TaS_2.

4.2.1 CDW-PLD formation in 3R-TaS_2. $(N_2H_4)_{4/3}$.
We have discussed, in section 3.3, the formation of superlattices in intercalate complexes involving alkali or silver ions, where it is important to consider the order-disorder transition of intercalant ions, in addition to the possibility of the host lattice forming charge density waves. Intercalant metal ion ordering is a property strongly related to the small size of the metal ions and hence high diffusivity, and the charge these ions carry which determines the Coulomb forces. These factors are absent in the organic intercalates except in smaller molecules such as ammonia and hydrazine. The advantages this offers are that with the organic intercalates, superlattice formation is usually still CDW induced, but with a modified Fermi surface due to charge transfer, and without the complexity of having to take into account ordering by intercalants. Furthermore with the layers separated by a large distance, the Fermi surface may be better approximated to that of a two-dimensional solid. Although this is an idealised view, we should nevertheless remember the different effects organic and alkali metal intercalation can have in extreme cases.

In addition to a wealth of structural information available for the hydrazine intercalation complexes of 1T-TaS_2, their electronic properties have also been studied in some detail. Both the 4/3 and 2/3 stoichiometry complexes show strong CDW-PLD distortions at room temperature and below, as shown in Fig.26(a) in the form of resistivity anomalies (Sarma et al., 1982). As in the pure 1T metals, there is a good correspondence between the observation of superlattice formation made by transport and by electron diffraction experiments. Electron diffraction measurements on $TaS_2.(N_2H_4)_{4/3}$ show a commensurate $8/\sqrt{7}$ a x $8/\sqrt{7}$ a superlattice at room temperature and another commensurate 3a x 3a superlattice at 80K (Tatlock and Acrivos, 1978).

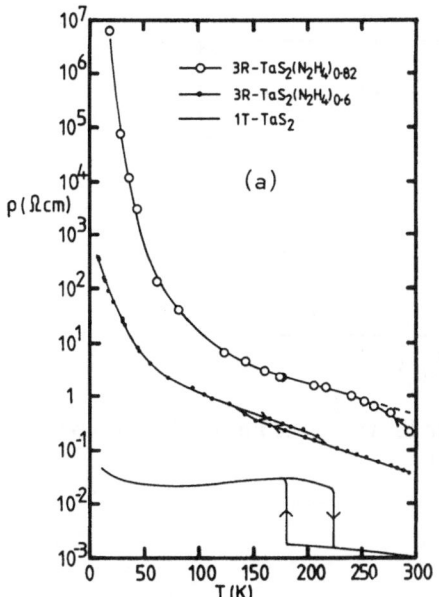
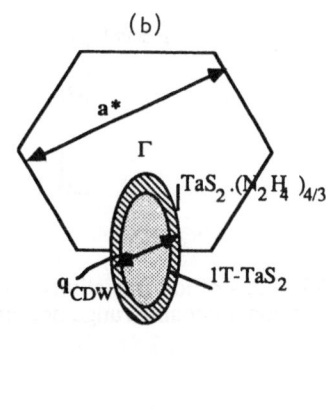

Figure 26. (a) Plot of log (resistivity) per unit thickness as a function of temperature for $1T\text{-}TaS_2$ before and after intercalation with hydrazine; (b) Fermi surface cross-sections of $1T\text{-}TaS_2$ and $3R\text{-}TaS_2$ $(N_2H_4)_{4/3}$. The increase in area after intercalation is an indication of a charge transfer of 0.33 electron to each $1T\text{-}TaS_2$ unit. $\mathbf{a^*}$ is the reciprocal lattice vector of the undistorted crystal, and $\mathbf{q_{CDW}}$ is the CDW wave vector.

These correspond to the resistivity anomaly seen at 280K which is probably the transition temperature between these two CCDW superlattice states (Sarma et al., 1982). Ghorayeb (1985) also found a further transition to a lower resistivity state at around 380K which may be the transition from the $8/\sqrt{7}$ $\mathbf{a}$ x $8/\sqrt{7}$ $\mathbf{a}$ commensurate superlattice to an incommensurate structure. Tatlock and Acrivos consider that the superlattices seen are associated with the electron-phonon driven CDW-PLD seen in $1T\text{-}TaS_2$ but modified by the increased band filling. Assuming a two-dimensional Fermi surface, the number of electrons in the conduction band is then proportional to the area of the Fermi surface. For $1T\text{-}TaS_2$ the electron Fermi surface will be enlarged if there is charge transferred from the hydrazine molecules, as shown in Fig.26(b). In the wavevector space, the above superlattices have distortion vectors equal to $0.331\mathbf{a^*}$ at 300K and $0.333\mathbf{a^*}$ at 80K, compared with $0.277\mathbf{a^*}$ for pure $1T_3\text{-}TaS_2$, where $\mathbf{a^*}$ is the reciprocal lattice vector of the undistorted structure. This is strong evidence for charge transfer, and simple estimates give a value of about 0.33 electrons to one TaS_2 unit, or a loss of 0.25 electron per N_2H_4 molecule. This degree of charge transfer, however, is much smaller than the 0.7 electrons estimated from transport measurements in $TiSe_2.(N_2H_4)_{2/3}$ (Sarma et al.,1982), and it is not understood why there is such a large difference in the charge transfers in these two materials. It should be said that in making the above estimates, the Fermi surface has been assumed to scale without changing shape, i.e. the rigid band approximation is used, and this may not be exact when the amount of charge transfer is appreciable.

4.2.2 <u>Localisation of carriers in the Peierls gap</u>. The amplitudes of the CDW-PLD's in the hydrazine intercalation complexes are very large, and Sarma et al. (1982) obtained a value of about 1eV for the energy gap or pseudo-gap near E_F from their optical transmission results. Such a strong distortion may be associated with the low wavelength "simple" commensurate superstructure which has a large lock-in energy as in (9). It turns out that intercalate complexes such as $1T\text{-}TaS_2$ with hydrazine is a very good system in which to observe variable range hopping conduction in two dimensions. The upturn in resistivity curves towards lower temperature is due to strong scattering of carriers by random potentials, and the reasons for this are: (a) electrons are confined to move in the layers and interlayer tunneling is prevented by a large intersandwich separation; (b) there is a strong random potential setup by the $(N_2H_4)^+$ ions

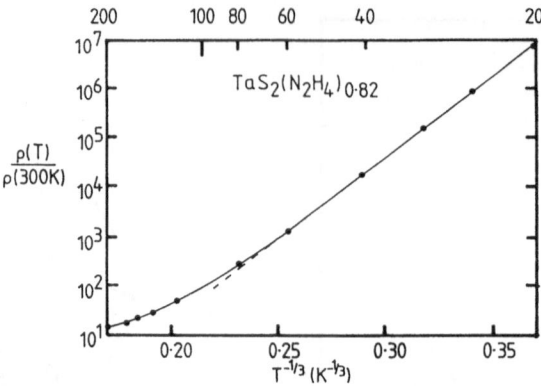

Figure 27. Plots of log (resistivity) versus $T^{-1/3}$ for 1T-TaS$_2$. The linear relationship confirms two dimensional variable range hopping conduction.

situated within the intersandwich space, the number of singly occupied sites is estimated to be 1/450 of Ta atoms in the intercalated samples, and only 1/2000 Ta atoms in the unintercalated crystals (Klipstein et al.,1985); (c) after the formation of CDW-PLD's at low temperatures, the number of carriers remaining is very small, only about 1.3×10^{20} /eV.cm^{-3}, compared with 2×10^{22} /eV.cm^{-3} in the metallic state, favouring Anderson localisation. Plots of logarithm of resistivity versus $T^{-1/3}$, Fig.27, indeed show reasonably good straight lines over five orders of magnitude variation in resistivity, providing evidence for two-dimensional variable range hopping conduction.

5. TWO-DIMENSIONAL SUPERCONDUCTIVITY

Two-dimensional (or single layer) superconduction has always been an intriguing experimental possibility in layer structure materials. In transition metal dichalcogenides, the d conduction band and the generally strong electron-phonon coupling, as is evident from the tendency to form CDW-PLD's, provide the bases for superconductivity. Thus the metallic systems, whether they are the pure group V compounds, or the intercalated group IV and VI complexes, can in general, become superconductors at low temperatures. There are of course, other processes of low temperature ordering such as the formation of CDW-PLD's and magnetic ordering, which compete with the superconducting transition, and they drive T_c to a low value or transform the material into an insulator. For this reason, the 1T structure group V compounds are not superconductors, and even among the trigonal prismatic coordinated forms, the Ta compounds are generally low T_c superconductors. There has been much interest in the study of the relationship between CDW-PLD's and BCS superconductivity. The suppression of a CDW-PLD, which can be achieved by the application of pressure or by intercalation, generally enhances T_c. We are interested in two dimensional superconductivity and there is a good deal of experimental evidence which supports this idea, which is summarised below.

(a) Heat capacity measurements on 2H-TaS$_2$ intercalated with the long chain n-octadecylamine indicates the presence of superconductivity (Gamble and Geballe, 1976). In this compound the layers have been separated to a distance of ~60Å and it would be reasonable to assume that the Josephson coupling between the layers will be weak or absent.

(b) For many long-chain amine intercalated compounds, the superconducting transition temperature T_c is found to be relatively insensitive to the lengths of the molecules and to lie around 3K (see Di Salvo et al., 1971 and also Coleman and Hillenius, 1981) for molecules

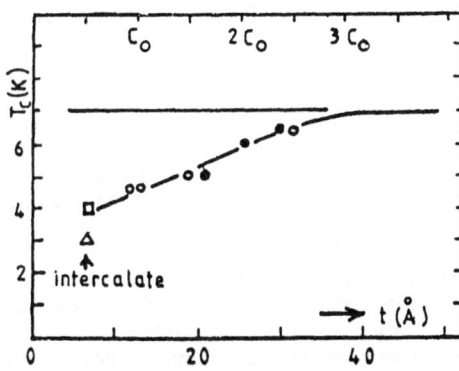

Figure 28. Superconducting transition temperature (T_c) for 2H-NbSe$_2$ as a function of crystal thickness. The point that corresponds to "intercalate" is taken to be an average value of T_c's when 2H-NbSe$_2$ is intercalated with long-chain molecules.

varying in length from 9 to 18 carbon atoms. If Josephson coupling was playing a role in which Cooper pairs tunnel between layers then T_c would be expected to fall with increasing layer separation.

(c) Superconductivity is unaffected by the presence of paramagnetic chromocene ions in the inter-sandwich sites. Experiments made on 2H-TaS$_2$ intercalated with chromocene and cobaltecene show that the T_c's are nearly identical (2.9K and 3.2K) (Dines, 1975; Gamble and Thompson, 1978). Here the cobalt and chromium are present as ions Co^{3+} and Cr^{3+} (d^4), with the cobaltocene complex being diamagnetic, but the chromocene complex being paramagnetic. Since local moments are generally thought to be responsible for the breaking of Cooper pairs (spin disorder scattering), this result again argues against Cooper pairs tunnelling across the van der Waals gap occupied by the intercalate molecules.

(d) The presence of superconductivity in a single layer of 2H-NbSe$_2$. Frindt (1972) measured the thickness dependence of the superconducting transition temperature of 2H-NbSe$_2$ and the results are shown in Fig.28. It can be seen that T_c remains constant down to thicknesses of the order of 40Å and then falls slightly, but even a crystal a unit cell thick (~12Å) has a T_c of nearly 5K. This extrapolates to a T_c of ~4K for a single layer.

The intercalation of 2H-TaS$_2$ with molecules such as pyridine, methylamine, dimethylamine and ethylene diamine can result in a substantial increase in the superconducting transition temperature, T_c, from 0.8K to temperatures in the region of 3-6K, following the suppression of the CDW-PLD's in the unintercalated material. Although the idea of two-dimensional superconductivity is an attractive one, it should be understood that this really is too crude to account for a more general situation. In reality, what we have is superconductors which can have varying degrees of anisotropic properties. This anisotropy manifests itself in the ratio of the upper critical field, and in the ratio of coherence length, measured parallel and perpendicular to the layers. Thus not surprisingly, after intercalation, there is usually a large increase in the observed anisotropy of the upper critical field $H_{c2//}/H_{c2\perp}$, parallel and perpendicular to the layers. Coleman et al. (1983) have followed the superconducting critical field behaviour in the field range 0-26.7T, and the temperature dependent critical field anisotropy can reach values approaching 60 for $T/T_c < 0.7$. They conclude that their results with pyridine and methylamine are consistent with a coherence length perpendicular to the layers $\xi_\perp$ that is less than the layer spacing (9-12Å for the molecules listed), and that their results best fit the theory of dimensional crossover developed by Klemm et al. (1975) for highly anisotropic superconductors, based on Josephson tunnelling between layers rather than the anisotropic Ginsberg Landau theory. The crossover of course is an indicator of a transition to a two-dimensional behaviour.

6. INTERCALATION WITH FIRST ROW TRANSITION METALS

6.1 Preparation, Structure and Phonon Spectra

Complexes of the group V dichalcogenides intercalated with 3d transition metals can only be prepared at high temperatures, by iodine transport techniques, at the same time that the layer compound is grown. Single crystals can be readily grown with the trigonal prismatic coordination host or with the octahedral coordination host. However, as 3d ions prefer to occupy octahedral sites, intercalation with trigonal prismatic hosts presents no complication, but with octahedral hosts there may be problems associated with the 3d ions substituting the host metal atoms. In a sense these are ternary compounds since the 3d ions can not be removed and the bonding between layers is greatly strengthened with the crystals unable to cleave easily. However, the 3d metal ions occupy interlayer sites and transfer electrons to the Nb or Ta d band, and descriptions given for the more conventional intercalate complexes are still applicable.

Two possible stoichiometries are normally obtained with M_xTX_2, either x=1/4 or 1/3 which form ordered superlattices as follows.

	x=1/4	x=1/3	ordered structure
superlattice	2a x 2a	√3a x √3a	in-plane
arrangement	-M-T-M-T-M-	M-T-M	inter-plane
	(continuous)	(chains)	

The presence of the intercalant superlattice is expected to reconstruct the Brillouin zone and give rise to new $q=0$ phonons which may be Raman-active. This has been observed in $Ag_{1/3}TiS_2$ by Leonelli et al. (1980), and in many 3d intercalation complexes by Pereira (1984). We show, as an example of zone folding and new Raman modes in Fig.29, the Raman spectra of $2H-TaSe_2$ and its $Fe_{1/4}$ intercalate complex. The insert shows the Brillouin zone of the pure compound, the larger hexagon, and the reconstructed zone, reduced to 1/4 of the original area.

We observe, as in the Raman spectrum of $2H-MoS_2$ in Fig.7 that, there are three easily identifiable modes, one due to the weak interlayer motion at 23 cm^{-1}, and two due to the intralayer atomic motion at 210 and 237 cm^{-1}. However, there is an additional mode in $2H-TaSe_2$ due to the presence of a soft phonon which is at ~140 cm^{-1} at 295K, but softens as the temperature is lowered towards the onset temperature of the CDW. By comparison, the Raman spectrum of $Fe_{1/4}$ $2H-TaSe_2$ shows no such CDW phonon mode, indicating the absence of a CDW-PLD. The intralayer modes are little affected by the presence of the intercalant, but there are now new modes at about 100 cm^{-1} which have been brought to Γ from M by folding the Brillouin zone as shown in the insert. Furthermore, the interlayer bonding force has been increased by about four times as the frequency is now doubled. The physical picture of the new folded zone modes is that in the enlarged unit cell containing four units of $TaSe_2$, there are basically two configurations of atomic motion that retain the centre of symmetry (required for Raman activity). These are, one when all four units are moving in phase, and two when only two units are in motion while the other two are stationary. The former has frequencies which are little modified from those of the pure crystal, and the latter has frequencies which are roughly halved.

The phonon modes associated with the movement of the Fe^{2+} in the layers cannot be determined by Raman experiments since the Fe^{2+} is in a centre of symmetry position, so to determine these it will be necessary to make infra-red or neutron diffraction measurements.

6.2 Transport and Magnetic Properties

The interest in this system is the fact that the 3d ions have strong local moments, but are separated, as imposed by the underlying host structure, by distances too large for direct exchange interactions. This gives a very simple system in which to study magnetic interactions either via the conduction electrons (R.K.K.Y.), or by superexchange involving the appropriate empty orbitals on the chalcogen atoms.

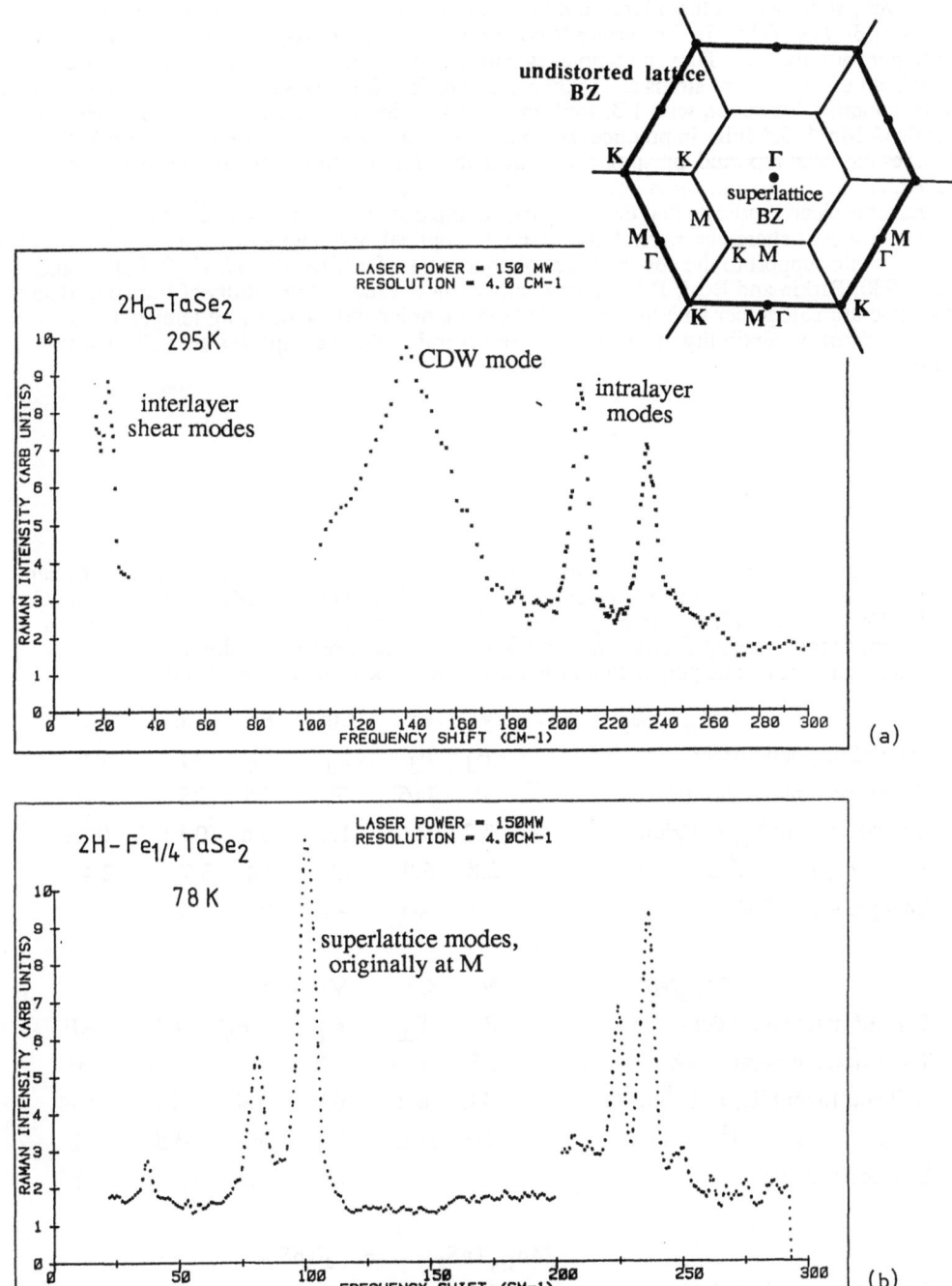

Figure 29. Raman spectra of (a) 2H-TaSe$_2$, and (b) 2H-Fe$_{1/4}$TaSe$_2$. The insert shows Brillouin zone of the pure crystal (thick lines and labels) and the effect of folding in a 2a x 2a superlattice (thin lines and labels) of the intercalate complex.

With transport properties, we are concerned primarily with the conduction band of the host crystal, namely the d_{z^2} band, and the rigid band model can be used to describe changes in electronic properties following intercalation. Evidence from local magnetic moments (Parkin and Friend, 1980) suggests that the 3d ions are in high ionisation states, with nominal charge state of either 2+ or 3+ (see Table 6). For group V compounds, the d_{z^2} conduction band contains one electron per formula unit. Thus in complexes with $1/3Cr^{3+}$ and $1/3V^{3+}$, the additional electrons coming from the intercalant atoms can in theory completely fill the conduction band, making them a semiconductor. However, with $1/3Mn^{2+}$ and $1/3Ni^{2+}$, the conduction band is only 5/6 full, and with $1/4Mn^{2+}$, 3/4 full. In practice, no semiconducting behaviour of the Cr^{3+} and V^{3+} complexes has been reported, presumably because the ideal 1/3 stoichiometry has not been achieved, or the charge transfer is not exactly three electrons for these ions. Nevertheless, Hall measurements clearly indicate that the intercalate complexes with $1/3Cr^{3+}$ and $1/3V^{3+}$ have the smallest number of charge carriers. Other transport, optical, and photoemission experiments also give reasonable support to the above values of charge transfer (Friend et al.,1977, Parkin and Friend, 1980; Parkin and Beal, 1980; Barry and Hughes, 1982). The results of transport data at room temperature, together with the types of magnetic order and the ordering temperatures determined from susceptibility measurements for several of these complexes are collected in Table 6.

Table 6. Magnetic transition temperature, type of order and room temperature Hall coefficient and electrical resistivity of the intercalation complexes, $M_{1/3}TaS_2$ and $M_{1/3}NbS_2$ (M = V, Cr, Mn, Fe, Co, Ni), $Mn_{1/4}TaS_2$ and $Fe_{1/4}NbSe_2$. F and AF correspond to ferromagnetic and antiferromagnetic orderings. The symbols // and $\perp$ give the direction of the easy axis of magnetization parallel and perpendicular to the c axis (Parkin and Friend, 1980).

$M_{1/3}TaS_2$:	M = V	Cr	Mn	Fe	Co	Ni
Type of magnetic order	$F_\perp$	$F_\perp$	$F_\perp$	$F_{//}$	AF	AF
Transition temperature (K)	35	116	70	35	35	120
Hall coefficient $R_H \times 10^3 cm^3 C^{-1}$	3.2	2.6	1.0	0.6	0.83	0.82
Resistivity $\rho \times 10^4$ ohm cm	2.8	5.9	1.3	2.4	3.7	2.4
Charged state of M	+3	+3	+2	+2	+2	+2

$M_{1/3}NbS_2$:	M = V	Cr	Mn	Fe	Co	Ni
Type of magnetic order	$F_\perp$	$F_\perp$	$F_\perp$	$AF_{//}$	AF	AF
Transition temperature (K)	55	115	40	45	25	90
Hall coefficient $R_H \times 10^3 cm^3 C^{-1}$	4.0	6.2	0.97	1.4	1.7	0.94
Resistivity $\rho \times 10^4$ ohm cm	3.0	11.6	1.5	3.2	3.5	2.2
Charged state of M	+3	+3	+2	+2	+2	+2

	$Mn_{1/4}TaS_2$	$Fe_{1/4}NbSe_2$
Type of magnetic order	$F_\perp$	$AF_{//}$
Transition temperature (K)	80	160
Hall coefficient $R_H \times 10^3 cm^3 C^{-1}$	0.7	0.095
Resistivity $\rho \times 10^4$ ohm cm	1.9	2.9
Charged state of M	+2	+2

7. CONCLUSION

The electronic properties of transition metal dichalcogenides and their intercalate complexes have been described in relation to the structural properties and the degree of electron filling of the conduction band. We have examined the conditions with which to apply the rigid band model. Superlattice formation, which is observed in a variety of compounds and intercalate complexes, is analysed in several separate examples. We show that $1T$-VSe_2 is a very special material electronically and deserving further studies.

I like to record my special thanks to Abe Yoffe, without whose help and constant encouragement and support, this paper would not have materialised. It is also a pleasure to thank Elizabeth Marseglia and Richard Friend for their useful advice on numerous occasions.

8. REFERENCES

Acrivos, J.V., 1979, "Intercalated Layer Materials", Vol. 6, ed. Levy, F., Dodrecht Reidel Ch.2, 33.

Anderson, O., Karschnick, F., Manzke R. and Skibowski, M., 1985, Sol. St. Comm. 53, 339.

Bagnall, A.G., Liang, W.Y., Marseglia, E.A., and Welber, B., 1980, Physica 99b, 343.

Barry, J.J and Hughes, H.P., 1982, J. Phys. C 15, L797.

Basu, S. and Worrell, W.L., 1979, "Fast ion Transport in Solids", ed. Vashishta, P., Mundy, J.N. and Shenay G.K., Elsevier, N. Holland.

Bayliss, S.C., Ghorayeb, A.M., Guy, D.R.P., 1984, J,Phys C, 17 , L533.

Bell, M.G. and Liang, W.Y., 1976, Adv. Phys., 25, 53.

Bernard, L., Glausinger, W. and Colombetm P. 1985, Sol.St. Ionics 17, 81.

Boebinger, G.S., Wakefield, N.I.F., Marseglia, E.A., Friend, R.H., and Tatlock, G.J., 1983 Physica 117B, 608.

Bonino, F., Lazzari, M., Vincent, C.A. and Wandless, A.R., 1980, Sol. St. Ionics 1, 311.

Brec, R. and Ouvrard, F., 1983, Sol,. St. Ionics 9,10, 481.

Brouwer, R. and Jellinek, F., 1976, 5th Int. Conf. on Solid Compounds of Transition Elements, Upsala, Sweden.

Brouwer, R. and Jellinek, F., 1980, Physica 99B, 51.

Chan, S.K. and Heine, V., 1973, J. Phys. F3, 795.

Chen, C.H., Gibson, J.M. and Fleming, R.M., 1982, Phys. Rev. B26, 184.

Chianelli, R.R., Scanlon, J.C. and Rao, B.M.L., 1978, J. Chem. Society, 125, 1563

Clarke, R. and Uher, C., 1984, Adv. Phys. 33, 469.

Coleman, R.V.and Hillenius, S.J., 1981

Dines, M.B., 1975, Science 188, 1210.

Dines,M.B., 1978 Inorg. Chem 17, 762 763; Mat. Res. Bull 10, 287.

Di Salvo, F.J. and Geballe, T.H., 1971, Science 174, 493.

Di Salvo, F.J., Moncton, D.E. and Waszczak, J.W., 1976, Phys. Rev. B14, 4321.

Di Salvo, F.J., Moncton, D.E., Wilson, J.A. and Mahajan, S., 1976b, Phys. Rev. B14, 1543.

Di Salvo, F.J., 1977, "Electron Phonon Interactions and Phase Transitions", ed. Riste, T. Plenum - NY, 107.

Doran, N.J., Ricco, B., Schreiber, M., Titterington, D. and Wexler, G., 1978, J. Phys. C 11, 689.

Dresselhaus, M.S. and Dresselhaus, F., 1981, Adv. Phys. 30, 139.

Dresselhaus, M.S., Dresselhaus, G., Fischer, J.E. and Moran, M.J., 1983, Materials Research Society, Intercalated graphite: North Holland N.Y. 20.

Fazekas, P. and Tosatti, E., 1979,Phil. Mag. B39, 229.

Folinsbee, J.T., Jericho, M.H., March, R.H. and Tindal, D.A., 1981, Can. J. Phys. 59, 1267.

Friend, R.H., Beal, A.R. and Yoffe, A.D., 1977, Phil. Mag. 35, 1269.

Friend, R.H., Jerome, D., Frindt, R.F., Grant, A.J. and Yoffe, A.D., 1977, J. Phys. C 10, 1013

Frohlich, H., 1955, Proc. Roy. Soc., A223, 296.

Friend, R.H., Jerome, D., Schliech, D.M. and Molinie, P., 1978, Sol. St. Comm. 27, 169.

Friend, R.H. and Jerome, D., 1979, J. Phys. C 12, 1441.

Friend, R.H., Jerome, D. and Yoffe, A.D., 1982, J. Phys. C 15, 2183.

Frindt, R.F., 1972, Phys. Rev. Lett. 28, 299.

Fung, K.K., McKernon, S., Steeds, J.W. and Wilson, J.A., 1981, J. Phys. C 14, 5417.
Gamble, F.R. and Geballe, T.H., 1976, "Treatise on Solid State Chemistry, Inclusion Compounds", Plenum N.Y. 3, Ch. 3.
Gamble, F.R. and Thompson, A.H., 1978, Sol. St. Comm. 27, 379.
Gerhards, A.G., Roede, H., Haange, R.J., Boukamp, B.A. and Wiegers, G.A., 1984, Synthetic Metals 10, 51.
Ghorayeb, A.M., Coleman, C.C. and Yoffe, A.D., 1984, J. Phys. C17, L715.
Ghorayeb, A.M., 1985, PhD dissertation, Cambridge.
Guo, G.Y. and Liang, W.Y., 1986, J.Phys. C, 19, 995.
Hallak, H.A. and Lee, P.A., 1983, Sol. State Comm. 47, 503.
Hambourger, P.D. and Di Salvo, F.J., 1980, Physica 99B, 173.
Hibma, T., 1980, Sol. St. Comm. 33, 445.
Hibma, T., 1982, Intercalation Chemistry, ed. Whittingham, M.S. and Jacobson, A.J. Academic Press.
Hughes, H.P. and Pollak, R.A., 1976, Phil. Mag. 34, 1025.
Hughes, H.P., 1977, J. Phys. C 10, L319.
Hughes, H.P. and Friend, R.H., 1978, J. Phys. C 11, L103.
Hulliger, F., 1976, "Structural Chemistry of Layer-type" Phases, ed. Levy, F., Dodrecth Reidel, Vol. 5.
Johnson, M.T., Starnberg, H.I. and Hughes, H.P., 1986, J. Phys. C, 19, L451.
Johnson, W.B. and Worrell, W.L., 1982, Synthetic Metals 4, 225.
Johnston, N.A. and Scholl, C.A., 1984, J.Phys. C, 17, L73.
Klemm, R.A., Luther, A. and Beasley, M.R., 1975, Phys. Rev. B12, 877.
Klipstein, P.C., Bagnall, A.G., Liang, W.Y., Marseglia, E.A. and Friend, R.H., 1981, J. Phys. C 14, 4067.
Klipstein, P.C., Guy, D.R.P., Marseglia, E.A., Meakin, J.I., Friend, R.H. and Yoffe, A.D., 1986, J. Phys. C (in press).
Klipstein, P.C., Pereira, C.M. and Friend, R.H., 1984, NATO ASI., ed. Acrivos, J.V.,
Leonelli, R., Plischke, M. and Irwin, J.C., 1980, Phys. Rev. Lett. 45, 1291.
Levy, F., 1979, "Intercalated Layer Materials", ed. Levy, F., Dodrecht Reidel, Vol. 6.
Liang, W.Y., 1984, "Physics and Chemistry of Electrons and Ions in Condensed Matter", NATO ASI 130, Dodrecht Reidel, 459.
McKinnon, W.R., Dahn, J.R., Levy-Clement, C., 1984, Sol. St. Comm. 50, 101.
McMillan, W.L., 1975, Phys. Rev. B12, 1187 & 1197.
McMillan, W.L., 1976, Phys. Rev. B14, 1496.
Marseglia, E.A., 1983, Int. Rev. Phys. Chem. 3, 177.
Monceau, P., 1985, "Electronic Properties of Inorganic Quasi-one Dimensional Materials II", 139-268, publ. D. Reidel.
Moncton, D.E., Axe, J.D. and Di Salvo, F.J., 1977, Phys. Rev. B16, 801.
Moncton, D.E., Di Salvo, F.J. and Davey, S.C., 1979, Bull. Am. Phys. Soc. 24, 446.
Motizuki, K. Suzuki, N., Yoshida, Y. and Takaoka, Y., 1981, S. State Comm. 40, 995.
Murphy, D.W., Di Salvo, F.J., Hull, G.W. and Waszczak, J.V., 1976, Inorg. Chem. 15, 17.
Myron, H.W., Rath, J. and Freeman, A.J., 1977, Phys. Rev. B15, 885.
Parkin, S.S.P. and Beal, A.R., 1980, Phil. Mag. B42, 627.
Parkin, S.S.P. and Friend, R.H., 1980, Physica 99B, 219; Phil. Mag. B41, 65, 95.
Peierls, R.E., 1955, Quantum Theory of Solids, OUP.
Pereira, C.M., 1984, Ph.D. Dissertation, University of Cambridge.
Pereira, C.M. and Liang, W.Y., 1985, J. Phys. C18, 6075.
Py, M.A. and Haering, R.R., 1983, Can. J. Phys. 61, 76.
Riekel, C., Hohlwein, D. and Schollhorn, R., 1976, Chem. Comm. 863.
Rouxel, J., 1978, "Intercalated Layer Materials", ed. Levy, F., Dodrecht Reidel, Vol. 6, 210.
Sarma, M., Beal, A.R., Nulsen, S. and Friend, R.H., 1982, J. Phys. C15, 477, 4367.
Schollhorn, R., Sick, E. and Weiss, A., 1973, Zeit fur Naturforsch 28b, 168.
Schollhorn, R., 1980, Physica 99B, 89; 1982, Intercalation Chemistry, ed. Whittingham, M.S. and Jacobson, A.J., Academic Press N.Y., Ch. 10, 315.
Scholz, G.A. and Frindt, R.F., 1980, Mat. Res. Bull. 15, 1703.
Shen, T.H., and Liang, W.Y., 1983, J.Phys C, 16, L883.
Shen, T.H., 1985, Private Communication.
Smith, N.V., Kevan, S.D. and Di Salvo, F.J., 1985, J. Phys. C18, 3175.

Subba Rao, G.V. and Shafer, M.S., 1979, "Intercalated Layer Materials", ed. Levy, F, Dodrecht Reidel, Vol. 1, $\underline{6}$, 99.

Suter, R.M., Shafer, M.W., Horn, P.M. and Dimon, P., 1982, Phys. Rev. B$\underline{26}$, 1495.

Suzuki, N.;, Yamamoto, A. and Motizuki, M., 1984, Sol. State Comm. $\underline{49}$, 1039.

Suzuki, N., Yamamoto, A. and Motizuki, M., 1986, J. Phys. Soc. Jap. (in press).

Tatlock, G.J., 1976, Commun. Phys. $\underline{1}$, 87.

Thompson, A.H., 1978, Phys. Rev. Lett. $\underline{40}$, 1511.

Tigchelaar, D., Wiegers, C.A. and Van Bruggen, C.F., 1982, Rev. Chim. Min. $\underline{19}$, 352.

Traum, M.M., Smith, N.V. and Di Salvo, F.J., 1974, Phys. Rev. Lett. $\underline{32}$, 1241.

Tsutsumi, K., 1982, Phys. Rev. B$\underline{26}$, 5756.

Wada, S., Alloul, H. and Molinie, P., 1978, J. Physique Lett. $\underline{39}$, L243.

Wertheim, G.K., Di Salvo, F.J. and Chiang, S., 1975, Phys. Lett. $\underline{54A}$, 304.

Wertheim, G.K., Di Salvo, F.J. and Chiang, S., 1976, Phys. Rev. B$\underline{13}$, 5476.

White, R.M. and Lucovsky, G., 1977, Nuovo Cim. B$\underline{38}$, 280.

Whittingham, M.S., 1977, "Solid Electrolytes", ed. Van Gool, W. and Hagenmuller, P.,

Wiegers, G.A., 1980, Physica $\underline{99}$B, 151.

Williams, P.M., Parry, G.S. and Scruby, C.B., 1974, Phil. Mag. $\underline{29}$, 695; 1975, Phil. Mag. $\underline{31}$, 255.

Wilson, J.A. and Yoffe, A.D.,1969, Adv. Phys., $\underline{18}$, 193.

Wilson, J.A., Di Salvo, F.J. and Mahajan, S., 1974, Phys. Rev. Lett. $\underline{32}$, 882; 1975, Adv. Phys. $\underline{24}$, 117.

Wilson, J.A. and Mahajan, S., 1976, Commun. in Phys. $\underline{2}$, 23.

Yoshida, Y. and Motizuki, K., 1980, J. Phys. Soc. Jap. $\underline{49}$, 898.

REACTIVITY AND PHASE TRANSITIONS IN TRANSITION METAL

DICHALCOGENIDES INTERCALATION CHEMISTRY

R. Brec and J. Rouxel

Laboratoire de Chimie des Solides
U.A. 279, 2, rue de la Houssinière
44072 Nantes Cedex, France

When an intercalation compound forms, ions or molecules are accepted by a host lattice. This results primarily in geometrical effects directly manifested through lattice parameter variations. At the same time an electronic exchange takes place between host and guest species. In addition the intercalation reaction is a reversible one. It is possible to return to the initial state through appropriate electrical, chemical, or thermal actions. This means that no strong bonds have been created between guest and host, or broken in the host structure.

Many tunnel structures allow one to practice intercalation chemistry. However, in each case there will be a suitable pathway for a particular type of ion or molecule but not for each kind of ion at the same time. This can clearly be an advantage when researching size selectivity effects in catalysis by open structures, but it is a drawback when one wants to understand the phenomenon in its generality. Then the ideal host structure would be a deformable one, adapting itself to each type of ions. This is the case of low dimensional solids. They are built up from a stacking of slabs or a juxtaposition of fibers. During the intercalation process such units are pulled apart in agreement with the weak interslabs or interfibers bonds (Van der Waals gap).

Among low dimensional solids, transition metal dichalcogenides (T.M.D.C.) represent a rather large family with a nice variety of geometrical and electronic situations. When considering the complete series of their alkali metal intercalates it becomes possible to recognize some general factors affecting reactivity and phase transitions in intercalation chemistry. We shall consider successively some electronic and geometric factors but it will rapidly appear clear that all these factors are strongly correlated.

1. Electronic Transfer and Reactivity

The electronic exchange between guest and host appears to play an essential role concerning the ability to intercalate and the stability of the intercalation products. It can impede any intercalation process even if convenient empty sites are available.

Figure 1 represents the band structure models that apply to transition metal chalcogenides. The main features are the following. Between a valence band built up from s and p anionic levels, and antibonding states provided by the s and p cationic levels, the d levels of the cations determine physical properties. These orbitals have been split by the crystal field which leads us to introduce the two classes of T.M.D.C. (Fig. 2). In both cases the slabs are made of three atomic layers with two anionic layers framing a cationic one, but the coordination of the metal in the slabs can be either octahedral like in TiS_2 or trigonal

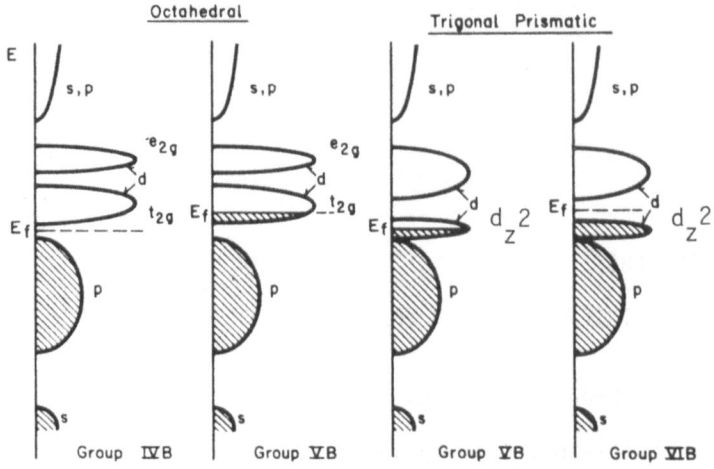

Figure 1 - Band schemes for IVB, VB and VIB transition metal layered dichalcogenides (after T. Hibma /14/)

prismatic like in 2H ṄbS$_2$. Representing the anionic layers by A, B, C and the cationic planes by a, b, c, the CdI$_2$ structure of TiS$_2$ will be described by the AbC, AbC sequence with a (AC)$_n$ hexagonal close packing of the anions. The 2H ṄbS$_2$ structure will appear as AbACBC, the anionic stacking consisting of blocks of two AA or CC superposing anions layers. In both cases (TiS$_2$ and NbS2) octahedral sites exist in the Van der Waals gap between C and A (TiS$_2$) or A and C (NbS$_2$). MoS$_2$ shows a first difference in respect of stacking of trigonal prismatic slabs, the atomic sequence being AbABaB with a B sulfur layer above the b cationic layer and a similar arrangement for a and A layers. More generally gliding motions of the slabs one over the other explain clearly the large number of polytypes that can be observed and are referred to in the RAMSDELL notation /1/. 2H means two slabs with hexagonal symmetry, 3R is for three slabs with rhombohedral symmetry etc...

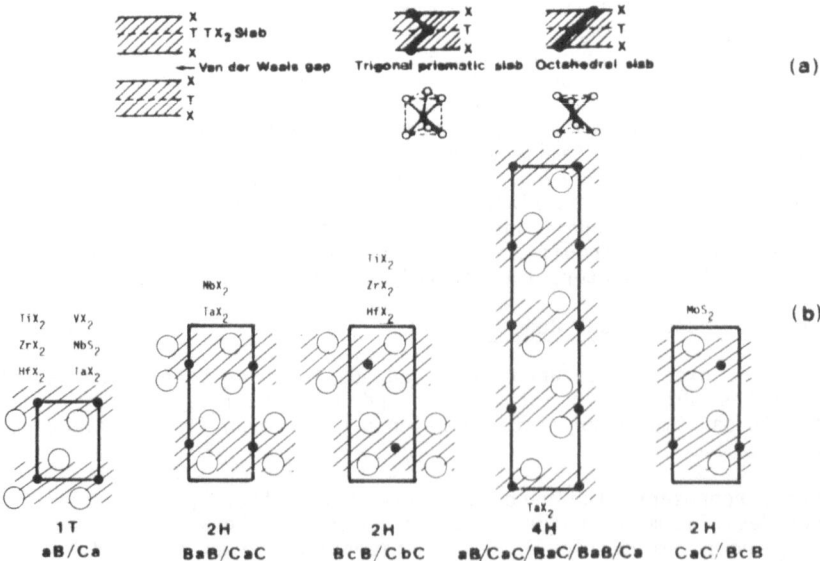

Figure 2 - (a) Basic feature for a representation of lamellar dichalcogenides
(b) Examples of some structural types

In the case of octahedral chalcogenides the lowest band built by the d cationic levels is a broad t_{2g} band able to accomodate six electrons, whereas a narrow a'_1 band formed by the d_{z^2} orbital and then a e' group (d_{xy}, $d_{x^2-y^2}$) are successively found in the case of trigonal prismatic chalcogenides. It is easier to intercalate in octahedral chalcogenides with the broad t_{2g} band (TiS_2, ZrS_2) than in trigonal prismatic chalcogenides (NbS_2, TaS_2) with a half filled narrow d_{z^2} band. Intercalation compounds formed with molybdenum dichalcogenides appear to be most unstable : this can be related to the fact that the d_{z^2} band of these trigonal prismatic chalcogenides is already filled up and the electron has to be accommodated on a higher level. The case of TaS_2 is of particular interest. This compound can be prepared with octahedral slabs (1T form) or with trigonal prismatic slabs (2H form). In both cases similar octahedral sites are present in the Van der Waals gap. The geometrical conditions for intercalation are identical. Differences are to be attributed to the band structure. Indeed it appears that there is no problem in intercalating up to x = 1 in the broad band 1T form (all available sites occupied which means a geometrical limitation), but it is rather difficult to reach the same limit in the narrow band trigonal prismatic structure of 2H TaS_2 /2/. Electrochemically it seems possible to force intercalation up to x = 1 /3/ but longly annealed samples reach an equilibrium at x = 0.72 which agrees well with previous solid state chemistry work /4/.

Electrointercalation which is often time a very useful technique to bring about intercalation, may also give some indication about the role of the band structure. Indeed it is worthwile to notice that an intercalation process with a complete delocalization of the transferred electron in a band of the host will be influenced not only by the number of empty levels but also by the structure of the band itself, i.e. by the density of states. A pseudo plateau in the potential variation will be observed when many electrons can be accommodated on levels that are very close in energy which is the case of a high density of states. The observed plateau would not mean a two-phase region but illustrates the fact that the difference in energy of electrons in the alkali metal, i.e. before intercalation, and in the host, i.e. after intercalation, remains quite constant.

The above considerations refer to a situation corresponding to a complete delocalization of the transferred electron which is the case usually encountered in T.M.D.C. However as far as the electronic exchange is considered three different situations have to be taken into account in order to cover all the experimental observations. They correspond to three different steps in the electronic delocalization. In fact, in addition to the previous case, the electron can be localized on discrete atomic levels or it can be partially delocalized on molecular levels of a discrete atomic entity of the structure. The first case corresponds to a reduction of a given ion in the host network, for instance M^{3+} to M^{2+} in MOX phases. The transition metal oxyhalogenide compounds present layered structures belonging to four structural types. The layered phases of the FeOCl type are made of slabs built up with two alternate metal-oxygen layers sandwiched by a double layer of halogen atoms. The metal ion presents a coordination of six (4 oxygen and 2 halogen atoms) and each distorted octahedral arrangement share two edges to constitute the (2D) layers (Fig. 3).

The lamellar oxyhalides chemistry differs from that of the lamellar chalcogenides by a substitution reaction which easily allows one to replace the outer layers of the slabs (i.e. the halogen layers) by other ion layers such as OH^-, NH_2^-... The structure thus lends itself not only to reversible topochemical reactions by intercalation, but also to irreversible topochemical reactions involving the complete replacement of the external layers of the sheets leaving only the inner layers unchanged.

From intensity potential curves of lithium electrochemical intercalation in several MOX phases (Fig. 4), it can be observed that only for FeOCl a reoxidation peak is recorded, demonstrating the reversibility of the intercalation. For that phase, the two waves at -270 mV (reduction) and -80 mV (oxidation) indicate easy reduction of Fe^{3+} and the weak difference between cathodic and anodic peaks means a fast electrochemical process.

Intercalation of lithium in FeOCl corresponds to an electronic transfer clearly seen at 77 K, with occurrence of Fe^{2+} cations (δ = 1.37mm.s^{-1} and Δ =

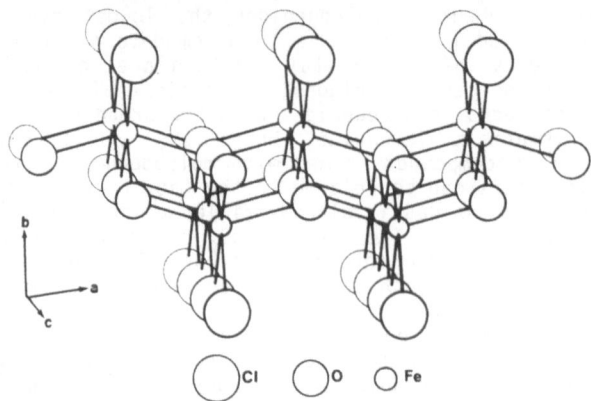

Fig. 3 - Perspective structure of FeOCl compound (after T.R. HALBERT /30/)

2.28mm.s^{-1}) quite distinct of the remaining Fe^{3+} cations (δ = 0.49 mm.s^{-1}, Δ = 1.02 mm.s^{-1}), the number of the reduced iron sites being equal to the number of Li$_2^+$ ions.With respect to increased temperature, one observes a rapid decrease of Fe^{2+} peak intensity, complete disappearance taking place above 110 K. The only doublet then recorded (δ = 0.42 mm.s^{-1}, Δ = 0.75mm.s^{-1}) has an isomer shift larger than that of iron in FeOCl (δ = 0.38mm.s^{-1}) showing an increased electronic density on the d orbitals. This can be interpreted by an electronic delocalization corresponding to a classical hopping mechanism, associated with a small polaron exchange between Fe^{3+} and Fe^{2+}. In such localized systems the ability of intercalating will depend on three factors which are the rigidity of the MOX lattice, the stability of the +3 oxidation state of the metal and the ionization potential of the guest /5/. The rigidity of the MOX lattice of course

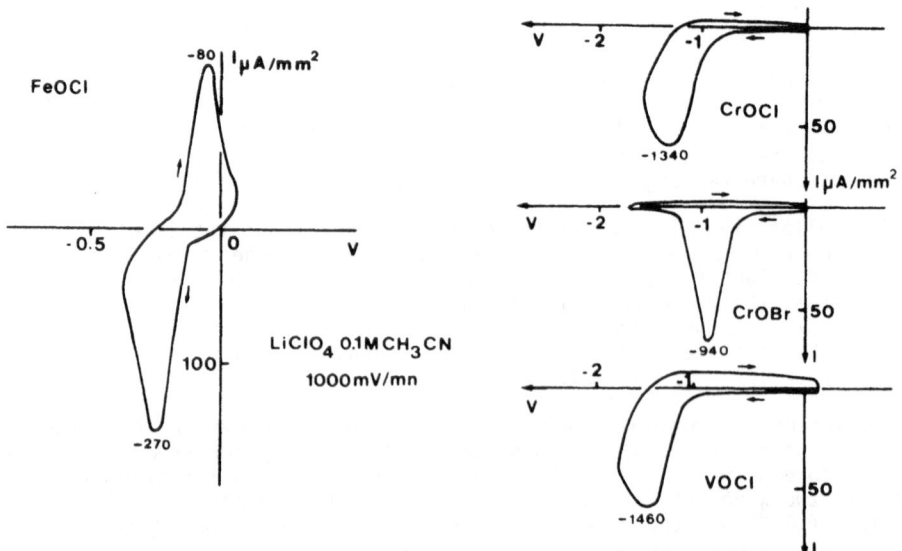

Figure 4 - Intensity potential curves for several oxyhalogenides

means the more or less easy separation of the slabs. Indications concerning this behaviour are to be found in thermal dilatation coefficients of the crystalline parameters /6/ measured with the help of a very precise X-ray camera /7/. A very interesting point in Table I is the value observed for the α_b thermal coefficient in the direction perpendicular to the slabs. This value is much higher in the case of FeOCl (α_b = 21.7) than in the case of VOCl (α_b = 15). In addition the α_a and α_c coefficients (associated with the slabs) express a higher rigidity of the MOX slabs in the case of CrOCl. This last point is probably related to a half filled t_{2g} electronic population around Cr^{3+} ions in pseudo octahedral sites. The higher value of α_b in the case of FeOCl agrees with the fact that molecular intercalation which needs a large Δb separation is easier in FeOCl than in other oxyhalides. Until now only electron donors have been intercalated in agreement with the reduction of the M^{3+} cation. Again $FeOCl_2$ represents the most favorable case as shown in Table II which gives the M^{3+}/M^{2+} redox potentials. This observation will have also drastic effects on the type of molecules that could be intercalated in the various hosts. Intercalation in FeOCl may occur for molecules presenting a pK_a factor at least equal to 2.8, whereas the corresponding values are 5.2 and 10.5 respectively for VOCl and CrOCl /8/. Similar considerations have also been introduced for transition metal dichalcogenides and MPS_3 derivatives /9/.

The last possibility corresponds to a partial delocalization among a polyatomic entity existing in the structure. CHEVREL phases /10/ built up from Mo_6 octahedral clusters enclosed in pseudo cubes of eight chalcogen atoms illustrate that situation with transferred electrons that are delocalized among a finite number of directly bonded metal atoms. Indeed through an intercalation process extra metal atoms can be seated between the Mo_6X_8 units but their number depends on the number of electrons that can still be accommodated on the molecular levels of the Mo_6 clusters /11/. One can intercalate twice as many univalent cations than divalent cations of the same size and changing the cluster electronic structure (substituting molybdenum) one changes the possible intercalation chemistry.

2. Geometrical Factors

The filling of a Van der Waals gap between slabs introduces geometrical aspects that can be described by considering both local structural effects, i.e. 1) symmetry of the occupied site, 2) ordering in the Van der Waals gap, and 3) a more global effect which is the Δc parameter expansion in the direction perpendicular to the slabs.

Let us consider at first the structural aspects. The alkali metal can occupy either octahedral or trigonal prismatic sites between the slabs of a T.M.D.C.. It stems from three factors : the size of the alkali metal, the amount of intercalation, the nature of the slabs of the host structure /12/. An octahedron can accommodate higher charges on the anions than a trigonal prism does. Thus the general ionization scheme xA^+, MS_x^- explains why, for a given alkali metal, the octahedral form may appear for the higher values of x, whereas the trigonal prismatic forms are to be obtained for the lower values of x. With a bigger alkali metal, the sulfur layers are more distant and the trigonal prismatic structure is favored. The last factor involves the covalency of the host structure : ZrS_2, more ionic, favours the formation of octahedral species as compared to TiS_2 (for example $KZrS_2$ is octahedral but $KTiS_2$ is trigonal prismatic). According to these remarks a general diagram concerning ionicity-structure relationship in the intercalation compounds has been proposed /13/. By plotting the $r_{A^+}/r_{C^{2-}}$ ratio versus a function related to stoichiometries and to the fractional ionicities of the M-S and A-S bonds, it is possible to draw an unambiguous limit between octahedral and trigonal prismatic regions (Fig. 5). Such a diagram can be used in order to predict the structural types and the range of compositions to be expected in a given series.

In the Van der Waals gap the A^+ ions repel each other as a first type of ion-ion interaction . For particular compositions such as 0.25, 0.33, 0.50 this should lead to an ordering between occupied and empty sites, at least at

Table I - Crystalline parameters and thermal coefficients in some oxyhalides. Mean thermal dilatation coefficients between 78 and 300 K ($\times 10^6$ K^{-1})

	a (Å)	b (Å)	c (Å)	γ (°)	α_a	α_b	α_c	α_v
FeOCl								
300 K	3.7729(5)	7.9104(3)	3.3026(4)	90	-	-	-	-
78 K	3.7667(5)	7.8733(3)	3.2961(4)	90	7.6(6)	21.7(6)	9.0(7)	38(2)
VOCl								
300 K	3.7734(5)	7.926(1)	3.2992(5)	90	-	-	-	-
78 K	3.7660(5)	7.900(1)	3.2890(5)	90	8.1(5)	15.7(5)	13.5(5)	37(1)
VOBr								
300 K	3.778(1)	8.411(1)	3.405(1)	90	-	-	-	-
78 K	3.772(1)	8.377(1)	3.401(1)	90	6.7(5)	18.6(5)	5.0(5)	30(1)
CrOCl								
300 K	3.855(1)	7.737(1)	3.166(1)	90	-	-	-	-
78 K	3.852(1)	7.714(1)	3.163(1)	90	3(1)	15(1)	5(1)	23(1)
CrOBr								
300 K	3.861(1)	8.191(4)	3.230(1)	89.94	-	-	-	-
78 K	3.858(1)	8.163(4)	3.224(1)	89.67	3(1)	20(1)	8(1)	31(3)
VOCl + p-methyl-pyridine								
300 K	3.792(5)	27.646(11)	3.225(5)	90	8(1)	39(2)	15(1)	62(4)
78 K	3.785(5)	27.409(11)	3.236(5)	90	-	-	-	-

Between brackets, estimated standard deviations.

TABLE II - Redox potential (volts) of some M^{3+}/M^{2+} couples

Fe^{3+}/Fe^{2+}	0.77
V^{3+}/V^{2+}	-0.25
Cr^{3+}/Cr^{2+}	-0.41

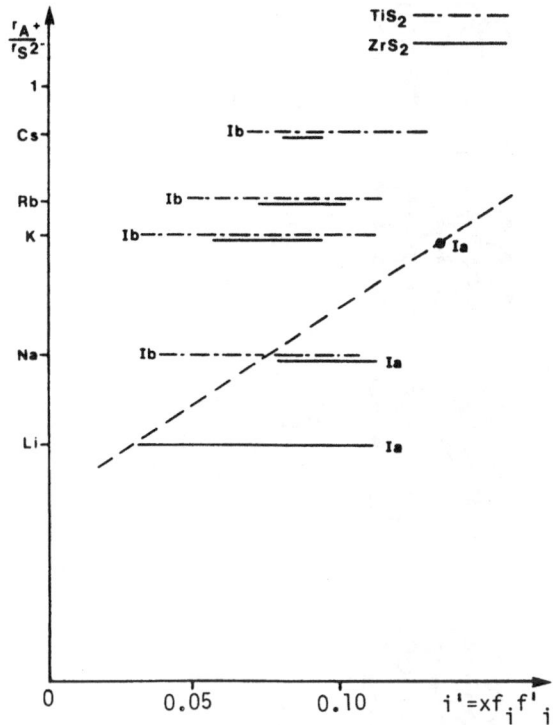

Figure 5 - Ionicity-Structure diagram for intercalates in CdI$_2$ like host structures. x is the stoichiometry, f_i and f'_i are the ionicities of A-S and M-S bonds respectively.

sufficiently low temperatures. The existence of such ordered structures has been proved in a TiS$_2$ single crystal electrochemically intercalated with sodium /14/.

Depending on the coordination type of the ions, there are three possible two dimensional sublattices for the intercalate ions in the layered dichalcogenides:a triangular lattice for the octahedrally coordinated ions, a honeycomb lattice for the trigonal prismatically coordinated ions, and a puckered honeycomb lattice for the tetrahedrally coordinated ions such as silver and monovalent copper (Fig. 6). Numerous differently ordered structures can be imagined on these sublattices.

In Na$_x$TiS$_2$, three types of superlattices were observed in the trigonal prismatically coordinated sodium ions domain (Fig. 7). Patterns A and B correspond to hexagonal (2x2) and ($\sqrt{3}$x$\sqrt{3}$) lattices, pattern C being a superposition of three rotationally equivalent rectangular (2x$\sqrt{3}$) lattices.

Let us consider now the Δc parameter expansion in the direction perpendicular to the slabs of the host structure. It is often considered as the proof that an intercalation process has really taken place. But one should notice that (i) there may be no Δc at all in a given intercalation process and (ii) this parameter expansion has not a purely geometrical origin but comes also from electronic effects. Concerning the geometrical contribution to Δc it depends on the size of the A$^+$ ions and also on the width of the Van der Waals gap. For a

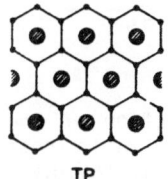
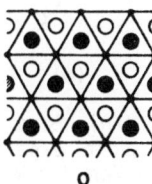
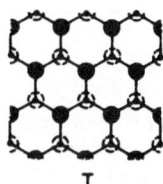

Figure 6 - Alkali metal sublattice as found in the Van der Waals gap. TP (honeycomb lattice), 0 (triangular lattice) and T (puckered honeycomb lattice) (after T. Hibma /14/).

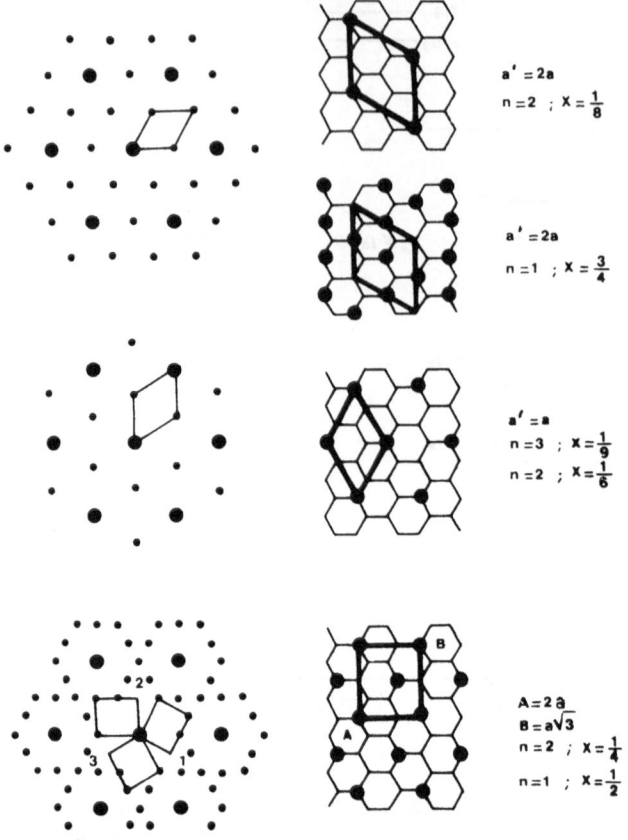

Figure 7 - Superstructure patterns in reciprocal space, and the corresponding unit cells in Na_xTiS_2. n is staging, x sodium concentration, a', A, B are the new supercell parameters (after T. Hibma/14/)

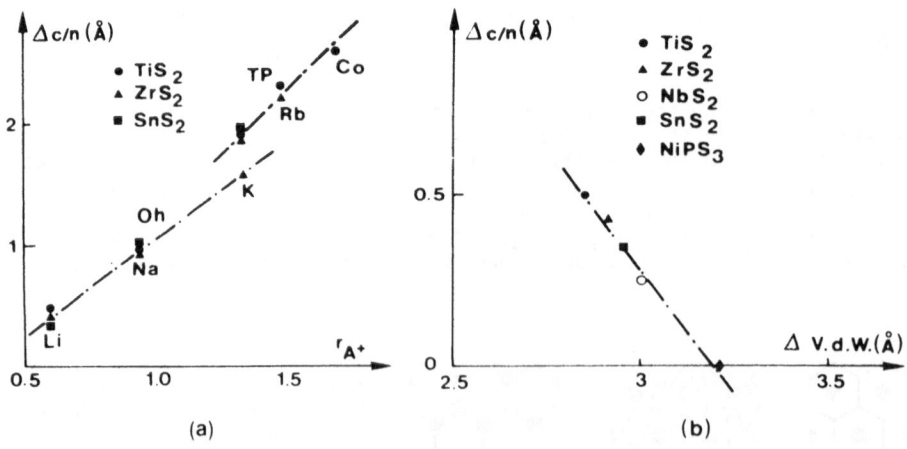

Fig. 8 - (a) c parameter expansion for <u>one</u> slab (Δ c/n) as a function of the alkali metal size (r_A+) and its site symmetry. (b) Δc/n expansion in Li host intercalates as a function of the initial width of the Van der Waals gap(ΔV.d.W.)

given structural type and the same amount of intercalation, Δc is proportional to the radius of the A^+ ions (Fig. 8a). Then, if we consider similar Van der Waals gaps such as the ones we have between sulfide layers in T.M.D.C., and a similar population of these Van der Waals gaps by the same A^+ ions which means the same electronic transfer to the host, Δc appears to be directly proportional to the initial size of the Van der Waals gap represented here by the sulfur-sulfur distance on each side (Fig. 8b). One can extrapolate a critical value of about 3.20 Å for which no Δc would be observed. Indeed, there is no Δc expansion when lithium is intercalated in this type of material /15/. Now the electronic transfer from guest to host contributes to Δc through two opposite effects, i.e. an increasing repulsion between slabs that are now negatively charged and an attraction between these slabs and the A^+ positive layers. If the A^+ ion is small, it does not sufficiently separate the slabs to minimize the repulsion and in that case the slab to slab repulsion may be the most important factor. In the case of the Li-TiS$_2$ system, for example, it has been said that the lattice expansion on intercalation is associated with the charge donation to the host rather than the propping of the layers by Li$^+$ ions /16/.

Such a situation is also highly favorable to obtain a dispersion of the slabs, like a dispersion of polymers, in a solvent. This was nicely illustrated for the first time in the case of Li$_2$Mo$_6$Se$_6$. When exposed to highly polar solvents such as dimethylsulfoxide or N-methylformamide, Li$_2$Mo$_6$Se$_6$ and Na$_2$Mo$_6$Se$_6$ can be dissolved /17/. The absence of swelling for the M$_2$Mo$_6$X$_6$ compounds with M = K, Rb, Cs and X = Se, Te, as well as the presence of macroscopic rod particles with Na$_2$Mo$_6$Se$_6$ in contrast to Li$_2$Mo$_6$Se$_6$ is clearly the result of a competition between solvation energy and a lattice energy which, as we mentioned above, will increase with an increasing size of the cationic species. Indeed coming back to these considerations, if the alkali metal is bigger, the balance between slab to slab repulsion and coulombic attraction between A^+ layers and negatively charged slabs is obviously in favor of the latter term as shown by the fact that true layer oxides do not exist due to the strong repulsion between O^{2-} layers but Na$_x$MO$_2$ derivatives exist that are isostructural with A$_x$MS$_2$ intercalates /18/.

After intercalation, a given host structure may relax. Δc will change with time which is equivalent to annealing at room temperature. This generally results in a structural contraction which is rapidly reached after heating the samples. Annealing the samples, for 15 days at 250°C for example, is a good precaution to take to insure good reproducibility in physical measurements that are sensitive to Δc (^{7}Li NMR in particular).

A last point that has to be underlined in this section concerns the fact that layered dichalcogenides are often time non stoichiometric due to the presence of some extra metal atoms in the Van der Waals gap. This may be a severe drawback concerning the reactivity as far as intercalation processes are concerned. Indeed the metal atoms in excess (Ti$_{1+x}$S$_2$ for example) will link the slabs together and impede the diffusion of intercalated species. Most of the advantages of layered structures associated with the elasticity and local deformation of the slabs may be lost. This can go as far as a structural change (3R structural type for Nb$_{1+x}$S$_2$ instead of 2H for stoichiometric NbS$_2$).

3. Strongly Coupled Electronic-Geometrical Effects

In the above section we have shown that Δc, the parameter expansion perpendicular to the slabs, cannot be reduced to a purely geometrical effect. In fact Δc is deduced from X rays measurements that are average long range observations, which imply a good transfer of information over the whole structure. An electronic delocalization acts as a good information transmitter. When the electron transferred from guest to host is accommodated on a particular site of a non-metallic host structure, the situation is rather different. There, the host may be able to accept small amounts of a small cation without any important parameter expansion. If the system becomes metallic, immediately the usual expansion is observed (see in section 4 the case of the Li-ZrSe$_2$ system).

If Jahn-Teller distortions are involved, inducing strains in the host lattice, largely biphased systems $A_\varepsilon MS_2$-$A_1 MS_2$ will be observed. This is for example the case of the Li-VSe$_2$ system /19,20/. The cost in elastic energy associated with the intercalation reaction is particularly high in that case and this implies a need for a sufficient amount of intercalation to have a gain in electronic energy able to overcome that cost. One could also suppose that the slabs do not allow existence of distorted and undistorted sites simultaneously (no dynamic Jahn-Teller effect).

Crystal field considerations have also to be taken into account in the case of the MPS$_3$ series of host structures.

4. Phase Transitions

Three types of transitions have been observed up to now in T.M.D.C. intercalation systems. They concern (i) the host structure through a shifting of its slabs, (ii) an occupancy of different sites by the alkali metal which may be related to the first point, (iii) electronic transitions induced by the electronic transfer.

A different arrangement of the slabs with respect to each other can be observed according to temperature or according to the rate of intercalation. At room temperature NaTiS$_2$ obtained by the liquid ammonia technique has a 3R structure but when obtained at higher temperature by solid state techniques /21/ it appears as a 2H polytype (Fig. 9). A shifting of the slabs is also involved when the A^+ ions move from trigonal prismatic sites to octahedral sites as mentioned above.

Such a transition is repeated many times when cycling electrochemically a given system (Na-TiS$_2$ for example). In that case it may lead to a complete disorganization of the stacking in the host structure. Finally we come to some one-dimensional amorphous situations (along the direction perpendicular to the slabs) with no visible plateau in the electrochemical curve. This increases the reactivity of the corresponding systems which may be considered as pseudo-monophased systems, allowing their use as intercalation cathodes in batteries. On the other hand it is a drawback in the sense that it makes it difficult to ascertain phase limits through electrochemistry.

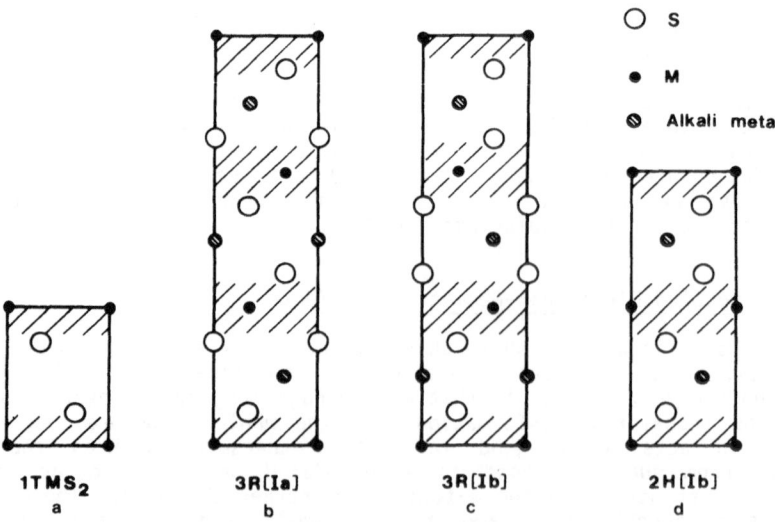

Figure 9 - NaTiS$_2$ structural types (a) Host structure, (b) 3R first stage (or Ia) with "octahedral" Na$^+$ ions, (c) 3R first stage (or Ib) with "trigonal prismatic" Na$^+$ ions, (d) 2H with "trigonal prismatic" Na$^+$ ions.

A change of sites is also possible without modifying the anionic organization in the host structure. In the Li-ZrS_2 system a phase transition takes place at 300°C for Li_xZrS_2 compositions corresponding to $0.30 < x < 0.50$. We move from a 3R octahedral structure at room temperature to a spinel structure /22/. Now lithium is in tetrahedral coordination. This transition does not need much energy because the anion stacking is not modified. It is ABC, cubic face centered, in both cases. Such a transition with no change in the anionic stacking can also be induced according to an increased ionicity of the structures. This last point is clearly apparent when comparing Li_xZrS_2 and the intercalation-substitution compounds $Li_x(Y_xZr_{1-x})S_2$. $LiZrS_2$ is of the NiAs type but in the latter case an NaCl structure is reached for the higher value of x (x = 1), following a spinel and a 3R octahedral model, which is in agreement with ionicity considerations /23/.

A critical discussion of the phase transitions in the sodium-TiS_2 system stems from a combined study by X rays and ^{23}Na NMR /24/, and band structure calculations /25/. In the sodium rich region one finds a phase with sodium in trigonal prismatic sites (Ib phase with $0.40 < x < 0.64$) and a phase with sodium in octahedral sites between TiS_2 slabs (Ia phase with $0.75 < x < 1$). When going from the Ib to the Ia phase the cell volume remains nearly constant with a 3.9 % decrease of the c parameter and a 2.2 % increase of the a parameter. Such a transition costs elastic energy because of the strong anisotropy of the structure which corresponds to a much larger stiffness in the slabs planes (a parameter) than along the c direction. To overcome this extra cost in elastic energy the transformation must lead to a lowering of the electronic energy. To characterize the Ib → Ia phase transition ^{23}Na NMR knight shifts were measured. Given the isotropic and anisotropic parts of the Knight shift as K_{iso} and K_{ax} respectively, it appears that the sodium Knight shift undergoes a 48 ppm increase when passing from Ib to Ia phase (Fig. 10). In addition, K_{ax} is negligible in the TP phase but not in the octahedral one. This implies an increase of the density of states during the transition and an anisotropic charge distribution around sodium in the octahedral phase. These statements have been confirmed by band structure calculations performed as a function of a parameter θ which characterizes the TiS_2 slab deformation at the transition (Fig. 11). To do electronic energy calculations, a TiS_2 slab is considered and characterized by its thickness dt the a parameter coinciding with the length of a side of S_3^{-6} triangle. Within each TiS_2 sheet , strong ionovalent bondings exist between Ti and S and it

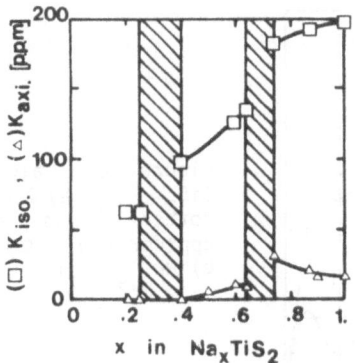

Figure 10 - Isotropic (K_{iso}) and anisotropic (K_{ax}) parts of the Knight shift versus x in the Na_x TiS_2 system.

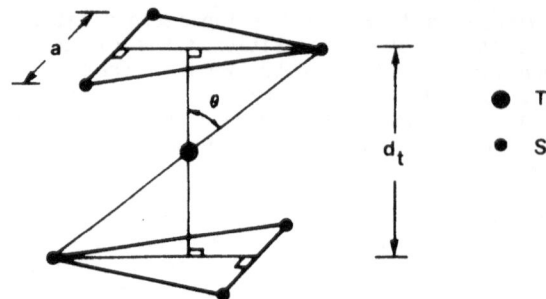

● Ti
● S

Figure 11 - The θ angle is a measure of a TiS$_2$ slab deformation during intercalation

is thus reasonable to assume that the Ti-S band remains quasi constant during the course of intercalation. This permits to calculate the structural parameter θ and to examine how the electronic structure is affected by its change.

Figure 12 shows the t$_{2g}$ subbands of a phase Ia calculated for a TiS$_2$ slab of θ = 54.9° as found in Na$_{0.5}$TiS$_2$ and a phase Ib for θ = 57.6° as found in Na$_{0.9}$TiS$_2$. We observe two important differences at the bottom portion of the t$_{2g}$ subbands[2]: i) in phase Ib, the lowest energy point at M(e$_M$) lies lower in energy that the lowest energy point at Γ(e$_Γ$). ii) In phase Ia however, e$_Γ$ lies lower in energy than e$_M$. These two differences are critical since it is only the bottom portion of the t$_{2g}$ subband that becomes filled by sodium intercalation.

As a function of θ (Fig. 13), the e$_M$ and e$_Γ$ levels vary and e$_Γ$ becomes more stable for θ > 56.4°. This critical value corresponds exactly to the structural discontinuity associated with the transition (a is calculated to be 3.50 Å to compare to 3.45 Å and 3.53 Å for the boundary values in Ib and Ia phases).

The crossing of levels along with the ionic interactions of Na$^+$ cations with sulfur layers of TiS$_2$ slabs is responsible for the transition. Band structure calculations confirm also the increase in density of states that was suggested by K$_{iso}$ measurements. /25/

Furthermore the analysis of the contribution of the d orbitals to the e$_M$ and e$_Γ$ levels shows a contribution of all d orbitals to the e$_M$ level whereas e$_Γ$ has no d$_{z2}$ contribution, which explains the variation of K$_{axial}$ Knight shift at the transition

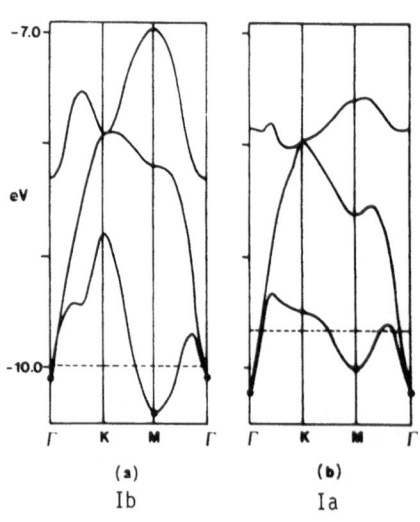

Figure 12 - Calculated t$_{2g}$ subbands of a TiS$_2$ slab. a) θ = 54.9° b) θ = 57.6°. The dotted lines are the Fermi levels appropriate for a) Na$_{0.5}$TiS$_2$(Ib phase), b) Na$_{0.9}$TiS$_2$(Ia phase)

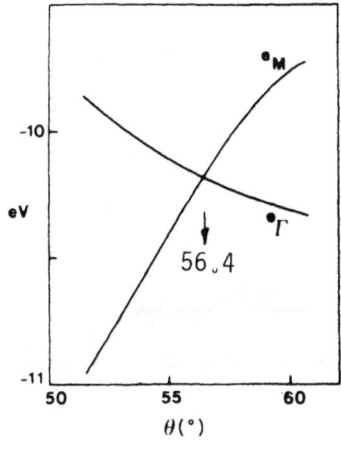

Figure 13 - e_M and e_Γ levels as a function of θ, defined in Fig.11.

Let us finally consider the Li-ZrSe$_y$ system ($1.85 < y < 1.95$). ZrSe$_y$ is a semi-conductor for high values of y. Experimental data indicate that Li$_x$ZrSe$_{1.95}$ intercalation compounds show a phase transition when the lithium concentration reaches the value of x = 0.40. A continuous filling of the octahedral sites of the Van der Waals gap is observed all along the composition domain ($0 < x < 1$) according to a classical CdI$_2$-NiAs evolution, and the transition is not structurally apparent. It is indeed a purely electronic transition observed through electrical and magnetic measurements /26/: Curie-Weiss paramagnetism for x < 0.40, and temperature independent paramagnetism for x > 0.40 ; in the ^{77}Se NMR experiments, no ^{77}Se shift below x = 0.40 and above, increasing shift respective to lithium concentration indicating increase metallic character.

In the metallic region, a study of the ^{77}Se NMR lineshape versus temperature (Fig. 14) shows, at room temperature, a Lorentzian shape which changes into two peaks below 250 K. The splitting of the peaks is proportional to the magnetic field which excludes any magnetic dipolar interaction between nuclei, but results from a hyperfine magnetic coupling with the conduction electrons /26/.

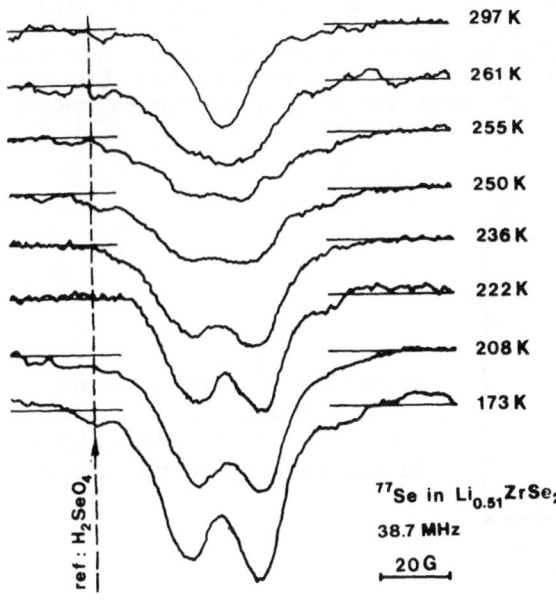

Figure 14 - ^{77}Se NMR lineshape versus temperature in a Li$_{0.51}$ZrSe$_2$ sample for various temperatures.

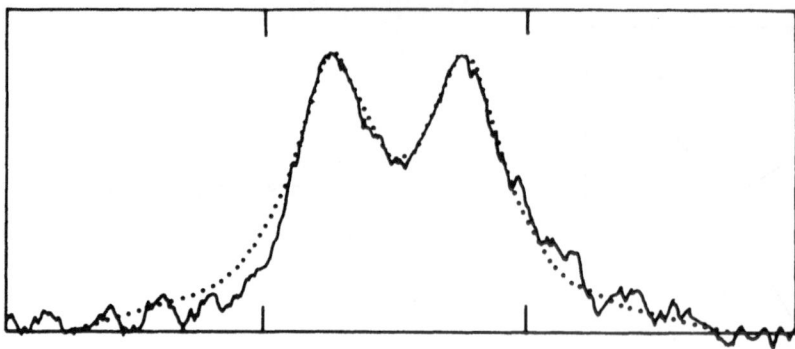

Figure 15 - Agreement with a lineshape that would be produced by a lithium short range ordering.

The low temperature ^{77}Se lineshape (Fig. 15) suggests a local order of lithium ions in the network resulting in a modulation of the electronic charge density at the Fermi level induced by the screening of the potential of the Li^+ ions.

In the semiconducting domain (x = 0.29 as shown in Fig. 16), at low temperature, the NMR line has three peaks. The simulation is in agreement with a statistical distribution of Li^+ ions leading to Se atoms having respectively 0, 1 or 2 Li^+ ions as nearest neighbors (no peak for 3 nearest neighbors). It is again a consequence of the modulation of the electron density at the Fermi level. A computer simulation has been made in order to illustrate the distribution of the lithium ions on their triangular lattice, taking into account the NMR results, i.e. the absence of three Li^+ nearest neighbors of a selenium atom /27/. Isolated groups of lithium are formed when increasing x up to the value of 0.42 which represents a critical percolation rate for the lithium sublattice (Fig. 17). It seems that the transition is certainly related to that percolation value illustrating a strong ion-electron coupling in this system. It is possible to consider that the potential of the lithium ions is responsible for a localization of the Anderson type leading to semi-conductor behavior for x< 0.40.

The semiconducting to metal transition can be seen only on carefully annealed samples during 15 days at 300°C, which is not the case in electrochemical studies /28/. In these experiments a mixture of grains more or less intercalated is generally observed with a too high value for the c/a ratio (due to the fact that small grains are more intercalated). The importance of annealing has been beautifully demonstrated by BUTZ et al. with the help of perturbed angular correlation measurements /29/.

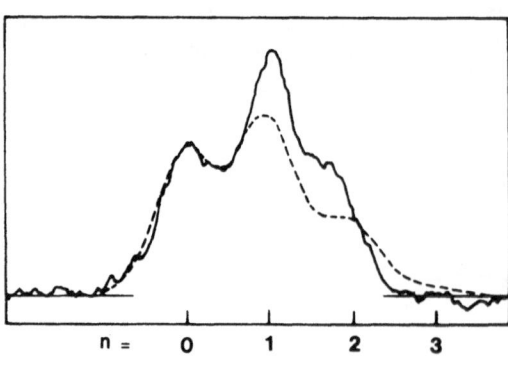

Figure 16 - ^{77}Se NMR lineshape for $Li_{0.29}ZrSe_2$. n is the number of Se first lithium neighbors. Solid line is experimental curve.

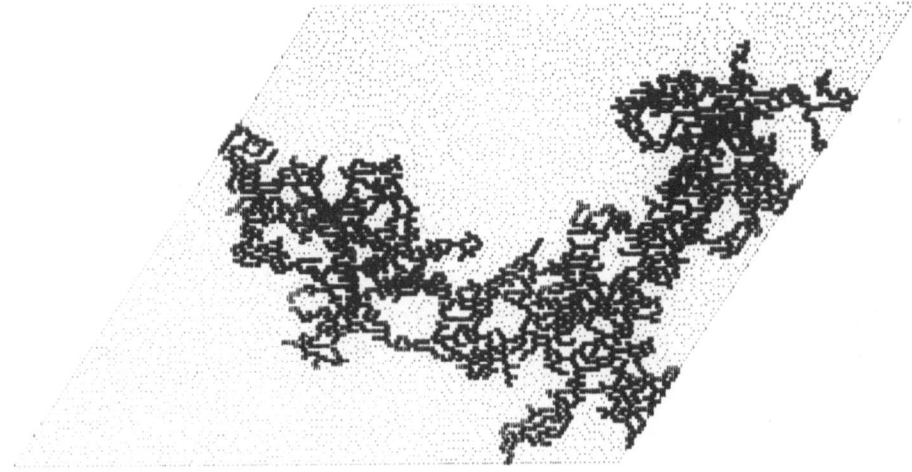

Figure 17 - Percolation of lithium ions obtained for x=0.42. (/27/)

The transition is also strongly influenced by the stoichiometry of the diselenide. In the above experiments powder of composition $ZrSe_{1.95}$ has been used. Increasing the non-stoichiometry, more and more d^1 electronic centers are present in the slabs as Zr^{3+} ions. Conductivity measured on powder samples of $ZrSe_y$ (1.88 < y < 1.94) shows a change of 10^7 (Fig. 18). In the 1.85 < y < 1.90 range the phase is metallic and becomes semi-conducting for y > 1.90. When intercalating such powders it is observed that a transition takes place towards much lower values of x, but is no more linked to a semi conductor metal transition.Figure 19

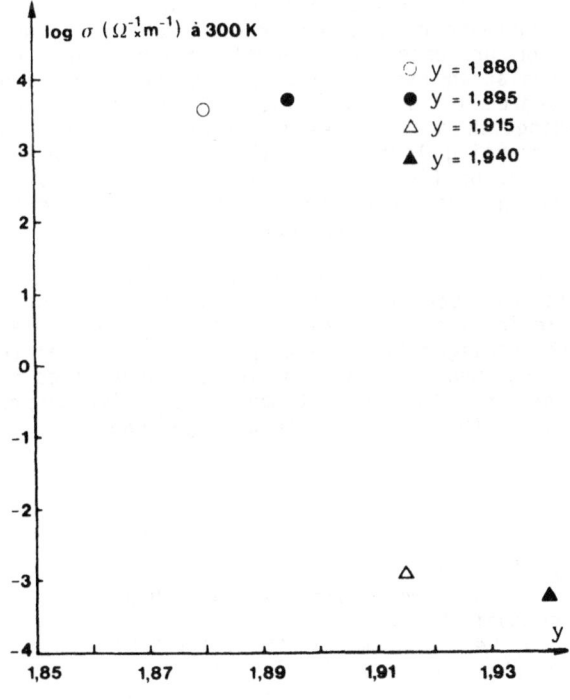

$\log \sigma \ (\Omega^{-1} \times m^{-1})$ à 300 K

○ y = 1,880
● y = 1,895
△ y = 1,915
▲ y = 1,940

Figure 18 - Conductivity measured on powder sample on several $ZrSe_y$ samples (1.80 < y < 1.94).

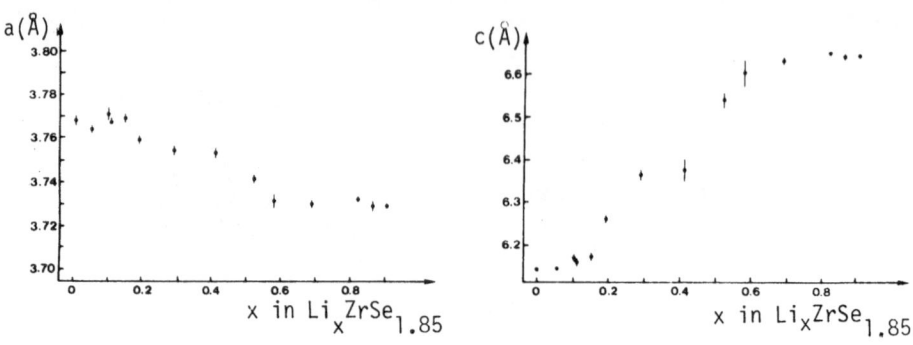

Figure 19 - Lattice parameter variation vs x for $Li_x ZrSe_{1.85}$ intercalates.

shows the c parameter expansion in the case of $Li_x ZrSe_{1.85}$. However again, breaks in the a and c parameters are found near x = 0.40, showing the influence of the percolation on the structure.

An important conclusion which stems from that study is that a semiconducting structure may accommodate a certain quantity of lithium without significant parameter expansion. Electronic delocalization, when it happens, acts as an information transmitter and the parameter expansion appears then at once. All the $ZrSe_2$ single crystals that have been tested were already metallic (because of a large non stoichiometry, for example $ZrSe_{1.88}$). The parameters expand then from the very beginning of intercalation.

5. Conclusion

In a first approach one would probably associate the ability to practice intercalation chemistry to the presence of convenient empty sites in a given structure. A careful discussion of the many systems that have been investigated up to now clearly emphasized the importance of the electronic exchange between guest and host species. Intercalation processes are largely governed by this exchange which may impede any intercalation even if the geometrical conditions are favorable. According to the degree of delocalization of the transferred electrons we have been led to distinguish among three situations corresponding respectively to the fact that the accepting level is a discrete atomic one, a molecular level of a polyatomic entity, or finally is situated in a conduction band. Even the structural expansion involves purely geometrical effects induced by the size of the intercalated species and electronic factors.

Finally intercalation chemistry appears as a competition between a cost in strain energy, needed for example to open the Van der Waals gaps of a two-dimensional structure, and a gain in electronic energy. Of course other factors are to be considered such as the configurational entropy of the intercalated species, but those are by far the most important contributions. There lies the driving force for intercalation chemistry. Sometimes it may be convenient to pre-intercalate a smaller ion that can be further exchanged to a bigger one.

6. References

1. L.S. Ramsdell : Am. Miner. 32, 64 (1947)
2. J. Rouxel, P. Molinie, L.H. Top : J. Power Sources 9, 345 (1983)
3. S.M. Whittingham : personal communication
4. W.P.F.A. Omloo, F. Jellinek : J. Less Com. Metals 20, 121 (1970)
5. J. Rouxel, P. Palvadeau : Rev. Chim. Min. 19, 317 (1982)

6. P. Palvadeau, J.P. Venien, G. Calvarin : C.R. Acad. Sci. 292II, 1259 (1981)
7. J.F. Berar, G. Calvarin, D. Weigel : J. Appl. Cryst. 13, 201 (1980)
8. F. Kanamaru, M. Shimada, M. Koizumi : Chem. Lett., 373 (1974)
9. J.P. Audiere, R. Clement, Y. Mathey, G. Mazieres : Physica B99, 133 (1980)
10. R. Chevrel, M. Sergent : J. Sol. State Chem. 3, 807 (1971)
11. R. Schollhorn, M. Kumpers, D. Plorin : J. Less Com. Metals 58, 55 (1978)
12. A. Leblanc, M. Danot, L. Trichet, J. Rouxel : Mat. Res. Bull. 9, 191 (1974)
13. J. Rouxel : J. Sol. State Chem. 17, 223 (1976)
14. T.J. Hibma : In Intercalation Chemistry, ed. by M.S. Whittingham and A.J. Jacobson (Academic Press, 1982)
15. R. Brec, A. Le Mehaute, G. Ouvrard, J. Rouxel : Mat. Res. Bull. 12, 1191 (1977)
16. A.H. Thompson, C.R. Symon : Solid State Ionics 3-4, 175 (1981)
17. J.M. Tarascon, F.J. Di Salvo, C.H. Chen, P.J. Caroll, M. Walsh, L. Rupp : J. Sol. State Chem. 58, 290 (1985)
18. C. Fouassier, C. Delmas, P. Hagenmuller : Mat. Sci. Eng. 31, 2 (1977)
19. D.W. Murphy, F.J. Di Salvo, G.W. Hull and J.V. Waszczak : Inorg. Chem. 15, 17 (1976)
20. M.S. Whittingham, F. Gamble : Mat. Res. Bull. 10, 363 (1975)
21. H.J.M. Bouwmeester, E.J.P. Dekker, K.P. Bronsema, R.J. Haange, G.A. Wiegers : Rev. Chim. Min. 7, 171 (1982).
22. P. Deniard, P. Chevalier, L. Trichet, J. Rouxel : Synt. Met. 5, 141 (1982)
23. O. Abou Ghaloun, P. Chevalier, L. Trichet, J. Rouxel : Rev. Chim. Min. 17, 368 (1980)
24. P. Molinie, L. Trichet, J. Rouxel, C. Berthier, Y. Chabre, P. Segransan : J. Phys. Chem. Solids 45, 105 (1984)
25. M.H. Whangbo, L. Trichet, J. Rouxel : Inorg. Chem. 24, 1824 (1985)
26. C. Berthier, Y. Chabre, P. Segransan, P. Chevalier, L. Trichet, A. Le Mehaute : Solid State Ionics 5, 379 (1981) and C. Berthier, Y. Chabre, P. Segransan, P. Deniard, L. Trichet, J. Rouxel : Nato ASI, 130, 561 (1985)
27. P. Deniard : personal communication
28. J.R. Dahn, W.R. Mc Kinnon, C. Levy-Clement : Solid State Com. 54, 245 (1985)
29. T. Butz, A. Lerf, J.O. Besenhard : Rev. Chim. Min. 21, 556 (1984)
30. T.R. Halbert : In Intercalation Chemistry, ed. by M.S. Whittingham and A.J. Jacobson (Academic Press, 1982)

REVIEW ON STRUCTURAL AND CHEMICAL PROPERTIES

OF TRANSITION METAL PHOSPHORUS TRISULFIDES MPS$_3$

R. Brec

Laboratoire de Chimie des Solides
U.A. 279, 2, rue de la Houssinière
44072 Nantes Cedex, France

1. INTRODUCTION

Layered phases are rather attractive compounds because they present many unusual physical and chemical properties that are not commonly encountered in three dimensional materials. Let us mention charge density waves occurrence, 2D anisotropic magnetic behaviour, strong anisotropy of conductivity for example. These specific properties are the consequence of space oriented and mainly covalent bondings taking place in the slabs that compose the structure. Transition metal dichalcogenides (T.M.D.C.) represent a significant and well known series of layered materials. In these, strongly bonded slabs composed of two anionic planes enclosing a cationic one, develop in infinite sheets. Stacking of such cation-chalcogen sandwiches leaves empty spaces between the layers otherwise weakly bonded to each other.

The two dimensional compounds may undergo intercalation reactions involving host-guest redox processes. This type of appealing chemical properties arise not only because of the weak van der Waals forces linking the structure slabs together, but also and mainly because of the existence, in the structure, of oxidizing centers.

Transition metal phosphorus trisulfides MPS$_3$ are semiconducting two dimensional phases that present most of the above mentioned low dimensional compounds properties. It renders this family very attractive, and the more so since it constitutes a broad class of phases, spanning from vanadium to zinc for the first row transition metal derivatives for instance. In addition, the MPS$_3$ present very unusual intercalation-substitution and intercalation-reduction behaviour. This article will review the physical and structural characteristics of the (2D) MPS$_3$ transition metals in relation with their substitution and intercalation properties. In this latter case, the problem of the electronic transfer upon lithium intercalation will be discussed. Some factors affecting capacity and energy of the MPS$_3$ positives in liquid electrolyte room temperature lithium battery will be outlined.

2. PREPARATION AND STRUCTURES

2.1. Synthesis Conditions

Several MPS$_3$ phases were first reported by M.C. FRIEDEL /1,2/ who obtained FePS$_3$ by heating some phosphorus pentasulfide with iron, and then by L. FERRAND /3-5/. However, the first extensive studies were done in the early seventies by HAHN et al. /6,7/. Hypothiophosphates are prepared, by direct combination of the

Table I - Single crystal microprobe elemental analysis (in %)

MPS_3	$V_{0.78}PS_3$		$MnPS_3$		$FePS_3$		$CoPS_3$		$NiPS_3$		$ZnPS_3$		$CdPS_3$	
	obs	calc	obs	calc	obs	calc	obs	calc	obs	calc	obs	calc	obs	calc
M	23.8	23.8	30.3	30.2	30.7	30.5	31.3	31.6	31.4	31.7	32.5	33.9	39.0	40.1
P	18.1	18.6	17.7	17.0	17.5	16.9	17.2	16.7	16.6	16.6	16.9	16.1	13.9	12.9
S	58.1	57.6	52.2	52.8	52.2	52.6	51.4	51.7	52.2	51.7	50.0	50.0	46.2	46.0
	Ref /15/		Ref /12/		Ref /12/		Ref /24/		Ref /12/		Ref /14/		Ref /12/	

elements, in a 400-750°C temperature range /6-8/. In order to obtain crystals large enough to enable good physical characterization of the materials, vapour-phase growth techniques can be used. Several works, particularly those of NITSCHE et al. /9,10/, WOLD et al. /11/, explored the various ways in which the ternary MPS_3 compounds can be prepared with, as transport agent, either iodine /9,10/ or chlorine /11/. This latter element is believed to act also as a mineralizing agent speeding the synthesis process.

If the above methods allow to obtain single crystals having the expected MPS_3 composition (see structure determinations and table I of element analysis below) it does not seem so easy to prepare large amount of all the phases with a homogeneous bulk, which is a mandatory requirement for any electrochemical industrial application.

$MnPS_3$, $NiPS_3$ and $CdPS_3$ are obtained in large quantities (> 50 g) /12/ with a very good homogeneity, very little sulfur excess being observed in the reaction tubes. Precautions have to be taken in the case of $CoPS_3$ for which a several month reaction at 540°C proves necessary. Occurrence of CoS_2 was detected in uncomplete reaction /13/, and in this case, successive firing and grinding should improve further the homogeneity of the bulk.

Well crystallized and homogeneous agregates (≈ 20 g) of $ZnPS_3$ can be prepared (see /14/ for example), but in some instances, $Zn_4(P_2S_6)_3$ was found to form /12/.

Recently, it was shown that VPS_3 was not the phase resulting from previously reported synthesis conditions /6/, but that a non-stoichiometric phase $V_{0.78}PS_3$ was obtained instead /15/. To date, it is not known whether there exists a composition domain for this mixed valence vanadium phase. In particular, the existence of a $V^{3+}_{2/3} \square_{1/3}PS_3$ phase, of the $In^{3+}_{2/3} \square_{1/3}PS_3$ type has not been demonstrated yet ($\square$ represents vacancy).

The synthesis of $FePS_3$ still raises some problems. The single crystals growing on the samples'surface exhibit a good stoichiometry as inferred from microprobe analysis (table I) and crystal structure calculations (see below). However, it has been pointed out in several examples /8,16,17/ that rather large quantities of unreacted phase (seemingly sulfur) remained in the reaction tubes. It follows that synthesis of the pure iron derivative in large quantities has not been reported so far. Occurrence of some iron sulfide was found in some instance in the bulk product /12/ that could explain the non-stoichiometric reaction. Possible formation of Fe_xPS_3 (x < 1) with iron of mixed valence (like in the case of vanadium) is another possibility that should be further explored. The observed very good stoichiometry of $NiPS_3$ (and also $CoPS_3$) can easily be explained from their known band structure (see below) showing the low lying electron accepting levels to be transition metal d orbitals. In the case of transition metal dichalcogenides (TiS_2 for instance), non-stoichiometry of the $M_{1+x}S_2$ type can be achieved by reduction of the M^{IV} host cation, whereas in the MPS_3, considering the high stability of the $(P_2S_6)^{4-}$ anion, a cationic excess would imply reduction of M^{II} leading to species unstable at the preparation temperature. The only possible non stoichiometry scheme left, as exemplified by the vanadium derivative, is a cationic deficiency compensated by a higher oxidation state of the metal provided it be stable in a sulfur environment, which is not the case for nickel and cobalt. Hence the high stoichiometry of the phases. This reasoning holds for MPS_3 with d filled cations.

With four components, synthesis of homogeneous substituted MPS_3 phases appears more delicate to achieve. Homocharge cationic substitution within the MPS_3 family has not been systematically studied and only the examples of $Zn_{1-x}Fe_xPS_3$ /18/ and $Cd_{1-x}Ni_xPS_3$ /19/ are known to some extent. If the synthesis is started from the elements, because of difference in stability and sublimation temperature, homogeneous products may be difficult to obtain. The best procedure consists in using chosen amounts of each MPS_3 and $M'PS_3$ compound and in heating the carefully ground and mixed phases at moderate temperature ($\approx$ 500°C).

The substituted layered materials of the types $M^{III}_{1/2}M'^{I}_{1/2}PS_3$ and $M^{III}_{1-x}M'^{I}_{2x}PS_3$ can only be prepared from the elements since corresponding single cation trithiophosphates do not exist. In the phases for which microprobe elemental analysis was made, i.e. $Cr_{1/2}Cu_{1/2}PS_3$ /20/, $Cr_{1/2}Ag_{1/2}PS_3$ /21/, $V_{1/2}Ag_{1/2}PS_3$ /22/ and $In_{1/2}Ag_{1/2}PS_3$ /23/, stoichiometry can be inferred from table II, analytical data. This is also the case for stoichiometric $Mn^{II}_{1/2}Ag^{I}_{1.0}PS_3$ /24/ of the second group.

For most of the above compounds, use of too high temperature may lead to some decomposition products and temperature appears to be a very critical factor if one wants to obtain pure samples, especially for large quantities for which synthesis have not been performed to date.

2.2 Structural Characteristics

It is one of the particularities of the MPS_3 compounds that they can be looked at in different ways according to the properties and behaviour one wants to shed light on. This fact is well illustrated through the names given to these phases labelled either metal phosphorus trichalcogenides or metal thiophosphates. In the first case, the phases are taken as double chalcogenides of metal and phosphorus, and using the formula $M_{2/3}(P_2)_{1/3}S_2$, they can be considered as layered disulfides in which a third of the cation sites are substituted by phosphorus atom pairs P_2 (fig. 1(a)). In the second case, the phase is viewed more as a salt constituted from M^{2+} cations and $(P_2S_6)^{4-}$ anions (fig. 1(b)).

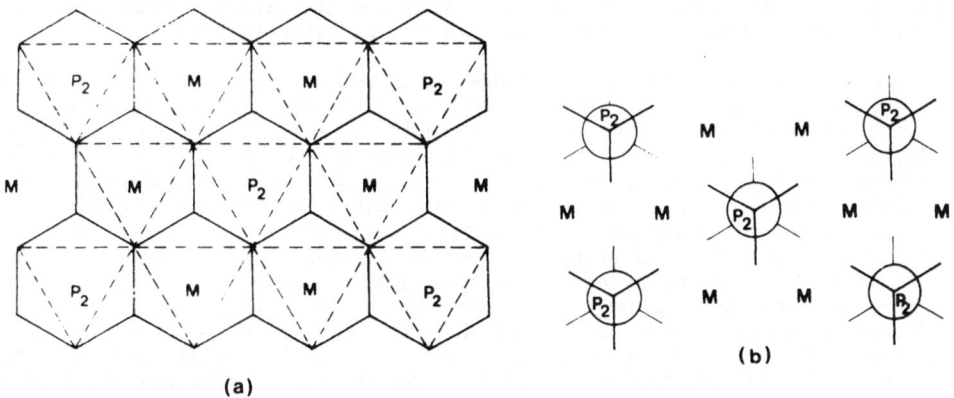

(a)   (b)

Figure 1 - (a) Octahedral cation ordering within a MPS_3 layer showing the honeycomb metal network. (b) MPS_3 structure considered as built from M^{2+} and $(P_2S_6)^{4-}$ ions.

Table II - Single crystal microprobe (Δ) and microcrystalline sample chemical (*) analysis of substituted MPS_3 (in %)

$M^{III}_{1/2}M'^I_{1/2}PS_3$	$Cr_{1/2}Cu_{1/2}PS_3^{\Delta}$		$Cr_{1/2}Ag_{1/2}PS_3^{\Delta}$		$V_{1/2}Ag_{1/2}PS_3^{\Delta}$		$In_{1/2}Ag_{1/2}PS_3^{\Delta}$	
	obs	calc	obs	calc	obs	calc	obs	calc
M^{III}	13.9	14.0	12.1	12.5	12.3	12.3	22.9	22.6
M'^I	17.2	17.2	27.4	26.1	26.3	26.1	23.6	24.1
P	15.2	16.8	14.6	14.9	14.6	15.0	12.8	13.0
S	52.8	52.0	47.5	46.5	46.5	46.6	40.1	40.3
	Ref /20/		Ref /21/		Ref /22/		Ref /23/	

$M^{II}_{1-x}M'^I_{2x}PS_3$	$Mn_{0.87}Cu_{0.26}PS_3^{*}$		$Mn_{1/2}AgPS_3^{*}$		$Cd_{1/2}AgPS_3$ $Cd_{0.87}Cu_{0.26}PS_3$
	obs	calc	obs	calc	
M^{II}	24.3	25.0	10.7	10.5	No analytical data available
M'^I	8.2	8.6	40.8	41.1	
P	16.2	16.2	11.3	11.8	
S	49.6	50.2	34.6	36.6	
	Ref /24/		Ref /24/		

2.2.1 Transition Metal MPS_3

Because all the MPS_3 structures are built from a close packed stacking of sulfur anions in the ABC sequence ($CdCl_2$ type), the MPS_3 phases could be expected to retain the trigonal symmetry and hexagonal cell generally encountered for such atomic arrays. Actually, the symmetry is monoclinic, but there is a close relationship between the hexagonal cell and the monoclinic one /6/ (fig. 2). For the latter, one observes that the β angle varies with the M cation (fig. 3), the ideal value for an undistorted cell being calculated for $\beta = 107.16°$. Two compounds, $FePS_3$ and $CoPS_3$ present such a value and can indeed be perfectly indexed with a hexagonal cell /8,25/, although the atomic array remains a monoclinic one as inferred by the crystal structure determinations.

A detailed single crystal structure calculation was first performed by W. KLINGEN et al. /6,7/ on the iron derivative. More recently, complete structure determinations were done on the whole MPS_3 family (M = V, Mn, Fe, Co, Ni, Zn, Cd) /14,15,26/. They showed that all the MPS_3 phases studied crystallize in the monoclinic symmetry, space group C2/m, the structure being constructed from an essentially sulfur cubic close-packed array. Every other layer of octahedral sites are completely filled by the transition metal ions and the phosphorus atoms as P_2 pairs in a 2/1 ratio (fig. 4), ordering being observed between the two species (fig. 1). To the exception of the vanadium derivative (see below) the structure refinements indicate, in agreement with single crystal analysis, occurrence of stoichiometric phases and no interslab cation in the van der Waals gap. For $CoPS_3$ and $NiPS_3$ some metal-phosphorus mutual substitution was found probably due to stacking faults. Polytipism was detected, although not studied, in the special case of $NiPS_3$.

For the first transition metal row phases, it appears that the van der Waals gap width remains fairly constant at 3.24(2) Å. It is much wider

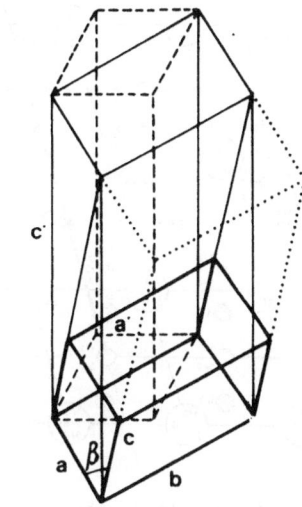

CELL
---- hexagonal
····· orthohexagonal
—— monoclinic

Figure 2 - Drawing showing the relationship between the undistorted hexagonal and distorted monoclinic cell of the MPS_3 (after W. Klingen /6/)

than in the MS_2 compounds. Different atomic radius cations are contained in the MPS_3, and the slabs size varies accordingly. To it corresponds a $(P_2S_6)^{4-}$ structural change, the flat (PS_3) pyramid constituting the anions remaining identical but the link between two of them through a phosphorus-phosphorus bond being more or less elongated to accommodate the various M cations. Thus, the P-P distance is found to shift from 2.148 Å in $NiPS_3$ (smallest cation) to 2.222 Å in $CdPS_3$ (largest cation), a good correlation being observed between the P-P bond length and the cation size (fig. 5) /27/. This rather unexpected breathing of the (PS_3) groups in the $(P_2S_6)^{4-}$ anionic units, along with their possible tilting

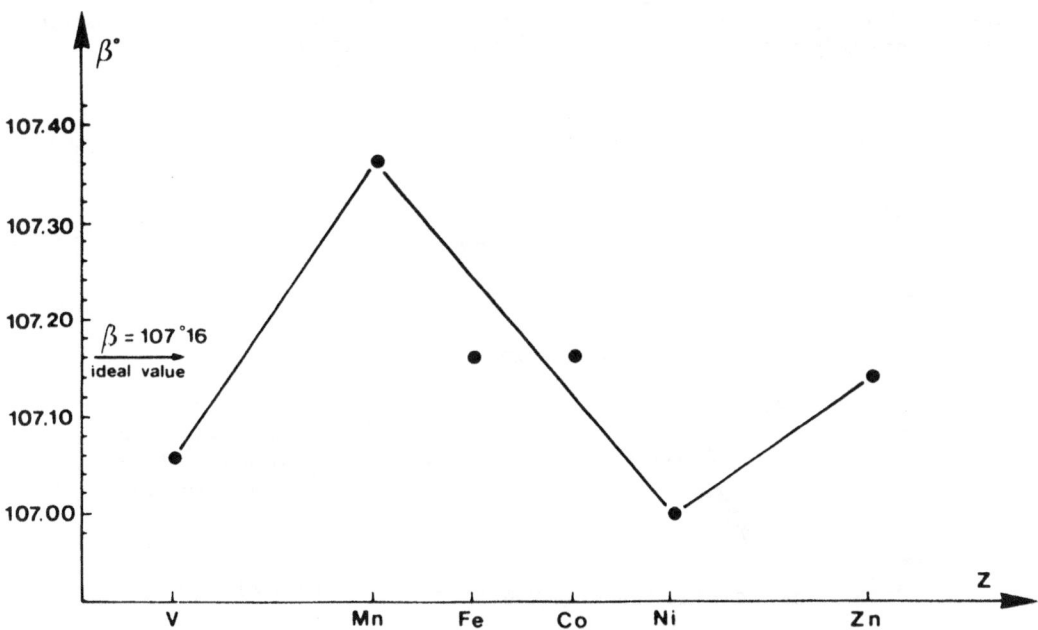

Figure 3 - MPS_3 monoclinic cell β value variation with respect to the M^{2+} cation. $FePS_3$ and $CoPS_3$ with an ideal β value have spectra indexable with a hexagonal cell

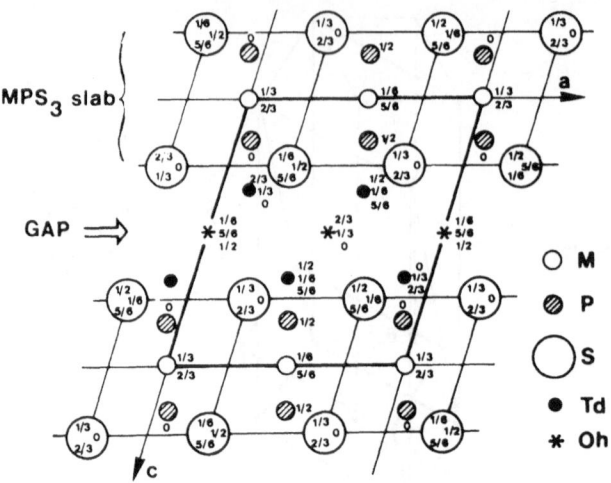

MPS$_3$ slab

GAP $\Rightarrow$

○ M
◍ P
◯ S
● Td
✳ Oh

Figure 4 - MPS$_3$ structure projection along the b axis showing slabs and van der Waals gap sites. T$_d$ and O$_h$ sites for intercalants are shown

respective to the P-P axis (see below the M$_{1-y}$M'$_y$PS$_3$ structures) illustrate the resilience of these groups and may explain many properties exhibited by this family.

Another trait revealed by the structure analysis lies in the equivalent temperature factor range values. Unexpected high values are calculated for those anions with spherical electron d distributions (Mn^{2+}, Zn^{2+} and Cd^{2+}) (fig. 6) for which one expects no crystal field stabilization. These values decrease steadily from MnPS$_3$ to NiPS$_3$ as the stabilization increases respective to the electronic configuration to reach the lower value that corresponds to maximum stability for NiPS$_3$ (t$_{2g}$^{6} e$_g$^{2}). In this series, the thermal local motion of the ions translates straightforwardly the crystal field stabilization effect /27/.

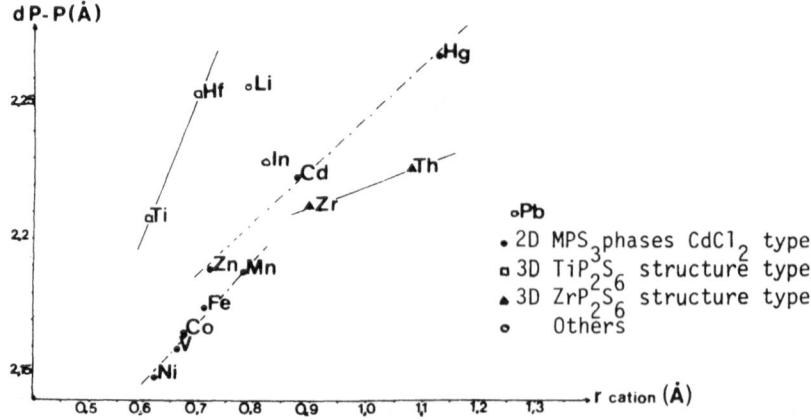

Figure 5 - P-P bond length versus ionic radius in some MPS$_3$ and related phases

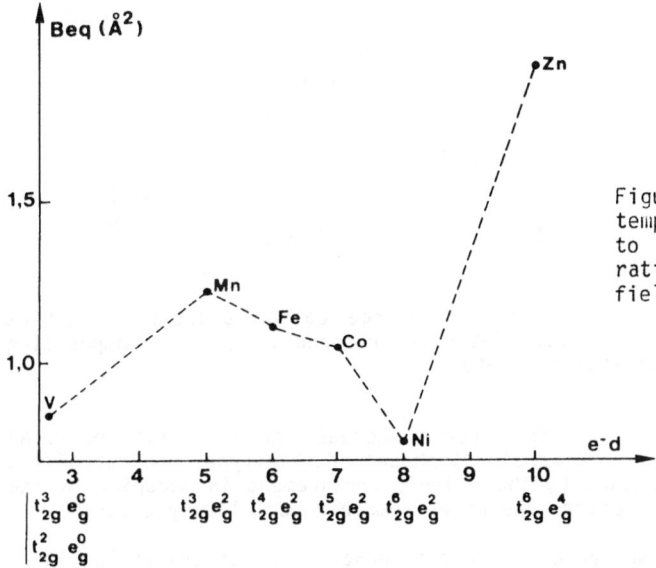

Figure 6 - Cations equivalent temperature factors respective to their electronic configuration evidencing the crystal field stabilization

In 1969 W. KLINGEN /6/ prepared, from the atomic ratio V/P/S = 1/1/3, a vanadium phase to which he attributed the formula VPS_3 since X-Ray analysis yielded cell parameters close to those expected for this type of compound. Considering the great stability of V^{3+} in a sulfur octahedral environment, a compound such as $V^{3+}_{2/3} \square_{1/3} PS_3$ would be expected to form instead of $V^{2+}PS_3$, the more so since a phase such as $In^{3+}_{2/3} \square_{1/3} PS_3$ can easily be obtained /28/. G. OUVRARD et al. /15/ showed recently that, in fact, the vanadium derivative corresponds to the non-stoichiometric phase $V_{0.78}PS_3$. The phase has the same structure as the stoichiometric MPS_3, and, because identical $(P_2S_6)^{4-}$ anionic species are found, the compound charge balance is achieved through vanadium of mixed valence occurrence and the phase is to be written $V^{2+}_{0.34}V^{3+}_{0.44} \square_{0.22} PS_3$. So far, it is the first and only MPS_3 compound of mixed valence. Neither $V_{2/3}PS_3$ nor VPS_3 seem to form and since it is an intermediary composition that is observed, this may be attributed to the gain in stability acquired through an electronic hopping between V^{2+} and V^{3+} (small polaron stabilization).

2.2.2 $M_{1-y}M'_yPS_3$ Substituted Phases

At least when the atomic radii of M and M' cations are close to each other, easy substitution is achieved. As an example, the parameters and volume variations of the series $Zn_{1-x}Fe_xPS_3$ and $Zn_{1-x}Ni_xPS_3$ $(0 \leqslant x \leqslant 1)$ are given on fig. 7, where it can be seen that the 'Vegard's law' is well followed /12/. Large cationic radii differences may create structure distortion preventing complete and homogeneous substitution, but this has not been explored so far. In particular, the possibility of a cationic ordering would deserve some attention.

The MPS_3 structural type that also occurs when replacing one M^{II} by $(2/3 M^{III} + 1/3 \square)$ suggested the possibility to replace M^{II} by a couple $(1/2 M^I + 1/2 M^{III})$. A series of such phases was obtained and their structures determined for some of them with $M^I = Ag$, Cu and $M^{III} = Cr$, V, In by P. COLOMBET et al. (21-23,29). The four phases $Cr_{1/2}Cu_{1/2}PS_3$, $Cr_{1/2}Ag_{1/2}PS_3$, $V_{1/2}Ag_{1/2}PS_3$ and $In_{1/2}Ag_{1/2}PS_3$, for which full structure determinations were made, show layered arrangements closely related to the MPS_3 structural type. In the three first

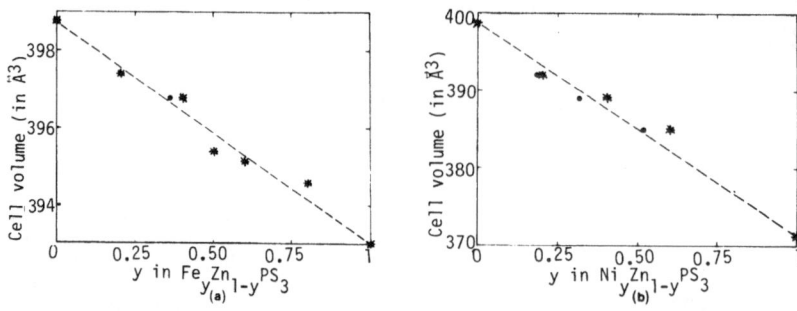

Figure 7 - Cell volume variation in the homocharge cation substituted series $Fe_yZn_{1-y}PS_3$ (a) and $Ni_yZn_{1-y}PS_3$ (b). Stars correspond to nominal composition and black circles to elemental analysis data

compounds the same ABC anion stacking occurs whereas for the last one a AB stacking is observed. Like in two dimensional phases of this type, one out of two octahedral sites sheets determined by the anionic arrangement is occupied by the two cations and the phosphorus pairs, the other remaining completely empty.

Except for $Cr_{1/2}Cu_{1/2}PS_3$ (see below), the octahedral sites of the filled layer are occupied by $M(1/3)$ $M'(1/3)$ and $P_2(1/3)$ in an ordered way. For $In_{1/2}Ag_{1/2}PS_3$, to that ordering corresponds a trigonal space group ($P\bar{3}1c$), the cations forming a triangular lattice (fig. 8(a)). In the case of $Cr_{1/2}Ag_{1/2}PS_3$ and $V_{1/2}Ag_{1/2}PS_3$ which have cell parameters very similar to that of the MPS_3 but with a $P2/a$ space group, one observes on the contrary occurrence of zigzagging (M^{III} and Ag^I) chains (fig. 8(b)).

In $Cr_{1/2}Cu_{1/2}PS_3$, structural determination showed the same type of cation triangular arrangement as in $In_{1/2}Ag_{1/2}PS_3$. However, at the copper site, an important anisotropic thermal factor revealed a possible statistical occupancy of several positions by copper atoms within the slabs octahedra. Subsequent EXAFS analysis /30/ indicated the occurrence of octahedral (Cu_2S_6) entities resembling the (P_2S_6) anionic groups with copper in a triangular environment, along with a few unusual (CuS_6) octahedral groups. The actual formulation of the compounds is thus to be written : $Cu_{(1/2-2y)}(Cu_2)_y \square_y Cr_{1/2}PS_3$.

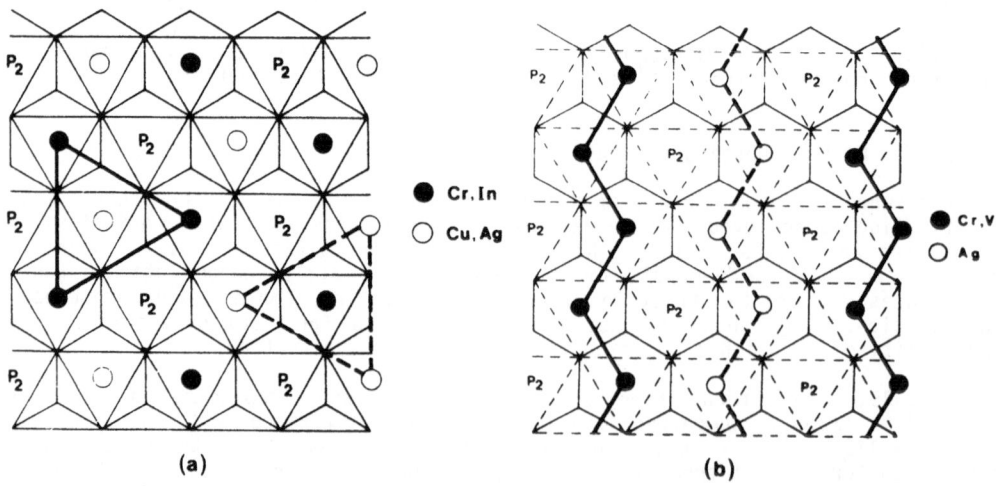

(a)   (b)

Figure 8 - Triangular (a) and linear (b) cationic arrangement in $M^I_{1/2} M^{III}_{1/2} PS_3$ compounds

Table III - Relationship between cationic radius ratio and sublattice type in substituted $M^I_{1/2}M^{III}_{1/2}PS_3$ phases (D and d are mean M^{III}-S and M^I-S distances)

Compound	D (Å)	d (Å)	D/d	Sublattice
$Ag_{1/2}Cr_{1/2}PS_3$	2.81	2.43	1.16	chains
$Ag_{1/2}V_{1/2}PS_3$	2.78	2.47	1.13	chains
$Cu_{1/2}Cr_{1/2}PS_3$	2.65	2.45	1.08	triangles
$Ag_{1/2}In_{1/2}PS_3$	2.78	2.65	1.05	triangles

The origin of the ordered distribution in the $M^I_{1/2}M^{III}_{1/2}PS_3$ series was discussed and correlated to the cationic radii (or M-S distances) ratio r^I/r^{III} /29/. For ratios close to one, a separated triangular arrangement of both cations is found, whereas for ratio quite departing from this value, zigzag chains of each cation are formed (table III). This can be explained from two conflicting factors related to coulombic and strain forces. Depending on how the structure can gain the highest energy, one arrangement will be favored over the other. Considering ionic charges, coulombic repulsions between the most ionized cations M^{3+} will tend to separate them. According to Pauling rules, the repulsive interaction produces a superlattice in which each cation maximizes the distance to its neighbors. This is achieved through a triangular lattice in which the $M^{III}S_6$ nearest neighbors do not share any edge or vertex (fig. 8(a)). The (P_2) pairs also order in triangular lattice probably for the same reason, leaving the same triangular arrangement for the M^+ cations. This situation happens for r^I/r^{III} ratio around unity, but when it becomes more different, strong distortions take place in the (P_2S_6) octahedra in order to accommodate the very different cationic radii. Clearly to minimize now this strain energy, the cations order as zigzagging chains (fig. 8(b)). Certainly, the breathing of the (P_2S_6) entities as discussed in the previous subsection plays an important role to locally adjust the anionic lattice from larger to smaller neighboring octahedra. A distortion involving bond angles and distances can also be observed on fig. 9 for $Ag_{1/2}V_{1/2}PS_3$ where it can be seen that towards the vanadium side, sulfur-sulfur distances are about 3.5 Å whereas towards the silver side it is around 4.0 Å. This implies very strong distortion of the thiophosphate group underlining its great adaptability to various cation sizes in the same slab /22/.

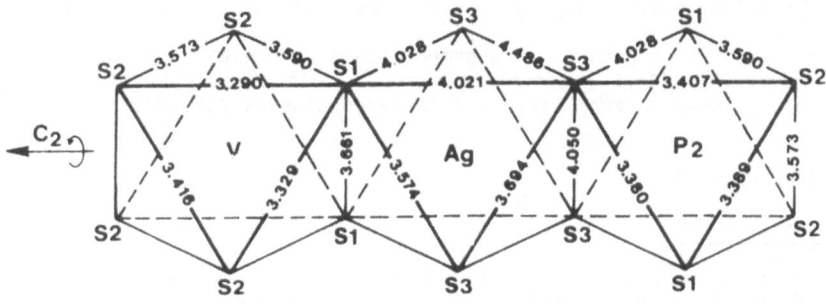

Figure 9 - Sulfur-sulfur distances encountered in (VS_6), (P_2S_6) and (AgS_6) octahedra of $Ag_{1/2}V_{1/2}PS_3$ revealing the (P_2S_6) group adaptation to cation size

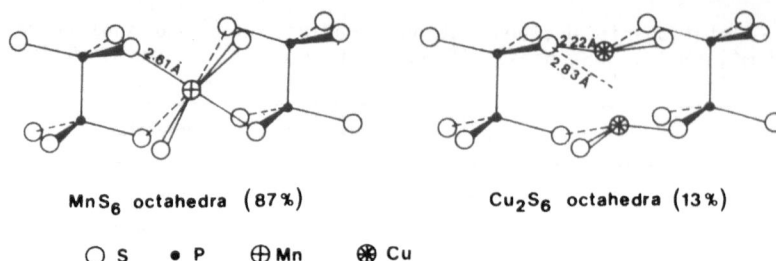

MnS$_6$ octahedra (87%)   Cu$_2$S$_6$ octahedra (13%)

○ S   • P   ⊕ Mn   ✳ Cu

Figure 10 - Schematic representation of (MnS$_6$) and (Cu$_2$S$_6$) octahedral sites in Mn$_{0.87}$Cu$_{0.26}$PS$_3$

A subclass of heterocharge cation substitution is exemplified by the families M$^{II}_{1-x}$Cu$^I_{2x}$PS$_3$ (M = Mn, Cd, x = 0.13) /31/ and M$^{II}_{1/2}$Ag^{I}PS$_3$ (M = Mn, Cd) /32,33/. Only for Mn$_{0.87}$Cu$_{0.26}$PS$_3$ has an extended X-Ray diffraction and EXAFS studies determined the structure of the phase. In this one, in a MPS$_3$ type of structure is found a disordered distribution of 87 % of (MnS$_6$) octahedra together with 13 % (Cu$_2$S$_6$) octahedra containing Cu$^+$ ions triangularly coordinated (fig. 10). Probably for stability reasons, the MPS$_3$ network can only accept a substitution site ratio of about 13 %. One of the interesting feature of this phase, putting aside the occurrence of (Cu$_2$S$_6$) entities resembling the (P$_2$S$_6$) ones, is the very strong thermal motion of the copper ions. Electrical conductivity measurements made on single crystals confirmed the insulating character of the phase (R // planes ≈ 10^9 Ω.cm), consistently with the MPS$_3$ general behavior. However, the value of the conductivity measured in the (ab) plane increases by four orders of magnitude between 300 and 500 K, suggesting the taking place of a thermally activated transport process of an ionic type, the electronic absorption spectra remaining unchanged /33/. A Raman study performed on Mn$_{0.87}$Cu$_{0.26}$PS$_3$ to elucidate the atomic motion in the phase established that, respective to temperature, the relative intensities of two signals attributed to copper translation modes were drastically modified (ν_1 and ν_2 on fig. 11), one of the peaks disappearing completely at low temperature. In the absence of any other spectrum modification, this reveals the existence of a distribution of Cu$^+$ on two types of sites whose population changes with temperature. The closely related frequencies of the signals (46 & 57 cm^{-1} at 300 K) along with the absence of important perturbation of the (P$_2$S$_6$) entities indicate comparable environment for copper. The dynamic proposed model is shown to correspond hence, at low temperature, to a structure of the MPS$_3$ type with all copper trapped in the intralamellar sites. At high temperature, almost all the copper ions are located in neighbor tetrahedral sites of the van der Waals gap, giving actually a new structural type, having lost its low dimensional character. Cd$_{0.87}$Cu$_{0.26}$PS$_3$ develops the same kind of behavior.

The same type of structural arrangement seems to prevail for the other phases, in particular with (Ag$_2$S$_6$) octahedral groups, in agreement with strong local distortions as inferred from Raman and Infrared studies /32,33/.

3. PHYSICAL PROPERTIES

3.1 Magnetism in MPS$_3$

The reciprocal susceptibility vs temperature variation for manganese, iron, cobalt and nickel thiophosphates /8,11,34/ characterizes low-dimensional magnetic systems, and a Curie-Weiss law is only obtained at quite high temperatures. All phases order antiferromagnetically at low temperature. Table IV summarizes the

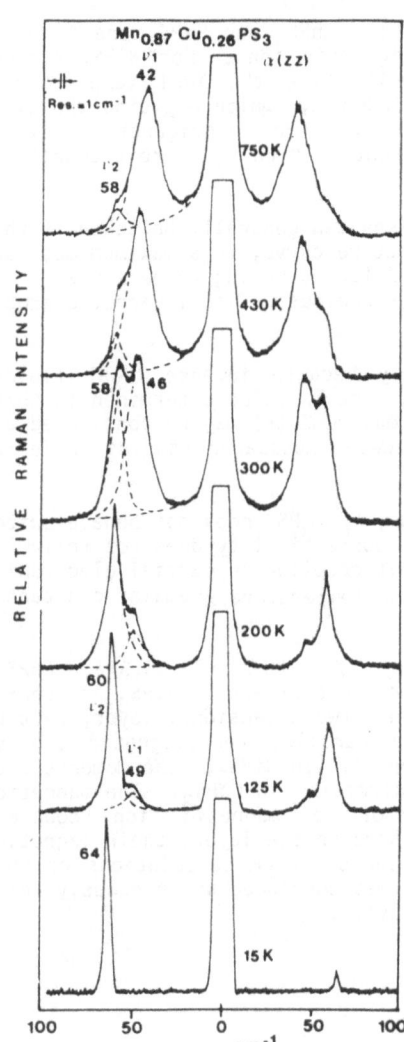

Figure 11 - Raman spectra (-100, +100 cm^{-1} of Mn$_{0.87}$Cu$_{0.26}$PS$_3$ obtained from microsingle crystals in the (ZZ) configuration in the 15-750 K temperature range (after P. Poizat /33/).

Table IV - Magnetic data relative to the MPS$_3$ phases

Compound	μ_{calc} (μ_B) (spin only)	μ_{eff} (μ_B)	θ_p (K)	T_{max} (K)	T_N (K)
MnPS$_3$	5.92	5.98	-217	110	82 *
FePS$_3$	4.90	4.94	104	123	{ 116 * / 117.4 o
CoPS$_3$	3.87	4.90	-116	127	122 •
NiPS$_3$	2.83	3.9	-712	300	155 *

Néel temperature determined by :	{ * N.M.R. of ^{31}P	Ref /35/
	• neutron diffraction	Ref /25/
	o Mössbauer spectroscopy	Ref /38/

main magnetic constants. The observed moments for $MnPS_3$ and $FePS_3$ are in good agreement with the spin only moment for M^{2+} high spin ions. For $NiPS_3$ strong short distance magnetic correlations, even well above the Néel temperature, account for the very wide minimum observed around 300 K, which explains the high value of μ_{eff}. For $CoPS_3$, because Co^{2+}, with a $3d^7$ configuration and a fundamental state $4T_1$ in octahedral field, presents a high spin orbit coupling, one observes a high μ_{eff}.

While magnetic ordering temperatures in 3D phases can generally be taken at the maximum of the magnetic susceptibility vs temperature curve, this maximum must be considered prudently in low dimensional systems. Actual ordering temperatures were found much lower using, in an N.M.R. study, ^{31}P nucleus as local magnetic probe /35/.

At T_N, the ^{31}P NMR line disappears suddenly because intense local fields broaden and shift the phosphorus resonance peak. The T_N value determined in this way (table IV) (or through neutron diffraction measurements) may be considered as closer to the true ordering temperature (see however Mössbauer studies on $FePS_3$ below).

Because of the mixed valence state of vanadium, $V_{0.78}PS_3$ does not behave quite like the other MPS_3. High temperature reciprocal susceptibility does not follow a Curie law, probably in part because of spin orbit coupling or partial electronic delocalization contribution and, below the ordering temperature, χ remains constant, consistently with a spin-flop mechanism /29/.

Magnetic structures were determined for $NiPS_3$ /8/, $MnPS_3$ /13,36,37/, $CoPS_3$ /13/ and $FePS_3$ /37/. The distribution of the magnetic carriers in $NiPS_3$ and $CoPS_3$ are given in fig. 12(a). It can be described, in a two dimensional layer, by the presence, for the two compounds, of double parallel ferromagnetic chains antiferromagnetically coupled to each other (type I). In $NiPS_3$, the moments are oriented in the $\vec{c}$ direction. It is the $\vec{a}$ direction in $CoPS_3$. The magnetic structure of $MnPS_3$ (fig. 12(b)) consists of a magnetic ion coupled antiferromagnetically to its three nearest neighbors in the layer, their magnetic moments pointing perpendicularly to the layer planes. From calculations of the exchange integrals in some MPS_3 phases /36/, it was concluded unambiguously that the compounds have a magnetic 2D anisotropic behavior.

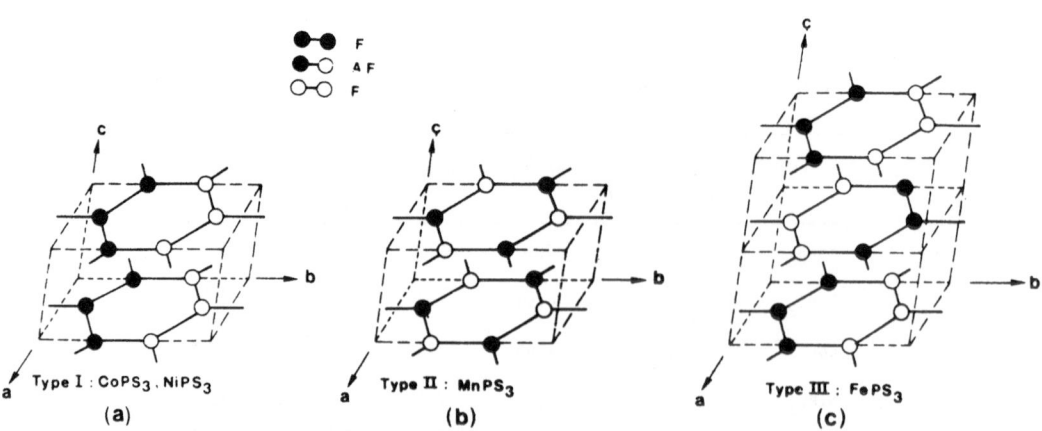

Figure 12 - Magnetic structures presented by the transition metal thiophosphates. Black and white circles are up and down atomic spins

In FePS$_3$, each Fe^{2+} ion is ferromagnetically coupled in a layer with two of the three nearest neighbors, and antiferromagnetically with the third one, giving the same structure as in CoPS$_3$ and NiPS$_3$. Contrary to type I structure, the chains of a plane are antiferromagnetically coupled to neighbor interplanar ones (type III), implying a doubling of the c parameter of the magnetic cell (fig. 12(c)). Like in the case of MnPS$_3$ and NiPS$_3$, calculations are consistent with a magnetic moment perpendicular to the layers.

In FePS$_3$ and FePSe$_3$, the Fe^{2+} ions participate in competitive ferromagnetic and antiferromagnetic couplings. For the selenide, a slightly negative θ value is recorded (-4 K), whereas a large positive value (θ = 104 K) is observed for the sulfide, in spite of interplanar antiferromagnetic coupling in this latter phase. The origin of the increase of θ must hence originate in shorter Fe-Fe distances in FePS$_3$, increasing the direct cation-cation ferromagnetic interaction, which is however not large enough to allow the existence of ferromagnetic layers.

Interestingly, the MPS$_3$ magnetic structures resemble the structural arrangement found for the M$^I_{1/2}$M$^{III}_{1/2}$PS$_3$, the triangular and chain cation sublattices in a layer corresponding to up and down magnetic moments arrangement.

A Mössbauer study of FePS$_3$ respective to temperature showed near ordering temperature, the coexistence of paramagnetic and antiferromagnetic phases, implying that the magnetic hyperfine field makes a sudden drop to zero at different temperatures for different parts of the sample within the transition region /38/. The existence of these two kinds of microdomains in FePS$_3$ layers raise the problem of the actual ordering temperature. P. JERNBERG et al. define, from their Mössbauer analysis, T_N as the temperature of equal paramagnetic and antiferromagnetic intensities (T_N = 117.4 K). This is to be compared to T_N taken at the temperature of the steepest rise of the magnetic parallel component (T = 120.8 K) /38/, to the minimum of $x^\perp$ measured on powder (T = 123 K) /8,11/, or from the N.M.R. study of ^{31}P (116 K) /35/.

The magnetic ordering in FePS$_3$ is found to take place between 120 and 114 K, and this phenomenon would deserve further study to determine how the ordered and unordered domains are distributed within the samples. This recalls, in Li$_x$MPS$_3$ intercalates, the occurrence of microdomains constituted of unreduced and reduced M cations (see lithium intercalation in MPS$_3$).

3.2 Magnetism in M$^I_{1/2}$M$^{III}_{1/2}$PS$_3$

Among the heterocharge cation substituted MPS$_3$, the phases that present an interest from a magnetic point of view and that have effectively been studied /29,21,22/, are Cr$_{1/2}$Cu$_{1/2}$PS$_3$, Cr$_{1/2}$Ag$_{1/2}$PS$_3$ and V$_{1/2}$Ag$_{1/2}$PS$_3$.

The first one presents a triangular cationic arrangement, leading to expected (2D) behavior at high temperature. A positive θ_T temperature of 31.5 K is attributable to a weakly positive intralayer interactions, adjacent layers being antiferromagnetically coupled at low temperature. Above 60 K, the magnetic susceptibility obeys a Curie-Weiss law allowing to calculate an effective moment (4.08 μ_B) very close to the spin-only value for Cr^{3+} and confirming the occurrence of Cu$^+$ in the compound.

To the rather classical (2D) magnetism of Cr$_{1/2}$Cu$_{1/2}$PS$_3$ corresponds, because of the one-dimensional and zigzagging arrangement of the magnetic element in Cr$_{1/2}$Ag$_{1/2}$PS$_3$ and V$_{1/2}$Ag$_{1/2}$PS$_3$, an original (1D) magnetic behavior. The magnetic susceptibility of Cr$_{1/2}$Ag$_{1/2}$PS$_3$ for instance (fig. 13) shows, around 200 K, a very flattened maximum corresponding to the one-dimensional antiferromagnetic character of the interactions, in agreement with the chromium arrangement. Below 50 K, the susceptibility increases, probably because of paramagnetic impurities and/or chromium situated at the chains tips. Then, and contrary to what one would expect from a paramagnetic contribution, a sharp maximum is recorded at 7 K corresponding to a 3D antiferromagnetic ordering.

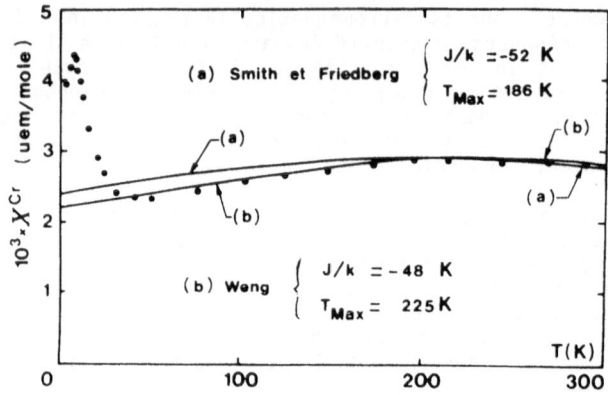

Figure 13 - Variation of the susceptibility per chromium mole versus temperature of $Ag_{1/2}Cr_{1/2}PS_3$ compound. J/k values were calculated by fitting experimental data with Smith and Friedberg and Weng models for magnetic chains (after P. COLOMBET /21/).

3.3 MPS$_3$ Electronic Band Structures

Knowledge of the nature of the electron accepting levels in host structures reducible by electron donors is important for both academic and technological reasons. Many X-Ray photoemission and absorption and optical measurements were carried out /39-43/, leading to proposing band models for several MPS$_3$.

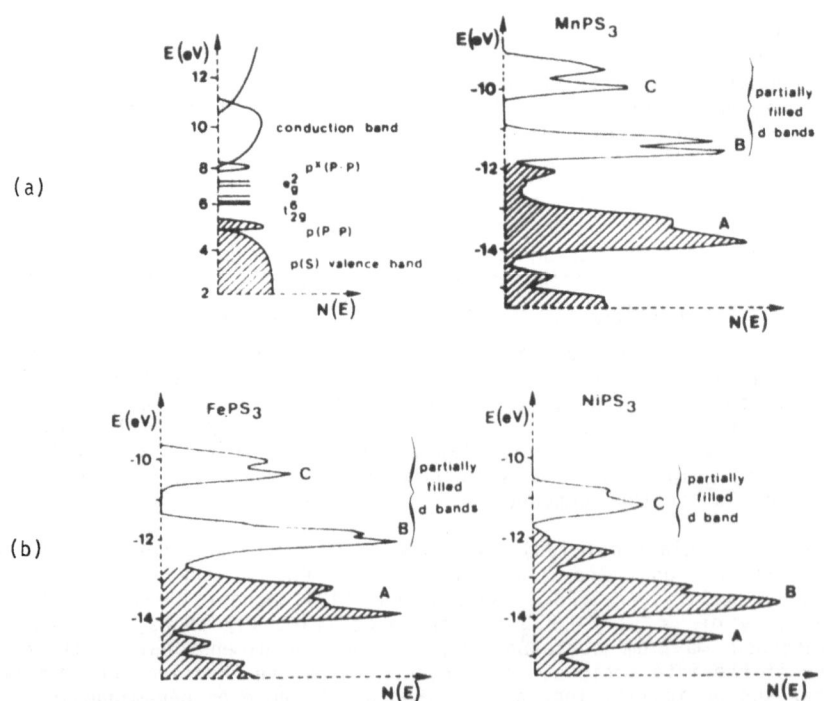

Figure 14 - Electronic band model of the MPS$_3$ layer structure compounds suggested by KHUMALO et al./39/. (a) Electronic band structure for MnPS$_3$, FePS$_3$ and NiPS$_3$ calculated using the extended Hückel method (only low lying orbitals are represented) (b)

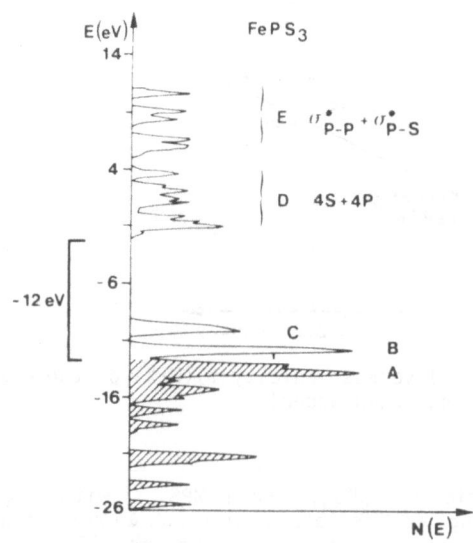

Figure 15 - Total density of states of FePS$_3$. Low lying partially filled predominantly d orbitals are situated at about 12 eV below empty 4s and 4p empty molecular orbitals /44/

Recently, the electronic band structure of an FePS$_3$ slab was calculated employing the tight-binding band scheme, based upon the extended Hückel method /44/, and later on, this calculation was carried out for NiPS$_3$, MnPS$_3$ and CdPS$_3$ /19/.

The electronic band model of MPS$_3$ layer structure compounds, suggested in particular by F.S. KHUMALO et al., were obtained considering little mixing of the sulfur (S)p with the metal (M)d states in the valence or conduction band states, and discrete localized 3d states. The semi-empirical model obtained resembles that of Wilson-Yoffe in the case of layered dichalcogenides for ionic extreme. The KHUMALO model is given on fig. 14(a) where it can be compared to the density of states calculated for several MPS$_3$ (fig. 14(b)).

Figure 15 shows the total density of states of an FePS$_3$ slab calculated by employing the special k-point method. Orbital component analysis of that density indicates that : 1) Peak A, largely sulfur 3p orbital in character is identified as the valence band of $(P_2S_6)^{4-}$ ions. 2) Peak B is largely composed of metal 3d orbitals and considered as the t_{2g} subband. 3) Peak C, in which sulfur 3p orbital character is nearly as large as the metal 3d contribution, is assigned as the e_g subband. At higher energy, 4) the major orbital components of the D peaks are metal 4s and 4p orbitals while those of the E peaks are associated with the σ^*_{P-P} and σ^*_{P-S} orbitals of $(P_2S_6)^{4-}$ ions. 5) the density of state peaks that occur in the energy region below peak A are largely associated with the 3s orbitals of sulfur and phosphorus. It is clear from fig. 15 that the metal 3d-block bands (B and C) which are partially filled (d^6 ions), must be responsible for the electron acceptor capability.

In fig. 14 where are represented the low lying density of states bands of NiPS$_3$ the component analysis shows that
1) peak A is identified as the valence band of $(P_2S_6)^{4-}$ ions,
2) peak B is considered as the t_{2g} subband, above which a small peak is a bonding contribution of the P-P type. All these bands are filled.
3) At higher energy peak C is assigned to subband e_g and is thus the first low lying partially filled level (d^8 ion). As pointed out for FePS$_3$, the accepting levels in NiPS$_3$ for electron reduction are constituted by the Ni^{2+} d orbitals.
These results are well verified experimentally through the observation that ZnPS$_3$ and CdPS$_3$ that contain d^{10} ions and for which accepting levels are 4s and 4p molecular orbitals too high in energy, do not intercalate electron donors like lithium /44,19/.

The MPS$_3$ electronic band structure determinations solved a few questions concerning some physical characteristics of the materials.In effect, and with the

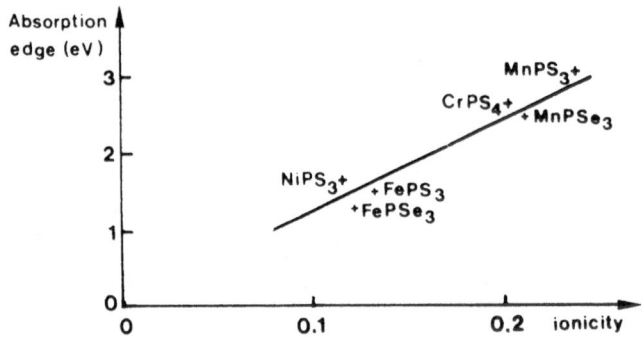

Figure 16 - Absorption edge (in electron-volt) versus ionicity f_i(M-X) of the M-X bonds for some MPX_3 compounds (parent $CrPS_4$ has been added)

noticeable exception of non-stoichiometric $V_{0.78}PS_3$, some MPS_3, with high resistivity ($MnPS_3$, $NiPS_3$) were appearing as insulators although a low optical absorption edge had led to classify them as broad band semiconductors /17/. The analysis of the total density of states of $FePS_3$ made by M.-H. WHANGBO et al. /44/ shows that the e_g subband has strong sulfur lone pair character and is therefore very compatible with the valence band allowing an intense transition : valence band (A peak) $\rightarrow e_g$ subband (C peak) (fig. 15), probably responsible for the apparent fundamental absorption edge of $FePS_3$. The calculated gap of 1.8 eV is in reasonable agreement with the absorption edge energy of 1.5 eV. The true band gap, located between the valence (highest occupied) and conduction (lowest unoccupied) band of $(P_2S_6)^{4-}$ ions should have a value of several electron-volts, nearly independent of the nature of the metal.

Recent calculations showed /19/ that increase of the metal electronegativity lowered the e_g subband much more than the valence band, decreasing the absorption edge accordingly. This explained why the absorption edge energy of the MPS_3 could be rather well correlated to the ionicity of the M-S bond, i.e. the electronegativity of M (fig. 16). Incidentally, this underlines the usefulness /45/ of the electronegativity concept in dealing, qualitatively, with compounds' characteristics and properties.

4. MPS_3 SUBSTITUTION INTERCALATION PROPERTIES

4.1 Molecular intercalation

Like in transition metal dichalcogenides, organic molecules intercalate in transition metal phosphorus trisulfides. However, if we consider intercalation of n-alkylamine in MPS_3, M being a d filled cation (Zn, Cd) /46,47/, the electronic transfer from host to guest is difficult to understand if one considers, in agreement with the MPS_3 band structures, the electron accepting levels as constituted by the 4s and 4p molecular orbitals of the cation in MPS_3. In the case of n-alkylamine intercalation in $MnPS_3$ /47/ or $NiPS_3$ /48/ for example, a charge transfer would be easier to imagine since these host structures possess low lying partially filled d bands. Nevertheless in these cases, intercalation reaction is not a simple one, since it has been shown that traces of water are necessary for the reaction to take place, as rigourously dried amines did not react at all. The water molecules would allow the following reactions :

$$RNH_2 + H_2O \rightleftharpoons RNH_3^+ + OH^-$$

$$NiPS_3 + 2x\ OH^- \longrightarrow (NiPS_{3-x}O_x)^{2x-} + x\ H_2O + x\ S$$

$$2x\ RNH_3^+ + y\ RNH_2 + (NiPS_{3-x}O_x)^{2x-} \longrightarrow (RNH_3^+)_{2x}(RNH_2)_y NiPS_{3-x}O_x^{2x-}$$

In support of the scheme, it was observed that an alcoolic LiOH allowed intercalation of lithium with release of an equivalent amount of sulfur in the solution.

A reaction of the type suggested by R. SCHOLLORN /49,50/ where oxidation of amines allows reduction of the host lattice and formation of protonated amine molecules then intercalated as ammonium cations is another possible mechanism. However, in the case of d filled cation MPS_3, we have again difficulties in determining the electron transfer nature. Actually P.J.S. FOOT et al, have shown, in the case of $CdPS_3$ alkylamine intercalates, that an amount of Cd^{2+} equivalent to the alkylammonium ions present in each adduct was found in the liquid amines after reaction. In that case, (and in the examples given by organometallic intercalates studied below), the formation of the ammonium cation is followed by a cationic exchange according to :

$$Cd^{2+}PS_3 + 2x\ RNH_3^+ \xrightarrow{RNH_2} Cd^{2+}_{1-x}PS_3\ (RNH_3^+)_{2x},\ (RNH_2)_y,\ xCd^{2+}$$

several amine molecules constituting a solvation shell around the protonated amine.

The last possibility for the intercalation process could be that of electrostatic interaction between the cation host and molecules as suggested in considering the MPS_3 phases as ionic compounds formed from M^{2+} and $(P_2S_6)^{4-}$ ions. In that case the reaction would be assimilated to a solvation one.

4.2 Organometallic Molecules Intercalations

Intercalations of organometallic molecules like $[Co(\eta C_5H_5)_2]$, $[Cr(\eta C_6H_6)_2]$ in MPS_3 structures were performed by R. CLEMENT /51-53/. In first experiments, reaction for example of $CdPS_3$ with toluene solutions of cobaltocene were reported /53/ to yield an intercalate with formula $CdPS_3[Co\eta(C_5H_5)_2]_{0.36}$. The I.R. spectra of this compound clearly established that the intercalated guest metallocene is fully cationic. Although, from the guest point of view, this result is consistent with the strongly reducing character of the metallocene, the question arises about the donated electrons, like in the above cases of the n-alkylamine intercalates. The d orbitals of some sulfur atoms have been suggested as possible locations for the transferred electrons, since a strong splitting upon intercalation of the asymmetric P-S stretching band is observed. However this does not seem consistent with $CdPS_3$ density of states that shows the accepting levels to be Cd 5s orbitals /19/, the more so if one considers that $CdPS_3$ does not intercalate lithium (see below). It is to be pointed out that this type of reaction made in the beginning of MPS_3 molecular intercalation history has not been repeated since, and that reactions in alcoholic or aqueous medium of organometallic salt have been done instead (see cationic substitution section). Using such a method, an intercalation by cation transfer is observed for many phases (excepted $NiPS_3$) according to :

$$MPS_3 + 2x(CoCp_2^+I^-) \xrightarrow[H_2O\ or\ C_2H_5OH]{solvent} M_{1-x}(CoCp_2)_{2x}PS_3, y\ sol + x(M^{2+}2I^-)z\ sol$$

(with cobaltocene iodide for instance).
This type of reaction may be the actual reaction taking place for all MPS_3 although occurrence of $MnPS_3,(Co(\eta C_5H_5)_2)_x$ and $MnPS_3,(Cr(\eta C_6H_6)_2)_x$ intercalates for example does not raise any conceptual problem.

4.3 Cationic Substitution-Intercalation

The above reactions introduce a more general property presented by the MPS_3 family, i.e. cationic substitution. This behavior has been extensively studied by R. CLEMENT /51/ who showed that many ionic salts (G^+X^-) could lead readily, in a polar medium and at room temperature, to substituted intercalates according to :

$$M^{2+}PS_3 + 2xGX \xrightarrow{solvent} M^{2+}_{1-x}(G^+)_{2x}PS_3\ (solvent)\ y + x(M^{2+}2X^-)$$

with for instance : $G = Cs^+,\ Rb^+,\ K^+,\ (C_2H_5)_4N^+,\ Co(\eta C_5H_5)_2^+,\ pyH^+\ \ldots,\ X = Cl^-,\ I^-,\ \ldots$

Between 10 and 20 % of M can thus be removed from the MPS$_3$ slabs, the substituting cation being found, with its solvation shell, in the gap leading to corresponding increase of the basal spacing. Reaction takes place between the solution containing the dissolved salt and a suspension of the thiophosphate. Similar phenomena have never been reported in the field of other layered materials and R. CLEMENT et al. /54/ ascribed it to lower lattice energy of the MPS$_3$ due to high lability of the metal M in the +II oxidation state and large size of the $(P_2S_6)^{4-}$ ion, allowing an heterogeneous equilibrium, in solution, between MPS$_3$, and Mn^{2+} and $(P_2S_6)^{4-}$ solvated ions. Thus, although the MPS$_3$ materials apparently remain solid during the intercalation reaction, the microscopic process would involve a locally highly solvated transition state accounting for the high rate of ion exchange at room temperature. Interestingly (see table V), whereas substitution is easily achieved in the case of MnPS$_3$, CdPS$_3$ and ZnPS$_3$, i.e. for those M^{2+} ions not stabilized by ligand fields, it is more difficult for more stabilized Fe^{2+} ($t_{2g}^4 e_g^2$) and basic pH and strong complexing agents (E.D.T.A.) are needed to reach sizeable substitutions. It is remarkable that Ni^{2+} that has in NiPS$_3$ structure the lowest thermal factor value (0.75 Å^2 as compared to 1.9 Å^2 for Zn^{2+}) and higher crystal field stabilization energy could not, to date, be substituted at all. This is a nice example of the relation between the liability and the crystal thermal motion of cations in a structure, themselves being linked to the ligand field stabilization.

Table V - Substitution conditions for several MPS$_3$ revealing the influence of the crystal field on the liability of the M cations (below, cations equivalent temperature factors and d configurations)

MnPS$_3$	MnPS$_3$ + 2xKCl $\xrightarrow[20°C]{water}$ Mn$_{1-x}$K$_{2x}$PS$_3$, nH$_2$O + x(Mn^{2+}2I$^-$)
CdPS$_3$	CdPS$_3$ + 0.5 KCl $\xrightarrow[20°C]{water + EDTA}$ Cd$_{0.75}$K$_{0.5}$PS$_3$,H$_2$O + 0.25 (Cd^{2+}2Cl$^-$)
ZnPS$_3$	ZnPS$_3$ + 2x(CoCp$_2$I) $\xrightarrow[20°C]{ethanol}$ Zn$_{1-x}$(CoCp$_2$)$_{2x}$PS$_3$,nC$_2$H$_5$OH + x(Zn^{2+}2I$^-$)
FePS$_3$	No exchange in water or C$_2$H$_5$OH up to 100°C FePS$_3$ + 2x(CoCp$_2$I) $\xrightarrow[20°C]{water + EDTA, basic pH}$ Fe$_{1-x}$(CoCp$_2$)$_{2x}$PS$_3$,nH$_2$O +.. + .. x(Fe^{2+}2I$^-$)
NiPS$_3$	no exchange

M	Mn	Fe	Co	Ni	Zn	Cd
B$_{eq}$ (Å^2)	1.22	1.11	1.05	0.75	1.90	1.74
d configuration	d^5(t$_{2g}^3$e$_g^2$)	d^6(t$_{2g}^4$e$_g^2$)	d^7(t$_{2g}^5$e$_g^2$)	d^8(t$_{2g}^6$e$_g^2$)	d^{10}(t$_{2g}^6$e$_g^4$)	d^{10}(t$_{2g}^6$e$_g^4$)

because of the low oxidation state of the cations in the MPS_3, and since the reduction of these host materials was expected to follow intercalation, some time elapsed before alkali metals were shown to intercalate chemically and electrochemically the thiophosphates /55,56,35/. $NiPS_3$, $FePS_3$ and $CoPS_3$ react readily with butyl-lithium whereas $CdPS_3$ and $ZnPS_3$ do not. $MnPS_3$ was found at first not to intercalate, probably because of a very slow kinetics, but recent experiments carried out for several weeks finally indicated large lithium intercalation /12/. These results confirm, as found from electronic band structure, the role of electron acceptors played by the d orbitals cations in the MPS_3 series.

Considering the structure of the MPS_3, filling by the lithium cations of all octahedral sites in the van der Waals gap should lead to the intercalation product $Li_{1.5}MPS_3$ (i.e. $Li_1M_{2/3}(P_2)_{1/3}S_2$ by analogy with Li_1TiS_2). Reversible intercalation effectively occurs for composition as high as $Li_{1.4}MPS_3$ /56/. Much higher lithium content /55/ ($Li_{4.5}NiPS_3$ for example) does not seem to be attributable to intercalation, but rather to irreversible reaction between lithium and host.

Because the van der Waals gap of the MPS_3 is so large, the cell parameters remain unchanged. This is conveniently illustrated on fig. 17, where has been reported the c parameter expansion calculated for one slab ($\Delta c/n$) versus the van der Waals gap width for several layered intercalates Li_1MS_2. Effectively, Δc is expected to be near zero for the thiophosphates, as can be also deduced from octahedral size considerations /56/. In the absence of cell parameters evolution, lithium intercalation in the MPS_3 family proved difficult to demonstrate. The first clue was given through X-Ray study of the products of the reaction with water /35/. It was shown that lithium intercalated $NiPS_3$ reacted readily with athmospheric humidity to yield a phase with a layer expansion of 5.8 Å, typical of a double layer of water molecules solvating the lithium ions. The X-Ray powder spectra of the products of the reaction showed a mixture of hydrated intercalate and $NiPS_3$ indicative of a possible diphased system, although a reaction of the intercalate with water leading to the formation of lithium hydroxide, hydrogen and original $NiPS_3$ is another possibility.

Infrared spectra (700 - 30 cm^{-1}) of several lithium intercalates (chemically prepared) Li_xMPS_3 with M = Fe, Ni and $0 \leqslant x \leqslant 1.5$ were carried out later on /57/. The intercalates showed essentially new absorption bands at 336 cm^{-1} and 310 cm^{-1}, characteristic of Li^+ in an octahedral sulfur environment, proving the nature of the intercalation sites (fig. 18). Considering the infrared bands of the host lattice, $FePS_3$ shows quite small frequency and intensity changes for the stretching and deformation modes of the (PS_3) group, for the ν_{P-P} mode and for

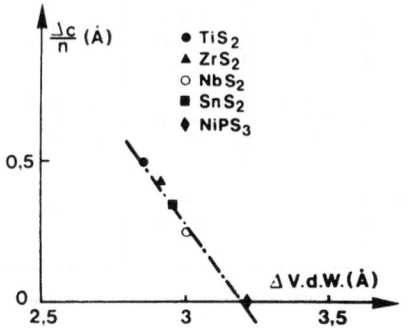

Figure 17 -   c parameter expansion per slab in some Li_1MS_2 layered phases plotted against the change in the Van der Waals gap width

111

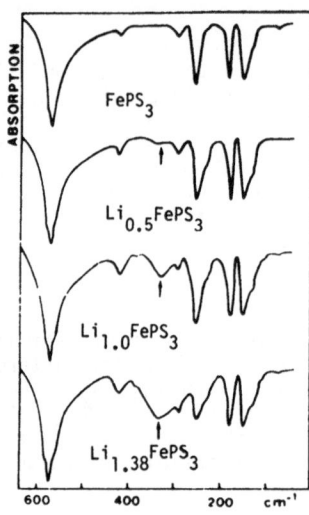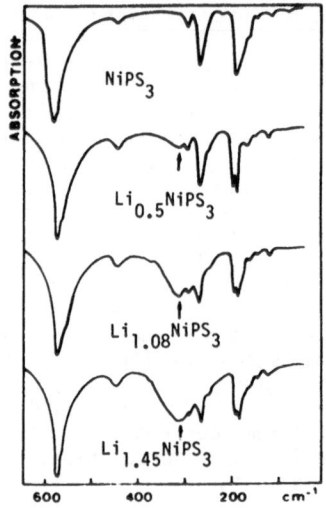

Figure 18 - Infrared spectra of Li_xFePS_3 and Li_xNiPS_3 (0 < x < 1.5) /57/

the low frequency translational motion of Fe^{2+} ions. On the other hand, in Li_xNiPS_3, as soon as x is equal to 0.5, some bands, characteristic of the deformation modes of the PS_3 groups, are split into doublets at 194-186 cm^{-1} for instance (see table VI), suggesting that distortions within the layers have

Table VI - Infrared band wavenumbers in $FePS_3$, $NiPS_3$ and several Li_xFePS_3 and Li_xNiPS_3 compounds at 300 K with 0 < x < 1.5 /57/

$FePS_3$	$Li_{0.5}FePS_3$	$Li_{1.0}FePS_3$	$Li_{1.38}FePS_3$	$NiPS_3$	$Li_{0.5}NiPS_3$	$Li_{1.08}NiPS_3$	$Li_{1.45}NiPS_3$	Assignments
580 sh	580 sh	580 sh	580 sh	589 sh				
571 vs	571 vs	572 vs	572 vs	575 vs	572 vs	575 vs	575 vs	ν_d (PS_3)
560 sh	560 sh	560 sh	560 sh	571 sh	560 sh	560 sh	560 sh	
445 m	445 m	445 bd,m	445 bd,m	440 m	440 m	445 bd,m	445 bd,m	ν P-P
–		384 w	384 w	–	–	385 w	385 w	$\nu_s PS_3$
–	336 m	336 bd,s	336 bd,s	–	310 m	310 bd,s	309 bd,s	ν Li^+-S_6
295 m	296 m	295 m	295 m	289 m	290 m	288 m	288 m	δ_s PS_3
257 s	255 s	255 s	252 s	265 s	265 s	265 s	264 s	δ_d PS_3
					246 m	246 m	246 m	
230 vw	230 sh	230 sh	230 sh	214 vw				
					194 s	194 s	193 s	ρ_r (PS_3)
183 s	181 s	181 s	181 s	187 s	187 s	186 s	185 s	+
				156 m	158 m	158 m	158 m	δ_s PS_3
153 s	153 s	152 s	150 s					
131 sh	130 sh	130 sh	130 sh	139 w	144 w	145 w	148 w	T' (M^{2+})
	–	108 w	108 w	109 w	118 w	120 w	123 w	
76 vw		75 vw	75 vw	72 vw				

Intensities are reported using the notations, vs : very strong, s : strong, m : medium, w : weak, vw : very weak, sh : shoulder, bd : broad.

occurred. Also, the low frequency band at 139 and 109 cm^{-1} in $NiPS_3$ assigned to translational motions of Ni^{2+} ions are shifted to higher values (144 and 118 cm^{-1} respectively), indicating strengthened Ni-S interactions. From x = 0.5 up, the ν_{P-P} mode broadens markedly and shows a maximum at 445 cm^{-1} for $Li_{1.08}NiPS_3$ and $Li_{1.45}NiPS_2$, in agreement with the successive occupancy of the two distinct types of octahedral sites in the van der Waals gap, and as evidenced by the electrochemical studies.

Recording of two chemically deintercalated nickel compounds ($Li_{0.02}$&$Li_{0.09}NiPS_3$) infrared spectra showed the disappearance of the band assigned to Li-S interaction whereas the ν_{P-P} mode was restored to its slightly lower original frequency. However, the splitting of the host lattice modes remain as in the previous intercalates, indicating that the geometrical distortions within the layers created during the intercalation are non reversible.

The open-circuit voltage discharge curve of a $NiPS_3$/P.C. $LiClO_4$/Li cell /58,59/ indicated a single phase region for x < 1.5 and a multiple region for x > 1.5 (fig. 19). In the single phase region itself, a small break occurs for 0.5 Li. Because the MPS_3 structures present two octahedral groups differing from the nature of the second neighbors (respectively 2d and 4h sites in the Wyckoff notation of the C2/m space group), the first domain with 0 < x < 0.5 corresponds to the filling of octahedral site (2d) and the second (0.5 < x < 1.5) to the filling of the second type of sites. This was clearly demonstrated through neutron diffraction studies /35/.

A quasi equilibrium discharge curve was also obtained using $V_{0.78}PS_3$ with vanadium of mixed valence to probe the non-stoichiometry of the phase /15/. Several breaks in the curve E(volt) = f (x in $Li_xV_{0.78}PS_3$) can be observed (fig. 20). One of them, at around x = 0.22 and considering the developed formula of the phase ($V^{2+}_{0.34} V^{3+}_{0.44} \square_{0.22} PS_3$) was attributed to the filling of the 0.22 empty sites of the structure slabs. Another possibility is that, in the beginning of intercalation, the lithium ions remain located in the van der Waals gap at octahedral sites rendered potentially favorable by the charge distribution in the compound sheets. The alkali ions could be located next to the voids of the slab up to the $Li_{0.22}V_{0.78}PS_3$ composition, hence the discharge curve break.

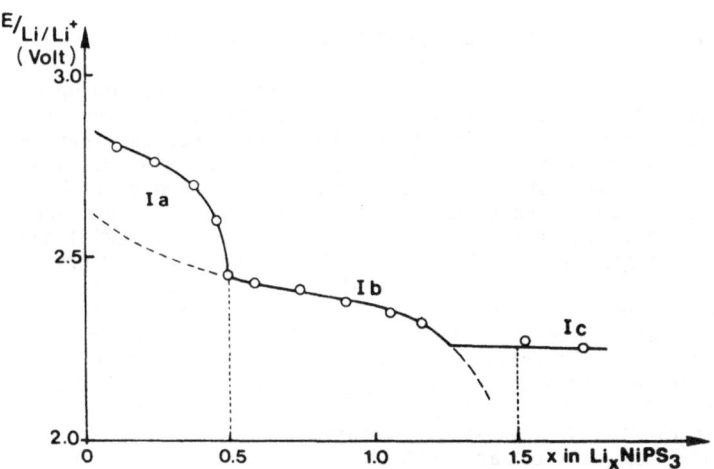

Figure 19 - Open potential circuit variation versus x in $Li_x NiPS_3$ /8/

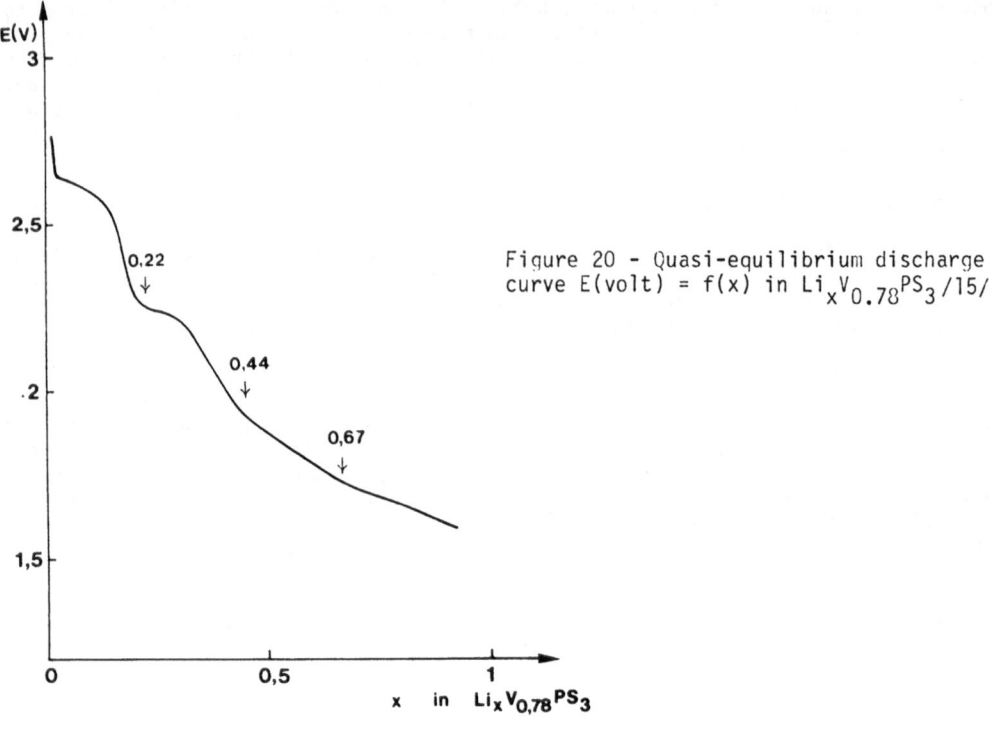

Figure 20 - Quasi-equilibrium discharge curve E(volt) = f(x) in $Li_xV_{0.78}PS_3$ /15/

The weaker slope change for 0.44 lithium atoms was interpreted as corresponding to the reduction of V^{3+} into V^{2+}.

Such quasi thermodynamic studies of lithium intercalation in layered structures, could probably help solve the copper distribution in a phase like $Cr_{1/2}Cu_{1/2}PS_3$. As explained above, this compound contains (Cu_2S_6) and (CuS_6) octahedral groups so that the formulation of the phase is $Cu^I_{1/2-2y}(Cu^I_2)_yCr^{III}_{1/2}\square_yPS_3$ and so far, y has not been determined.

5. THE LITHIUM INTERCALATION AND THE ELECTRONIC TRANSFER

Like in the Li_xTX_2 systems, the complete ionisation of lithium upon intercalation was determined though N.M.R. of ^7Li in Li_xNiPS_3 /57/, implying electron donation to the host. According to the electronic band structure, the transferred electron must be accommodated on the molecular orbitals of mainly d character of the transition metal. Most of the property changes recorded on Li_xMPS_3 intercalates agree well with these conclusions. In particular, the impossibility to intercalate in MPS_3 with filled d orbital cations ($ZnPS_3$, $CdPS_3$) and the magnetic susceptibility decrease in Li_xNiPS_3 /8,35/ (fig. 21). However, among the thiophosphate intercalates, an example seemed to contradict the above oxidoreduction process, since the lithium intercalation in $FePS_3$ was reported to alter the magnetic susceptibility very little /35/. More recent results /61/ indicate that a slow transfer kinetics may be responsible for the unchanged magnetic properties. Actually, magnetic moment decrease is indeed observed in Li_xFePS_3 phases (fig. 22) /60/ provided enough time (up to several weeks) elapse between intercalation and physical measurements. Finally, it appears clearly that, contrary to earlier arguments, the redox process in the MPS_3 is a classical one corresponding to reduction of the cation. In view of this conclusion, a very low oxidation state metal is expected for lithium content as high as 1.42 per MPS_3 mole. It is hence possible to probe past physical properties of some Li_xMPS_3 intercalates and to analyse how they match the redox

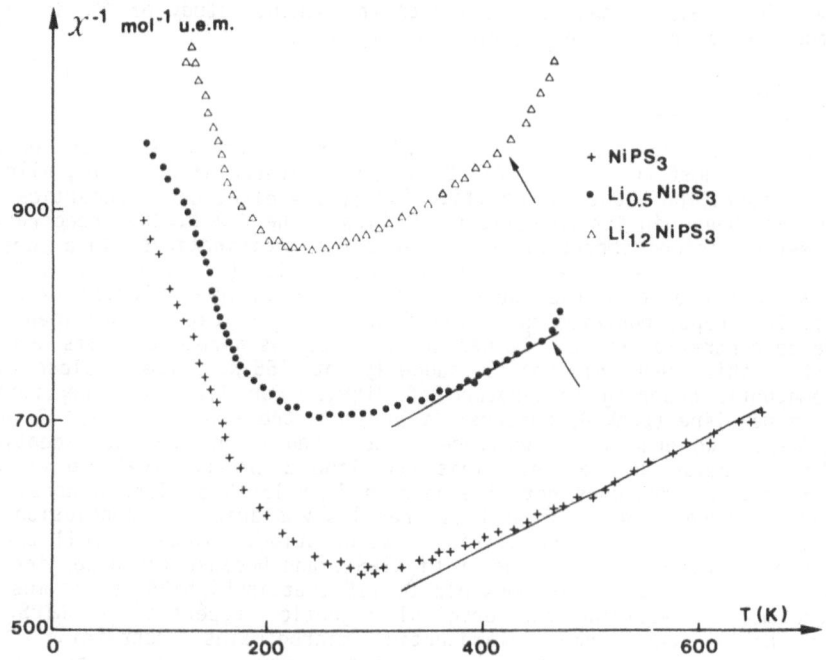

Figure 21 - Inverse molar susceptibility of Li_xNiPS_3 (x = 0, 0.5 and 1.2) vs T. The arrows indicate irreversible decomposition temperature /34/

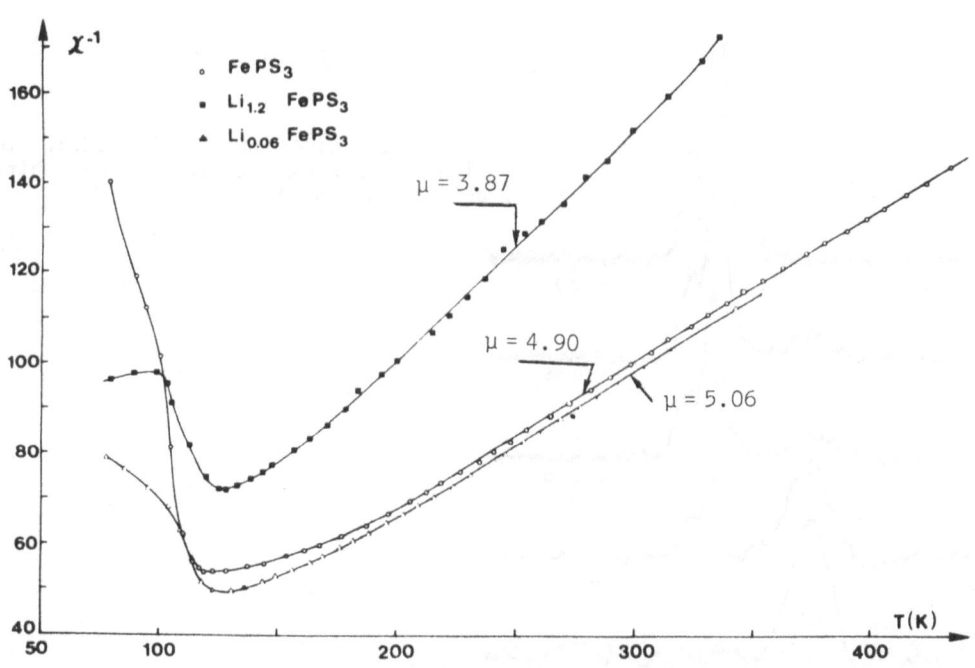

Figure 22 - Reciprocal molar susceptibility of Li_xFePS_3 phases versus temperature o : x = 0, ■ : x = 0.5, Δ : x = 0.06 (de-intercalated phase)

mechanism. Two examples can be taken with the N.M.R. study of ^{31}P in Li_xNiPS_3 /36,61/ and Mössbauer spectroscopy in Li_xFePS_3 /62/.

5.1 The Li_xNiPS_3 Series

Systematic studies of the Li_xMPS_3 bulk electronic properties, such as Knight shift of the ^{31}P host nuclei as a function of the intercalation ratio, allowed us to obtain information about the modification of the electronic properties of the host from the change in its magnetic properties. The ^{31}P N.M.R. spectra of the Li_xNiPS_3 series showed important and interesting modifications. In a pure $NiPS_3$ matrix, the ^{31}P line shape appears as a Pake doublet (peaks A on fig. 23) that does not seem to change in the early stage of lithium intercalation (x < 0.5 in Li_xNiPS_2). Its shape, position and relaxation time T_1 (≈ 16 ms) are identical in the whole concentration range. It was shown that, as a result of its broadening and shift, this peak disappears suddenly at 155 K, i.e. close to the antiferromagnetic ordering temperature of $NiPS_3$. For lithium concentration of x ≈ 0.5, a new line (peak B) appears. It presents the same shift as ^{31}P in $CdPS_3$ and a T_1 value two orders of magnitude longer than T_1 of peak A, identical to that in the cadmium derivative. This new line B progressively grows at the expense of line A, and does not disappear at T_N ≈ 155 K as line A does. It is even still observed at 4 K. From these results was drawn the conclusion that a diamagnetic phase occurred upon lithium intercalation. Since no cell change is recorded upon lithium intercalation in Li_xNiPS_3, and because the accepting level is the e_g band of Ni^{2+}, it is possible to say that in Li_xNiPS_3 there must exist some microdomains retaining the original magnetic properties of $NiPS_3$ (with unreduced Ni^{2+}) and some diamagnetic microdomains containing reduced Ni^0 atoms. The reduction process may hence be written $xLi + NiPS_3 \rightarrow Li_x^+ Ni_{1-x/2}^{2+} Ni_{x/2}^0 PS_3$. The unchanged properties of the matrix in the $0 < x < 0.5$ concentration range is a curious phenomenon. It may, in part, be explained by the detection threshold of the N.M.R. technique and also by the fact that xLi are needed to produce $x/2 Ni^0$, lowering further this threshold. Also, it was found that an electronic transfer slow kinetics takes place during

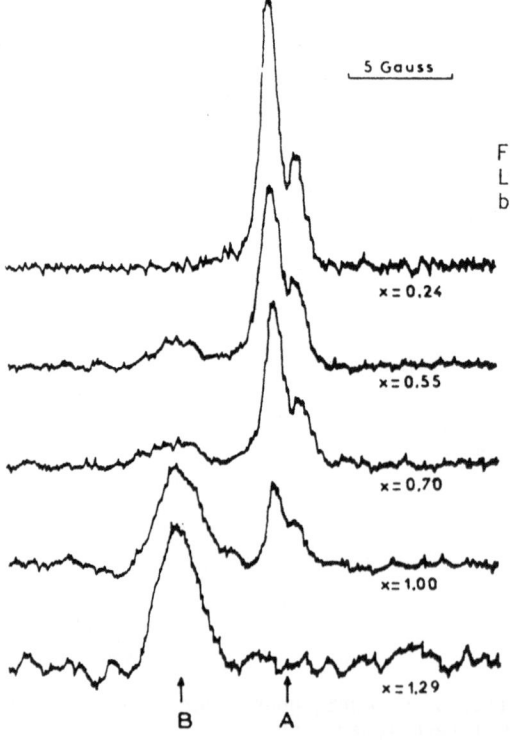

Figure 23 - Lineshape of ^{31}P N.M.R. in Li_xNiPS_3 (0.24 ≤ x ≤ 1.29). The shift between line A and B is 0.046 %

5 Gauss

x = 0.24

x = 0.55

x = 0.70

x = 1.00

x = 1.29

B   A

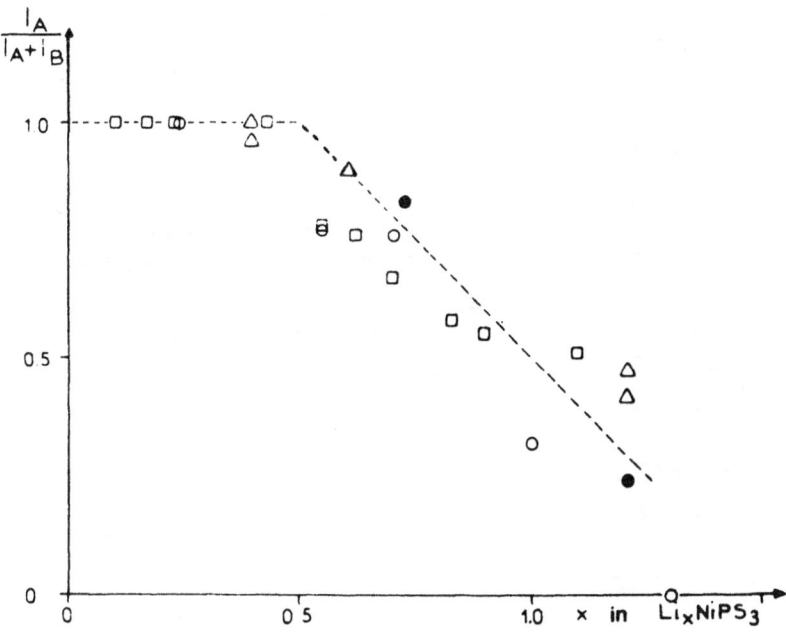

Figure 24 - Relative intensity of the $^{31}P(A)$ line corresponding to the magnetic A phase versus x in Li_xNiPS_3. (o, ▫ chemically intercalated powder respectively at 68°C and ambient temperature, ●, △ electrochemically intercalated powder, respectively in presence of electrolyte and dried under vacuum /8/

lithium intercalation in $FePS_3$ (see Mössbauer studies below), and this phenomenon may very well also play a role in the case of intercalation in $NiPS_3$. Another possibility is that the reduced nickel be statistically scattered within the structure, gathering in detectable microdomains only when reaching some critical concentration around x ≃ 0.5.

All these factors may explain the break observed around $Li_{0.5}NiPS_3$ (fig. 24). Clearly more studies should be dedicated to intercalation in $NiPS_3$, particularly in situ ones, to solve the problem of early intercalation reaction.

5.2 The Li_xFePS_3 Series

Hopefully, although some shade remains, Li_xFePS_3 intercalate behavior appears to be more easily explained. In that series of phases, thanks to Mössbauer spectroscopy, the reduced cationic species may be directly observed and this was recently done by G.A. FATSEAS et al. /62/. The spectra obtained at room temperature for a Li_xFePS_3 series (0 < x < 1.42) are given in fig. 25. For x = 0, a quadrupole doublet is observed (site Fe(A)) and for x ≠ 0, the lithium intercalates give rise to a second iron doublet attributed to a second site of reduced iron called Fe(B), whose relative intensity increases with x. Table VII, where both Fe(A) and Fe(B) hyperfine parameters are given, shows that the site characteristics remain very much the same in the whole concentration range, only their relative intensities being altered. This compares well with the observations made for Li_xNiPS_3. Also, a decrease instead of an increase of isomer shift is calculated for the reduced iron species. This unexpected result has been linked to a large increase of the 4s electron population and possible stronger overlap distortion of the core orbitals and/or more extended 3d orbitals taking place on Fe(B). Since it can be seen that, at least in the first steps of the intercalation, around two lithium are needed to produce one Fe(B) site, a complete reduction of Fe^{2+} is inferred, in agreement with the reduction process in $NiPS_3$.

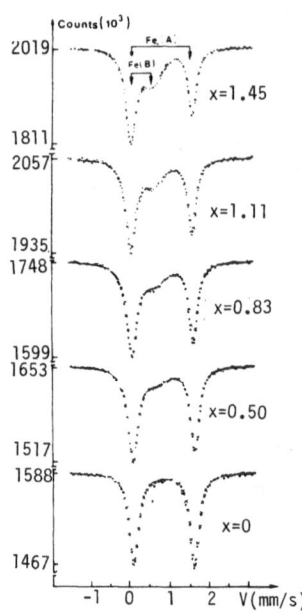

Figure 25 - Mössbauer spectra for Li_xFePS_3 at room temperature. The Fe(A) doublet, present in all spectra, is that of the host compound $FePS_3$. The Fe(B) doublet is present only in intercalated compounds /62/

It is to be pointed out that, in the Mössbauer experiments, samples did not show any spectra change up to x = 0.5, unless repeated spectra recording spread in time on the same capsules showed progressive appearance and intensity increase of Fe(B) site /60/, fig. 25 corresponding to final spectra intensities. The same ^{31}P N.M.R. study on Li_xFePS_3 intercalates as that done on Li_xNiPS_3 has been performed thoroughly recently /63/. Relative intensity variation of site A versus x in Li_xFePS_3 is shown on fig. 26, where the Mössbauer results have been added.

Table VII - Hyperfine Mössbauer parameters for Li_xFePS_3 at room temperature
δ : isomer shift, ΔE : quadrupole splitting, FWHM: full width at half maximum, % : relative intensity for each iron site

x	site	δ (mm/s)	ΔE (mm/s)	FWHM (mm/s)	%
0	Fe(A)	0.86	1.52	0.29	100
	Fe(B)	-	-	-	-
0.50	Fe(A)	0.87	1.55	0.27	75
	Fe(B)	0.47	0.52	0.64	25
0.83	Fe(A)	0.87	1.56	0.27	65
	Fe(B)	0.45	0.53	0.68	35
1.11	Fe(A)	0.86	1.55	0.26	60
	Fe(B)	0.44	0.52	0.54	40
1.45	Fe(A)	0.86	1.55	0.25	52
	Fe(B)	0.43	0.48	0.51	48

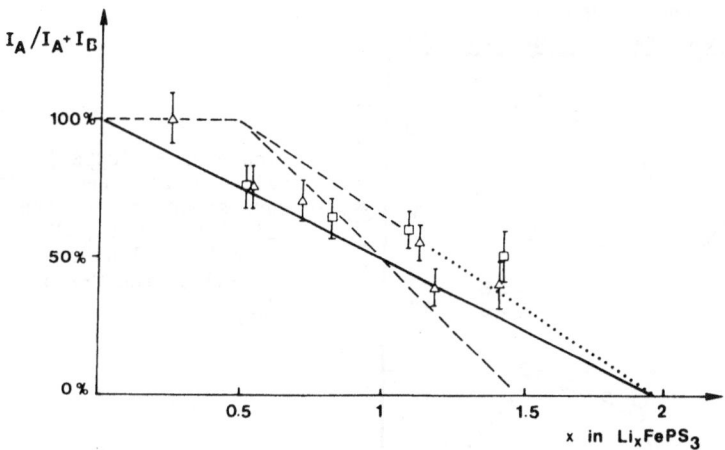

Figure 26 - Relative intensity of the $^{31}P(A)$ lines (Δ) and of Fe(A) site ($\square$) in Li_xFePS_3 The solid line corresponds to a continuous cation reduction, the broken one to a variation of the Li_xNiPS_3 type, and the dotted one to another possible reduction process /63/

Both results indicate progressive reduction of iron, and extrapolation of the straight line going through $I_A = 100$ % ($x = 0$) and $I_A = 0$ % ($x = 2$) assuming an utter reduction of Fe^{2+} into Fe^0 is not contradictory with the experimental data.

Indeed, these conclusions make the MPS_3 behave in a classical way with respect to reduction mechanism of intercalation in transition metal chalcogenides. In return, since the original phases contain low oxidation state cations, this implies occurrence of at least kinetically stable highly reduced transition metals within the intercalated host. Certainly further studies are needed to better characterize these new and unusual species that make the MPS_3 look more like organometallic phases than usual solid state sulfides.

6. THE MPS_3 AS CATHODIC MATERIALS

Besides their fundamental interest, it is their use as positive electrodes in lithium batteries /64,65/ that constitutes the other attractive side of the transition metal thiophosphates. They have been studied in systems using liquid electrolyte constituted by a polar organic solvent having dissolved a lithium salt, most usually $LiClO_4$. It appears, from the few known examples, that the discharge current obtained depends on the electrolyte used. This could be the result of intercalating solvent molecules in the coordination sphere of the lithium into the host structure, although this has not been structurally demonstrated so far.

The influence of a certain number of factors acting on the performance of the MPS_3/Li systems have been studied /65/, with respect to the main requirements to be considered for such secondary systems. Thus a high formation energy for the intercalate, related to the high potential of the host compound is sought. Calculation of the $|\Delta G|$ free energy values for the MPS_3 showed the highest value for $NiPS_3$ which is also the material that gave the best electrochemical yield in galvanostatic discharges. The $|\Delta G|$ variation versus Z demonstrates the strong influence of the transition metal present in the MPS_3, i.e. the influence of the M-S bond ionicity. Actually it was possible to observe that the variation of the

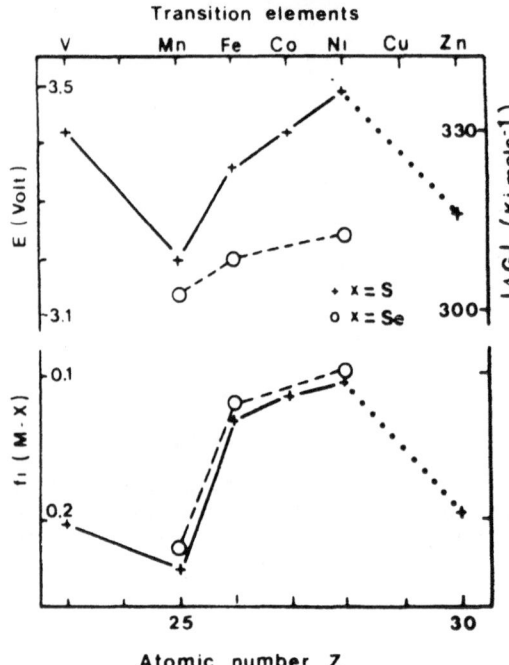

Figure 27 - (upper) Absolute value of the free energy change corresponding to the reaction $xLi + MPX_3 \rightarrow Li_xMPX_3$ (lower) Calculated ionicity f_i(M-X) of the M-X bond versus Z /45/

M-S bond ionicity $f_{i(M-S)}$ versus Z is fairly well correlated with that of $|\Delta G|$ (fig. 27). This result is quite consistent with the nature of the accepting electronic levels constituted by the molecular d orbitals of the transition metal. In effect, going from the left to the right of the periodic table, one expects a lowering of the t_{2g}-e_g band energies as the atomic core charge, hence the electronegativity, increases. This, in turn, augments the potential difference (E_{INT}), thus $|\Delta G|$, between the Li^+/Li potential and the accepting d level (fig. 28). Clearly, for a given type of structure, the more covalent host should potentially give the best positive. This the more so as the site potential for lithium intercalation will be minimized, authorizing, under high discharges rates, a better lithium diffusion between the structure layers.

These conclusions are well confirmed by the correlation (see fig. 14) between the optical absorption edge and the ionicity f_i values of the M-S bond, and the fact that the best MPS_3 positives are those with the smallest optical absorption

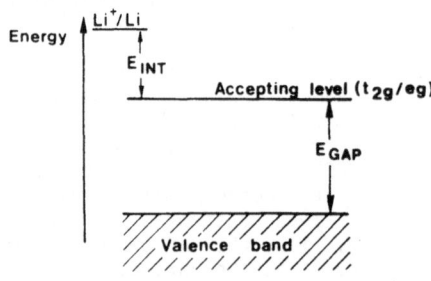

Figure 28 - Scheme showing the relation between energy (E_{INT}) and the energy gap between electron accepting level (constituted by partially filled d bands in the MPS_3) and valence band in intercalation systems

edge (i.e. the nickel, cobalt and iron derivatives). It was suggested /45/ that, if a compound like $CuPS_3$ existed, (it is not known so far), with a very low Cu-S bond ionicity, it should, nevertheless, be the best positive material of the thiophosphate family.

One may wonder why, because of the small ionicity difference between $NiPS_3$ and $FePS_3$, the nickel phase proves the best intercalation cathode. Setting apart the possible stoichiometric and purity problems encountered in the preparation of $FePS_3$, a convincing explanation of such a difference can be put forward from structural and chemical properties of this compound. Since, with a $t_{2g}^4 e_g^2$ configuration, Fe^{2+} is moderatly stabilized in its octahedral sulfur site, it can be exchanged by more ionic cation. In the lithium batteries studied, $FePS_3$ is in contact with a lithium salt in a polar solvent. The conditions for a substitution-intercalation are gathered for lithium to substitute iron ions, liberating Fe^{2+} cations in the electrolyte. This would lead to discharge of these ions on the anode and to a drop of the voltage. With unexchangeable Ni^{2+}, a higher potential will be maintained, as it is effectively observed (see fig. 29). It is also the case of Co^{2+} ($t_{2g}^5 e_g^2$) that shows in $CoPS_3$ a polarisation similar to that of $NiPS_3$.

Structural changes taking place upon intercalation like phase transitions and parameter expansions also play an important role in the reversibility of the intercalation secondary systems. No such changes seem to occur during lithium intercalation in the MPS_3, because of the favorable structural features of these phases.

However, although the van der Waals gap presents the ideal size to accommodate the Li^+ ions without expansion, several other factors, with opposite effects, act upon the bond lengths, thus the parameters values. We mention the increase in the negative charge borne by the MPS_3 slabs, leading to coulombic attraction between sulfur anions and intercalated Li^+ (decrease of c and increase of a). Correlatively, since the M^{2+} cation is reduced, one expects larger (MS_6) octahedra with increased parameters. Both effects may either compensate or give rise to local distortions not observed from X-Ray measurements that are average long range observations. In any case, unchanged macroscopic cell volume remains a strong advantage of the MPS_3 as positive materials.

The lack of good electronic conductivity of the trithiophosphates, which is certainly an a priori unfavorable factor, is overcome by the use of an electronic conductor like carbon black, added to the electrode, and the increase

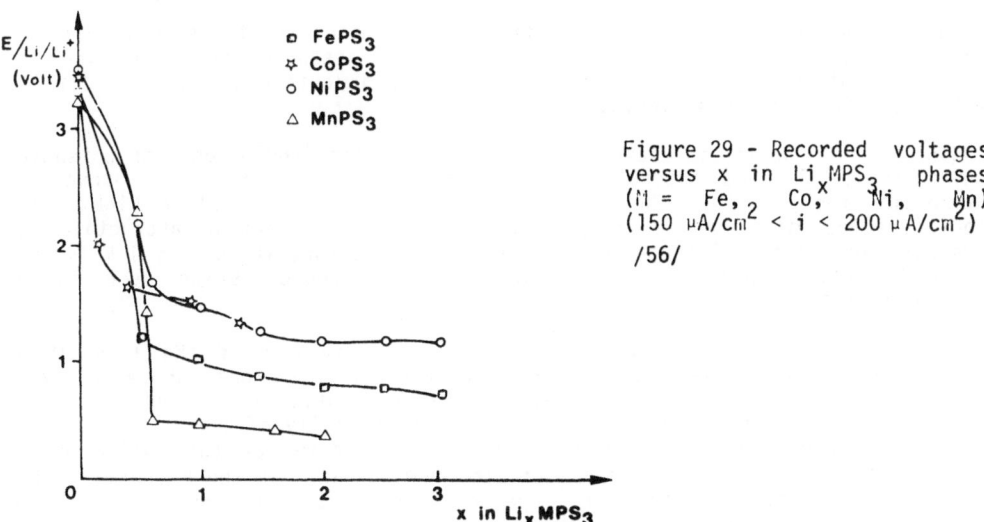

Figure 29 - Recorded voltages versus x in Li_xMPS_3 phases (M = Fe, Co, Ni, Mn) ($150\ \mu A/cm^2 < i < 200\ \mu A/cm^2$) /56/

in conductivity attributable to the electronic transfer that improves the functioning of the MPS$_3$ positives. On the other hand, the number of available sites per molecular weight and the corresponding oxidizing centers give to the MPS$_3$ capacities similar to the well known TiS$_2$ positive.

7. CONCLUSIONS

Because of the stability of transition cations M^{II} in an octahedral sulfur environment, the MPS$_3$ phases constitute a large family that spreads over a wide area of the periodic table, from vanadium to zinc inclusively. This is an important difference with parent transition metal dichalcogenides MX$_2$ where the presence of M^{IV} cations limits the occurrence of layered materials to IV, V and VID elements. Thus, change in the size of the cation, its electronegativity and electronic configuration can be used to study the structural, chemical and electrochemical variation of the phases.

Because of the stability of many M^I and M^{III} cations in an octahedral sulfur environment, a new class of substituted phases M$^I_{1/2}$M$^{III}_{1/2}$PS$_3$ has recently appeared. With respect to the cationic radius size ratio, they present different original cationic ordering, in particular with one-dimensional magnetic chains. Probably, this new area will present as many possibilities as the MPS$_3$ family itself and will likely develop greatly in the future, giving nice examples of (1D) magnetic chains trapped in (2D) slabs.

With the help of crystal and band structure calculations, fine correlations are drawn between intercalation reaction and electronic transfer processes, allowing a better understanding of the factors influencing intercalation in two-dimensional materials. In relation to the composition and symmetry of the phases, lithium cation ordering can be inferred from discharge curves. It was shown that the ability to practice intercalation chemistry is bound not only to the presence of convenient empty sites in the structure, but also, since the driving force for such intercalation lies in the electronic exchange between guest and host, to corresponding oxidizing centers provided by partially filled transition cation d bands. Hence, MPS$_3$ materials with d filled cations lack the electronic factor to intercalate reducing species, that is without cationic exchange.

Comprehension of the redox processes is an important step in deepening fundamental knowledge of intercalation chemistry. It is necessary in order to master the factors involved in the electrochemical mechanism and develop applications.

The unexpected and exotic substitution-intercalation reactions underline the originality of the MPS$_3$ and open new fields of research for these types of compounds, at the same time as they may have important effects in lithium batteries using organic solvents.

Since the M^{II} cation is reduced during intercalation, and beyond its potential applications, lithium intercalation-reduction in host structures proves an interesting tool, within the scope of soft chemistry, to obtain kinetically stable species with very low oxidation states that may present attractive new physical properties. Although not thoroughly done so far, it is very likely that the whole MPX$_3$ group can see its cations in highly reduced states upon lithium intercalation.

It must be pointed out that the MPS$_3$ are a subdivision of the large M-P-S group where are found numerous and appealing new low dimensional phases /67-70/. As far as transition metals are concerned, let us mention for example the group where M is a VD cation. In these materials, the possibility of having a wide range of oxidation states not only for the cations, but also for the sulfur anions through catenation allows occurrence of many new phases. In these ones, sulfur catenation leads not only to (S$_n$)$^{2-}$ species (1 < n < 3), but even to infinite sulfur chains (S^0) as found in the tunnel structure Ta$_4$P$_4$S$_{29}$ /70/.

Since lithium intercalation in chalcogenides and chalcogenophosphates mostly involves the cations as reduction centers, use of extra oxidizing centers is a way to increase capacity of lithium batteries using these materials as secondary positive. Because the new M-P-S compounds present anionic species with high oxidation states, reduction both of cationic and anionic species may be envisioned and prove another way of obtaining highly efficient secondary electrodes.

8. REFERENCES

1. M.C. Friedel : C.R.Acad.Sc. 119,269 (1894)
2. M.C. Friedel: Bull. Soc. Chim. Fr. 11, 115 (1894)
3. L. Ferrand: Bull. Soc. Chim. Fr. 13 115 (1895)
4. L. Ferrand: Ann. Chim. Phys. 17, 388 (1895)
5. L. Ferrand: C.R. Hebd. Acad. Sc. 122, 621 (1896)
6. W. Klingen: Thesis, Höhenheim (W.G.) (1969)
7. W. Klingen, R. Ott, H. Hahn: Z. anorg. allg. Chem. 396, 271 (1973)
8. G. Ouvrard: Thesis, Nantes (France) (1980)
9. B.J. Curtis, F.P. Emmenenegger, R. Nitsche: R.C.A. Review 647 (1970)
10. R. Nitsche, P. Wild: Mat. Res. Bull. 5, 419 (1970)
11. B.I. Taylor, J. Steger, A. Wold: J. Solid State Chem. 7, 461 (1973)
12. G. Ouvrard, R. Brec: unpublished results.
13. G. Ouvrard, R. Brec, J. Rouxel: C.R. Acad. Sc. 294II, 971 (1982)
14. E. Prouzet, G. Ouvrard, R. Brec: Mat. Res. Bull. 21, 195 (1986)
15. G. Ouvrard, R. Fréour, R. Brec, J. Rouxel: Mat. Res. Bull. 20, 1053 (1985)
16. A.H. Thompson, M.S. Whittingham: Mat. Res. Bull. 12, 741 (1977)
17. R. Brec, D. Schleich, A. Louisy, J. Rouxel: Ann. Chim. Fr. 3, 347 (1978)
18. J.P. Odile, J.J. Steger, A. Wold: Inorg. Chem. 14, 2400 (1975)
19. H. Mercier, E. Canadell, Y. Mathey: Inorg. Chem. to be published (1986)
20. A. Leblanc-Soreau, J. Rouxel, C.R. Acad. Sc. 291C, 263 (1980)
21. P. Colombet, A. Leblanc, M. Danot, J. Rouxel: Nouv. J. Chim. 7, 333 (1983)
22. S. Lee, G. Ouvrard, P. Colombet, R. Brec: Mat. Res. Bull. to be published (1986).
23. Z. Ouili, A. Leblanc, P. Colombet: J. Solid State Chem. to be published (1986).
24. Y. Mathey, R. Clément, J.P. Audière, O. Poizat, C. Sourisseau: Solid State Ionics 9&10, 459, (1983)
25. G. Ouvrard, R. Brec, J. Rouxel: C.R. Acad. Sc. 294C, 971 (1982)
26. G. Ouvrard, R. Brec, J. Rouxel: Mat. Res. Bull. 20, 1181 (1985)
27. R. Brec, G. Ouvrard, J. Rouxel: Mat. Res. Bull. 20, 1257 (1985)
28. S. Soled, A. Wold: Mat. Res. Bull. 11, 657 (1976)
29. P. Colombet, A. Leblanc, M. Danot, J. Rouxel: J. Solid State Chem. 41, 174 (1982)
30. Y. Mathey, H. Mercier, A. Michalowicz, A. Leblanc: J. Phys. Chem. Solids 46, 1025 (1985)
31. Y. Mathey, A. Michalowicz, P. Toffoli, G. Vlaic: Inorg. Chem. 23, 897 (1984)
32. O. Poizat, C. Sourisseau: J. Solid State Chem. 59, 371 (1985)
33. O. Poizat: Thesis, Paris (1985).
34. R. Brec, D.M. Schleich, G. Ouvrard, A. Louisy, J. Rouxel: Inorg. Chem. 18, 1814 (1979)
35. Y. Chabre, P. Segransan, C. Berthier, G. Ouvrard: In Fast Ion Transport in Solids, Ed. P. Vashishta, J.N. Mundy, G.K. Shenoy, 221 (North-Holland 1979)
36. G. Le Flem, R. Brec, G. Ouvrard, A. Louisy, P. Segransan: J. Phys. Chem. Solids, 43, 455 (1982)
37. K. Kurosawa, S. Saito, Y. Yamaguchi: J. Phys. Soc. Jap. 52, 3919 (1983)
38. P. Jernberg, S. Bjarman, R. Wäppling: J. Mag. Mag. Mat. 46, 178 (1984)
39. F.S. Khumalo, H.P. Hugues: Phys. Rev. B 23, 5375 (1981)
40. M. Piacentini, F.S. Khumalo, C.G. Olson, J.M. Anderegg, D.W. Lynch: Chem. Phys. 65, 289 (1982)
41. M. Piacentini, F.S. Khumalo, G. Levêque, C.G. Olson, D.W. Lynch: Chem. Phys. 72, 61 (1982)

42. M. Piacentini, V. Grasso, S. Santangelo, M. Fanfoni, M. Modesti, A. Savoia: Solid State Comm. 51, 467 (1984)
43. M. Piacentini, V. Grasso, M. Fanfoni, S. Modesti, A. Savoia: Nuov. Cim. 4D, 444 (1984)
44. M.H. Whangbo, R. Brec, G. Ouvrard, J. Rouxel: Inorg. Chem. 24, 2459 (1985)
45. R. Brec. G. Ouvrard, A. Louisy, A. Le Méhauté, J. Rouxel: Solid State Ionics, 6, 185 (1982)
46. M. Hanggo, T. Noguchi, S. Nakashima, A. Mitsuishi: 9nth Int. Conf. Raman Spect. (ICORS) Tokyo (1984)
47. S. Yamanaka, H. Kobayashi, T. Tanaka: Chem. Lett. 329 (1976)
48. P.J.S. Foot, N.G. Shaker: Mat. Res. Bull. 18, 173 (1983)
49. R. Schöllhorn, H.D. Zagefka: Angew. Chem. 89, 193 (1977)
50. R. Schöllhorn, H.D. Zagefka: Angew. Chem. Int. Ed Eng. 16, 199 (1977)
51. R. Clément, M.L.H. Green: J. Chem. Soc. Dalton Trans. 10, 1566 (1979)
52. R. Clément: J. Chem. Soc. Chem. Com. 647 (1980)
53. J.P. Audière, R. Clément, Y. Mathey, C. Mazière: Physica 99B, 133 (1980)
54. R. Clément, O. Garnier, J. Jegoudez: Inorg. Chem. to be published (1986)
55. A.H. Thompson, M.S. Whittingham: Mat. Res. Bull. 12, 741 (1977)
56. A. Le Méhauté, G. Ouvrard, R. Brec, J. Rouxel: Mat. Res. Bull. 12, 1191 (1977)
57. M. Barj, C. Sourisseau, G. Ouvrard, R. Brec: Solid State Ionics 11, 179 (1983)
58. A. Le Méhauté: C.R. Acad. Sci. 287C, 309 (1978)
59. A. Le Méhauté: Thesis, Nantes (1979)
60. G.A. Fatseas, G. Ouvrard, M.-H.Whangbo, R. Brec: Solid State Ionics, to be published (1986)
61. C. Berthier, Y. Chabre, M. Minier: Solid State Comm. 28, 327 (1978)
62. G.A. Fatseas, M. Evain, G. Ouvrard, R. Brec, M.-H. Whangbo: Phys. Rev. B to be published (1986)
63. Y. Chabre, G. Ouvrard, R. Brec: Solid State Ionics, to be published (1986)
64. A.H. Thompson, M.S. Whittingham: U.S. Patent 4, 049, 879 (1977)
65. R. Brec, A. Le Méhauté: Fr Patents 7, 704-518, 7,704-519 (1977)
66. J. Rouxel, R. Brec: Ann. Rev. Mater. Sci. 16, 137 (1986)
67. P. Grenouilleau, R. Brec, M. Evain, J. Rouxel: Rev. Chim. Min. 20, 628 (1983)
68. R. Brec, P. Grenouilleau, M. Evain, J. Rouxel: Rev. Chim. Min. 20 295 (1983)
69. R. Brec, G. Ouvrard, M. Evain, P. Grenouilleau, J. Rouxel: J. Solid State Chem. 7, 174 (1983)
70. M. Evain, M. Queignec, R. Brec, J. Rouxel: J. Solid State Chem. 56, 148 (1985)

RECENT DEVELOPMENTS IN NEW HOST STRUCTURES

R. Brec

Laboratoire de Chimie des Solides
U.A. 279, 2, rue de la Houssinière
44072 Nantes Cedex, France

The reverse of the intercalation process involves oxidizing an intercalation compound to generate the host frame into which the ions were inserted. This frame was reduced during intercalation, and it is in turn oxidized by a more oxidizing agent that removes the mobile intercalated species (for example an alkali metal). Many reagents are sufficiently oxidizing to do this reaction, but it is necessary to use proper solventsto avoid hydrolysis for instance. Also, during the reaction $Li_xMX_n \rightarrow xLi^+ + xe^- + MX_n$, the oxidized and reduced forms of the reagent should be soluble in the non-reactive solvent.

From the position of a given host structure relative to various reagents, in a redox scale (Fig. 1), we can determine whether a particular phase will intercalate or deintercalate /1/

Starting from an $A_xM_yS_2$ ternary sulfide obtained at high temperature, it may be possible to remove the alkali metal (A) through low temperature deintercala-

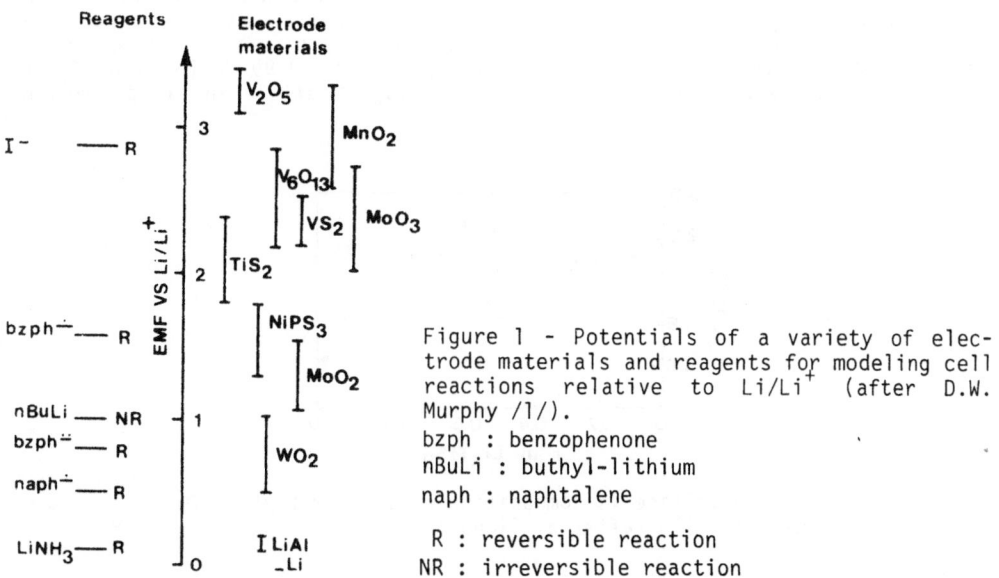

Figure 1 - Potentials of a variety of electrode materials and reagents for modeling cell reactions relative to Li/Li^+ (after D.W. Murphy /1/).
bzph : benzophenone
nBuLi : buthyl-lithium
naph : naphtalene

R : reversible reaction
NR : irreversible reaction

tion. This produces a new M_yS_2 binary phase, either previously unknown or of a new structural type. This method is useful in generating new highly reactive solids for intercalation cathodes.

To allow such a reaction to occur, the ternary sulfides must possess two essential properties : they must have mobile cations and easily oxidizable centers in the structure. When total deintercalation is achieved, the oxidized phases may retain their original structural frames with sites previously filled by the mobile cations now empty. Such open structures are often metastable and can be reduced again through reintercalation. Electrochemically, the ternary sulfide can be described as a discharged cathode that can be recharged (chemically or electrochemically) by a redox deintercalation process.

1. VS_2 Synthesis

An example of this type of reaction was first given by D. MURPHY /2,3/ who started with the vanadium derivative $LiVS_2$. This material is prepared by high temperature synthesis and has a TiS_2 structure with lithium ions located in the octahedral sites of the Van der Waals gap. With V^{3+} in an octahedral sulfur environment, this phase is stable whereas VS_2 cannot be prepared by direct combination of the elements. Complete delithiation of $LiVS_2$ can be achieved using iodine in acetonitrile according to the equation : $LiVS_2 + \frac{1}{2} I_2 \rightarrow LiI + VS_2$.

Although VS_2 presents a TiS_2 structure, several monoclinic distortions occur during deintercalation. These phase transitions in the Li_xVS_2 series were related to electronic instabilities in the VS_2 layers and may take the form of a charge density wave as observed in many MX_2 layered compounds (including VSe_2).

2. The Cubic TiS_2

Recent deintercalation reactions have been performed from the cubic defect spinel $CuTi_2S_4$, where Cu and Ti ions are respectively in tetrahedral and octahedral coordination in a cubic close-packed sulfur anion array /4,5/.

The reaction involves oxidation of the high temperature stable spinel $CuTi_2S_4$ with Br_2 in acetonitrile at room temperature according to :

$$CuTi_2S_4 + nBr_2 \rightarrow Cu_xTi_2S_4 + yCuBr + zCuBr_2$$

It was shown that, to achieve complete copper removal and stability of the oxidized phase, a slight titanium excess was needed, so that the copper free material corresponds to $Ti_{2.05}S_4$ ($C-TiS_2$). The cubic unit cell parameter decreases approximately linearly with x and goes from 9.99 Å for $CuTi_2S_4$ to 9.75 Å for $C-TiS_2$ (7 % volume decrease). Cubic TiS_2 is stable in air at ambient

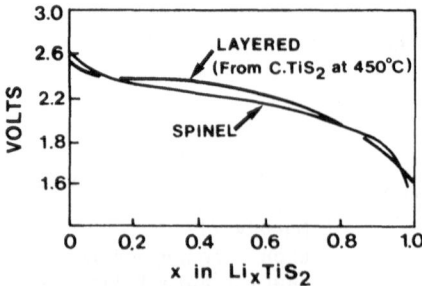

Figure 2 - Voltage vs composition curves for Li/Li$^+$/2D and C-TiS$_2$ cells (after S. Sinha /5/)

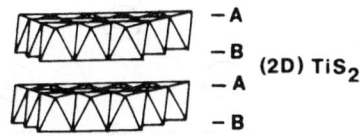

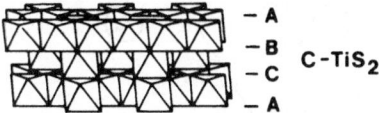

Figure 3 - Comparison of the
layered- and C- TiS$_2$ structures
(after S. Sinha /5/).

temperature and converts irreversibly to a layered structure above 400°C. It can be reversibly intercalated by lithium which goes into octahedral sites of the structure to produce the series Li$_x$Ti$_2$S$_4$ (0 < x < 2), the intercalates remaining single phased.

The experimental voltage vs composition curve for a Li/Li$^+$/C-Li$_x$TiS$_2$ cell was fit to a lattice gas model (E = E$_0$ + 6 Ux + kT Log $\frac{x}{1-x}$) yielded E$_0^x$ = 2.3 V and U = 3.0 kT. The E.M.F. vs x curve for C-Li$_x$TiS$_2$ is very similar to that of layered Li$_x$TiS$_2$ (E$_0$ = 2.3 V, U = 2.5kT), although in this latter case (Fig. 2) a deviation from the fit was thought possibly connected to two dimensional lithium ordering, which is not the case for C-Li$_x$TiS$_2$.

A comparison of the structures of layered and C-TiS$_2$ may be useful. Starting from layered TiS$_2$, layers are first shifted to form a CdCl$_2$ structure, from which one quarter of the Ti atoms are displaced from each layer and ordered in the octahedral sites in the gap between the layers such that each Ti has an octahedron of next nearest Ti atoms (Fig. 3). In layered TiS$_2$, excess Ti occupies octahedral sites between the layers and in C-TiS$_2$ it seems that, from structure calculations, it is located in the tetrahedral sites.

For layered TiS$_2$, the presence of excess Ti between the layers slows intercalation considerably, yet C-TiS$_2$ may be considered as having a layer structure with a quarter of the interlayer octahedral sites filled by Ti. It appears that any unfavorable effect of interlayer Ti is compensated by the intralayer vacancies which allow three dimensional diffusion.

3. The Li$_x$FeS$_2$ System

Li$_2$FeS$_2$ is a double sulfide obtained by synthesis at high temperature between Li$_2$S and FeS. According to X-Ray results /6/ the unit cell of the phase is hexagonal, and the phase has a trigonal symmetry. The sublattice hexagonal parameters (a = b = 3.908(2) Å and c = 6.279(4) Å) suggest a CdI$_2$ structural type, the low c/a ratio (1.61) hinting at the occupancy of tetrahedral sites by some cations.

Recent structure determinations /7/ taking into account the strongest super-structure (2a, 2b, c) showed the phase to be built from a hexagonal close packing of the anions. Iron ions are found to be located statistically on two tetrahedral sites of the slabs (66 % of Fe and 33 % of Fe') (Fig. 4), constituting infinite (2D) (FeS$_2$) slabs. The lithium is scattered on half of the Oh sites and some of the four tetrahedral available sites per molecular weight (Li$_{Oh}$ Li$_{Td}$ Fe$_{Td}$ S$_2$).

Electrochemical oxidation of Li$_2$FeS$_2$ allowed removal of lithium down to FeS$_2$, and from open potential circuit analysis /8/, several domains in the oxidized

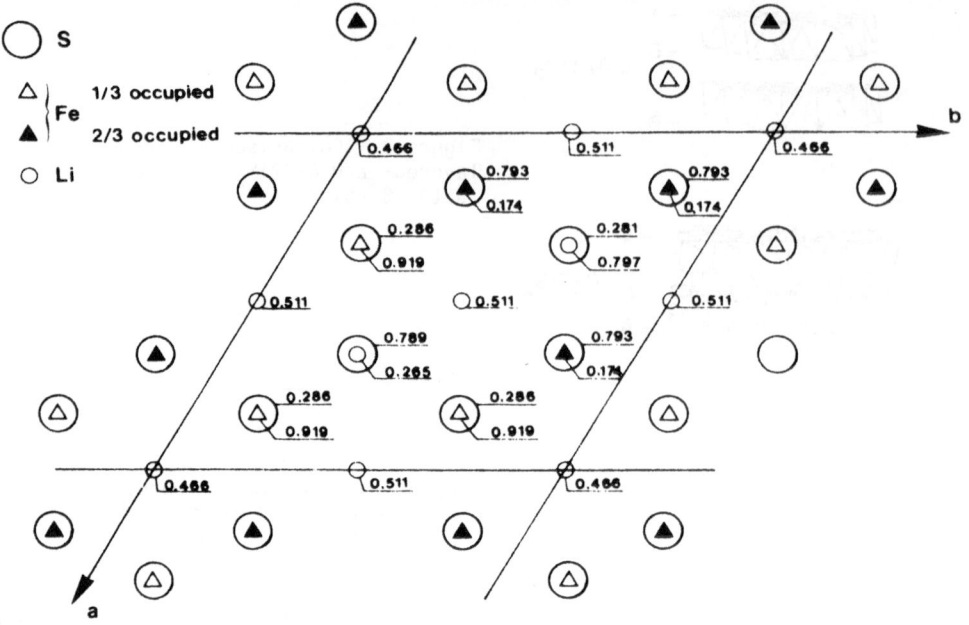

Figure 4 - Projection on the ab plane of the Li_2FeS_2 structure.

Li_xFeS_2 (0 < x < 2) system were characterized (Fig. 5). There is a lithium rich phase Li_xFeS_2 (1 < x < 2) that can be considered single phased in spite of an electrochemical transition detected around $Li_{1.5}FeS_2$, and a biphased region that spreads from Li_1FeS_2 to FeS_2 and which corresponds to the occurrence of both these phases. For Li_1FeS_2 the phase retains the original hexagonal axes (a = 3.87 Å and b = 6.21 Å) whereas the highly oxidized FeS_2 spectrum presents a monoclinic cell deriving from the original one, suggesting a new structure for the phase. A similar monoclinic unit cell was also reported for delithiated Li_xVS_2 phases /2/.

FeS_2, the recharged state of Li_2FeS_2 can then be reintercalated electrochemically to its initial composition. This reversible behavior with respect to lithium is an appealing property of the material from an electrochemical point of view, since it corresponds to a capacity about double that of TiS_2.

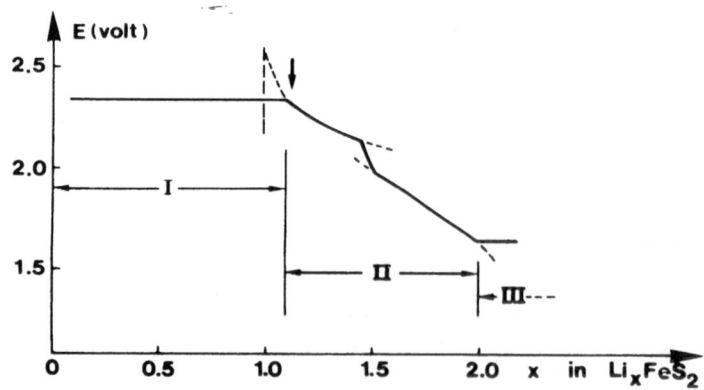

Figure 5 - Quasi equilibrium potential of the Li_xFeS_2 system (0 < x < 2).

The formation of this new iron disulfide raises the question of its electronic structure, since only pyrite or marcassite are the known stable phases for FeS_2.

In the starting material Li_2FeS_2, magnetic and Mössbauer studies indicate the occurrence of Fe^{2+} high spin in tetrahedral environment. After deintercalation to Li_1FeS_2, Mössbauer spectra (see table I) are consistent with oxydation of iron according to : $Li_2Fe^{2+}S_2^{-2} \rightarrow Li + Li_1Fe^{3+}S_2^{-2}$. Removal of the last lithium poses a problem, since it is difficult to imagine a further oxidation of Fe^{3+}.

From a schematic band model, it is expected that, in the original Li_2FeS_2 phase, the level of the d bands will be situated well above the anionic valence band (Fig. 6a). Oxidation to Li_1FeS_2 can be assumed to give a model of the type drawn in Fig. 6a, where the level of the d bands is strongly lowered and close to the anionic filled band. Further oxidation, if considered to take place at the iron site, would necessarily push the d band well into the anionic one, meaning that the S^{-2} ions become reductors of the highly oxidized iron. This means actually, that beyond the Li_1FeS_2 composition, the S^{-2} anions must be oxidized, holes being created in the anionic band and $(S_2)^{-2}$ polyanions must form according to : $Li^+Fe^{3+}S_2^{-2} \rightarrow Li + Fe^{3+}S^{-2} (S_2)^{-2}_{1/2}$.

In the absence of usable structural data because of amorphization of the deintercalated materials, only an infrared study, backed by Mössbauer work, can help to follow the perturbation of lattice phonons and to define the redox process of the Li_xFeS_2 systems.

The infrared spectra (550-200 cm^{-1}) of Li_2FeS_2 and of some oxidized Li_xFeS_2 compounds with x = 1.64, 0.98, 0.72, 0.28, 0.14 and $\simeq$ 0.0 are shown on figures 7 and 8a /9/. Similarly, the infrared spectra of three reintercalated Li_xFeS_2 samples with x = 0.69, 0.93 and 1.5 are presented in figure 8b. All the observed band maxima and proposed assignments are reported in table II.

3.1 Li_2FeS_2 and $Li_{1.64}FeS_2$ Compounds (Fig. 7a)

The infrared spectra of both systems are quite simple and do not show tremendous changes with respect to the lithium content. One observes mainly two broad absorption maxima at about 320 and 420 cm^{-1} (these bands split into several components at low temperature). The lower frequency and more intense IR bands are assigned to asymmetric stretching modes of $Li-S_6$ entities. The higher frequency bands are thus attributed to lithium motions in "tetrahedral" positions ; this result is in complete agreement with the structural determinations. Moreover, one notes a relative intensity decrease of the band at $\simeq$ 420 cm^{-1} for x = 1.64 : this

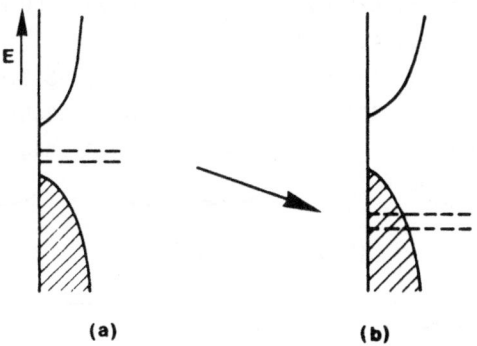

Figure 6 - Schematic band model for Fe^{II} (a) and higher oxidation state (b) in sulfur environment.

TABLE I - Mössbauer data for various compositions in the Li_xFeS_2 system.

x in Li_xFeS_2	Isomer shift mm.s^{-1}			Quadrupolar splitting mm.s^{-1}			Relative intensities %		
Iron site	1	2	3	1	2	3	1	2	3
2	0.50	0.49		0.90	1.61		77	23	
1.78	0.45	0.38		0.84	1.39		64	36	
1.31		0.35	0.18		1.24	0.37		73	27
1.10		0.36	0.20		1.18	0.35		55	45
0.85		0.32	0.16		0.93	0.49		60	40
0.57		0.32	0.23		0.94	0.53		65	35
0.21		0.32	0.26		0.85	0.56		50	50

TABLE II - Infrared band wavenumbers (cm^{-1}) and proposed assignments in Li_2FeS_2, some oxidized samples (x = 1.64, 0.98, 0.72, 0.28, 0.14 and ≈ 0.0) and in three reduced phases (x = 0.69, 0.93 and 1.5) at 300K.[†]

Li_2FeS_2				416(s)	392(s)	350(s)	321(vs)	274(sh)
$Li_{1.64}FeS_2$				420(m)	396(w)	351(s)	320(vs)	273(sh)
$Li_{0.98}FeS_2$				423(w)	388(w)	350(vs)	320(sh)	
$Li_{0.72}FeS_2$	520(sh)	494(m)	435(w)	420(vw)	386(sh)	349(vs)	320(m)	
$Li_{0.28}FeS_2$	524(sh)	477(m)	425(w)?	425(w)?	392(m)	348(vs)		
$Li_{0.14}FeS_2$		470(vs)	425(m)		395(vw)	350(m)		
$Li_{0.0}FeS_2$		469(vs)	430(vs)					280(w)
$Li_{0.69}FeS_2$		489(s)	434(m)		390(w)	343(vs)		308(m)
$Li_{0.93}FeS_2$		495(m)	430(sh)		388(w)	345(vs)		308(sh)
$Li_{1.15}FeS_2$		496(sh)			388(sh)	344(vs)		306(sh)
Assignments	$(S-S)^{2-}$							(S-Fe-S)
		$(S-S)^{2-}$		$(Li-S_4)$			$(Li-S_6)$	

[†] sh=sharp, m=medium, w=weak, vs=very strong

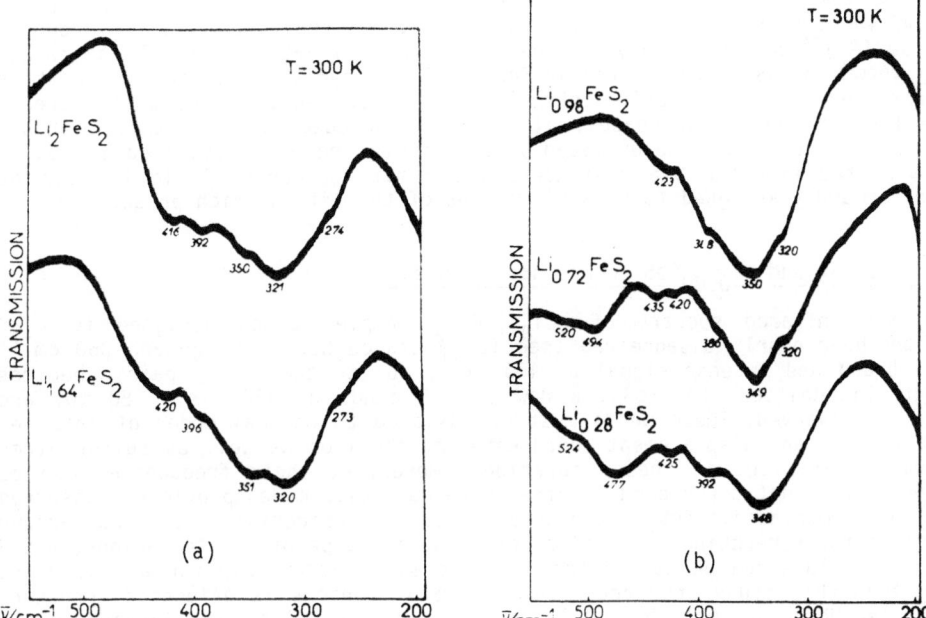

Figure 7 - Infrared spectra at 300 K of Li_xFeS_2 compounds (a) : x = 2 and 1.64 (b) : x = 0.98, 0.72 and 0.28 /9/

suggests that during the first oxydation step of Li_2FeS_2, lithium ions are more stabilized in "octahedral" positions ; nevertheless, in agreement with X-ray data, the crystal structure is not markedly modified since all the bandwave-numbers are similar. Because these spectra are not temperature dependent and these compounds are single phased, we are dealing with a statistical disorder in lithium position occupancies which may be the origin of local distortions. In the frequency region explored in this work, iron translational modes are not expected to be found since they must appear at much lower frequencies.

3.2 $Li_{0.98}FeS_2$, $Li_{0.72}FeS_2$ and $Li_{0.28}FeS_2$ Compounds (Fig. 7b)

For a $Li_{0.98}FeS_2$ sample, one notes a very simple spectrum which exhibits essentially an intense broad band maximizing at 350 cm^{-1} with weaker components at 423, 388 and 320 cm^{-1}. The bands at 350 and 320 cm^{-1} correspond to asymmetric stretching modes of lithium ions in "octahedral" sites and their relative intensities can be explained by a new distribution of lithium ions in $Li-S_6$ environments. The infrared spectrum contains only an additional weak band at 423 cm^{-1} which may come from some lithium remaining on "tetrahedral" sites.

We can thus say that most of lithium ions have been removed from the "tetrahedral" sites upon deintercalation from Li_2FeS_2 to Li_1FeS_2. At this stage of the delithiation process, one must assume, in agreement with Mössbauer results /8/, the formation of Fe^{3+} ions since strong electronic perturbations within the sulfur layers are not spectroscopically confirmed.

In sharp contrast with the above results, the infrared spectra of $Li_{0.72}FeS_2$ and $Li_{0.28}FeS_2$ derivatives are quite complex and exhibit intense new bands respectively at 494 cm^{-1} and 477 cm^{-1}, i.e. in the frequency region of $(S_2)^{2-}$ pairs stretching vibrations. According to STEUDEL's relation /10/ : $d_{(S-S)Å}$ = $2.57-9.47.10^{-4}.\nu(S-S)cm^{-1}$, distances equal to 2.10 Å and 2.12 Å can be proposed for these $(S_2)^{-2}$ anions ; they can be compared with S-S interatomic distances already known in MnS_2 (2.09 Å) and in FeS_2 (2.17 Å) pyrite compounds /11/. It is worth noticing that the relative intensities of these new bands seem to increase as the lithium content decreases. It can be concluded that in this biphased domain, the sulfur atoms are oxydized progressively and are stabilized in $(S_2)^{-2}$ pairs with the expected S-S distances. The remaining bands (Table II) can be straightforwardly assigned to Li-S vibrations of the lithium rich phase.

3.3 $Li_{0.14}FeS_2$ and $Li_{0.00}FeS_2$ Compounds (Fig. 8a)

On the infrared spectrum of a $Li_{0.14}FeS_2$ sample, bands assigned to Li-S vibrations have nearly disappeared (see for instance bands at 425 and 350 cm^{-1}) while a broad and intense signal at 470 cm^{-1}, due to the $(S_2)^{-2}$ pairs, becomes dominant. In addition, for FeS_2, a new set of bands at 430 (vs), 280 (w) and 223 cm^{-1} is observed. These bands must be assigned to optical modes of this new iron disulfide, and displacement coordinates of the iron as well as sulfur atoms are probably involved in these vibrations. Moreover, their frequencies can be compared, but do not correspond, to the infrared active modes previously observed in pyrite and marcassite FeS_2 compounds /12,13/. Considering the band around 430 cm^{-1} as the stretching vibration of a second type of $(S_2)^{-2}$ anions, a S-S distance of 2.16 Å can be calculated, a value still acceptable for a (S_2) pair. Clearly from that study, the occurrence of $(S_2)^{-2}$ anions is evidenced. Presence of Fe^{3+} (see Mössbauer work below) leads to the charge balance put forward above : $Fe^{3+}S^{-2}(S_2)^{-2}_{1/2}$.

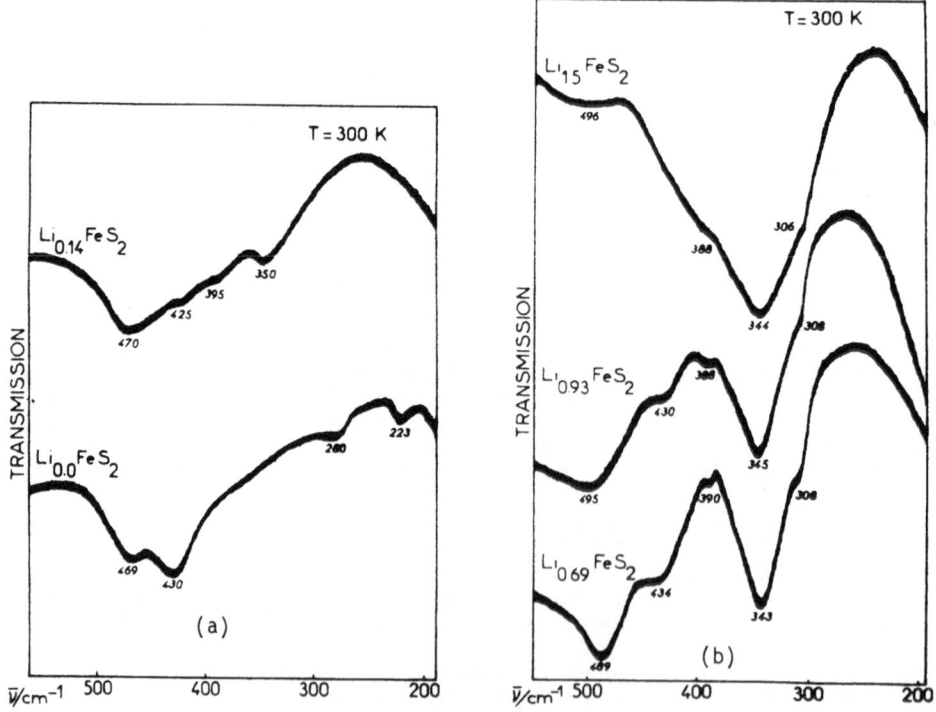

Figure 8 - Infrared spectra at 300 K of Li_xFeS_2 compounds (a) : x = 0.14 and 0.00 (b) : x = 0.69, 0.93, 1.5 obtained by reintercalation.

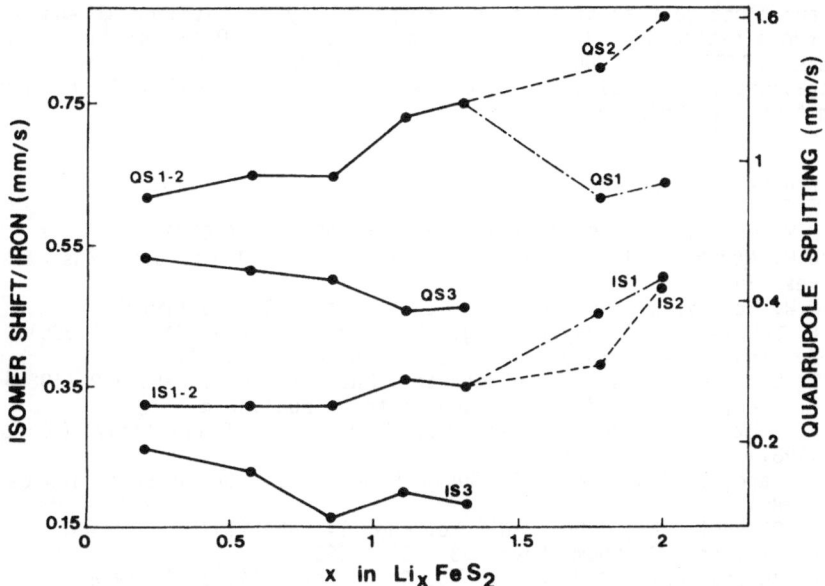

Figure 9 - Variation of isomer shift and quadrupole splitting versus x in Li_xFeS_2.

3.4 $Li_{0.69}FeS_2$, $Li_{0.93}FeS_2$ and $Li_{1.5}FeS_2$ Reintercalated Samples (Fig.8b)

The infrared spectra of these chemically reintercalated phases can be nicely compared with those of the previously discussed Li_xFeS_2 oxidized forms with x = 0.72, 0.98 and 1.64, respectively. Some differences are clearly evidenced. First of all, we observe two absorption bands at 490 cm^{-1} and 430 cm^{-1} due to the stretching vibration of $(S_2)^{-2}$ pairs; the intensities of these bands have almost vanished when the lithium content is equal to 1.5; secondly, a very intense signal at 344 cm^{-1} with weaker components at 388 cm^{-1} and 308 cm^{-1} increases and becomes dominant on the $Li_{1.5}FeS_2$ spectrum. These results show that lithium ions are mainly accomodated on "octahedral" $Li-S_6$ sites and the $(S_2)^{-2}$ have been almost completely reduced.

The fact that, for the $Li_{1.5}FeS_2$ composition, the 1.5 lithium ions seem to be located in octahedral sites indicates a structural change respective to the deintercalated phase with the same composition. Moreover, it appears difficult to reintercalate lithium chemically beyond the $Li_{1.5}FeS$ composition, as though no other octahedral sites were available. A shift of the iron ions from tetrahedral to octahedral sites might take place and explain the structure difference. Coordination change is not new in lithium intercalated phases (see for example the case of $Li_xFe_3O_4$). Actually it is confirmed by the Mössbauer study of the Li_xFeS_2 systems /14/.

Considering the isomer shift of iron in the original phase, two iron sites attributable to Fe^{2+} in tetrahedral sulfur environments are found, in agreement with the structural determination (Table II), with however a somewhat different occupancy ratio. This difference seems to be related to the synthesis conditions of the samples, different preparations leading to various filling ratios /14/.

Mössbauer study of deintercalated Li_xFeS_2 samples shows (Fig.9) that, respective to lithium content decrease, the iron isomer shift decreases as would be expected for an oxydation of Fe^{2+} into Fe^{3+}. In the vicinity of $Li_{1.3}FeS_2$, the two Fe(1) and Fe(2) sites give two new Fe(1-2) and Fe(3) sites whose isomer shifts are close to that of iron III, respectively in tetrahedral and octahedral sulfur coordination (I.S.(1-2) = 0.35 and I.S.(3) = 0.18 mms^{-1}), in agreement with the FATSEAS and GOODENOUGH /15/ relation. Finally, at low lithium concentration, the t(1-2)/t(3) occupancy ratio is found around 1, in accord with the limit of $Li_{1.5}FeS_2$ found for chemical reintercalation.

4. References

1. D.W. Murphy, P.A. Christian : Science 205, 651 (1979).
2. D.W. Murphy, J.N. Carides, F.J. Di Salvo, C. Cros, J.V. Waszczak : Mat. Res. Bull. 12, 825 (1977).
3. D.W. Murphy, J.N. Carides : J. Elec. Soc. 126, 349 (1979).
4. R. Schöllhorn, A. Payer : Ang. Chem. Int. Ed. Eng. 24, 67 (1985).
5. S. Sinha, D.W. Murphy : Solid State Ionics 20, 81 (1986).
6. R. Brec, A. Dugast, A. Le Mehaute : Mat. Res. Bull. 15, 619 (1980).
7. R. Brec, G. Ouvrard, L. Blandeau : to be published.
8. A. Dugast, R. Brec, G. Ouvrard, J. Rouxel : Solid State Ionics 5, 375 (1981).
9. P. Gard, C. Sourisseau, G. Ouvrard, R. Brec : Solid State Ionics 20, 231 (1986).
10. K. Steudel : Ang. Chem. Int. Ed. Eng. 14, 655 (1975).
11. N. Elliott : J. Chem. Phys. 33, 903 (1960).
12. E. Anastassakis, C.H. Perry : J. Chem. Phys. 64, 3604 (1976).
13. H.D. Lutz, G. Kliche, H. Haenseler : Z. Nat. 36a, 184 (1981).
14. L. Blandeau, G. Fatseas, G. Ouvrard and R. Brec, J. of Solid State Chem., to be published 1986.
15. G.A. Fatseas, J.B. Goodenough : J. Solid State Chem. 41, 1 (1982).

BAND STRUCTURE CHANGES UPON LITHIUM INTERCALATION

OF 1T- AND 4Hb-TaS$_2$

A.M. Ghorayeb[§], W.Y. Liang and A.D. Yoffe

Cavendish Laboratory
Madingley Road
Cambridge CB3 OHE
U.K.

The band structure of the layered transition-metal dichalcogenides (TMD's) is now fairly well understood (Liang, 1986, this volume). Of particular importance is the lowest-lying part of the conduction band which is known to contain a considerable d_z2 character: its energy position seems to be associated with its electronic filling as well as with the c/a ratio of the compound in question. Experimental evidence for such dependence is hereby given for the first time, through room-temperature optical transmission spectra of lithium-intercalated 1T- and 4Hb-TaS$_2$.

It is known that 1T-TaS$_2$, where the Ta atoms are octahedrally coordinated by the surrounding S atoms, exhibits a nearly-commensurate $\sqrt{13}$ x $\sqrt{13}$ in-plane superstructure at room temperature (Scruby et al., 1975; Wilson et al., 1975). The 4Hb-polytype, which consists of an alternation of trigonal prismatic and octahedral layers, is particularly interesting in that its differently-coordinated sub-layers behave almost independently from each other. This is deduced from structural (Wilson et al., 1975; Tatlock, 1976) and optical (Beal, 1978) data.

In this work, the intercalation of 1T- and 4Hb-TaS$_2$ with lithium was effected by immersing the samples in a solution of butyl-lithium (BuLi) in hexane. Figures 1 and 2 show room-temperature optical density spectra, taken before and after intercalation with Li, of 1T- and 4Hb-TaS$_2$ respectively. Figure 3 shows our proposed energy band scheme for (a) 1T-TaS$_2$ and (b) 1T-LiTaS$_2$, taking into account the effect of the charge-density wave (CDW) distortion which causes a reduction in the density of states at the Fermi level. This is represented by the splitting of the d_z2 band into two sub-bands, d_z2_1 and d_z2_2. The absorption bands A and B in the spectrum of pure 1T-TaS$_2$ (figure 1) are thought to correspond to $d_z2_1 \rightarrow d_z2_2$ and $d_z2_1 \rightarrow$ higher d transitions respectively, while C and D would respectively represent p $\rightarrow d_z2_2$ and p $\rightarrow$ higher d transitions. For 1T-LiTaS$_2$, the absorption band A" indicates that CDW's are still present in the intercalated sample, in agreement with the structural results of Clark and Williams (1977) who report the presence of a 3 x 3 in-plane superlattice at room temperature in the intercalated complex. An important feature in the spectrum of 1T-LiTaS$_2$, however, is the

─────────
[§] Present address: Université Pierre et Marie Curie, Laboratoire de Physique des Solides, Tour 13, 2ème étage, 4 place Jussieu, 75252 PARIS CEDEX 05, FRANCE.

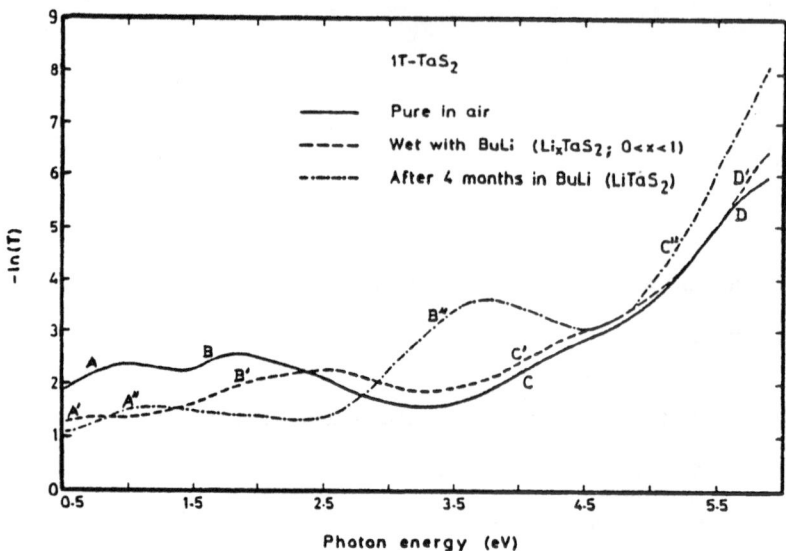

Fig. 1. Room-temperature optical density spectra of
1T-TaS$_2$ and its lithium-intercalation complexes.

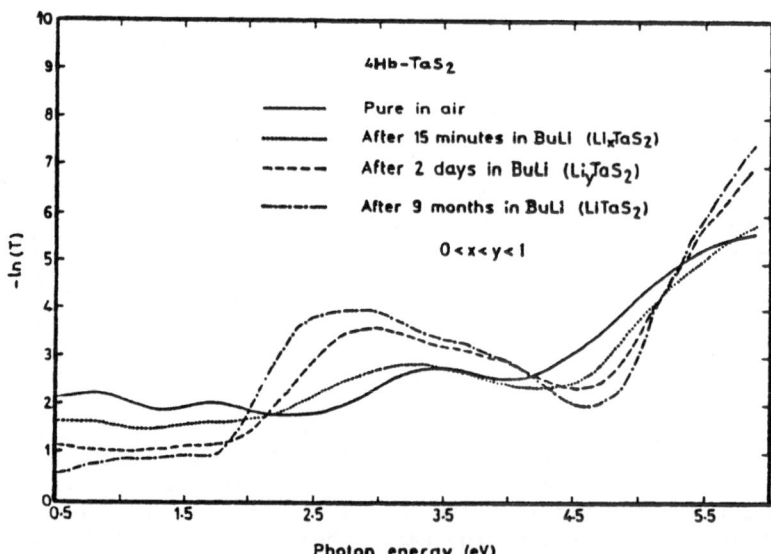

Fig. 2. Room-temperature optical density spectra of
4Hb-TaS$_2$ and its lithium-intercalation complexes.

energy position of the absorption band B" which represents $d_{z^2}1 \rightarrow$"d" transitions and which indicates a considerable lowering of the d$_{z^2}$ band with respect to its position in the pristine compound. The filling of this band with electrons donated from Li, as well as the increased c/a ratio after intercalation, contribute to this band lowering which was suggested by Liang (1986, this volume) and which is here experimentally evidenced for the first time. The lowering of the d$_{z^2}$ band is expected to lead eventually to a change of coordination from octahedral to trigonal prismatic. However, this change is believed not to have yet occurred here since the spectrum of 1T-LiTaS$_2$

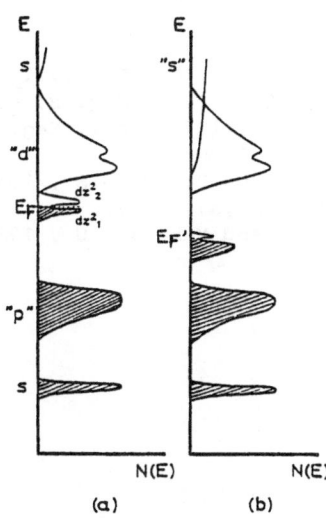

Fig. 3. Schematic density-of-states diagram for
(a) 1T-TaS$_2$ and (b) 1T-LiTaS$_2$.

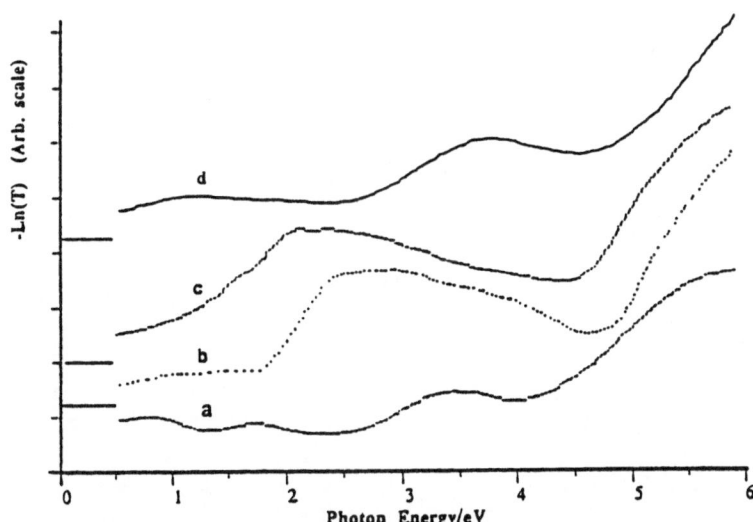

Fig. 4. Comparative representation of the room-temperature
optical density spectra of (a) 4Hb-TaS$_2$, (b) 4Hb-LiTaS$_2$,
(c) 2H-LiTaS$_2$ and (d) 1T-LiTaS$_2$.

shown in figure 1 is different from that of 2H-LiTaS$_2$ reported by Beal and
Nulsen (1981). Heating the intercalated complex above room temperature could
facilitate such a change.

As for the results on the 4Hb-polytype, their interesting aspect lies in
that the spectrum of 4Hb-LiTaS$_2$ may be regarded as a superposition of the
spectra of 1T-LiTaS$_2$ (shown in figure 1) and 2H-LiTaS$_2$ (reported by Beal and
Nulsen, 1981). A comparison of these spectra is shown here in figure 4. This
indicates that the nearly-independent differently-coordinated sub-layers in
4Hb-TaS$_2$ remain intact even after intercalation.

REFERENCES

Beal, A.R., 1978, J. Phys. C: Sol. St. Phys., 11:4583.
Beal, A.R. and Nulsen, S., 1981, Phil. Mag. B, 43:985.
Clark, W.B. and Williams, P.M., 1977, Phil. Mag., 35:883.
Liang, W.Y., 1986, see this volume.
Scruby, C.B., Williams, P.M. and Parry, G.S., 1975, Phil. Mag., 31:255.
Tatlock, G.J., 1976, Commun. Phys., 1:87.
Wilson, J.A., Di Salvo, F.J. and Mahajan, S., 1975, Adv. Phys., 24:117.

Reflectivity of $Cr_{1/3}TaS_2$ and its Kramers-Kronig Analysis

T. H. Shen, W.Y. Liang and A.D. Yoffe

Cavendish Laboratory
University of Cambridge
U.K.

As a member of the layered transition metal dichalcogenides (TMD'S), TaS_2 has attracted considerable research interest because of its unusual physical properties such as charge density waves and other quasi-two-dimensional behavour. Rich in polytypism, it is also very interesting following intercalation. Varieties of chemical species and substances have been intercalated (see e.g. SUBBA RAO et al [1]). Apart from the potential industrial application (see e.g. WHITTINGHAM [2]; JOHNSON et al [3]) employing some of the intercalation complexes such as lithium titanium disulphide, there is also the need to understand the various mechanisms involved in intercalation [4].

In this paper, we examine the optical properties of $Cr_{1/3}TaS_2$. The intercalate complexes are prepared from an initial charge by the iodine vapour transport method. X-ray results show that the complex has a $\sqrt{3}a_0$ by $\sqrt{3}a_0$ superlattice with a_0=3.31 Å [5]. This is taken to be an indication that the crystals are stoichiometric[6].

The reflectivity of the $Cr_{1/3}TaS_2$ was measured over the photon energy range from 1.7 eV to 14 eV. A specially designed and 'home made' near normal incidence visible-vacuum u.v. spectrometer allowed us to perform the experiment with satisfactory accuracy. The result is presented in Fig.1. Although the reflectivity spectrum below 6 eV has previously been studied by PARKIN and BEAL [7] and re-examined by BAYLISS [8], to the best of our knowledge, this is the first time the spectrum between 6eV to 14eV has been presented.

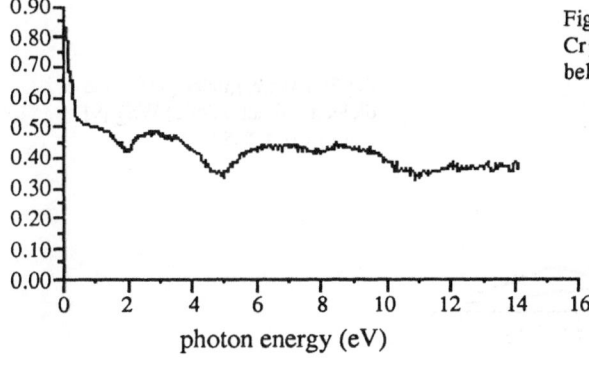

Fig 1: Reflectivity spectrum of $Cr_{1/3}$ TaS_2 crystal. The data below 1.7 eV is taken from [7]

A wide range of reflectivity data combining the infra red spectra [7] [8] and the vacuum u.v. has enabled us to perform Kramers-Kronig analysis and to deduce the absorption spectrum and dielectric functions for $Cr_{1/3}TaS_2$. The high energy tail is extrapolated with reflectivity propotional to (energy)$^{-4}$.

The calculated absorption coefficient is plotted in Fig.2. It features a broad band of absorption starting from about 2 eV until 4.5 eV with a maximum near 3eV followed by further absorption at higher energies. In Fig.2 we also plot the spectrum of its isoelectronic compound, $3R-WS_2$ [9]. The imaginary part of the dielectric function of both are plotted in Fig.3. The similarity suggests that the overall band schemes for these two materials are the same. Two major interband transitions can be easily identified at about 2.5 and 5.5eV from the imaginary part of the dielectruc function. Apart from the free carrier feature which is not suppressed completely upon intercalation, the curve for $Cr_{1/3}TaS_2$ shows little fine structure compared with $3R-WS_2$. and the corresponding peaks seem to be shifted to a slightly lower energy. The difference arising at higher energy could be due to two reasons. Firstly our extrapolation starts at 14eV , whilst that of WS_2 starts at 30 eV. The lack of information beyond 14eV will inevitably introduce errors especially at higher energy. Secondly, the intercalate could possibly modify the higher bands if some of its states have the same symmetry as the states of the host compound, causing hybridization in the conduction bands.

To summarise, we have presented the reflectivity spectrum of $Cr_{1/3}TaS_2$ over the range 1.7 to 14 eV. Combining these results with published data below 1.7 eV, a Kramers-Kronig analysis was performed. The deduced imaginary part of the dielectric function and the absorption coefficient were compared with an isoelectronic system. We believe that further systematic studies of the intercalated TaS_2 system, of its structure and its electronic states by methods such as Raman, photo-emission, optical and transport will provide useful

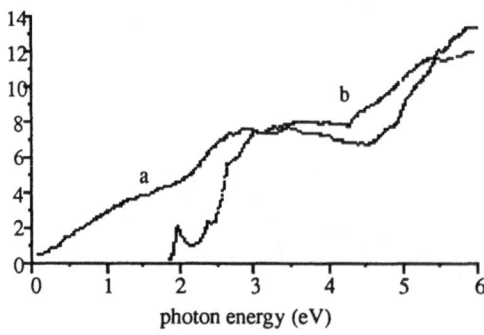

Fig 2: Absorption coefficient of a) $Cr_{1/3}TaS_2$ and b) WS_2 [9] in the unit of 10^5 cm^{-1}

photon energy (eV)

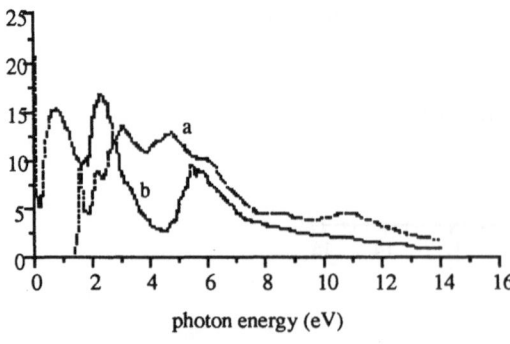

Fig 3: The imaginary part of the dielectric function of a) WS_2 [9] and b) $Cr_{1/3}TaS_2$

photon energy (eV)

information. *Ab initio* band structure calculations for the cases such as that of 1T- and 2H-LiTaS$_2$ will also help to clarify many of the problems involved in the intercalation proccess.

ACKNOWLEDGMENTS

We would like to thank Mrs. S. Nulson for growing the crystals. One of us (T.H.S) would also like to thank the organizors for the NATO grant provided to participate in the course. Thanks are also due to Miss R F Warren and Dr C M Pereira for technical help in preparing the manuscript.

References

1. G.V. Subba Rao and M.W. Shater: in Intercalated Layered Materials, ed. by F.A. Levy, Phys. and Chem. of Materials with Layered Structure, Vol 6 (D.Reidel, Dordrecht, Boston, LOndon, 1979) p99
2. M.S. Whittingham: J. Sol. St. Chem. 29, 303 (1979)
3. W.B. Johnson and W.L. Worrell: Synthetic Metals , 4, 225 (1982)
4. A.D. Yoffe: in Phys. and Chem. of Electrons and Ions in Layered Matter , Proc. NATO Davy Institute Summer School, September,1983, Cambridge, U.K., ed by J.V. Acrivos, N.F. Mott and A.D. Yoffe (D.Reidel, Dordrecht) p437
5. S. Nulsen: private communication, unpublished
6. A. Le Blanc-Soreau, J. Rouxel, M. Gardette, O. Gorochov:Mat. Res. Bull. 11, 1061 (1976)
7. S.S.P. Parkin and A.R. Beal: Phil. Mag. B42, No5,627 (1980)
8. S. Bayliss: Ph.D thesis,Cambridge,U.K. (1980)
9. A.R. Beal, W.Y. Liang and H.P. Hughes: J. Phys. C:Solid State Phys, 9,2449 (1976)

THE LMTO-ASA METHOD AND BAND STRUCTURES OF

LAYER-STRUCTURE TRANSITION METAL CHALCOGENIDES

G. Y. Guo, W. Y. Liang and A. D. Yoffe

Cavendish Laboratory
Madingley Road
Cambridge CB3 OHE, U. K.

In a band structure calculation, eigenvalues and eigenfunctions of the effective one-electron wave equation can not, in practice, be determined exactly and a number of approximate methods have been proposed. Band structure methods used in the early studies up to 1979, of layer compounds have been reviewed by Fong [1]. In the case of compounds there is a greater need to carry out selfconsistent (SC) calculations since there would be certain charge transfers between the constituent atoms. With the traditional methods described in [1], SC calculations could be a very formidable task. Recently, a number of more efficient linear band methods such as the linear augmented plane wave (LAPW) and the linear muffin-tin orbital (LMTO) methods have been developed. In particular, the LMTO method with atomic sphere approximation (ASA) [2,3] is probably the fastest method available at present, and gives satistfactory results, particularly for close-packed structure materials. This method uses a set of atomic orbital-like muffin-tin orbitals, which are energy-indepedent, as bases, and replaces the various integrals over a primitive cell by those over a set of atomic spheres. This is computationally fast since the eigen equation is linear in energy and its dimension is fairly small. The LMTO-ASA method has been successfully used to calculate the SC band structures and ground state properties of a number of close-packed elemental metals and compounds (see [3] chapter 1, and references therein). However, as the ASA approximation is designed for close-packed systems, this method may give rise to inaccuracy for systems with open-structures such as layer compounds. Nevertheless, with suitable modifications the method can still be applied to open structure crystals while exploiting its computational speed. One such modification is the introduction of so-called 'empty spheres' into the interstitial sites. This then gives band structures of open structure crystals, such as diamond and Si, with accuracy comparable to those obtained by the computationlly more demanding methods with less stringent assumptions about the form of the potential [4]. Temmerman and coworkers [5] first used the LMTO-ASA method to calculate the SC band structures of TiS_2 and $TiSe_2$. More recently, we carried out sevaral SC LMTO-ASA band structure calculations, based on local density functional theory, of layer-structure transition metal chalcogenides such as $PtSe_2$, $PdTe_2$ and Hf_2S. In our LMTO-ASA calculations, we introduced artificial 'empty-spheres' to interstitial octahedral sites within the so-called 'van der Waals' gap. Treating these empty spheres as 'real' atoms with zero nuclear charge, the resultant structure looks like that of a close-packed ternary compound. Obviously, this method can be easily extended to calculate the band structures of the intercalate complexes with simple structure such as $LiTiS_2$, $LiNbS_2$ and $LiTaS_2$.

Our calculations show that in layered Ni-group metal dichalcogenides there is a strong metal d-orbital and p-orbital mixing forming highly p/d hybridised valence

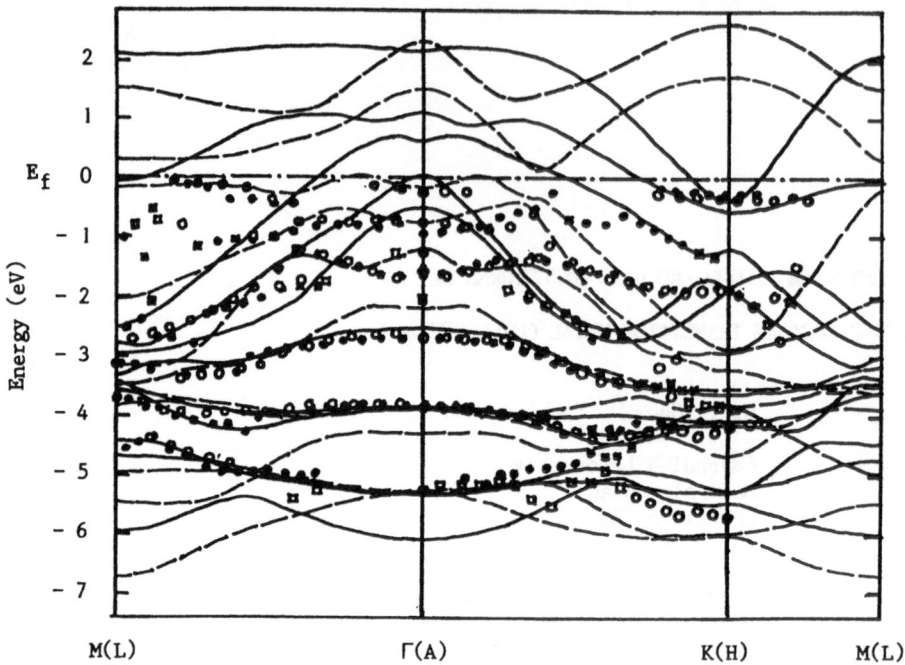

Fig. 1 Relativistic SC energy bands for PdTe$_2$: solid curves for wavevector $\underline{k}$ in the ΓMK plane (i.e. k_z=0), broken curves for wavevector $\underline{k}$ in the LAH plane (i.e. k_z=π/c) (refer to Liang in this volume for Brillouin zone notations). Experimental data derived from ARUPS [8] as follows: full and open circles represent data from He I and Ne I spectra respectively, and full and open squares similarly represent data from weak features (shoulders)

and conduction bands. This is in contrast to the band structures of group IVB, VB and VIB transition metal dichalcogenides in which the conduction and valence bands are based mainly on metal d and chalcogen p orbitals respectively. We found that there is a shift from semiconductor to metal in the progression from PtS$_2$ through PtSe$_2$ to PtTe$_2$. These have been explained qualitatively in terms of atomic orbital binding energies and local coordination of the constituent atoms [6,7]. Fig. 1 shows the relativistic SC band structure of PdTe$_2$, together with experimental points derived from photoemission measurements [8]. It is clear that there is a substantial agreement on both the position and the dispersion of energy bands obtained theoretically and experimentally. Finally, SC calculations on the dimetal chalcogenides Hf$_2$S and Hf$_2$Se (which has NbS$_2$ structure and detailed calculations will be reported elsewhere; also see Liang in this volume) predict that both are metals with Fermi levels cutting many Hf d-dominant bands and, as in HfS$_2$, the Hf d-dominant bands lie above the chalcogen p-dominant bands.

REFERENCES

1. Fong C Y (1979) <u>Electrons</u> and <u>Phonons</u> in <u>Layered Crystal Structures</u> ed. by Wieting T J and Schulter M (Dordercht: Reidel)
2. Andersen O K (1975) Phys. Rev. B12 3060
3. Skriver H L 1984 <u>The LMTO method</u> (Berlin: Springer)
4. Glotzel D, Segall B and Andersen O K 1980 Solid State Commun. <u>36</u> 403
5. Temmerman W M, Glotzel D and Andersen O K 1983 (unpublished)
6. Guo G Y and Liang W Y 1986 J. Phys. C: Solid State Phys. in press
7. Guo G Y and Liang W Y 1986 J. Phys. C: Solid State Phys. <u>19</u> 995
8. Orders P J, Liesegang J, Jeckey R C G, Jenkin J G and Riley J D 1982 J. Phys. F: Metal Phys. <u>12</u> 2737

ALUMINO-SILICATE CLAYS AND CLAY INTERCALATION COMPOUNDS

S.A. Solin

Department of Physics and Astromomy
Michigan State University
East Lansing, MI   48824-1116

INTRODUCTION

Background

Intercalation compounds formed from graphitic host materials and from layer dichalcogenides such as TaS_2 and $HfSe_2$ are quite familiar to most solid-state physicists and chemists who have extensively studied the properties of such materials over the past several decades [1,2]. There is, however, another general class of host materials which can form a wide range of intercalation compounds yet has until very recently received very little attention from the solid-state community, although this class is heavily studied by inorganic chemists, soil scientists, geologists, and minerologists. This "new" class of materials is that which includes alumino-silicate clays (often called sheet silicate clays) [3-5]. Although the term "clay" is often used to refer to minerals, the morphology of which is small particles of a typical size less than 1μm, I will use the term in this paper as synonymous with layered alumino-silicates. In this chapter, I will clarify the relationship between layered alumino-silicates and the more familiar layered solids and, in addition, highlight the key features of the former which make them interesting systems for the exploration of unusual physical phenomena.

Classification

It is useful to classify layered solids into three general groups according to the rigidity of their layers with respect to distortions involving displacements transverse to the layer planes [5]. Such a classification scheme is shown in Fig. 1a which depicts schematically the three general classes.

Class I contains only two compounds, graphite and boron nitride, which are the only layered solids that are composed of atomically thin sheets of atoms. As a result of this unusual structure, the individual layers of Class I solids are "floppy" with respect to the above described transverse distortions i.e. it is not energetically costly to transversally distort the layers. As examples of Class II materials, we cite the layer dichalcogenides as well as FeOCl and other compounds [1]. The layers in these Class II materials are often composed of three distinct planes of strongly bonded atoms and present a stiffer structure vis-á-vis transverse distortions than do the Class I solids. The third class of interest is that defined by the silicate clays, the layers of which are thick assemblies of as many as seven planes of strongly bonded atoms. It can be intuitively surmised that structures such as those of Class III compounds will be quite "rigid" against transverse layer distortions.

Each of the materials discussed above is considered a "layered compound" because the intralayer forces binding the atoms together are much stronger than the interlayer forces. As a result, foreign species can be readily inserted or intercalated into the galleries between the host layers which remain essentially undistorted. But each class of layered solid exhibits novel intercalation properties which are illustrated in Fig. 1b. The graphite intercalation compounds (GIC's) of Class I are famous for their ability to form stages [6,7] in which monolayers of guest intercalant are separated by n multilayers of host to form a stage-n compound as shown in Fig. 1b. In contrast, the Class II materials usually accept foreign species into random gallery sites until, at sufficient concentration, a saturated stage-1 compound is formed. The Class III silicate clays have

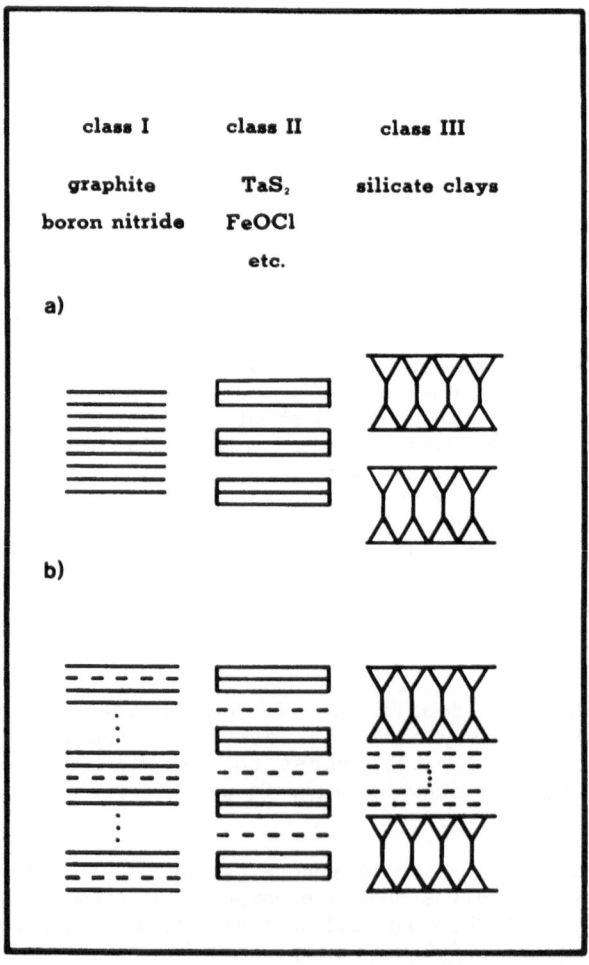

Figure 1. Schematic classifications of a) pristine layered solids and b) the intercalated forms of a). In b) the dashed (solid) lines represent guest (host) layers.

the unusual ability to form clay intercalation compounds (CIC's) in which monolayers of host are separated by multilayers of guest in a stacking sequence that is always stage-1 [3]. (Note that while Class I and II solids sometimes accommodate bi-layers or even tri-layers of intercalant, some CIC's readily accommodate 20 or more intercalant layers in their galleries [3,4].)

STRUCTURE OF THE CLAY HOST

Though the host structures of the Class I and Class II materials are familiar to this readership, the structures of Class III hosts are probably foreign and will thus be illucidated here. Figure 2 shows the structure of a typical and chemically relevant CIC known as montmorellonite [4].

The 2:1 alumino-silicate layer can be seen to be composed of two sheets of corner-connected SiO_4 tetrahedra bound at an interfacial oxygen/hydroxal plane to an octohedral sheet of AlO_6 edge-connected octohedra. Note that the free tetrahedral and octohedral sheets, both of

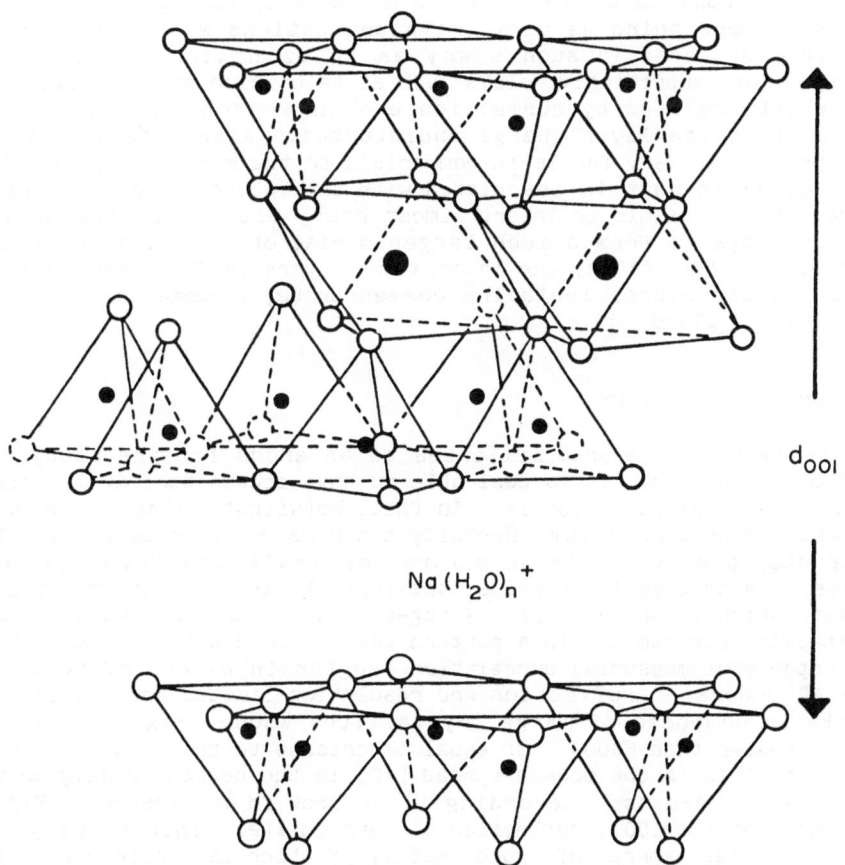

$$d_{001}$$

$$Na(H_2O)_n^+$$

Figure 2. Schematic illustration of the tetrahedral and octahedral sites in a 2:1 layered silicate. Open circles are oxygen, closed circles are cations in tetrahedral (Si, Al) and octahedral (Al, Fe, Mg, Li) positions. Hydroxyl groups (not distinguished from oxygen in the figure) are located in the second and third basal planes of oxygens.

which exist in nature, have slightly different in-plane lattice constants for their respective interfacial oxygen planes. This is a very important point which will enter into later discussion.

A key feature which distinguishes the clay host from a graphite host is that, unlike graphite which is amphoteric and can accommodate either donor or acceptor species [8] in its galleries, clay layers are either neutral (as in the mineral talc) or more commonly carry a fixed negative charge. This negative charge state is determined by the number of hydroxal units (solid circles in Fig. 2) which replace oxygens (open circles in Fig. 2) and by the substitution of other elements or vacancies at the Si tetrahedral and/or Al octohedral sites. It is the multitude of substitution/vacancy possibilities which generates the large, many-membered class of 2:1 clays since each set of substitutions/vacancies defines a distinct mineral [3,4].

In order to compensate for the negative layer charge and to guarantee overall charge neutrality, the galleries of a typical 2:1 clay contain cations (M^+ in Fig. 2) which may or may not carry waters of hydration. Note that the intercalation process in CIC's is really a process of ion exchange. Thus, a given number of say M^+ cations is exchanged with an equal number of other monovalent cations M'^+ or with half as many divalent cations M'^{++}, one third as many trivalent cations M'^{+++}, etc., or with combinations of these in such a way as to maintain overall charge neutrality. Not surprisingly, clays such as talc which have neutral layers cannot be intercalated by normal ion exchange processes since such processes leave the layer charge undisturbed. Also note that the ion exchange process is very facile in comparison to the more disruptive charge exhange mechanism that is associated with the intercalation of graphite. Therefore, it is possible to insert almost every element of the periodic table into clays to form a much larger myriad of CIC's than is possible with GIC's. But like GIC's, the intercalant layers in CIC's may acquire a number of 2D structures including commensurate, incommensurate, lattice gas, liquid, and glass.

DIFFRACTION AND MORPHOLOGY

If we are to use layered silicates as an arena for the study of 2D physical phenomena, then it is desirable to have access to clay material in the form of large single crystals. In fact, notwithstanding the negative images which the word "clay" normally conjures up, excellent crystalline forms of clay are available as natural minerals and from synthetic processes. As an example, consider the (00ℓ) X-ray diffraction pattern of vermiculite which is shown in Fig. 3 together with and a schematic diagram of its stacking structure. This pattern was acquired with a diffractometer that is capable of measuring correlation lengths in excess of 5000Å [9]. Yet the 14 orders of reflection are resolution-limited and indicate that the sample is composed of large crystallites whose c-axis correlation length is greater than 5000Å. Of equal importance to the study of physical phenomena in CIC's is the mosaic spread [10] in the degree of alignment of the crystallite c-axes. Accordingly, we show in the insert of Fig. 3 a rocking curve of the (007) reflection of vermiculite. This rocking curve indicates a mosaic spread of approximately 5° which is considerably higher than that of highly oriented pyrolytic graphite (HOPG) [11], but comparable to that of graphoil [12], a material with which much excellent solid-state physics has been carried out [13]. The 5° mosaic spread of vermiculite can also be achieved with self-supporting films that are prepared by the sedimentation from solution of small (~1μm) vermiculite particles onto a suitable (glass) substrate [14]. The resultant films are the morphological analog of HOPG.

LAYER RIGIDITY AND ITS CONSEQUENCES

The concept of layer rigidity was introduced in section 1. We now apply this concept to illucidate the range of physical behavior exhibited by intercalated layered solids which span the classes illustrated in Fig. 1.

Consider the mixed layered solids of the type $M_{1-x}M'_xL$, $0 \leq x \leq 1$, where M and M' represent cations and L represents the host layers and, for the purposes of this discussion, takes only the symbols V = vermiculite or C = graphite. Solids of this type are often referred to as ternary intercalation compounds [15]. Of initial particular interest here are the GIC's and CIC's with respective compositions $Cs_{1-x}Rb_xC_8$ and $Cs_{1-x}Rb_xV$ which can be synthesized over the full composition range by methods described in detail elsewhere [14]. If we adopt the simplified model in which the graphite layers are infinitely floppy with respect to transverse distortions while the vermiculite layers are infinitely rigid, as illustrated in Fig. 4, then the composition dependence, d(x) vs. x, of the X-ray-derived basal plane spacing of the layers would have the functional forms shown in that figure. The non-Vegard law [16] (i.e. non-linear) form of d(x) for graphite can be deduced from computer simulations [17] and has been measured experimentally [18],[14] (see discussion below). It is a consequence of the ability of the carbon layers to wrap or pucker around

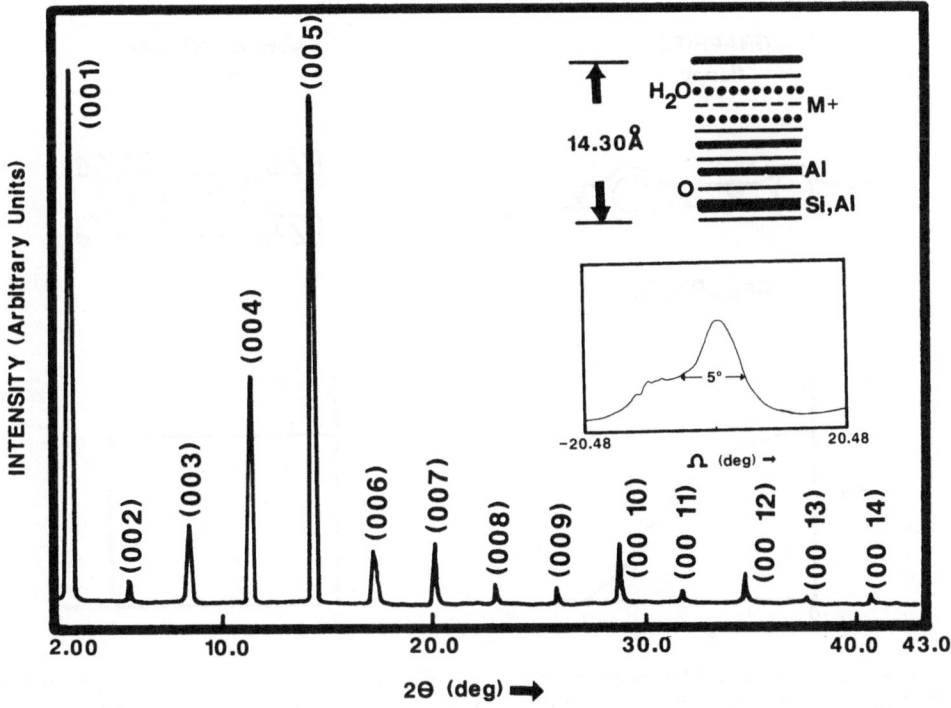

Figure 3. Room temperature (00ℓ) X-ray diffraction pattern of vermiculite recorded using Mo Kα radiation. The chemical formula of vermiculite is $[(Mg,Al,Fe)_3(OH)_2(Si_3Al)O_{10}Mg^{2+}(H_2O)_4]$. Inset: Rocking-curve of the (007) reflection. The diagram at the upper right of the figure represents the c-axis structure.

the large ion in the gallery [19]. In contrast, the step-function-like behavior which would be predicted for d(x) in the vermiculite case, reflects the assumption that as few as three noncolinear Cs ions could fully prop the clay layers apart if they were infinitely rigid.

In order to compare the composition-dependent basal spacings of different layered solids, we have defined a normalized basal spacing $d_N(x)$ given by

$$d_N(x) = \frac{d_{obs}(x) - d_{min}}{d_{max} - d_{min}}$$ (1)

where $d_{obs}(x)$ is the measured basal spacing of $M_{1-x}M'_xL$ and, if we take M as the larger ion, $d_{min} = d_{obs}(1)$ while $d_{max} = d_{obs}(0)$. Note that the function $d_N(x)$ takes on the values 1 and 0 at x = 0 and x = 1, respectively.

In Fig. 5 we show plots of $d_N(x)$ vs. x for the Cs-Rb-C [18] and Cs-Rb-V [14] ternary intercalation systems. As can be seen from that figure, a behavior opposite to that which was predicted on the basis of our simple model is observed and the basal spacing of $Cs_{1-x}Rb_xV$ drops precipitously at a lower value of x than does the basal spacing of $Cs_{1-x}Rb_xC$. The data of

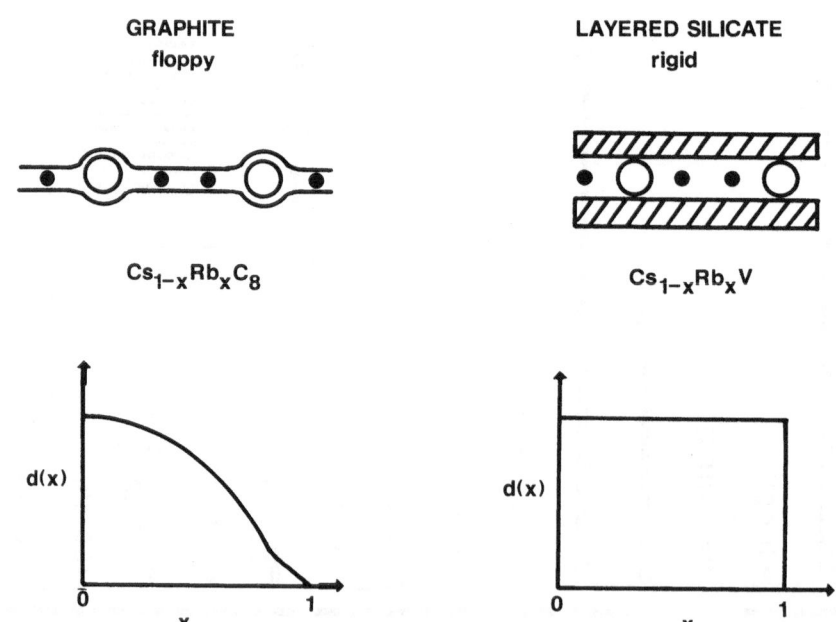

GRAPHITE
floppy

$Cs_{1-x}Rb_xC_8$

LAYERED SILICATE
rigid

$Cs_{1-x}Rb_xV$

Figure 4. A schematic representation of the transverse layer distortions at the large ion in the mixed layer intercalation compounds Cs-Rb-graphite, the layers of which are "floppy" and Cs-Rb-vermiculite, the layers of which are "rigid". Also shown are the composition dependences of the corresponding basal spacings, d(x) vs. x, deduced from the simple model discussed in the text.

Fig. 5 were obtained from (00ℓ) X-ray diffraction studies of the Cs-Rb-C GIC's [18] and Cs-Rb-V CIC's [14].

The anomalous behavior exhibited in Fig. 5 can be readily understood as follows. It is well known that the lattice mismatch between the tetrahedral and octahedral sheets in vermiculite generates an interfacial strain which is relieved in the binary CIC form, e.g. M-V, by alternate clockwise/counterclockwise rotations of adjacent SiO_4 tetrahedra about an

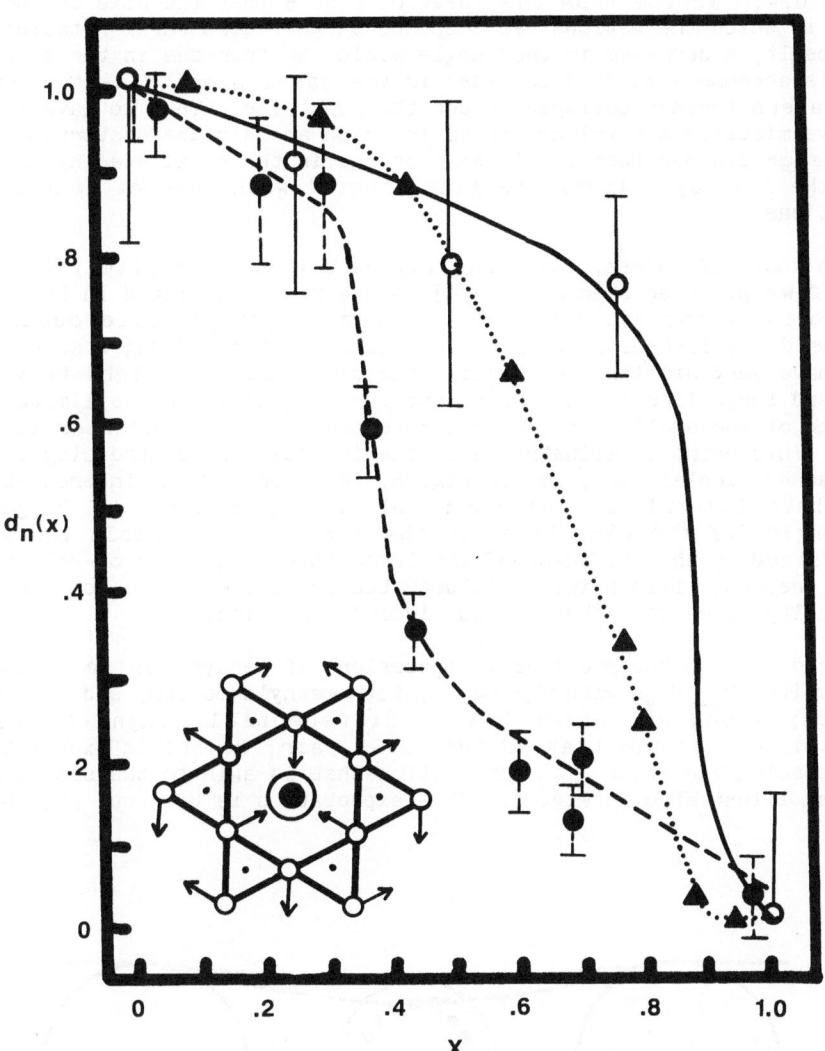

Figure 5. The composition dependence of the normalized basal spacing of $Cs_{1-x}Rb_xC_8$ (solid line), $Cs_{1-x}Rb_x$-vermiculite (dashed line), and $[Me_4N^+]_{1-x}[Me_3NH^+]_x$-vermiculite (solid line). Typical error bars are indicated in the figure. Inset: c-axis view of the tetrahedral rotations discussed in the text. The symbols have the following meanings: ◯ oxygen, • Si; large cation: ◯, small cation: ●.

Si-O axis which is parallel to the crystal c-axis [3]. This coupled rotational in-plane (longitudinal) distortion is shown in the inset of Fig. 5 and is very similar to the soft mode rotations of the TiO_6 octahedra near the ferroelectric phase transition in $BaTiO_3$ [20]. Moreover, the angle of the tetrahedral rotation depends on the size of the ion M which is encased by the hexagonal oxygen pockets of the basal surfaces of the bounding clay layers.

When two species M and M' occupy the gallery, the tetrahedral rotation angle becomes localized to that corresponding to the species at a particular site [14]. But the coupling between tetrahedra causes a free energy lowering switch from the Cs rotation angle to the Rb rotation angle at x ~ 0.4. Notice from the inset of Fig. 5 that the size of the oxygen pocket in which the cations rest depends on the tetrahedral rotation angle. As a result, a decrease in that angle yields an increase in the pocket size which is accommodated by a decrease in the basal spacing as the bounding clay layers further collapse around the guest ion. Thus we have in mixed-layer vermiculite a novel mechanism in which an in-plane distortion of the clay layer can manifest itself as a change in the basal spacing of the CIC while the 2:1 layer itself remains "rigid" with respect to transverse distortions.

We know, of course, that clay layers are not infinitely rigid, and that if we pillared them apart [21] in the manner depicted in Fig. 6, some electrostatic and van der Waals induced "sagging" would occur as we increased the lateral spacing of the pillars. For ion pairs such as Cs-Rb which have very similar ionic radii, the differential height between the small and large ions is only 0.65Å and the sagging effect is limited by the presence of the smaller ion according to the rotation mechanism discussed above. This point is illustrated by the intersection of the clay layer and the (dashed circle) small ion in Fig. 6. However, if the intercalating ion pairs have individual radii which yield a large differential height, then the sagging of the clay layer in the region of the small ion will be constrained by the stiffness of the layer itself, whether or not the small ion is present. This point is illustrated in Fig. 6 by the nonintersection of the clay layer with the (dotted circle) small ion.

According to the preceeding discussion, if we synthesize mixed layer vermiculites $M_{1-x}M'_xV$ with $M = Me_4N^+$, tetramethylammonium, and $M' = Me_3NH^+$, trimethylammonium, which have a differential height of 1.0Å or approximately double that of the Cs-Rb pair, we should expect to see a basal spacing variation with composition that is akin to the step-function behavior illustrated in Fig. 4. This expectation is born out [22] by the

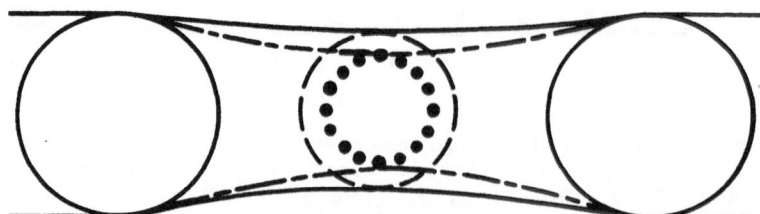

Figure 6. A schematic representation of the "sagging" of clay layers at the small ion (dashed and dotted circles) when the layers are pillared apart by laterally spaced large ions (continuous circles).

plot of $d_N(x)$ vs. x for $(Me_4N^+)_{1-x}(Me_3NH^+)_x-V$, $0 \leq x \leq 1$ which is shown as a solid line in Fig. 5.

CONCLUDING REMARKS

It is useful to summarize the distinctions between intercalation compounds based on graphite and clays, two materials which represent the extremes in the classification of layered solids. Graphite is rigid against longitudinal in-plane distortions and in the a-axis direction it is even "harder" than diamond (has higher phonon frequencies [23,24]). But it is very soft with respect to transverse distortions for which the layers are "floppy". In contrast, vermiculite and other clays are very soft with respect to longitudinal in-plane rotational distortions but are relatively rigid with respect to transverse distortions. It is thus intriguing that mixed layer CIC's can readily adjust their basal spacings with composition variations while the host layers remain transversely rigid.

ACKNOWLEDGEMENTS

I am grateful to several of my colleagues who collaborated with me in some of the research reported here and who have enlightened me on the subject of this paper. Thus, it as a pleasure to acknowledge T.J. Pinnavaia, N. Wada, I.D. Johnson, R. Raythatha, S. Lee, and W. Leonard. This work was supported in part by the U.S. National Science Foundation under grant DMR-85-17223.

REFERENCES

1. F. Levy (ed.): Physics and Chemistry of Materials with Layered Structures (Reidel, Dordrecht, 1979)
2. K. Nakao and S.A. Solin (eds.): Graphite Intercalation Compounds (Elsevier, Lausanne, 1985)
3. G.W. Brindly and G. Brown (eds): Crystal Structure of Clay Minerals and Their X-Ray Identification (Minerological Society, London, 1980)
4. R.E. Grimm: Clay Minerology (McGraw-Hill, New York, 1968)
5. S.A. Solin: J. Molec. Catalysis 27, 293 (1984)
6. S.A. Solin: Adv. Chem. Phys. 49, 455 (1982)
7. M.S. Dresselhaus and G. Dresselhaus: Adv. Phys. 30, 139 (1981)
8. J.E. Fisher: In Physics and Chemistry of Materials with Layered Structures, Vol. 6, ed. by F. Levy (Reidel, Dordrecht, 1979) p. 481
9. B.R. York, S.K. Hark, and S.A. Solin: Phys. Rev. Lett. 50, 357 (1983)
10. B.E. Warren: X-Ray Diffraction (Addison-Wesley, Reading, Mass., 1969)
11. A.W. Moore: In Chemistry and Physics of Carbon, Vol. II, ed. by P.L. Walker Jr. and P.A. Thrower (Marcel Dekker, New York, 1973) p. 69
12. W.N. Reynolds: Physical Properties of Graphite (Elsevier, Amsterdam, 1968)
13. S.K. Sinha (ed): Ordering in Two Dimensions (North-Holland, New York, 1980)
14. B.R. York, S.A. Solin, N. Wada, R. Raythatha, I.D. Johnson, and T.J. Pinnavaia: Solid State Comm. 54, 475 (1985)
15. S.A. Solin: this volume, p. 291.
16. R.M. Hazen and L.W. Finger: Comparative Crystal Chemistry (Wiley, New York, 1982)
17. S.D. Mahanti: Unpublished
18. D. Billaud and A. Hérold: Bull. Soc. Chim. Fr. 11, 2402 (1974)
19. N. Caswell, S.A. Solin, T.M. Hayes, and S.J. Hunter: Physica 99B, 463 (1980)

20. A.S. Chaves, R.S. Katiyar, and S.P.S. Porto: Phys. Rev. B10, 3522 (1974)
21. T.J. Pinnavaia: Science 220, 365 (1983)
22. S. Lee, W. Leonard, T.J. Pinnavaia, and S.A. Solin: To be published
23. S.A. Solin and A.K. Ramdas: Phys. Rev. B4, 1687 (1970)
24. S.A. Solin, R.J. Nemanich, and G. Lucovsky: Mat. Sci. and Eng. 31, 157 (1977)

ALKALI METAL INTERCALATION-DEINTERCALATION REACTIONS IN 2D OXIDES

Claude Delmas

Laboratoire de Chimie du Solide du CNRS
Université de Bordeaux I
351, cours de la Libération - 33405 Talence Cedex (France)

While much work has been devoted to intercalation reactions in graphite, transition metal dichalcogenides (TMDC) and clays, only a few studies have been reported on A_xMO_2 layer oxides (A : alkali ion, M : transition element x < 1)[1,2,3]. Stoichiometric AMO_2 oxides have been mainly studied from a crystal chemistry point of view[4,5]. Although A_xMO_2 layer oxides show close likeness with intercalated TMDC they cannot exist in the unintercalated form. This property comes from the large electronegativity difference between the oxygen and the M element.

In TMDC the lower electronegativity of the chalcogens (x_S = 2.5, x_{Se} = 2.4) in comparison with the oxygen one (x_O = 3.5) leads to high covalent character of the M-S bond. This gives rise to a small negative charge on the adjacent chalcogen layers and so that the Van der Waals bonding can stabilize the 2D structure.

In oxides the M-O bonds are more ionic, then the electrostatic repulsions between highly negative charged oxygen layers are very strong and prevent the existence of unintercalated layer oxides. Nevertheless if alkali ions are intercalated between the adjacent oxygen layers, strong ionic bonds appear and stabilize the A_xMO_2 layer oxides. In these materials x varies in the $x_\ell \leqslant x \leqslant 1$ range. The minimum intercalation rate x_ℓ is related to the nature of A and M elements but is always over 0.5[3].

CRYSTAL CHEMISTRY

The structure of these materials is built up of MO_2 sheets, made of edge-sharing MO_6 octahedra, between which alkali ions are inserted. As shown in Fig. 1[6] three packing types can be observed. The structures are symbolized by a letter which refers to the alkali ion environment, octahedral (O) or trigonal prismatic (P), followed by a figure indicating the number of $(MO_2)_n$ sheets within the unit cell[6].

Using an ionicity-structure diagram, like that proposed by J. ROUXEL for TMDC[7], it was possible to separate phases with octahedral and trigonal prismatic alkali ion environment[8].

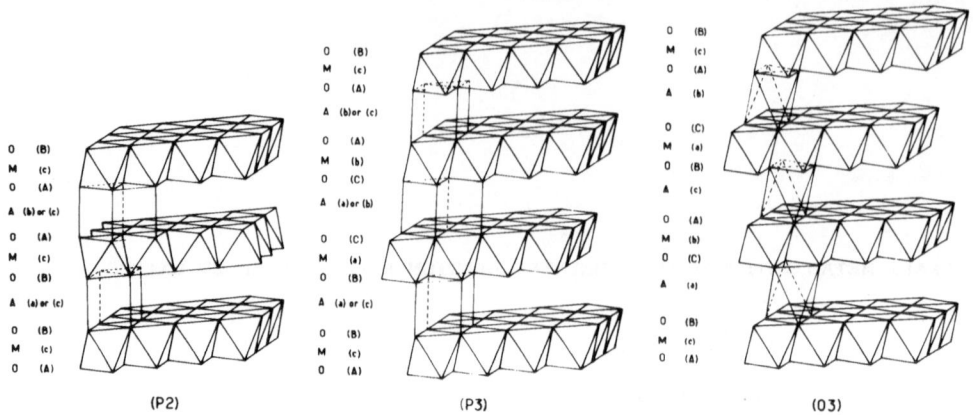

(P2)   (P3)   (03)

FIG. 1 - Various oxygen packing types in A_xMO_2 layer oxides.

INTERCALATION - DEINTERCALATION

When alkali metals are intercalated in a host structure, the electrons coming from the alkali metal ionization : $A \rightarrow A^+ + e^-$ must be trapped by the host. These latter can be injected in the conduction band or they can reduce the oxidation state of one of the atomic species constituting the host (if the electrons are localized in the material). Alkali metal intercalation-deintercalation reactions can be observed in transition metal layer oxides if there is no kinetic limitation due to low ionic or electronic conductivity. These reactions (soft chemistry) have been carried out either electrochemically or chemically by using oxidizing or reducing reagents. For the electrochemical reactions, the following electrochemical chain is used : $A/A^+ClO_4^-$ in propylene carbonate/A_xMO_2. Fig. 2 shows the voltage variation vs composition curves obtained from $NaMO_2$ (M = Ti, Cr, Co, Ni) phases

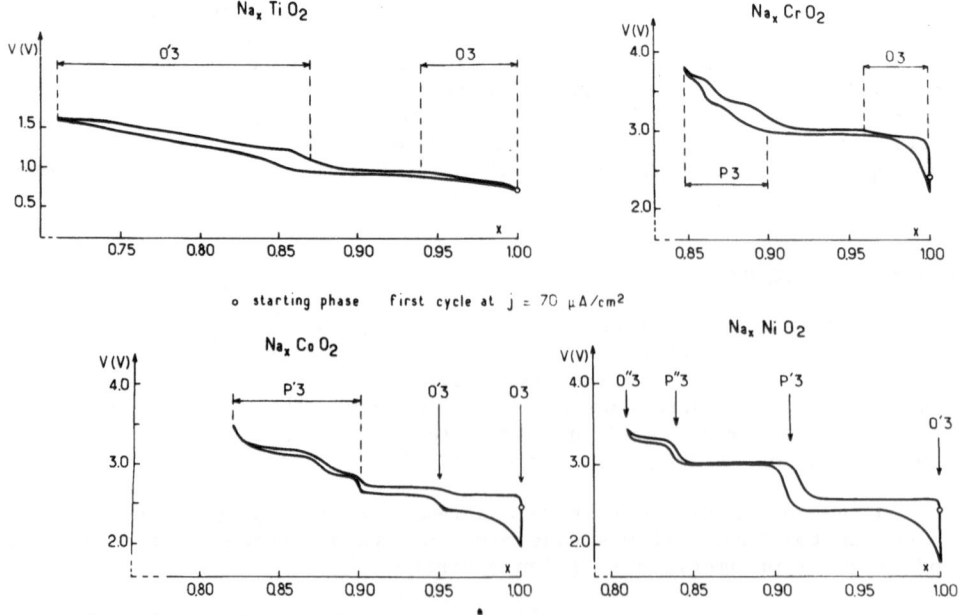

FIG. 2 - Electrochemical behavior of $NaMO_2$ phases (M = Ti, Cr, Co, Ni).

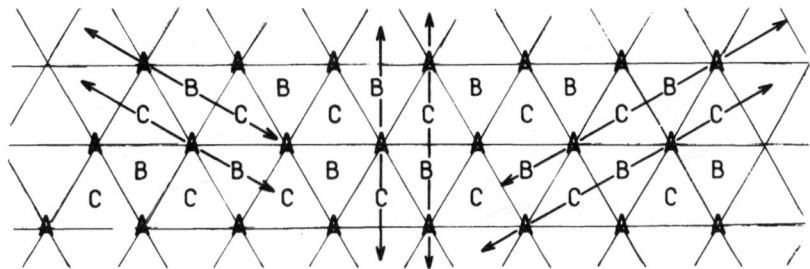

FIG. 3 - Triangular lattice with ABC notation and gliding directions.

Deintercalation reactions from $NaMO_2$ allow to obtain new Na_xMO_2 materials. According to x values, solid solutions (continuous variation of voltage vs composition) or defined compounds separated from one another by biphased domains (constant voltage vs global composition) can be observed. The voltage values (vs Na/Na^+) clearly illustrate the oxidizing character of tetravalent chromium, cobalt or nickel on one hand and the reducing character of trivalent titanium on the other hand.

Reversible structural transitions between phases with octahedral (O) or trigonal prismatic (P) alkali ion environment can be observed. As shown in figure 1, the structural difference between O3 and P3 phases consists of a gliding of the second and third $(MO_2)_n$ sheets within the unit cell. As in such a gliding process the M-O bonds are not broken, this transition is already possible at room temperature and is furthermore completely reversible.

Figure 3 shows a triangular lattice with ABC oxygen positions. Three equivalent gliding directions are possible. For each of them the new position can be deduced from the starting one by circular permutation of A, B, C.

This behavior is illustrated in Fig. 4 by the structural transformations between the P3 and O3 structural types. Such transformations occur for instance in the Na_xTiS_2 system[10].

It should be noticed that in intercalated layer oxides or in TMDC, each slab contains three successive layers : oxygen, M, oxygen or sulphur, M, sulphur. As the integrity of the slab is maintained during its gliding, all packings cannot be obtained by sheet gliding from a given one. It is particularly the case of P2 type phases (ABBA oxygen packing) which cannot be obtained from O3 or P3 only by sheet gliding. The structural transition needs MO_6 octahedron rotations around the hexagonal axis. As a result the M-O-M bonds should have to be broken and the reaction does not occur at room temperature. On the contrary, these restrictions do not apply to GICs as there is only one graphite layer per slab.

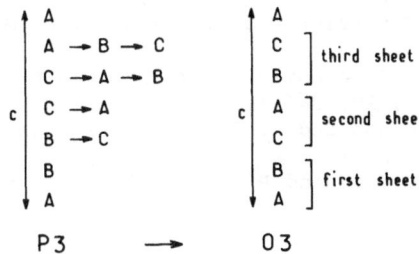

FIG. 4 - Transitions between P3 and O3 structural types.

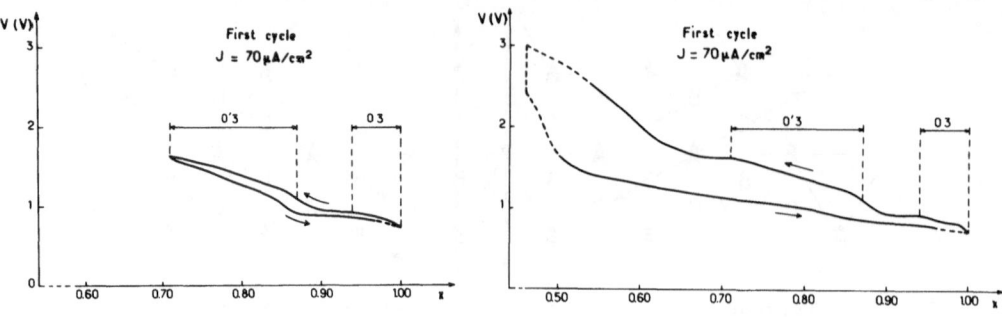

FIG. 5 - Reversible and irreversible transition in the $Na//Na_xTiO_2$ system.

Irreversible 2D → 3D Host Modifications

As shown in Fig. 5 the behavior of the Na_xTiO_2 system is strongly dependent on the deintercalation rate. A reversible charge-discharge reaction is observed for $0.71 \leqslant x \leqslant 1$. Contrarily, if the positive electrode is overcharged ($x < 0.71$), an irreversible reaction occurs. Fig. 5 shows unambiguously the shape difference between the discharge curves in such a case.

As above mentioned, intercalated layer oxides exist only for high amounts of intercalation. In the Na_xTiO_2 system, the limit is close to 0.71. When the cell is overcharged, Na^+ ions must be removed. To stabilize the structure, some titanium ions move from the TiO_2 sheet to the intersheet space.

Crystallographic formula of the charged phase can be written $(Na_xTi_y\square_{1-x-y})(Ti_{1-y}\square_y)O_2$. This material can no longer be considered as a layer compound, but rather as a tridimensional one. During reintercalation, there is no driving force which would lead to the migration of the titanium ions to their original positions. Consequently, the transition is irreversible.

REFERENCES

1. J.P. Parant, R. Olazcuaga, M. Devalette, C. Fouassier and P. Hagenmuller, J. Sol. State Chem., 3, 1 (1971).
2. C. Fouassier, G. Matejka, J.M. Réau and P. Hagenmuller J. Sol. State Chem., 6, 532 (1973).
3. C. Delmas, C. Fouassier and P. Hagenmuller, Mat. Sc. Eng., 31 297 (1977).
4. R. Hoppe, Bull. Soc. Chim. Fr., p 1115 (1965).
5. H. Brunn and R. Hoppe, Z. anorg. allg. Chem., 417, 213 (1975).
6. C. Fouassier, C. Delmas and P. Hagenmuller, Mat. Res. Bull., 10, 443 (1975).
7. J. Rouxel, J. Sol. State Chem., 17, 223 (1976).
8. C. Delmas, C. Fouassier and P. Hagenmuller, Mat. Res. Bull., 11, 1483 (1976).
9. C. Delmas, J.J. Braconnier, A. Maazaz and P. Hagenmuller Rev. Chim. Min., 19, 343 (1982).
10. J. Rouxel, M. Danot and J. Pichon, Bull. Soc. Chim. Fr., p 3390 (1971).
11. A. Maazaz, C. Delmas and P. Hagenmuller, J. of Incl. Phenomena, 1, 45 (1983).

ELECTRICAL AND OPTICAL PROPERTIES OF INTERCALATED

In-Se LAYERED MATERIALS

Christian Julien

Laboratoire de Physique des Solides, associê au CNRS
Université Pierre et Marie Curie
4, Place Jussieu, 75252 Paris Cedex 05, France

III-VI compounds such as InSe, GaSe or In_2Se_3 are layered semiconductors and are intercalated with lithium or silver by spontaneous reaction in n-butyllithium and an electrochemical process. These compounds appear to be good hosts for the insertion of Li at room temperature and Li intercalated InSe has been shown to function as a cathode in Li solid state batteries.

Electrical and optical properties of intercalated materials are investigated. The insertion reaction in n-butyllithium is characterized by the time resistivity variation of the host material. Electrical measurements on Li intercalated compounds show a change by 3-orders of magnitude of the conductivity with respect to non intercalated samples[1]. In indium selenide this change is mainly due to the increasing carrier concentration.

Electrochemical discharge of a Li/InSe cell has been cycled at a 25 hour rate, with current density 0.7 mA/cm^2; the chemical lithium diffusivity was found to be 10^{-11} cm^2/s at room temperature for lithium content x=1 and the stored energy was estimated as 220 Wh/kg. The continous decrease in EMF with increasing x is expected for a single phase system up to 0.8 Li.

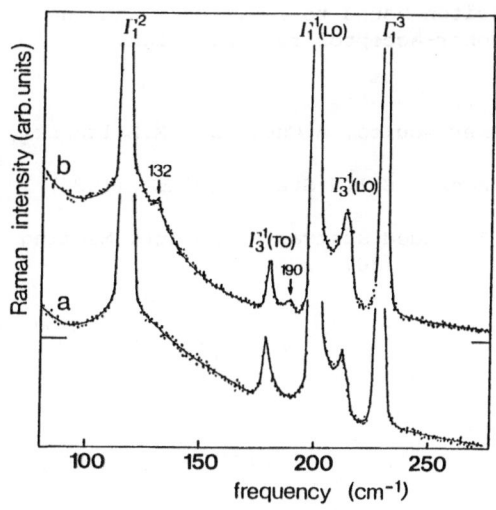

Fig.1 Raman spectra taken at 5K for (a) pure InSe and (b) Li intercalated InSe.

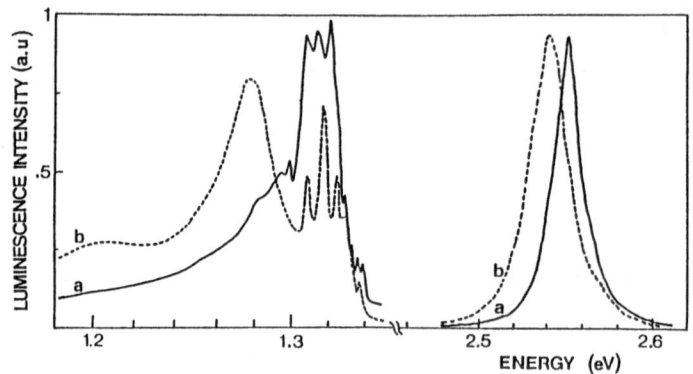

Fig.2 Luminescence intensity at 5K of the different recombination bands (a) pure InSe and (b) Li intercalated InSe

In the case of the silver intercalation, low values of the chemical
diffusivity less than $10^{-12} cm^2/s$ were measured and variations as a function
of the silver concentration have been studied in the range 0<x<0.03[2].

Optical properties have been mainly investigated by photoluminescence
(PL) and Raman spectroscopy (RS). Li intercalation affects the emission spectrum and phonon scattering spectrum[3]. In indium selenide, Li ions are intercalated on tetrahedral sites and optical measurements have been done with
in-plane van der Waals ordering at x=0.7.

The RS spectrum observed at 5K (fig.1) in the region of the first
order scattering of polar optical modes shows two new features appearing for
the intercalated samples. In the first, the RS intensity ratio of the 3D-LO
mode to the 2D-LO mode is enhanced. Secondly new weak bands appear in the
vinicity of the main peaks at 118 cm^{-1} and 182 cm^{-1}. These new peaks may be
due to the folding of the Brillouin zone resulting from the creation of a new
superlattice.

Figure 2 shows the PL spectra associated with the two first edges in
the absorption coefficient near 1.35 eV and 2.60 eV. The presence of alternating sheets of Li^+ ions induces changes in the electronic band structure
due to a complementary Coulomb interaction binding energy. Two dominant bands
appear at lower energies. The 1.278 eV line can be attributed to the phonon
assisted recombination of an indirect exciton bound to a Li-donor and the
1.206 eV broad band to phonon assisted donor-acceptor recombination.

REFERENCES

1. C.Julien, E.Hatzikraniotis, K.M.Paraskevopoulos, A.Chevy and M.Balkanski,
   Solid State Ionics, 18-19:859 (1986)
2. C.Julien, E.Hatzikraniotis and M.Balkanski, Solid State Ionics, (1986)
   in press
3. M.Balkanski, C.Julien and M.Jouanne, Extended abstracts 3rd Int. Meeting
   on Lithium Batteries (Kyoto, Japan, 1986), p.307

FREE-EXCITON PHOTOLUMINESCENCE: A PROBE FOR STUDYING THE INTERCALATION PROCESS IN THE III-VI LAYER COMPOUND InSe

K.M. Paraskevopoulos and E. Hatzikraniotis

Solid State Physics Department
University of Thessaloniki
540 06 Thessaloniki, Greece

The group of III-VI (III: Ga, In-VI: S,Se) layered compounds, with large structural anisotropy, seems to be a promising family of materials suitable for intercalation applications, mainly as electrodes in solid state batteries [1]. Each layer consists of four two-dimensional sheets of like atoms in the sequence X-M-M-X with mainly strong covalent bonding between the sheets, while the complete four fold layers are bound together by weak Van der Waals type forces, forming different crystalline structures (polytypes) with different stacking sequences [2]. Among other intercalants it is Li on which attention has been centered [3,4], although the question has been asked whether these compounds are really intercalated or whether new compounds with Li participation are formed [5]. In this paper is presented a new damageless method for the characterization of intercalated samples by studying the influence of the intercalant on the free exciton (FE) line of the host material.

The FE was excited in the photoluninescence spectrum with Ar (476 nm) or Kr (671 nm) laser beam incident on the sample in the $\vec{E} \perp \vec{c}$ geometry, where $\vec{c}$ is the hexagonal axis perpendicular to the layers. The samples used had mirror-like surfaces and were prepared by careful cleaving (because of the material's high plasticity) of single crystals grown from the nonstoichiometric melt with the Bridgman method [6]. Solutions of n-butyllithium in hexane are used for the intercalation of Li species between the layers of the material. The photoluminescence spectra were recorded after full re-

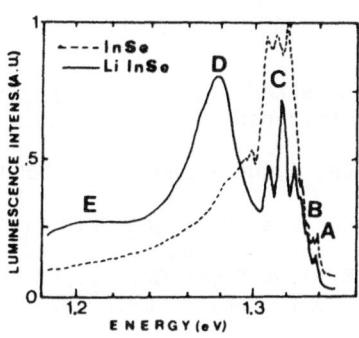

Fig. 1. The photoluminescence spectra of pure and intercalated InSe at T = 4.2K.

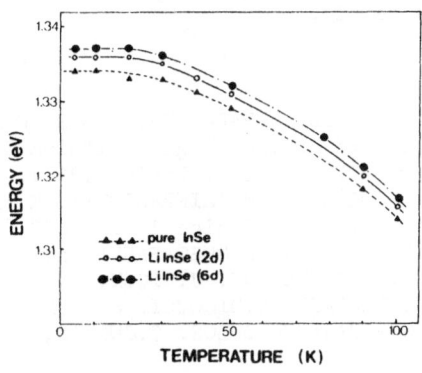

Fig. 2. Temperature dependence of FE energy for different times of intercalation.

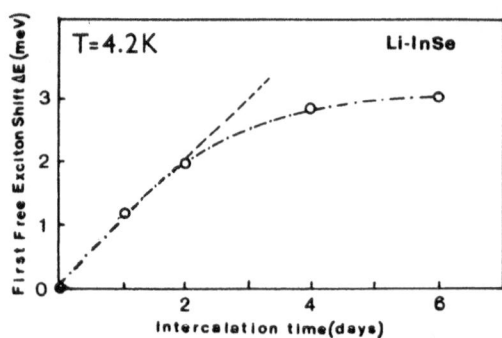

Fig. 3. Shift of the FE line with the
intercalation time.
------- is the slope at the origin.

laxation of the samples, in the temperature range 4.2-100K, with different values of the excitation power incident on many places on the same sample, in order to ensure that the measured spectra were characteristic of each sample.

The excitation spectra of the intercalated samples (fig. 1) are similar to the spectra of the pure InSe indicating that the characteristics of the host material are preserved [7] though additional features were also found. The band A (free exciton band) B (bound exciton) and C (donor-acceptor pair) are common in the two spectra while the presence of D and E is due to Li insertion. The analysis of the dependences of these peaks on temperature and incident excitation power leads to the conclusion that band D is due to plastic deformation and E_1, E_2 are attributed to Li donor centers and donor acceptor recombination centers [3,7].

The FE line in pure samples exhibits a temperature dependence, quite parallel to the energy gap one [8]. Also the intercalated samples present a parallel to the pure FE dependence on temperature (fig. 2). A shift is observed to higher energies with the increase of the intercalated quantity of the Li-species. This shift is characteristic for each intercalation dose and is the same for all the temperatures in the range 4.2-100K. By using the quantity

$$\Delta E = E_i - E_p \tag{1}$$

where E_i and E_p are the free exciton energies of intercalated and pure samples respectively, the characteristic of the material behaviour of ΔE whith intercalation time is found in fig. 3. From this plot it is clear that there is a

$$\Delta E \propto t^{1/2} \tag{2}$$

functional dependence where t is the time of intercalation.

Phenomena like the intercalant domain formation, playing a role in the kinetic processes [9], related with this observed dependence, will be discussed elsewhere [10].

This behaviour gives the opportunity to use the FE line as a probe for the degree of intercalation to be found in intercalated samples, by using this characteristic shift dependence. The above method is proposed as a means to probe the Li intercalation in a sample of InSe. This probe has still to be calibrated against an absolute measurement of the Li concentration in the sample.

REFERENCES

[1] C.JULIEN, K.PARASKEVOPOULOS, M.JOUANNE and M.BALKANSKI, Extended Abstract of 3rd Int.Meeting on Lithium Batteries (Kyoto,Japan 1986) p. 307.
[2] A.CHEVY, Thesis, Paris (1981).
[3] C.JULIEN, E.HATZIKRANIOTIS, K.PARASKEVOPOULOS, A.CHEVY and M. BALKANSKI, Solid State Ionics 18-19, 859 (1986).
[4] G.GUSEINOV, S.MUSTAFAEVA and E.ABDULLAEV, Phys.Stat.Sol. a88, K205 (1985).
[5] C.LEVY-CLEMENT, J.RIOUX, J.DAHN, W.McKINNON, Mat. Res. Bull. 19, 1629 (1984).
[6] C.DEBLASI, G.MICOCCI, S. MONGELLI, A. TEPORE, J. of Cryst. Gr. 57, 482 (1982).
[7] K.PARASKEVOPOULOS, C.JULIEN, M.BALKANSKI, C.PORTE, Phys. Scripta (1986) in press.
[8] J.CAMASSEL, P.MERLE, H.MATHIEU, A.CHEVY, Phys. Rev. B17, 4718 (1978).
[9] S.A. SOLIN, This Volume p. 173.
[10] K. PARASKEVOPOULOS et al (to be submitted).

SYNTHESIS OF GRAPHITE INTERCALATION COMPOUNDS

P.C. Eklund

University of Kentucky
Lexington, KY 40506

I. Introduction

The intercalation process involves a chemical reaction between a
layered host material and reagent which results in the insertion of new
atomic or molecular layers, termed intercalate layers, between the host
layers. In graphite, this reaction takes advantage of the weak (van der
Waals) bonds between carbon layers. The intralayer hexagonal
organization of the carbon atoms is not affected by intercalation.
Typical syntheses of graphite intercalation compounds (GICs) take place
at moderate temperatures ($T < 700°C$), and occur on a relatively short time
scale (several minutes to several days). Intercalation reactions are
known to occur between graphite and literally hundreds of reagents. The
compounds are, for the most part, very unstable in the open laboratory
environment, and they must therefore be handled with extreme care and
encapsulated in vacuum or in an overpressure of inert gas or reactant.
In many cases, the samples are sufficiently unstable to merit in situ
investigation of the physical properties.

Recent reviews of GIC properties by Dresselhaus and Dresselhaus [1]
and Solin [2] are available in the literature. Other views and articles
useful in grasping the scope of GIC-synthesis are: Herold (structure and
chemistry) [3], Ebert (general physical and chemical properties) [4],
Stumpp (metal chlorides and bromides) [5], Bessenhard et al.
(electrochemical synthesis) [6], Makrini et al. (rare earths) [7],
Lagrange et al. (Hg-amalgams) [8], Guérard et al. (3D structure and
epitaxial effects in stage 1 metal-GICs) [9] and Beguin et al. (alkali
metals co-intercalated with organic molecules) [10,11]. Fundamentally
new and exciting GICs are still being discovered and are discussed in the
proceedings of the most recent topical conference at Tsukuba (1985) [12].

Consistent with its amphoteric character, graphite reacts with
reducing or oxidizing agents. Charge transfer between C (carbon) and I
(intercalant) layers therefore plays an integral role in the reaction.
Depending on the reagent, one obtains either positively charged C-layers
(acceptor-type GICs) or negatively charged C-layers (donor-type GICs).
Graphite is somewhat unique among layered host materials (eg., transition
metal dichalcogenides, clays) because, for a given reagent, its
intercalation compounds can often exist in a variety of stoichiometries

associated with different stage indices. The stage index n refers to the number of carbon layers found between periodically inserted intercalate layers. In this paper, and in much of the current GIC research, there is a strong emphasis on samples which exhibit a single (or dominant, >90%) stage index. This is the case because many of the chemical and physical properties of interest are stage-dependent. The most difficult part of many GIC syntheses is then the determination of the growth conditions which lead to predominantly single stage material. In-plane order in the intercalate layer is also an important issue [13]. A GIC may exist in a single well-defined stage, but at the same time may exhibit a variety of in-plane intercalate layer structures (i.e. different superlattices, large concentrations of vacancies, island structures [14]) which also effect the sample stoichiometry. The relation of the growth conditions to a particular in-plane structure is more subtle, presumably because the free energy differences involved are small. Complete three dimensional crystal structures have been determined from some acceptor and donor GICs [1-3,9].

II. Common Reaction Techniques for GIC Synthesis

General Remarks

For many GIC syntheses, 99.9% pure chemicals and the rigid exclusion of O_2 and H_2O is sufficient to guarantee desired results. Of course, further consideration of reagent purity may be necessary, depending on the particular experimental probe. The selectivity of the intercalation process prevents the introduction of some impurities into the host. Reagents to be used in the intercalation reaction are, in some cases, best synthesized in an adjoining apparatus. They are then transferred to the reaction vessel by vacuum line techniques (i.e. distillation or sublimation) [15]. Most reactions can be carried out in Pyrex (Corning), quartz, Teflon (Dupont) or stainless steel reactors, depending on the compatibility with the reagent(s) at the temperature of the reaction.

Several forms of graphite are used in GIC research. Which form is used depends to some extent on the experiment to be performed, and to some extent on the kinetics of the particular reaction. Slow kinetics usually require smaller and thinner single crystals, or polycrystals in the form of flakes or powders. Small (~1 mm dia. by 10^{-2} mm thick) single crystals can be extracted from ore samples. Somewhat larger, and less perfect crystals referred to as "Kish" graphite may be obtained from slag skimmed from molten high carbon steel. Research and development in technologies seeking high strength/low mass/high electrical conductivity products based on carbon fibers have sparked recent research into the reaction of various reagents with pitch-based and gas-derived carbon fibers [16]. The most commonly employed graphitic host for fundamental studies of GIC's is a polycrystalline synthetic graphite called pyrolytic graphite. A particularly well-ordered form of pyrolytic graphite, referred to in the literature as "highly oriented pyrolytic graphite" (HOPG), Union Carbide), exhibits an average crystallite basal plane diameter of ~1 micron and c-axis alignment of better than 1 degree [17]. Large area HOPG plates can be readily cleaved with tape or razor blades to expose fresh c-faces. HOPG plates of various areas can be cut with an abrasive-type wire saw using either a diamond-impregnated wire or a smooth wire and glycerine /H_2O/abrasive slurry. The effects on the intercalation reaction of graphite flake or plate thickness [18], pyrolytic graphite plate size [19] and structural perfection of the host [20] have been investigated.

Two Zone or Vapor Phase Method

The two-zone method of preparation is useful whenever the reagent exhibits a high vapor pressure at moderate temperature. The reagent and graphite are held in opposite ends of a sealed reaction tube positioned in a two-zone furnace (Fig. 1). The furnace shown is also equipped with a vacuum muffle tube (quartz or alumina). We use evacuated muffle tubes in our laboratory when heating metallic reaction vessels. This is done to both reduce oxidation attack of the vessel walls and provide back-up vacuum in case of small leaks in the reaction vessel. The schematic temperature profile of the two-zone furnace is shown (Fig. 2). A reaction tube is shown with graphite (G) and reagent (R) suitably positioned in the flat zones of the profile. The stoichiometry of the GIC is controlled by independent regulation of the zone temperatures (T_G, T_R). For this reason the two-zone method is also referred to as the "two-temperature method". The zone containing the graphite is maintained at a higher temperature (T_G) than the zone containing the reagent (T_R) to prevent condensation on the GIC. T_R controls the vapor pressure of the reagent and T_G the reaction rate. The amount of intercalant uptake is observed to be a function of $(T_G - T_R)$ as shown in Fig. 3 for graphite-K [21]. These data are referred to as an "isobar" because T_G is varied and T_R is fixed (constant reactant vapor pressure). On the other hand, T_G can be fixed and T_R varied. This type of reaction scheme leads to the "isotherm" shown in Fig. 4 [22]. The plateaus in Figs. 3 and 4 correspond to single stage regions. It should be noted that the width of the plateau decreases with increasing stage index (lower intercalant uptake). This result is consistent with intuition--very high stage compounds are the most difficult to prepare.

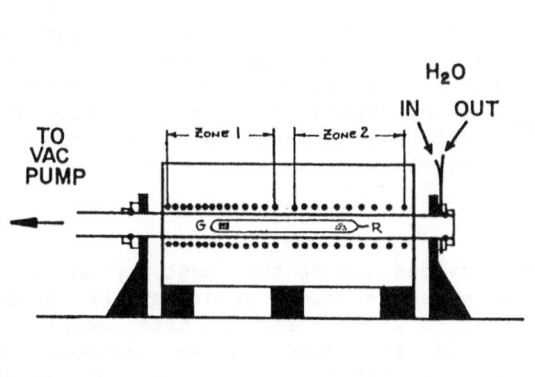

Fig. 1

Schematic representation of a two-zone vacuum furnace.

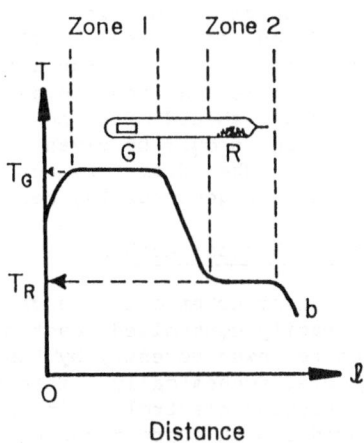

Fig. 2

Temperature profile of the two-zone furnace.

One Zone Method

This method differs from the method described above in that the intention is now to cause direct contact of the graphite with the liquid or solid reagent. The reaction tube is usually short and maintained at a nearly uniform temperature T_R in a single zone furnace. The one-zone method can be used to promote a more vigorous reaction. However, sample stoichiometry and stage can be controlled in the one-zone

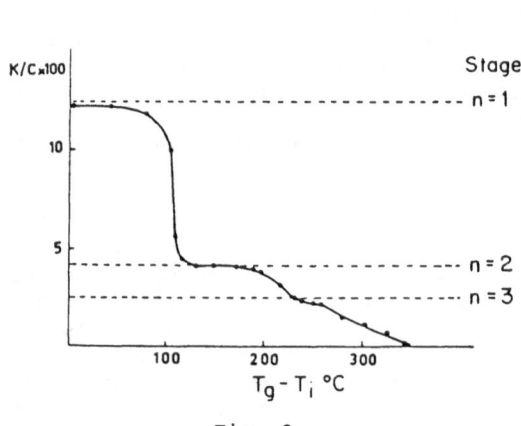

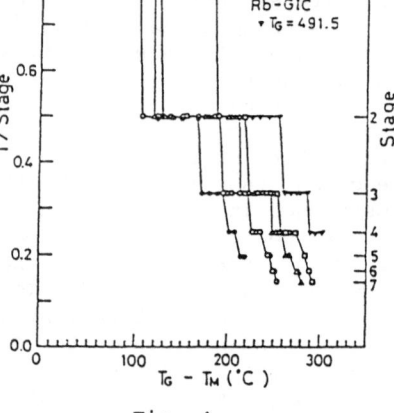

Fig. 3

Isobar for growth of graphite-K ($T_i = 250°C$) [21].

Fig. 4

Isotherm for growth of graphite-K [22], where T_m is the temperature of the metal intercalant.

method. If excess reagent is loaded into the reaction tube, then (by trial and error) a combination of parameters T_R and reaction time can be found which leads to fairly reproducible results. In another important variation of the one-zone method, a stoichiometric amount of reagent is introduced into the reaction tube and the reaction is run to completion. It is the author's opinion that, in both this and the preceding method, the temperatures and reaction times reported in the literature should often be considered as approximate points of departure. Attention should be given to all the synthesis conditions (i.e., tube length, tube diameter, graphite host dimensions, oven temperature profile, reagent purity, reaction time, etc.).

Electrochemical Method

Electrochemical intercalation provides one of the most reproducible and easily controlled reaction methods. This aspect of GIC synthesis has been reviewed recently by Bessenhard et al. [6]. The graphite sample to be electrochemically intercalated is suspended in an aqueous or non-aqueous electrolyte in a non-reactive metallic clip (i.e. Pt) which provides both support and electrical contact (Fig. 5). A fixed current I from a regulated supply is passed between the sample electrode (SE) and the counter electrode (CE) via the electrolyte. The current loop is completed in the external circuit. A third electrode, or reference electrode (RE) is present if quantitative cell potentials $V = (V_{RE} - V_{SE})$ are required. The polarity of the SE relative to the CE determines whether the graphite is reduced or oxidized. Initially it is assumed that ~ one electron flows in the external circuit for every charged species (cation or anion) intercalated into the graphite. Thus the reaction rate is proportional to the cell current, and the stoichiometry (within the context of a particular electrochemical model) is determined by the total charge Q passed in the electrochemical circuit. Neutral species may also co-intercalate. The charge/mass ratio Q/M, where M is the initial graphite mass, is thus a measure of stoichiometry (i.e. it is

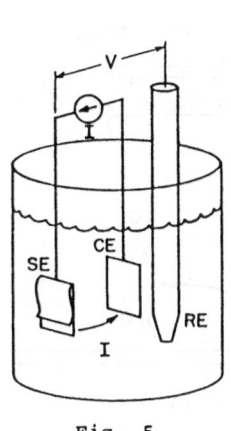

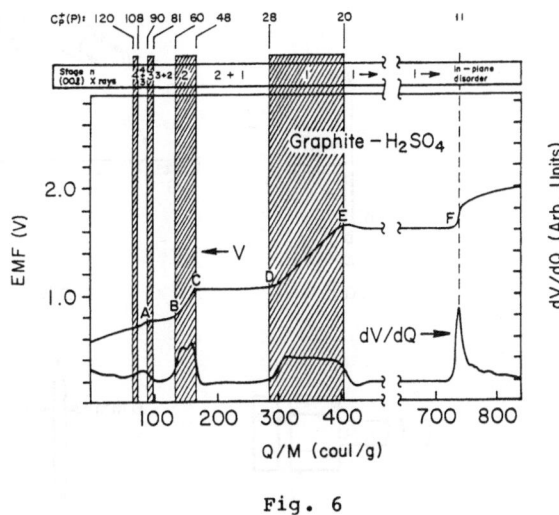

Fig. 5

Electrochemical Cell for intercalation from solution.

Fig. 6

Cell Potential vs. Q/M

proportional to the concentration ratio $C:X^{+(-)}$, where $X^{+(-)}$ represents
the charged intercalate species). In the case of graphite-H_2SO_4, the
material undergoes stage evolution $5{\to}4{\to}3{\to}2{\to}1$ with increasing Q/M. This
evolution of the anode (i.e. successive stages) exhibits the
staircase-like cell potential shown in Fig. 6. Assumptions involving a
particular electrochemical reaction mechanism are tested by a variety of
experimental probes [23]. A large variety of in situ measurements are
possible in an electrochemical cell [23,24].

III. Typical Reaction Vessels

In Fig. 7a-d we display an assortment of typical reaction vessels
used in our laboratory. The graphite and reactant are labeled G and R,
respectively. The vessels (7a-c) are made from pyrex or quartz. The
softening point of pyrex is ~400°C and is therefore used in lower-T
reactions than quartz. These glass reactors, in 1 mm or less wall
thickness, are reasonably transparent to Mo (Kα) x-rays, making them
convenient for in situ x-ray characterization of stage (00ℓ) and in-plane
order (hk0). Quartz is also a suitable ampoule material for in situ
neutron scattering studies. In (7a) or (7b) rectangular cross section
tubing can be used at the sample end for optical studies. Metal clips
fashioned from stainless steel or Pt, or bundles of quartz or pyrex wool,
can be used to stabilize the orientation of the sample. In (7c) we show
a large diameter tube containing a graphite plate too large to slide down
the connecting tube at the right. Fig. 7c is included to serve as a
reminder that pristine graphite is affected minimally by the momentary
hot gases generated during the sealing-off operation with a torch. In
Fig. 7d we show a thin wall (.010") stainless steel reactor which can be
loaded with graphite and reagent in an inert atmosphere glove box. One
end of the steel tube is first crimped and fused shut using a Heliarc
plasma torch. The tube is next thoroughly cleaned and taken into the
glove box for loading. The remaining open end is closed in the glove box
using a wrench to tighten a metal-to-metal compression seal (Swagelok).
The stainless steel reactor shown in Fig. 7d, depending on its length, is
suitable for either one-or two-zone reactions.

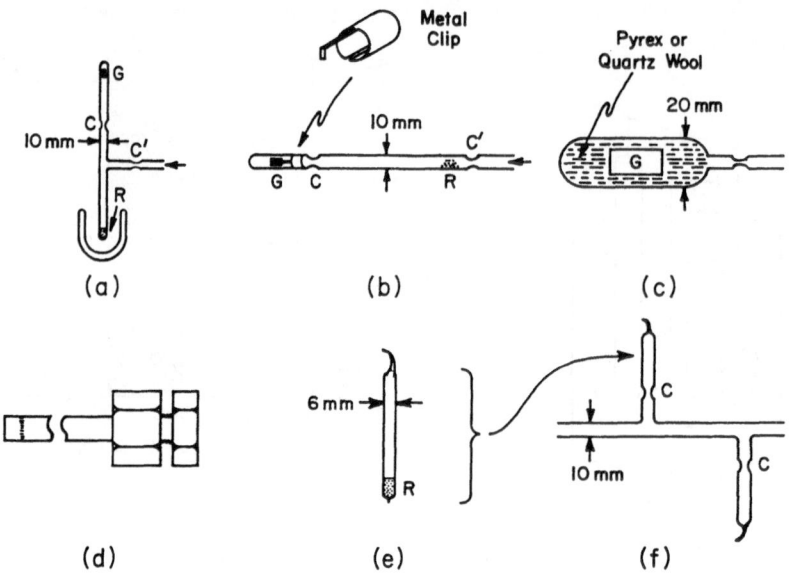

Figs. 7a-f

Intercalation Apparatus (see text).

The configurations shown in Fig. 7a and 7b, in the orientations shown, are most convieniently used for vacuum line distillation (7a) and sublimation (7b) of intercalant into the reaction tube. In (7a) a dewar surrounds the lower end of the reaction tube to promote condensation of the intercalant vapor entering through the slide arm. Once loaded, the reaction tube is sealed off with a torch. In (7b) the vapor from a solid heated to the right of C' is sublimed at R. This can be accomplished in many cases by the gentle action of a torch. For a one- or two-zone reaction, the tubes (7a) and (7b) are sealed-off at either C or C', respectively. An alternate way to introduce intercalant into the reactor tube is via a glass-encapsulated charge (Fig. 7e) equipped with an easily broken break-seal. A glass "tree" (Fig. 7f), a "branch" of which is shown sealed off in Fig. 7e, is used to prepare a series of several charges at one time. They are individually sealed off at the constrictions C and stored for future use.

Recently, there has been considerable interest in the ternary compounds made from binary alkali metal (M) GICs by successive intercalation (i.e. graphite-$M(NH_3)_x$, graphite-MH_x (chemisorbed hydrogen) and graphite-$M(H_2)_x$ (physisorbed molecular hydrogen). These reactions require the initial formation of the binary GIC, followed by the safe transfer of the sample to another apparatus in which the next step of the reaction occurs. In this step, the binary GIC reacts with the vapors of the second intercalant (e.g. NH_3 or H_2, etc.). In

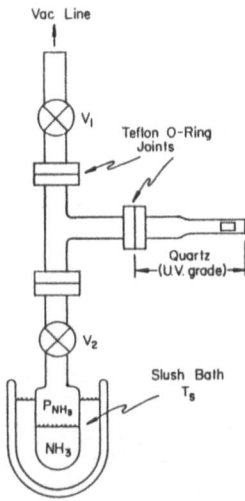

Fig. 8

Apparatus for the intercalation of NH_3 into KC_{24}. Thick-walled glass is necessary to withstand ~9 atm of NH_3 pressure at room temperature.

Fig. 8, we show schematically the apparatus used in collaboration with the Michigan State group to study optically the back-donation of charge during the intercalation of ammonia into KC_{24}. Purified NH_3 is contained in a reservoir surrounded by a dewar. Various slush baths of characteristic temperature T_S are placed in the dewar to achieve the desired equilibrium pressure of NH_3 gas. The stoichiometry (K/NH_3) of the sample is controlled by the vapor pressure [25] and the C:K ratio is determined by measurement of the weight uptake after formation of the binary compound. The sample is contained in the rectangular cross section quartz side arm. The remainder of the apparatus is fashioned from pyrex glass. The valves (Kontes) have teflon stems and the joints (Fisher-Porter) are·sealed with teflon o-rings. Attention must be given to the wall thickness of the glass apparatus which, if the liquid NH_3 is at room temperature, reaches ~9 atm. internal pressure.

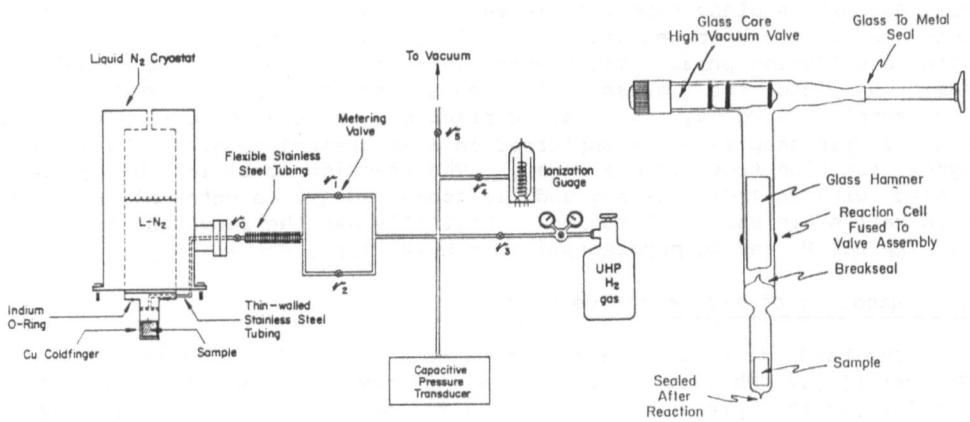

Fig. 9a

Schematic H_2 gas handling system.

Fig. 9b

Glass reactor for H_2 chemisorption in KC_8.

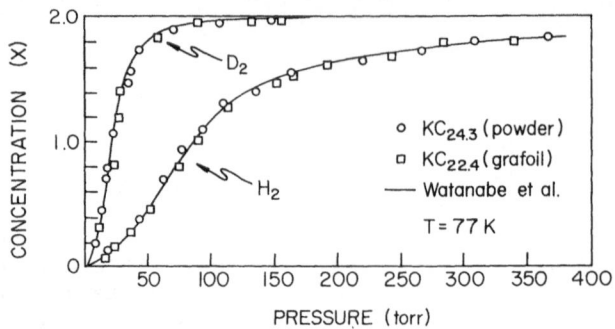

Fig. 10

Isotherms for H_2 and D_2 physisorption in stage 2 graphite-K (symbols [26], solid lines [27]). The ratio of $(H_2/K)=x$.

The apparatus used in our laboratory for the reaction of alkali metal GICs with hydrogen is shown in Fig. 9a. The amount of hydrogen uptake is determined from the decrease of hydrogen pressure in a known volume (V). The pressure decrease is measured using an electronic manometer (MKS-Bartron) at point P. For the chemisorption studies, KC_8 is prepared in a separate two-zone reaction and the KC_8 sample is sealed off from the supply of K. The other end of the KC_8 ampoule has a delicate break seal and is subsequently joined to the right half of the apparatus shown in Fig. 9b, with a glass hammer included. The glassware in Fig. 9b is then connected to the hydrogen apparatus at S via an o-ring seal and evacuated with a diffusion pump. The hammer is then used to break the seal and admit H_2 into the glass cell containing the KC_8. For the H_2 physisorption studies, which are carried out at cryogenic temperatures, stage 2 $C_{24}K$ samples are transferred in a He atmosphere glove box from a pyrex reaction tube into a dewar. The reaction cell in the dewar is sealed with an Indium o-ring and is connected to an external valve by thin wall stainless steel tubing. Fig. 10 shows the resulting isotherms (T=77K) for H_2 and D_2 physisorption in $C_{24}K$ [26].

IV. Handling of Air Sensitive GICs

The handling of air sensitive materials is discussed in detail by Shriver [15]. This text is a highly recommended sourcebook for those involved in the preparation and handling of GICs. A well-maintained, inert atmosphere glove box is a useful, if not necessary, apparatus for dealing successfully with the air sensitive reagents and GIC samples.

For those glove boxes not equipped with expensive D_2 and H_2O monitors, certain organo-metallic reagents can be used as indicators for the presence of small amounts of O_2, H_2O and solvents. Samples can be transferred there to small cells equipped with suitable windows or electrical feedthroughs. We have on occasion employed an optical cell with salt windows, a small attached 3/4 litre/sec ion pump and evacuation valve. After the loaded cell is removed from the glove box, the He atmosphere is removed with a diffusion pump through the evacuation valve and the ion pump is started. The cell is valved off from the diffusion pump and a vacuum of 10^{-5} - 10^{-6} torr is maintained by the ion pump. The cell is small enough to easily fit into spectrometers. Samples may also be transferred through glass break seals under diffusion pump vacuum into new cells suitable for various experimental studies. This procedure was described above in the remarks concerning the chemisorption of hydrogen into KC_8.

V. Brief Overview

We give a brief overview of the spectrum of GICs, most of which can be prepared with the techniques and apparatus described above. The purpose of this section is not to make an exhaustive list of GICs, these lists are available in the references cited in section I. Literally hundreds of atomic and molecular reagents react with graphite to form reasonably well-staged intercalation compounds. In many cases a range of stage index (n) is possible for a given reagent, depending on the growth conditions. The range n=1-3 can often be achieved, but there are many instances where the proper conditions have not yet been found to synthesize this range of compounds. A good example of this is graphite-Br (acceptor), which can readily be prepared in the stage n=2-4 range, but no stage 1 graphite-Br compound has been reported to date. A donor example of a reagent which reluctantly forms low stage compounds, is Na, which has been found not to form stage 1,2 compounds, in contrast to the other alkali metals. Generally sepaking, the higher stage (n>3) compounds are the most difficult to prepare.

Donors

Both binary (M_xC_n) and ternary ($M_xM'_yC_n$) donor compounds are well known. Binary donor GICs have been prepared from alkali metals (Li,K,Cs,Rb react readily, Na reacts reluctantly), rare earths (e.g. Eu,Sm,Yb) and alkaline earths (e.g. Ca, Ba, Sr). Full 3D crystal structures have been reported for all stage 1 examples of these compounds. Ternary donor GICs have been synthesized where one, or both, of the intercalated species are alkali metals (e.g. M=K,M'=Rb and M=K,M'=Hg). Donor ternary GICs exist in one of three classes. The examples given below apply to stage 1 compounds: (1) three-sublayer intercalant layer (e.g. -C-K-Hg-K-) and (2) alternating atomic intercalant layers (e.g. -C-K-C-Rb-) and (3) single, disordered layer (e.g. -C-(K,Rb)-). Another interesting class of ternary GICs involve the subsequent co-intercalation of a second non-metallic specie into a binary alkali metal compound to form $M_xZ_yC_n$, where for example M=K and $Z=NH_3$,H or benzene.

Acceptors

Acceptor GICs are usually produced from the reaction of graphite with molecular reagents. These intercalant species are large and therefore produce thick intercalate layers which lead to GICs with greater anisotropy than observed in the binary donor GICs. Most of the acceptors are based on halogens (e.g. Br_2,IBr,ICℓ) or halogen containing molecules metal chlorides- e.g. $NiC\ell_2$, $PdC\ell_2$, $A\ell C\ell_3$, $REC\ell_3$, $REC\ell_4$ (RE = rare earth), $SbC\ell_5$, $WC\ell_6$; metal bromides- e.g. $A\ell Br_3$, $AuBr_3$, UBr_5; fluorides- MF_5, e.g. M=As,Sb; MF_6, e.g. M=W,Re,Te,NO_2). Acceptors can also be obtained from the acids HNO_3, H_2SO_4 and $HC\ell O_4$. Recently, interesting new stage 1 ternary acceptor GICs $C_pZ_nY_m$ have been discovered [12] (e.g. Z = $FeC\ell_3$ and Y = N_2O_5, $CoC\ell_2$, $GaC\ell_3$) where the Z and Y species appear in alternate intercalate layers (-C-Z-C-Y-).

171

VI. Acknowledgements

This work was supported, in part, by the United States Department of Energy (#DE-FGO5-ER45151). The author would like to express special gratitude to W. Mateyka for his collaboration on the design and construction of the GIC glassware used in our laboratory. Many of my graduate students have spent long hours practicing the art of GIC synthesis-- discovering which methods succeed, and which fail. I salute their dedication and success.

References

1. S.A. Solin, Adv. Chem. Phys. 49, 455 (1982).
2. M.S. Dresselhaus and G. Dresselhaus, Adv. Phys. 30, 139 (1981).
3. A. Herold, in Physics and Chemistry of Materials with Layered Structures, Vol. 6, ed. F. Levy (Dordrecht, Reidel, 1979), p. 323.
4. L. Ebert, Am. Rev. Mat. Sci. 6, 181 (1876).
5. E. Stumpp, Mat. Sci. Engng. 31, 53, (1977)
6. J.O. Bessenhard, H. Moewald and J.J. Nickl, Syn. Met. 3, 187 (1981).
7. M. El Makrini, D. Guérard. P. Lagrange and A. Herold, Physica 99B 481 (1980).
8. P. Lagrange, M. El Makrini, D. Guérard and A. Herold, Syn. Met. 2, 191 (1980).
9. D. Guérard, P. Lagrange, M. El Makrini and A. Herold, Syn. Met. 3, 15 (1981).
10. F. Beguin, R. Setton, A. Hamwi and P. Touzain, Mat. Sci. Eng. 40, 167 (1979).
11. F. Beguin, R. Setton. L. Facchini, A.P. Legrand, G. Merle and C. Mai Syn. Met. 2, 161 (1980).
12. Proceedings of the International Symposium on Graphite Intercalation Compounds Syn. Met. 12 (1985).
13. M.S. Dresselhaus, Superlattices and Intercalation Compounds, in this volume, p. 1.
14. D.M. Hwang, X.W. Qian and S.A. Solin, Phys. Rev. Lett. 53, 1478 (1984).
15. D.F. Shriver, The Manipulation of Air-sensitive Compounds, (McGraw-Hill, New York, 1969).
16. M.S. Dresselhaus, Graphite Fibers, in this volume, p. 461.
17. A.W. Moore, in Physics and Chemistry of Carbon, Vol. 11, ed. P.L. Walker and P.S. Thrower (Dekker, New York, 1973) p. 69.
18. J.G. Hooley, Carbon 10, 155 (1972), Mat. Sci, Engng. 31, 17 (1977).
19. J.G. Hooley, W.P. Garby and J. Valentin, Carbon 3, 7 (1965).
20. M.B. Dowell, Mat. Sci. Engng. 31, 129 (1977).
21. D.E. Nixon and G.S. Parry, J. Phys. D, 1, 291 (1968).
22. R. Nishitani, Y. Uno and H. Suematsu, in Summary Report The Study of Graphite Intercalation Compounds eds. S. Tanuma and H. Kamimura, p. 25 (1984).
23. See for example, J.O. Bessenhard, H. Moewald, J.J. Nickl, W. Biberacher and W. Foag, Syn. Met. 7, 185 (1983).
24. P.C. Eklund, E.T. Arakawa, J.L. Zarestky, W.A. Kamitakahara and G.D. Mahan, Syn. Met. 12, 97 (1985).
25. S.K. Hark, B.R. York, S.D. Mahanti and S.A. Solin, Solid St. Commun. 50, 595 (1984).
26. G.L. Doll and P.C. Eklund, unpublished.
27. K. Watanabe, T. Kondow, T. Onishi and K. Tamura, Chem. Lett. 51 (1978).

KINETICS AND DIFFUSION IN GRAPHITE INTERCALATION COMPOUNDS

S.A. Solin

Department of Physics and Astronomy
Michigan State University
East Lansing, Michigan 48824-1116

INTRODUCTION

Serious and systematic studies of the kinetics of intercalation of graphite intercalation compounds (GIC's) were first carried out by Hooley [1,2] who investigated the bromination of various forms of graphite in the early 1960's. He found that the rate of intercalation was dependant on not only the intercalating species, but also on the microstructural morphology and macroscopic form of the graphitic host material. In addition, Hooley related the intercalation rate to the partial pressure of the vapor of the intercalating species and found a threshold pressure below which no intercalation occurred [1,2]. He also discovered that for pressures above the threshold pressure, there existed a dwell time which retarded the onset of intercalation.

The seminal approach established by Hooley in his classic studies of bromine intercalation remained essentially dormant until Dowell and Badorrek [3] emulated Hooley's efforts in their careful diffusion studies of the intercalation of highly oriented pyrolytic graphite (HOPG) by HNO_3 and $PdCl_2$ as well as by Br_2. They reported diffusion constants that were not atypical of gaseous diffusion in solids [4] and qualitatively associated those rates with the presumed two dimensional (2D) structure of the intercalating species, i.e. liquid, crystalline solid, etc. Moreover, they adopted a procedure which is now commonplace and analyzed their weight gain vs. time data according to the equations which govern ordinary 2D diffusion. These equations predict a $t^{1/2}$ [5] behavior which accounts for only a small portion of the weight-gain curve as can be seen from Fig. 1 which shows the time dependence of the weight gain for nitric acid intercalation.

The seminal experiments of Hooley [1,2] and of Dowell and Badorrek [3] were performed in an era of minimal theoretical interest in intercalation kinetics. Moreover, although it was known during this period of underwhelming theoretical interest that intercalant domain formation [6] played a role in the kinetic processes, the complex connection between domain formation and kinetics was, not surprisingly, unappreciated--especially the relationship of staging to the rate of intercalation and to domain formation.

Recently there has been an explosion of both theoretical [7-12] and experimental [13-17] interest in the kinetics of the intercalation of graphite as well as other layered solids. This new interest in intercalation kinetics is collateral with, if not driven by, a plethora of new insight into domain formation in GIC's. New electrical techniques have been established for studying 2D diffusion processes [16] and there have been several real-time X-ray diffraction studies of structural transformations [14,15,17] associated with intercalation. On the theoretical side, both analytic methods involving stochastic models [7] and computer simulation Monte Carlo methods [10-12] have been applied to the study of intercalation kinetics with great success. In addition, considerable fundamental insight has been gained by the recent extension of experimental kinetic studies to ternary graphite intercalation compounds [16]. Accordingly, the purpose of this paper is to highlight the recent theoretical and experimental developments in GIC intercalation kinetics and, where necessary, reinterpret earlier results in the light of modern concepts.

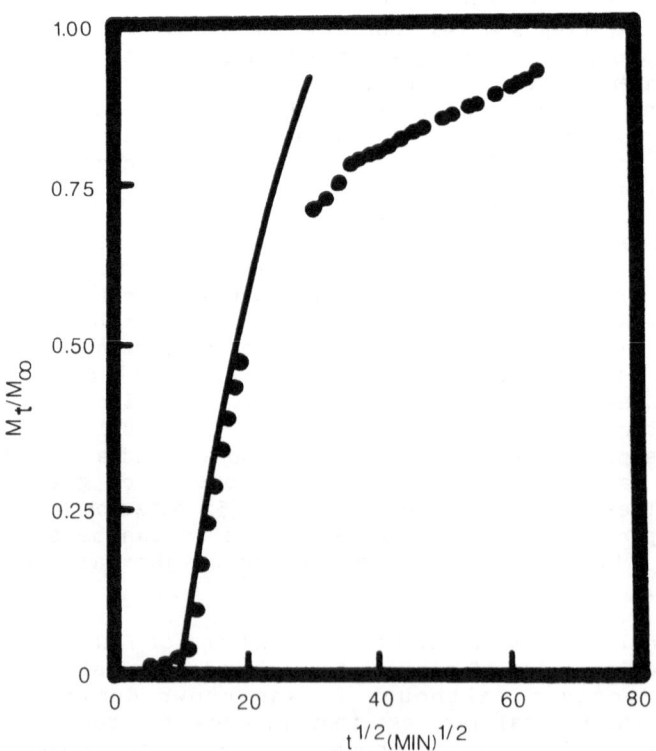

Fig. 1. Weight gain M_t/M_∞ as a function of the square root of time during the intercalation of highly oriented pyrolytic graphite by HNO_3 vapor in equilibrium with 98% HNO_3. The sample temperature was 30°C (from ref. 3).

DOMAINS, DOMAINS, DOMAINS

It is now recognized that domains play an important role in the kinetics of intercalation. The most famous domain structure which is generally believed to exist in GIC's is the so-called Daumas-Herold (DH) domain [6] which consists of three-dimensional (3D) stackings of planar islands of intercalant. Note that to date the specific planar structure of the island-like regions of which the DH domain is composed have remained unspecified. There is much evidence [18-20] that DH domains are present in large numbers in binary GIC's and they appear to be a prominent feature of ternary GIC's as well [16].

Clarke and coworkers [21] have investigated the in-plane structure of the islands in 3D DH domains and have interpreted their X-ray diffraction data on both donor [21] and acceptor [22] binary GIC's in terms of quasi-commensurate (QC) 2D regions of intercalant which are separated by domain walls. In the QC 2D domains, the ions or molecules of the intercalant occupy positions which are keyed to the hexagonal pockets in the graphite layers. Thus the coordinates of the ions can be referenced to the positions of the hexagonal sites in the graphite lattice even though the ions themselves may not lie exactly on those sites. This situation is similar to the thermal displacements of atoms in a crystalline solid and in principle gives rise to a corresponding Debye-Waller factor. The boundary of a QC 2D domain is defined by the region in which the above described coordinate referencing obtains. In the special case in which the intercalant forms 2D commensurate domains in which the ions or molecules lie (neglecting thermal motion) exactly on hexagonal sites and these regions are separated by domain walls in which the site occupancy is not preserved, the walls are referred to as discommensurations. Examples of a QC 2D domain and a 3D DH domain are shown in Figs. 2a and 2b, respectively.

KINETICS AND DOMAINS IN BINARY GIC'S

Two types of kinetic experimental studies have been carried out to date. In one case, the composition of the specimen is kept constant and the time dependence of structural changes induced by a pressure [17] or temperature [13] up or down quench is monitored by real-time diffraction measurements. In the second type of kinetic study, the composition is varied in response to a step-like change in the vapor pressure of the intercalant around the specimen [14-16] or a change in the specimen temperature at fixed vapor pressure and the time dependence of the resultant weight change, conductivity change, staging transition etc. is monitored.

Constant Composition Studies

About 25 years ago Lifshitz [23] analyzed the effect of multidegenerate ground states on domain-growth kinetics. A multidegenerate ground state is one in which many domains may form simultaneously on equivalent sublattices. When the system is quenched from a high temperature disordered phase into a single multidomain phase, the size of the average domain can be shown to follow a power-law form $L(t) \propto t^{\alpha}$, where α is a universal constant that depends on the ground state degeneracy, p [24]. Lifshitz showed that for $p > 3$, domain boundaries in 2D systems retard the growth of domains which, as a result, should exhibit sluggish growth kinetics. While it is clear that for $p = 2$, $\alpha = 1/2$ [24], the value of α for $p > 2$ is not well established, but there is evidence from Monte Carlo calculations [25] that for $p = 14$, $\alpha = 0.45$. In addition, p-state Potts model calculations [26] show that α ranges from the classical value of 1/2 to a value of 0.41, for $p > 30$.

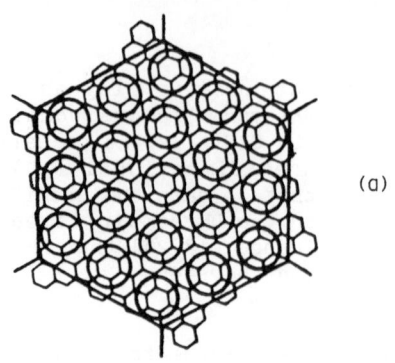

(a)

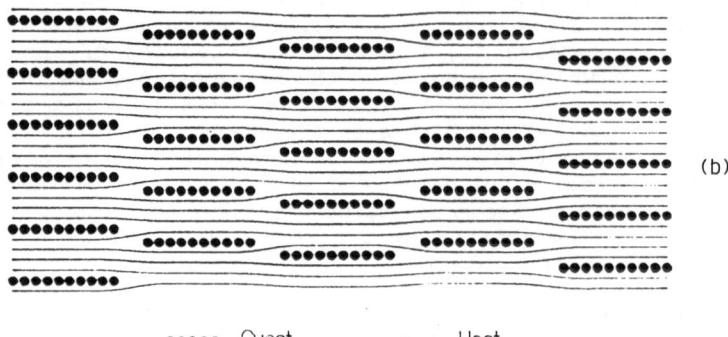

(b)

●●●●● Guest   ———— Host

Fig. 2. a) The domain model for a quasicommensurate domain in Cs-graphite (from ref. 21); b) A schematic representation of the Daumas-Hérold domains in a stage-4 graphite intercalation compound. The groupings of solid circles represent islands of intercalant (from ref. 11).

Homma and Clarke [13] realized that when graphite is intercalated with $SbCl_5$ to form a stage-6 GIC, the intercalant molecules (at room temperature) form commensurate 2D multidegenerate domains which have a $(\sqrt{7} \times \sqrt{7})R \pm 19.11°$ superlattice structure with a degeneracy of p = 14. Thus, by temperature quenching $SbCl_5$-graphite from its high temperature disordered phase (T > 450K) or its high temperature incommensurate phase (338K $\leq$ T $\leq$ 450K) into its room temperature single phase multidegenerate domain structure and monitoring the X-ray diffraction intensity of a reflection from the ordered superlattice domains, they were able to test the modern theories alluded to above. Such tests of the theories had been previously lacking because samples with high degeneracy were not readily available.

The results of the real-time X-ray studies of stage-6 $SbCl_5$-graphite are shown in Fig. 3 which displays the variation of the intensity of the (10) reflection from the $(\sqrt{7} \times \sqrt{7})R \pm 19.11°$ superlattice with time following a temperature quench to room temperature from the disordered (lattice gas) phase (Fig. 3a) and the incommensurate phase (Fig 3b). Both quench measurements exhibit a characteristic time t_p beyond which the ordering kinetics become very sluggish. Below t_p the intensity follows a power-law behavior $I \propto t^{2\alpha}$, whereas $I \sim \ln t$ for $t > t_p$. But it can be shown [13] that $I \propto L^2$ over the entire range of the measurement. Therefore, $L \propto t^\alpha$ and the data of Figs. 3a and 3b yield $\alpha = 0.52 \pm 0.05$ and $0.15 \leq \alpha \leq 0.25$, respectively.

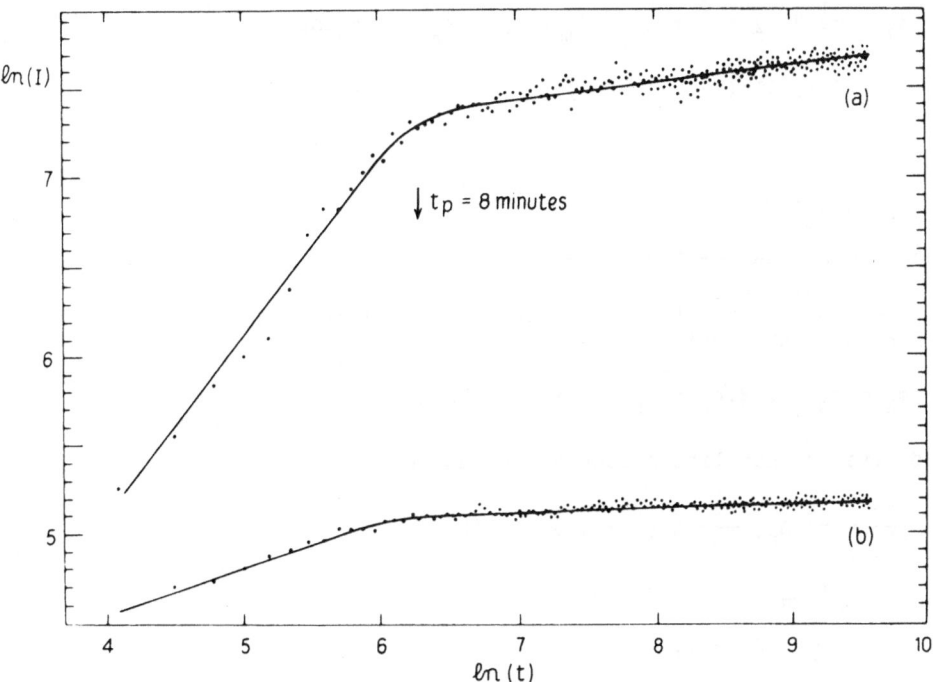

Fig. 3. The evolution of the (10) intensity from the ($\sqrt{7}$ x $\sqrt{7}$)R 19.11°
superlattice regions of stage-6 SbCl$_5$-graphite after quenching to room
temperature from (a) 470K and (b) 378K. The solid lines are fits to
scaling behavior discussed in the text (from ref. 13).

The domain size corresponding to t_p was found from the width of the
(10) X-ray reflection to be about 3000Å, an indication that "the crossover
at t_p is the result of pinning of the domain boundaries on defects
introduced into the graphite matrix by intercalation" [13]. Since the
value of L at the onset of the quench shown in Fig. 3b is close to the
3000Å pinning limit, a smaller value of α was anticipated and observed.
Thus the measurements of Homma and Clarke [13] confirm the power-law
scaling behavior of the time dependence of domain-growth below t_p. This
result is convincingly attributed to the interaction between
discommensuration domain walls. The anticipated sluggish kinetic behavior
for $t > t_p$ was also confirmed.

Variable Composition Studies

Theoretical and experimental efforts to understand variable
composition kinetics in GIC's have focused on the staging transformations
which accompany the change in composition. Miyazaki and Horie [7] have
used a stochastic model to determine the time evolution of the staging
transition which follows a step-function change in the chemical potential
of the intercalating species while the sample temperature is held constant.
In their model, the graphite host is treated as a set of c-axis columns,
the stacked galleries of which constitute cells which can accept single
islands of intercalant whose lateral size is fixed.

If we let ρ_n be the mean density of the guest atoms in the nth cell
within a single column, then $0 \leq \rho_n \leq 1$ and ϕ_1, the intra-column
contribution to the free energy of the system can be written as

$$\phi_1 = -\frac{1}{2} V_0 \sum_n \rho_n^2 + \frac{1}{2} \sum_{m,n} V_{mn}\rho_m\rho_n + k_B T \sum_n \{\rho_n \ln \rho_n$$

$$+ (1-\rho_n)\ln(1-\rho_n)\} - \mu \sum_n \rho_n \qquad (1)$$

where

$$V_{nm} = v_0 \ell^{-\alpha} \qquad (2)$$

is the Safran-Hamann [27] form of the electrostatic repulsive interaction, ℓ is the number of graphite layers between islands n and m, and μ is the chemical potential of the intercalant. Miyazaki and Horie [7] define the inter-column interaction, ϕ_2 as

$$\phi_2 = W_0 \sum_n \{-2(\rho_n + \rho_{n+1})^2 + (\rho_n + \rho_{n+1})^4\} \qquad (3)$$

and derive the non-linear Langevin equation

$$\frac{\partial \rho_n}{\partial t} = -\sum_m \Delta_{nm} \frac{\partial}{\partial \rho_m} \beta(\phi_1 + \phi_2) + f_n(t) \qquad (4)$$

where $\beta = 1/k_B T$,

$$\Delta_{nm} \begin{cases} = 2a + b, & n = m \\ = -a, & n = m \pm 1, \end{cases} \qquad (5)$$

and $f_n(t)$ is a Gaussian random force which satisfies the equation

$$\langle f_n(t)f_m(t')\rangle = 2\Delta_{nm}\delta(t - t'). \qquad (6)$$

Here a and b represent, respectively, the c-axis diffusion rate and island growth rate. Using this formalism, the isothermal transformation from a stage-3 to a stage-2 structure can be simulated to yield the results shown in Fig. 4a. Notice from that figure that the two prominent peaks at π/c and $2\pi/3c$ which correspond to stage-2 and stage-3 structures coexist and remain unbroadened throughout the transition.

Kirczenow [10,11] has also simulated the time evolution of a staging transition, but he used a 3D Monte Carlo technique [11] to study the isothermal evolution from the pristine host to stage-2. In his approach, the Hamiltonian is written in terms of the interactions of elementary islands (EI) and

$$H = \sum_m KN + \frac{1}{2} \sum_{m,n} NV_{nm} + E_d. \qquad (7)$$

Here K is the Helmholtz free energy of an island per EI, E_d is the elastic interaction between EI's, the summations are over the EI's, and V_{nm} has been defined in (2). The results of Kirczenow's 3D simulations are shown in Fig. 4b and again exhibit the coexistence of well-defined stages as the system evolves to the equilibrium stage-2 form.

The essential qualitative features of the staging kinetics simulations discussed above are beautifully displayed in the real-time X-ray studies by Misenheimer and Zabel [14] (MZ) and by Nishitani et al. [15] (NUS) the latter of whose results for a stage-3 to stage-2 transition in potassium-graphite are shown in Fig. 5. That figure clearly shows the coexistence of well-ordered regions (e.g. sharp X-ray reflections) of both stages throughout the transition, in qualitative agreement with the simulations discussed above. Note however, that Misenheimer and Zabel's more recent

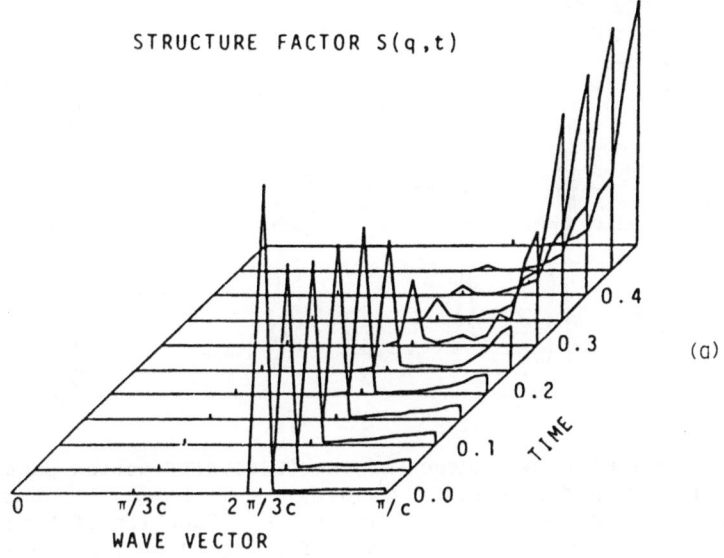

STRUCTURE FACTOR S(q,t)

(a)

WAVE VECTOR

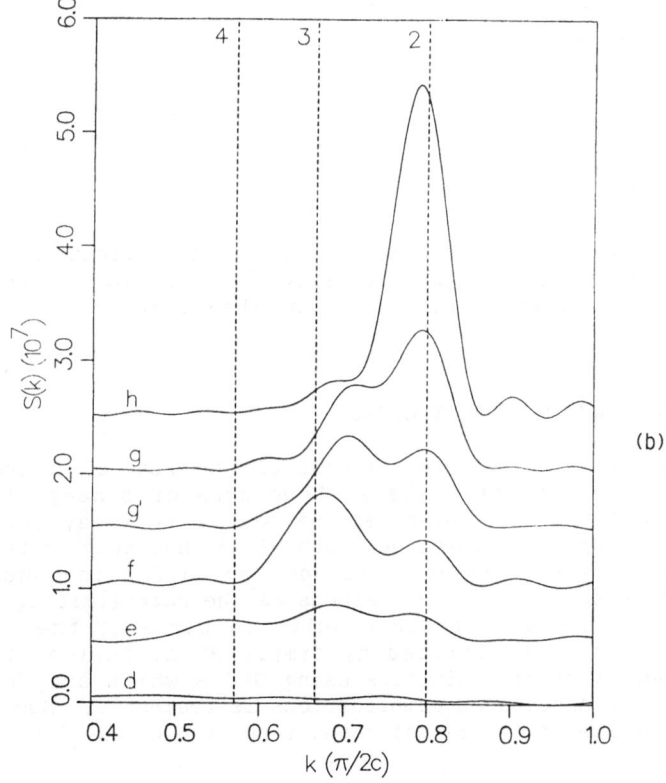

(b)

Fig. 4. Time evolution of the structure factor calculated
for   a) a stage-3 to stage-2 transition using a
stochastic model and computer simulation (from ref. 7)
and b) a pristine graphite to stage-2 transition
calculated using a 3D Monte Carlo computer simulation
(time increasing from d to h) (from ref. 11).

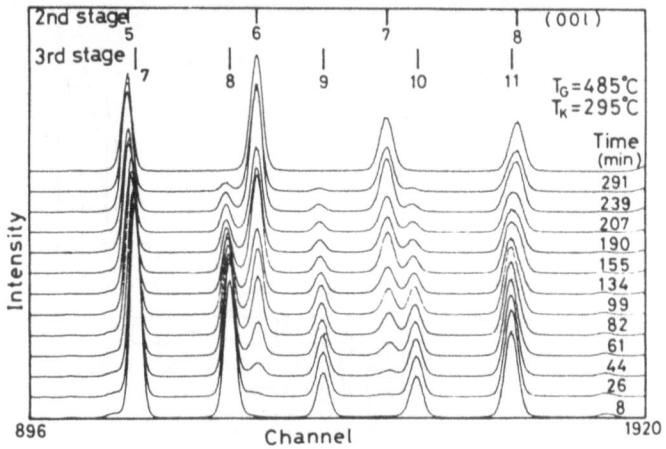

Fig. 5. The time dependence of the (00ℓ) diffraction pattern in the stage transformation from stage-3 to stage-2 potassium-graphite. The patterns are shown as a function of time after changing the potassium temperature, T_K, from that corresponding to the equilibrium condition for stage-3 to that for stage-2 (from ref. 15).

high resolution X-ray studies [28] of the (0 0 ℓ) reflections of potassium graphite as it undergoes a stage-5 to stage-4 transition do indeed reveal the line broadening and shift attributable to stage disorder and the presence of DH domains.

KINETICS AND DOMAINS IN TERNARY GIC'S

The complications associated with multidegenerate QC 2D domains which are prominent in the intercalate structures of binary GIC's make it difficult to quantitatively interpret the real-time X-ray studies of the kinetics of staging. For instance, both MZ and NUS studied the potassium-graphite system, but MZ found no evidence for DH domains and identified nucleation of (presumably QC) 2D islands as the rate-limiting process. In contrast, NUS interpreted their results in terms of the growth of DH domains which was rate limited by simple 2D diffusion of potassium. Clearly, a study of staging kinetics using GIC's which are free of QC 2D domains and their associated complications is desirable. Such a study has been recently carried out [16] using the ternary GIC $K(NH_3)_x C_{24}$ [29], $0 \le x \le 1$.

The potassium-ammonia intercalant layers in $K(NH_3)_{4.38} C_{24}$ are known to form a simple 2D [30] liquid which is thus free of QC 2D domains. But it is presumed that islands of this liquid participate in the formation of 3D DH domains [16]. Solin and coworkers [16] have studied the composition dependence of the c-axis resistance of the potassium-ammonia GIC's through the stage-2 to stage-1 transition and the kinetics of this resistance increase that results from a step-function increase in ammonia pressure

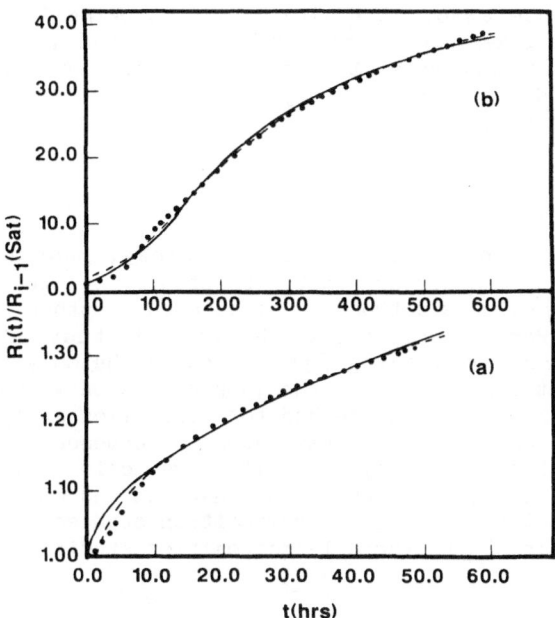

Fig. 6. Time dependence of the resistance ratio $g(t) = R_i(t)/R_{i-1}(sat)$ (solid circles). $R_i(t)$ is the time dependence of the c-axis resistance of $K(NH_3)_x C_{24}$ following an incremental ammonia pressure increase from $P_{i-1} = 1.5 \cdot 10^{-3}$ atm. to $P_i = 4.5 \cdot 10^{-3}$ atm (panel a) and from $P_{i-1} = 0.13$ atm. to $P_i = 0.22$ atm. (panel b). $R_{i-1}(sat)$ is the saturation value ($t = \infty$) of the resistance corresponding to P_{i-1}. Panels a) and b) correspond respectively to pressure increments which do not and do span a staging phase transition. The dashed (solid) line is a least squares fit with (without) a dwell time, t_d, and yields the following diffusion constants: panel a) solid line: $D = 9.06 \cdot 10^{-9} cm^2/sec$, dashed line: $D = 2.62 \cdot 10^{-8} cm^2/sec$; panel b) solid line: $D = 5.56 \cdot 10^{-9} cm^2/sec$, dashed line: $D = 4.72 \cdot 10^{-9} cm^2/sec$.

(chemical potential). They have solved the 2D diffusion equation for ammonia penetration into their rectangular parallelepiped specimens and used the solutions to calculate the time dependence of the c-axis resistance for a fixed pressure increment. The results of this approach are given in Fig. 6 which shows the resistance change for pressure increments which do (Fig. 6b) and do not (Fig. 6a) span the staging phase transition. From the theoretical fits (solid and dashed lines in Fig. 6) to the data, diffusion constants have been deduced (see the caption of Fig. 6).

Notice that although Fig. 6a has an S shape similar to that of Fig. 1, the full functional form has been accounted for without the need for breaking the curve up into three distinct regions as was done in the interpretation of the kinetic measurements of the bromination of HOPG. Thus it appears that QC 2D domains and other types of multidegenerate domains such as those associated with c-axis stacking faults (e.g. αβγδ, αβδγ etc.) which are likely to be present in the binary GIC's do indeed complicate the kinetic response to the intercalation phenomena as they do in SbCl$_5$-graphite and that earlier analyses of kinetic studies of binary GIC's may need to be refined.

CONCLUDING REMARKS

In this paper I have discussed only X-ray, mass transport, and resistance methods for monitoring kinetic processes and diffusion in GIC's. Space limitation precludes a thorough discussion of other powerful probes of such phenomena. Particularly noteworthy is the study of diffusion by quasi-elastic neutron scattering [31] which is the only technique known that can yield direct information on the jump diffusion vectors of lattice gas diffusion. Another important method for diffusiion analysis is nuclear magnetic resonance (NMR) [32], the temperature dependence of which yields a wealth of information on the type of diffusive motion being probed, e.g. rotational, translational, etc. While both neutron and NMR techniques have generally been applied to constant composition studies of GIC's, new NMR imaging methods applicable to variable composition studies are beginning to emerge [33].

ACKNOWLEDGEMENTS

It is a pleasure to acknowledge useful conversations with J. Heremans, Y.Y. Huang, X.W. Qian, D.R. Stump, and Y.B. Fan. This work was supported by the National Aeronautics and Space Administration under grant NAG-3-595 and in part by the National Science Foundation under grant DMR 85-17223.

REFERENCES

1. J.G. Hooley: Canad. J. Chem. 40, 749 (1962)
2. J.G. Hooley: Carbon 11, 225 (1973)
3. M.B. Dowell and D.S. Badorrek: Carbon 16, 241 (1978)
4. M.H. Jacobs: Diffusion Processes (Springer, New York, 1967)
5. J.G. Hooley and J.L. Smee: Carbon 2, 135 (1964)
6. N. Daumas and A. Hérold: C.R. Acad. Sci., Ser. C 268, 273 (1969)
7. H. Miyazaki and C. Horie: Syn. Met. 12, 149 (1985)
8. S.A. Safran: Phys. Rev. Lett. 46, 1581 (1981)
9. P. Hawrylak and K.R. Subasswamy: Phys. Rev. Lett. 53, 2098 (1984)
10. G. Kirczenow: Phys. Rev. Lett. 52, 437 (1984)
11. G. Kirczenow: to be published
12. S. Miyazima: Syn. Met. 12, 155 (1985)
13. H. Homma and R. Clarke: Phys. Rev. Lett. 52, 629 (1984)
14. M.E. Misenheimer and H. Zabel: Phys. Rev. B27, 1443 (1983)
15. R. Nishitani, Y. Uno, and H. Suematsu: Phys. Rev. B27, 6572 (1983)
16. Y.Y. Huang, D.R. Stump, S.A. Solin, and J. Heremans: to be published
17. B. Sundquist and J.E. Fischer: to be published
18. S.A. Solin: Adv. Chem. Phys. 49, 455 (1982)
19. R. Clarke, N. Wada, and S.A. Solin: Phys. Rev. Lett. 44, 1616 (1980)
20. R. Clarke and C. Uher: Adv. Phys. 33, 469 (1984)
21. M.J. Winokur and R. Clarke: Phys. Rev. Lett. 54, 811 (1985)
22. H. Homma and R. Clarke: Phys. Rev. B31, 5865 (1985)
23. I.M. Lifshitz: Zh. Eksp. Teor. Fiz 42, 1354 (1962) [Sov. Phys. JETP 15, 939 (1962)]
24. P.K. Wu, J.H. Perepezko, J.T. McKinney, and M.G. Lagally: Phys. Rev. Lett. 51, 1577 (1984)

25. P.S. Sahni, G.S. Grest, M.P. Anderson, and D.J. Srolovitz: Phys. Rev. Lett. 50, 263 (1983)
26. P.S. Sahni, G.S. Grest, M.P. Anderson, D.J. Srolovitz, and S.A. Safran: Phys. Rev. B28, 2693 (1983)
27. S.A. Safran and D. Hamann: Phys. Rev. B22, 606 (1980)
28. M.E. Misenheimer and H. Zabel, Phys. Rev. Lett. 54, 2521 (1985)
29. B.R. York and S.A. Solin: Phys. Rev. B31, 8206 (1985)
30. X.W. Qian, D.R. Stump, B.R. York, and S.A. Solin: Phys. Rev. Lett. 54. 1271 (1985)
31. H. Zabel, A. Magerl, A.J. Dianoux, and J.J. Rush: Phys. Rev. Lett. 50, 49 (1983)
32. C.P. Slichter: Principles of Magnetic Resonance (Springer, Berlin, 1978)
33. G.C. Chingas, J. Milliken, H.A. Resing, and T. Tsang: Syn. Metals 12, 131 (1985)

STRUCTURAL PROPERTIES AND PHASE TRANSITIONS OF GRAPHITE INTERCALATION COMPOUNDS

R. Moret

Laboratoire de Physique des Solides, associé au CNRS, Bâtiment 510
Université de Paris Sud, F-91405 Orsay, France

In this contribution I would like to present an overview of the structural properties and phase transitions of graphite intercalation compounds (hereafter noted
GICs). Because of their intrinsic anisotropy these materials provide very attractive opportunities for studying structural phase transitions as a function of dimensionality. Quite naturally structural studies have taken an important and pro-
ductive part in the renewed interest of the past ten years for GICs. Although several recent review papers [1,2,3,4] have dealt with the structural properties of
GICs numerous new results have enlarged our knowledge in the past 2 or 3 years.
However the aim of the present article is not to give a comprehensive review of
the subject but to focus on selected structural results which are representative
examples of the exciting, albeit complex, structural phenomena encountered in
GICs. I also hope to convince the reader that despite the important efforts made
recently, our knowledge and understanding are still poor in many cases and that
these studies are bound to develop rapidly in the next years.

I should point out that some topics which are closely related or even belong
to the structural properties will be treated in separate papers so that they will
not be considered here. This is the case of kinetics effects or of the structure
of ternary GICs. The subject matter of this chapter is organized as follows.
Section 1 deals with the general structural characteristics of GICs. In section 2
I introduce briefly the different experimental techniques used for structural studies. The last sections, 3 and 4, are devoted to a description of selected examples
of structures and phase transitions in the so-called donor and acceptor compounds,
respectively.

1. STRUCTURAL CHARACTERISTICS OF GRAPHITE INTERCALATION COMPOUNDS

Let us begin with a reminder of the basic structural properties of pristine graphite. Graphite is a natural compound which consists of honeycomb layers of carbon
atoms strongly linked by covalent bonds with only weak interlayer forces as evidenced by the in-plane and out of plane carbon-carbon distances (Fig. 1). These
layers form stacks of hexagonal symmetry with a sequence A B A B (the rhombohedral
sequence ABC ABC is rare and will not be considered here).

Several types of graphite samples have been used for intercalation. Natural
graphite single crystals can be found in the form of flakes which can be purified.
Their size is limited (to roughly 1 cm^2 × 0.2 mm) but they are the best choice for
detailed X-ray scattering experiments. A synthetic single crystal type of graphite
named Kish graphite has been used recently for intercalation although it is probably less perfect and less pure than natural graphite.Pyrolytic graphite and especially highly oriented pyrolytic graphite (HOPG) has been used extensively for
structural studies especially by neutron scattering because it can be available

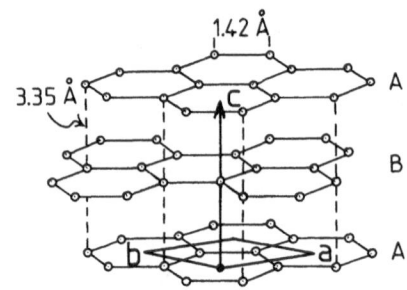

Fig.1
Structure of hexagonal graphite
showing the ABAB stacking of honeycomb carbon layers (Space group
$P6_{3/m} m_c$, a_G = 2.46 Å , c_G = 6.70 Å)

in large sizes (several cm³). It consists of compacted crystallites with a common stacking axis c but with random in-plane orientations. Therefore HOPG can be considered as a two-dimensional powder so that information on orientational effects in the basal plane is lost. Finally, intercalation compounds prepared from graphite powder or graphitized carbon fibers are important for applications but they can only give limited structural information.

It should be kept in mind that the choice of a particular graphite form introduces specific limits on the type of structural studies which are feasible. Also note that the properties of a GIC (and in particular the structural ones) can be influenced by the type of graphite which is used.

The intercalation consists in the introduction of layers of atoms or molecules between the graphite planes which retain their integrity because of the strong covalent bonding. The intercalants are ionized as a result of a change transfer to or from the graphite which leads to classify GICs as donor or acceptor compounds, respectively. Most GICs are highly ordered materials and they exhibit several types of ordering. This originates from their composite nature (host + intercalant) and from their layered structure which lead to a variety of intralayer and interlayer orderings and interactions. These structural features are now described separately, for clarity, though they usually interact.

1.1. Staging

This is the most peculiar type of ordering as it is rarely observed in other intercalation compounds. The stage (or stage index) of a GIC is the number n of graphite layers separating successive intercalant layers. The distinctive property of GICs is that a given intercalant is able to form compounds with different and well defined stages. Most GICs form stage 1 compounds (where graphite and intercalant layers alternate) as well as higher stage ones.

Stage determination is the first step in the structural characterization of GICs. It is usually done by scanning the (00ℓ) reflections whose Bragg angles Θ_ℓ are related to the stage n by the relation :

$$I_c = (n-1) c_0 + d_s = \frac{\lambda}{2\ell \sin \Theta_\ell}$$

where λ is the radiation wavelength and I_c, c_0, d_s are characteristic distances defined in Fig. 2. This method using X-ray or neutron scattering is straightforward in the case of single-staged samples. However the occurrence of stage disorder and/or mixed stages is common, especially for high stage materials in which case the (00ℓ) reflections can be broadened and displaced so that a more careful analysis of the whole (00ℓ) intensity distribution is required. Of particular importance is to distinguish between simple stage disorder which implies a limited correlation length due to random faults and mixed-stages where 2 (or more) stage units (possibly faulted) are distributed in the sample. This type of analysis has been developed recently [5] on the basis of the early works of Hendricks and Teller [6] , Kakinoki and Komura [7].

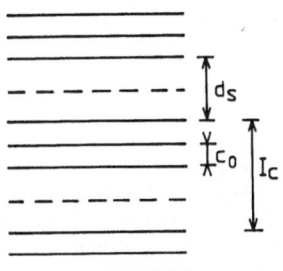

Fig.2
Model for a stage 3 GIC. Solid and dashed
lines represent carbon and intercalate layers
respectively.
d_S : intercalate sandwich distance
c_0 : carbon layer interlayer distance
I_c : repeat distance

An important aspect of the staging phenomenon is that, at least for stages higher than 1, the intercalate layers are divided into domains and that the carbon layers are pleated as shown schematically in Fig. 3. It results in an identical concentration of intercalant between every pair of carbon layers. This model was first suggested by Daumas and Hérold (DH) [8]. It provides a basis for the understanding of the intercalation process and of the staging transitions which can thus occur through a motion of the DH domains together with absorption-desorption of intercalant atoms or molecules. There seems to be a large range of values for the size of the DH domains going from less than 100 Å to 10000 Å [9] (although few data are available). It also appears that this size is mainly related to the nature of the intercalant and the quality of the graphite host. Thus the largest sizes have been observed in the case of molecular intercalants in single crystals. Actually the DH domain sizes are limited by those of the mosaic crystallites of the host graphite sample which are usually larger in single crystals than in HOPG.

After the pioneering work of Safran and Hamann [10] numerous theoretical models have been elaborated to account for the staging effect and to derive corresponding phase diagrams. The fundamental interactions which are included in these models are briefly described below. The elastic deformation of the host by the intercalant leads to a long-ranged effective in-plane interaction which is attractive. On the other hand the electrostatic and elastic interlayer interactions are both repulsive. In a simplified picture this explains the segregation of the intercalant into dense intercalate layers as is the case for staging, instead of an homogeneous intercalant distribution between every pair of carbon layers. Refinements

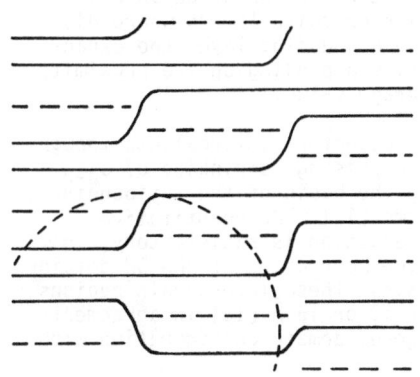

Fig. 3
Daumas-Hérold domain model for a stage
2 GIC. A faulted region is circled.

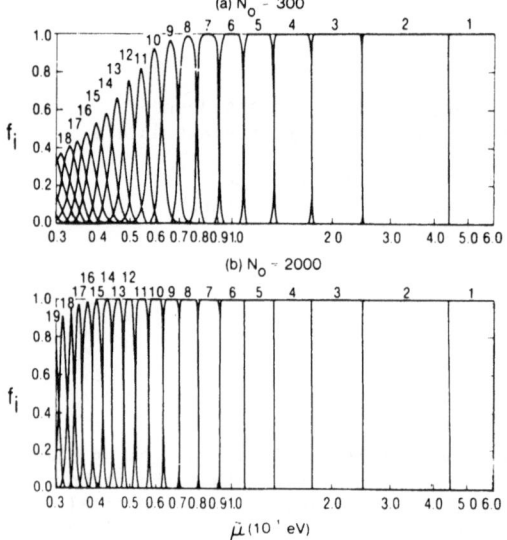

Fig.4
Theoretical phase diagram for K-graphite. f_i is the proportion of stage i indicated in the upper part, $\tilde{\mu} = \mu - \mu_\infty$ where μ_∞ is the lowest chemical potential for intercalation with infinite domains. N_O is the domain size. The stage disorder is found to increase with stage and to decrease with domain size (from ref. 13b).

of the models have shown that elastic forces coming from the graphite layer deformation stabilize a staggered arrangement of the intercalate domains as required by the DH model [11,12] . The finite DH domain size has been included very recently by Kirczenow [13] thus enabling to account for the occurrence of stage disorder (Fig. 4) as observed in some high resolution experiments. Other developments of the theory will be mentioned in section 3 together with experimental results on staging phase transitions.

1.2. Distortion of the Graphite Layers

Although the overall structure of the graphite layers is maintained upon intercalation several types of distortion have been identified. First there is a slight change of the in-plane C-C bond length from its value in pristine graphite d_{C-C} = 1.420 Å. From the repulsion of the charged carbon atoms and intercalant moieties one expects an increase of d_{C-C}. This is actually verified for donor GICs as measured by X-rays [14,15]. Furthermore the effect is found to be roughly proportional to the valence and inversely proportional to the stage and to the ionic radius of the alkali metal, alkaline earth or lanthanide. It should be noted however that the X-rays measurements give an average value of d_{C-C}. Therefore the observed stage dependence is likely due mainly to an average effect where the stage increase corresponds to an average over more and more interior graphite layers where distortions are smaller. Also consider that even within a bounding layer the expansion is certainly not homogeneous but that d_{C-C} varies depending on the proximity to an intercalate atom and again X-rays give an average value.

In contrast d_{C-C} decreases upon intercalation of acceptor intercalants though the effect is weaker than for donors [16]. This surprising shrinking of d_{C-C} has been explained from theoretical calculations on the basis of the antibonding character of the orbitals for second neighbor C atoms [17]. The quantitative agreement between theory and measured bond length reduction is satisfactory(Fig.5) Another type of carbon layer distortion occurs at the boundaries of the DH domains due to the dislocation loops which limit these domains. These interdomain regions are not well characterized and probably extend over 10 or 15 Å [16] in agreement with energy minimization [12] showing that a staggered domain configuration with such a distance between domains is favorable.

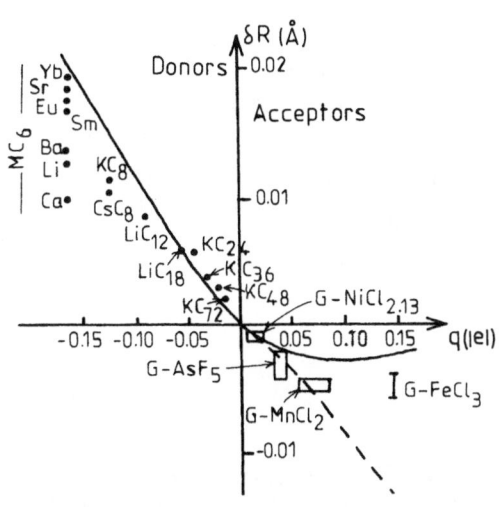

Fig.5
Variation of the calculated optimal
C - C bond length with charge trans-
fer (CT) for π (full line) and $\pi + \sigma$
(broken line) CT. From ref. 17b.

Finally, distortion arises because every C atoms of a given bounding layer does not have the same interaction with the intercalant atoms or molecules due to the different distances which are involved. This is exemplified in potassium-graphite where EXAFS studies have shown that the carbon layers are puckered in the vicinity of the K atoms [18].

1.3. Graphite Layer Stacking

Intercalation modifies the stacking arrangement of the graphite layers leading to various possible stackings which belong to 2 basic categories. The arrangement of the graphite layers about the intercalate one can be symmetric (A / A) or asymmetric (A / B). In donor and in several acceptor GICs the symmetric arrangement is obser- ved. This is the case for example of stage 1 (A/A/), stage 2 (AB/BC/CA/A) alkali metal [14], graphite-HNO_3 (A/AB/BA/) [19,20], graphite-Br_2 (A/AB/BC/CA/A) [21]. This situation implies a shift of the graphite layers bounded by a dislocation whose Burger vector component in the basal plane is of the type 1/3 ($\vec{a}_G + 2\vec{b}_G$).

On the other hand asymmetric arrangements are found in some acceptor GICs like stage 2 graphite $FeCl_3$ [22] or graphite $SbCl_5$ [1] (AB/AB/) where there is no la- teral shift of the graphite layers with respect to the pristine host.

However more complex situations may arise with relatively large molecular in- tercalants whose interactions with the bounding graphite layers are not compatible with the A, B, C hexagonal type of stacking (because of the relatively low symme- try of the molecular intercalant for example). This appears to be the case of graphite-HNO_3 (α-C_{5n} HNO_3) at low temperature as shown in section 4 [20]. One could also imagine that when the in-plane arrangement of the intercalant is disordered or incommensurate with the graphite layers then these layers might be uncorrelated, but this does not seem to have been observed so far.

1.4. Intercalate Layer Ordering

Depending on the nature of the intercalant, the stage, the in-plane concentration and external parameters (temperature, pressure), the intercalate layers in GICs exhibit various degrees of two-dimensional (2D) organization. This goes from long- range ordered simple structures (e.g. stage 1 alkali metal GICs at room tempera- ture, (RT)) to liquid or disordered states (e.g. graphite-HNO_3 at RT [19] or graphite-Bromine above 374 K [21,23] including various types of incommensurate and faulted structures. Some general considerations on the phenomena which deter- mine this structural variety are now presented.

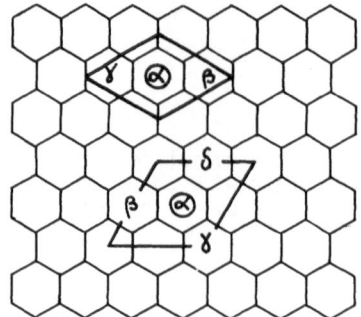

Fig. 6

Units cells of the p $(\sqrt{3} \times \sqrt{3})$ R 30°
and p (2×2) R0° structures projected
on a graphite layer. The in-plane stoichiometries are C_6M and C_8M respectively. The equivalent intercalate
sites are labelled α, β, γ and δ.

It is often possible to relate the structure of the intercalate layer to that
of the parent material. This has been done for instance for alkali metals compounds.
Stage 1 alkali metal GICs present simple in-plane structures characterized by pri-
mitive unit cells like p $(\sqrt{3} \times \sqrt{3})$ R 30° for C_6Li and p (2×2) R0° for KC_8, RbC_8 and
CsC_8 as shown in Fig. 6. It was shown that the geometry of the above structures
can be reproduced with little distortion using the atomic arrangement of the (111)
planes of the parent bcc alkali metal structures [2]. Thus the p $(\sqrt{3} \times \sqrt{3})$ R 30°
structure of C_6Li corresponds to one single (111) plane of solid Li while the
p (2×2) R0° structures of the heavier alkali metal compounds can be associated
with the projection of 3 successive (111) planes of the parent material.

In the case of Na such geometrical analogies fail either for the p (2×2) R0°
or p $(\sqrt{3} \times \sqrt{3})$ R 30° lattices (unless large distortions are allowed) which was related to the fact that Na does not intercalate easily into graphite.

For the higher stage K, Rb and Cs compounds the in-plane concentration is lower
(roughly $C_{12}M$) and the organization of the alkali atoms is complex and temperature dependent. However geometrical relations between the alkali metal areas in
the bcc metal and in the GICs were also suggested but with a more limited success
[2].

The intercalate layer structure of several acceptor compounds shows similarities with the structure of the intercalant (e.g Br_2, metal chlorides..). In the
case of bromine the molecular bromine arrangement in the (010) plane of orthorhombic solid bromine shows similarities with the RT commensurable layer structure
of the saturated variety of bromine GIC as studied by different groups [24] [25]
[26]. Also note that a comparison of the areas per Br_2 molecule indicate that the
arrangement is apparently denser in the GIC than in the solid. This lead to assume that in fact the Br_2 molecules are tilted with respect to the graphite plane by angles on the order of 10 or 20°, in agreement with EXAFS results [27].

Another phenomenon which influences the structure of the intercalate layer is
clearly the intercalant-graphite interaction. This effect is most simply seen in
the case of stage 1 alkali metal GICs where it is responsible for the small displacements needed to bring the atoms from their positions in the (111) planes of
the bcc structure to the centers of the carbon hexagons in the p $(\sqrt{3} \times \sqrt{3})$ R 30°
and p (2×2) R0° structures. The alkali metal-graphite interaction in stage 1 compounds was shown to be dominated by the competition between Coulomb attraction
(between the ionized alkali metals and the charged carbons) and core repulsion
[28] while the alkali metal-alkali metal interaction is essentially a repulsive
Coulomb interaction. Indeed, these simple assumptions result in a satisfactory
reproduction of the structural properties of these compounds [28].

The influence of the graphite bounding layers on the structure of the intercalate layer is more complex in higher stage alkali metals and also in molecular ac-

ceptor compounds. Generally speaking there is a tendency for the intercalate atoms to be driven towards the center of the carbon hexagons. The competition between this effect and the intralayer interaction is the origin of most structural phenomena in GICs. For instance the disordered or liquid GIC phases often exhibit positional and orientational correlations corresponding to a "lattice liquid". This is also responsible for the frequent incommensurate structures and the strong modulation effects which are observed when a soft intercalate layer is distorted by the more rigid carbon layer. Examples will be given in sections 3 and 4.

The intercalant-graphite interaction also leads to the existence of domain structures. Thus, when the point group symmetry of the intercalate layer structure is lower than that of its graphite bounding layers, 6mm, (which is frequent in acceptors) there will be different energetically equivalent orientations of the intercalate structure with respect to the graphite. This is illustrated in Fig. 7. That is, domains corresponding to each orientation will form at random during intercalation or at a phase transition and they will have equal concentrations on a macroscopic scale. Therefore a macroscopic hexagonal symmetry will appear even though the real symmetry can be much lower. In diffraction experiments it gives rise to intricate diffraction patterns with a superposition of the individual domain patterns which is often difficult to unravel. In some few cases electron diffraction has allowed to select only one orientation domain by taking advantage of the small electron beam size [29].

The number of these domains and their symmetry relations can be simply determined by the inventory of the symmetry elements of the 2D point group symmetry of the graphite layer which do not belong to that of the intercalate layer structure. In the case of Fig. 7 this gives 3 domains related by a 3-fold axis.

Another type of domain originates from the translational symmetry when the 2D unit cell of the intercalate structure is a multiple of the graphite one. If the multiplicity is N then N domains can be considered. They are of course fully identical, if taken separately, but in the real compound they do not coalesce since they are related by translation vectors (of graphite) which do not belong to those of the intercalate structure and they are joined by domain boundaries. Usually this has a domain size effect only which leads to reduction of the observed coherence length. However if the domains have similar sizes and shapes they can form a perio-

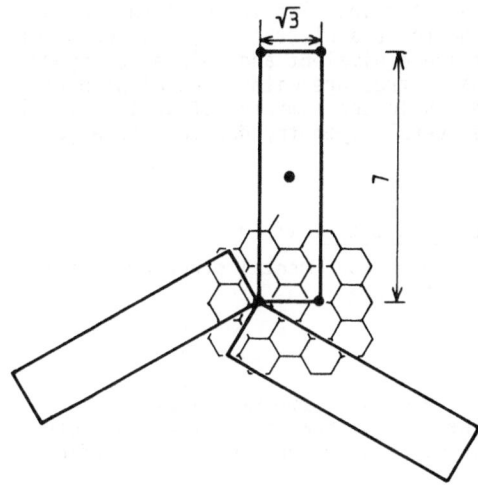

Fig.7
For the c $(7 \times \sqrt{3})$ structure (in graphite-bromine for example according to Erbil et al. [26] the point group is mm. These mirror planes coincide with those of the graphite layer (6 mm). Therefore the 3-fold axis is not common to the two point groups and it generates 3 types of domains.

dic domain pattern with very peculiar diffraction effects. This is the situation currently proposed to account [30] for the structure of some stage 2 heavy alkali metal GICs at low temperature with 7 different domains based on the p $(\sqrt{7} \times \sqrt{7})R19°$ structure and separated by discommensurations (section 3).

1.5. Intercalate Interlayer Ordering

Besides the staging phenomenon which reflects effective correlations of the intercalate layers in the c direction, these layers often exhibit correlations in the relative placement of the intercalant. This results in stacking correlations or even three-dimensional (3D) order of the intercalant. The stage 1 alkali metals are again typical and simple examples. In the p (2×2) R0° (p $(\sqrt{3} \times \sqrt{3})$ R30°) structure the multiplicity with respect to the graphite unit cell is 4 (3) and there exists 4 (3) equivalent domains labelled $\alpha, \beta, \gamma, \delta$ (α, β, γ). (Fig. 6 shows one site of each type). These domains are stacked according to periodic sequences in the successive intercalate layers leading to superstructures. As a consequence the period along c becomes a multiple of I_c, the repeat distance related to staging.

The origin of this type of stacking can be usually attributed to a steric effect in which close-packing and cohesion would be enhanced by avoiding the occupancy of similar sites in successive intercalate layers. In other words the elastic energy is minimized by this stacking. Although these arguments are valid it should be noted' that in some cases these conditions are not fulfilled. For instance in CsC_8 the stacking is $\alpha \beta \gamma$ [31] instead of $\alpha \beta \gamma \delta$ as in RbC_8 and KC_8 [32]. Even more striking is LiC_6 where the lithium atoms are stacked according to $\alpha \alpha$ [33], which should be ruled out following the above compactness arguments. However LiC_6 is special in its compactness due to the small size of the Li atom, its high compressibility and the possibility of a lithium-carbon hybridization leading to a weak covalent bonding. One could imagine that the $\alpha \alpha$ stacking is stabilized by bonding between the Li atoms and C hexagons in a symmetric arrangement leading to a $Li-C_6-Li-C_6-Li$ "chain" along c.

Intercalate layer stacking correlations are evidently stronger for stage 1 GICs and they are reduced as the stage increases because of the screening effect of the graphite layers. However such correlations have been identified in stage 2 alkali metal GICs at low temperature and in several cases of acceptor compounds (e.g. stage 2 and 4 graphite bromine [26,34] and graphite-HNO_3 at low temperature [20]. These correlations are also largely dependent on temperature and pressure and although very few data have been obtained so far the development of temperature and pressure studies of these correlations are promising as they offer means other than staging to vary the dimensionality of the phenomena. For instance an unexpected effect was observed in stage 2 graphite bromine where the correlations between the bromine layers disappear at low temperature [21] (see section 4).

Finally I mention that the intercalate layer stacking also leads to the formation of multiple domain structures. Considering for example the $\alpha \beta \gamma \delta$ stacking of RbC_8 it is easily seen that 6 other sequences are equivalent and related by symmetry (e.g. $\alpha \delta \gamma \beta$ ..). This results in a 3D domain structure with an equal proportion of each domain in the macroscopic sample. While the symmetry of an individual domain is orthorhombic (space group Fddd) the overall symmetry due to this effect is hexagonal as observed by diffraction [32].

2. EXPERIMENTAL TECHNIQUES FOR THE STRUCTURAL STUDIES OF GICs

Because of the distinctive properties of GICs their experimental structural studies present some special aspects which I indicate here.

2.1. Diffraction and Diffuse Scattering Methods

The diffraction from intercalation compounds like GICs is special because it reflects their layered structure and it combines the diffraction effects of both host and intercalant. In constrast with common solid structures where the reflec-

Table 1

Type of reflection	Structural information available from :	
	Geometrical analysis (position, symmetry, diffuse features)	Intensity analysis (integrated intensity, diffraction profiles)
00ℓ	Stage Repeat distance I_C Intercalate sandwich distance d_s	Stage disorder, staging transitions [35-41] Orientation of molecular intercalant (G -AℓCℓ$_3$ [42],G-HNO$_3$ [43]) Intercalant in-plane density
$(hkℓ)_G$	Graphite layer stacking (in simple cases) Stacking disorder Stacking period I_G (multiple of I_C)	Graphite layer stacking and stacking faults Contribution from intercalant structure or Fourier components of graphite potential [44-46]
$(hk0)_I$	2D unit-cell, space group of intercalant Diffuse scattering due to disorder (liquid halo, diffuse streaks)	2D structure of the intercalate layer Characteristics of phase transitions (liquid-solid, incommensurate, commensurate) [20-26-48]
$(hkℓ)_I$ (ℓ variable)	Stacking period of the intercalate layers when ordered	Correlations of intercalate layers as a function of T, P...

tions form an "homogeneous" set, in the case of GICs different types of information can be obtained from the analysis of selected reflections as summarized in Table 1. A few interesting points are underlined below.

A careful analysis of the (00ℓ) reflection profiles allows investigation of stage fidelity and stage disorder using for example algorithms based on the Hendricks and Teller work [6,7]. On the other hand the integrated intensity is used to determine the intercalate in-plane density or the orientation of molecular intercalants (when combined with the knowledge of the molecular shape). However it is worth noting that a reliable measurement of these quantities is often difficult because of the mosaic spread of the c axis in GICs which can reach a few degrees, particularly with HOPG samples. It was also shown that a triple-axis spectrometer is preferable to a double-axis one which can cause spurious broadening effects (Fig. 8) [44].

The reflections other than (00ℓ) corresponding to the reciprocal lattice points of the graphite sublattice (referred to as graphite reflections) such as $(10ℓ)_G$ or $(20ℓ)_G$ allow to determine the particular stacking of the graphite layers using standard crystallographic methods. Furthermore a contribution from the intercalate structure occurs except when it is fully incommensurate with the graphite. Of particular interest is the intermediate case where the incommensurate or partial order of the intercalant is modulated by the graphite. Then a Fourier series expansion of the intercalant density will have Fourier components for the graphite wave vectors, leading to a significant contribution to the graphite reflection inten-

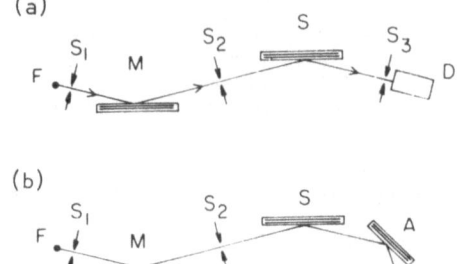

Fig. 8
Sketches of double-axis (a) and (b)
triple-axis diffraction geometries
F : X-ray source ; M : Monochromator ;
S : sample ; A : analyzer ; D : detector;
S_1, S_2, S_3 : slits. Double-axis geometry gives profiles which may be asymmetrically broadened in the case of
low absorbing samples (from ref. 44)

sity. This contribution was analyzed for the modulated liquid state of stage 2
alkali metal GICs to obtain either the fraction of registered alkali metal atoms
[44,45] or the Fourier coefficients of the graphite modulation [46].

The $(hk0)_I$ reflections from the intercalate structure contain most of the information on its in-plane ordering. The geometry of the $(hk0)_I$ pattern is often
quite intricate due to the existence of several domains, large unit cells or modulation satellites. Its study requires in most cases the use of single crystallites
using electron diffraction or for more quantitative results X-ray diffraction on
single crystals. In the later case it is highly recommended to begin with diffraction photograph obtained with standard photographic techniques (rotating crystal
or precession methods). The so-called "monochromatic Laue" technique (Fig. 9) is
also very powerful though less popular [47] . The development of 2D localization
detectors will provide even better quantitative results in the near future. From
the geometrical analysis of the $(hk0)_I$ pattern one can extract the 2D unit cell
and space group of the intercalate layer when it is ordered or identify diffuse
scattering features due to disorder. Diffractometer measurements are required for
more quantitative studies such as the intercalate structure determination
or the study of melting or commensurate-incommensurate transitions. Actually there

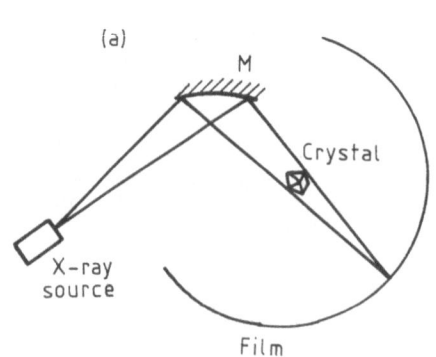

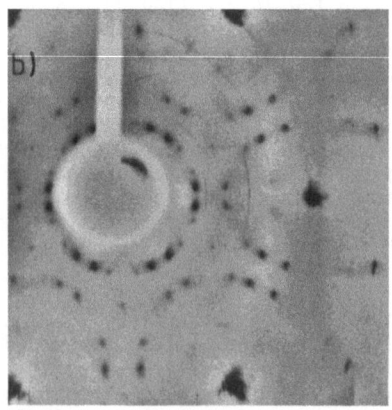

Fig. 9. a) Schematic diagram of the fixed-crystal fixed film method (or monochromatic Laue method). M : focussing monochromator.
b) X-ray photograph from KC_{24} at $T \simeq 100$ K. c is parallel to the beam. The five large spots are $10.\ell$ graphite reflections. Small ones are fundamental and satellite
reflections from potassium (from ref. 47).

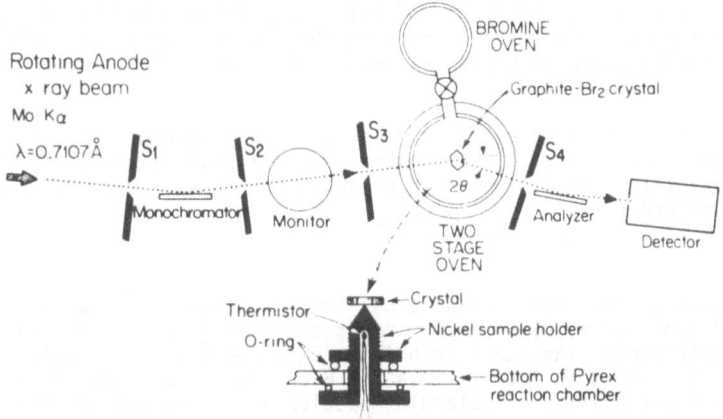

Fig. 10. Schematic diagram of the X-ray spectrometer used for structural studies of graphite-bromine (from ref. 26).

is a current trend towards this type of work on both alkali metals and molecular acceptor GICs.

The $(hk\ell)_I$ reflections with $\ell \neq 0$ allow study of the stacking of the intercalate layers. When there is no correlation between them the diffraction effects are similar to those of a 2D system (as if there was only one layer) and ℓ scans show a monotonic intensity variation due to the scattering factors and temperature effects dependence on $\sin \theta/\lambda$ [34] . This is often observed in high stage compounds or in some cases at high temperature in low stage ones. Long-range ordered stacking can be observed in stage 1 compounds; otherwise only partial or short range order is usually achieved.

The present development of structural studies of GICs is marked by the use of the most advanced scattering techniques. In situ X-ray scattering with the monitoring of the intercalant vapor pressure was applied successfully for alkali metals [35,37] and bromine [26].Such an experimental arrangement is shown in Fig. 10. High pressure scattering experiments is also a very active domain using standard hydrostatic pressure cells for neutrons or diamond-anvil cells for X-rays with MoKα radiation. As these measurements are often limited by the weakness of the scattered intensity, the development of synchrotron radiation with its high photon flux and good resolution will be essential. The study of kinetic effects during intercalation or at phase transitions is now possible as shown recently [35] [48].

2.2. Lattice imaging

High-resolution electron microscopy is a powerful method to determine the ultramicrostructure of crystalline materials (i.e. the structure at the atomic level). However its use for the study of GICs has been limited because they are often unstable in air and in the high vacuum chamber of the electron microscope. Therefore it has been applied most often to partly desorbed compounds (so called residue compound). The electron beam can also induce changes in the molecular nature of the intercalant. A typical case is graphite - $SbCl_5$ where the solid-glass transition observed with electron diffraction is not detected with X-rays [50]. This is tentatively attributed to a radiolysis process corresponding to the relation.

$$SbCl_6^- + SbCl_3 \rightarrow SbCl_5 + SbCl_4^-$$

and giving rise to disorder. Image simulation are in agreement with the existence of $SbCl_6^-$ and $SbCl_3$ species.[50]

Furthermore it should be kept in mind that the electron beam has a limited penetration (a few atomic layers). Therefore the ordering which is probed by the electron beam may be somewhat different from that of the bulk (as seen by X-rays or neutrons).

In spite of these problems electron diffraction and imaging are invaluable in the study of individual intercalate domains (see 1.4 [50,51], the determination of their sizes [52] , the characterisation of stacking faults in the graphite sublattice or stage disorder [53].

2.3. EXAFS

From the analysis of the characteristic oscillations near the X-ray absorption edge (e.g. the K edge) of a specific atom it is possible to get information on its short-range environment. The EXAFS technique (Extended X-ray Absorption Fine Structure) has been applied with success to a few GICs. It can be used to determine the bond lengths and orientation of molecular species as in graphite-bromine where the Br-Br distance was found to increase slightly from the normal value (2.28 Å) by 0.15 Å to match the spacing between adjacent hexagons [54]. As mentioned earlier (section 1.3) the interatomic distance between carbon and intercalate atoms can also be analyzed as in K-graphite where the C layers were found to be puckered in the vicinity of the K atoms [18].

3. STRUCTURE AND PHASE TRANSITIONS OF ALKALI METAL COMPOUNDS

Among GICs containing donor intercalants the alkali-metal compounds have been the subject of most structural studies. These have been stimulated by a confrontation with theoretical phase diagrams and predictions. Therefore a very large amount of data is available which is not possible to cover here in its entirety. I have chosen to analyze three main topics in which important contributions have been made recently using sophisticated techniques and arguments. These are in-situ X-ray studies of the staging and melting transitions in K-graphite (3.1), high-pressure phase transitions (3.2) and the structure and low-temperature behaviour of stage 2 heavy alkali metal GICs (3.3).

3.1. Staging and Melting Phase Transitions in Potassium-Graphite

The intercalation and the formation of different stages in a GIC like K-graphite can be studied as a function of the temperature of the graphite host (T_G) and the K vapor pressure P_K (or temperature T_K) in a two-zone furnace. Measuring the weight uptake as a function of T_G-T_K allows one to associate characteristic plateaus with stability regions for the different stages [55].

More detailed information on the nature of the staging transitions can be obtained from in-situ X-ray experiments by measuring the (00ℓ) diffraction spectrum, as done recently. Misenheimer and Zabel [56] first observed that (00ℓ) scans either reveal the existence of one single stage or the coexistence of 2 distinct stages with long-range order (resolution limited (00ℓ) profiles) in all cases. Since nonpure stages or stage disorder would give rise to (00ℓ) peak displacements or broadening the authors concluded that the transition from stage n to n$\pm$1 is a direct process with no intermediate or mixed structures even for higher stages.

A similar conclusion was drawn from the work of Nishitani, Uno and Suematsu [35] who studied the (00ℓ) reflections as a function of K vapor pressure and were able to construct a phase diagram for stages up to n = 7. Again they saw no evidence for stage disorder even for time-dependent measurements of the transition from stage 3 to stage 2 as shown in Fig. 11. In this case the intensity of stage 3 reflections decreases continuously while that of stage 2 reflections increases with no broadening or diffuse scattering effects.

However more recent high-resolution experiments by Misenheimer and Zabel [37] provided different results. Typical data are shown in Fig. 12 for a constant potassium temperature and 3 different values of T_G corresponding to the transition from stage 5 to stage 4. The reflections of Fig. 12 a and c are broader than the experimental resolution and their profile can be fitted using the Hendricks-Teller

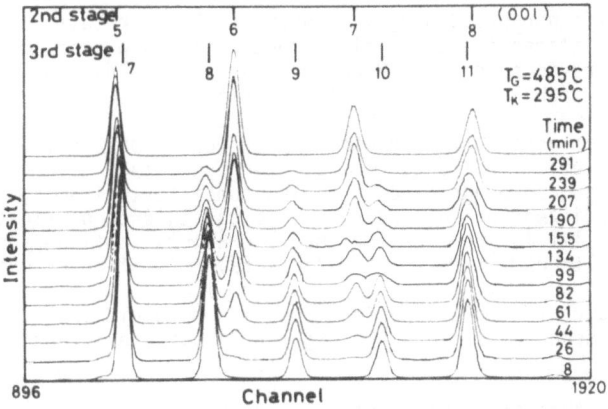

Fig. 11. Time dependence of the (00ℓ) spectrum during the stage 3 - stage 2 transformation in K-GIC. At time t = 0, T_K is changed from 290° to 295°C (from ref.35)

theory [6] (see section 1.1.) leading to admixtures of sizeable fractions of stage $n \pm 1$ packages in stage n compounds (e.g. 71% stage 5, 19% stage 4 and 10% stage 6 for the (005) peak of Fig. 12 a . In the stability region of a given stage for $n > 2$ the (00ℓ) peaks shift continuously while their position and width exhibit dramatic changes near the transitions (Fig. 13). However for the transition from stage 2

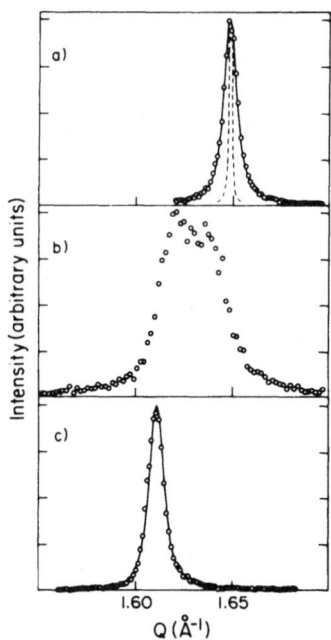

Fig. 12 Profiles of (00ℓ) in K-GIC
a) (005), stage 5, $T_G = 455°C$
b) Transition region, $T_G = 450°C$
c) (004), stage 4, $T_G = 444°C$ (from ref. 37).

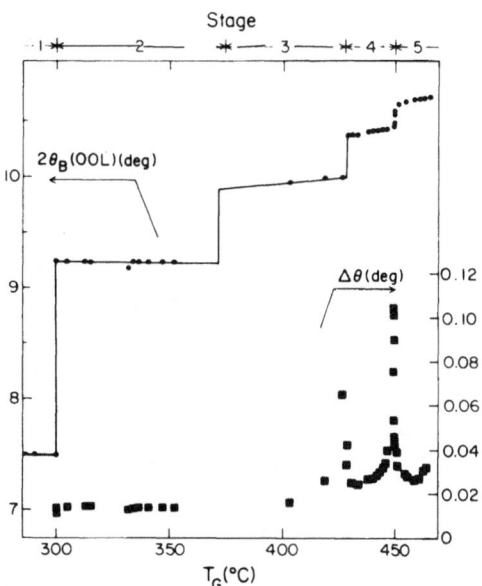

Fig. 13. θ_{Bragg} and half width at half maximum $(\Delta\theta)$ of (00ℓ) peaks during staging transitions in K-GIC ($T_K = 230°C$ (from ref.37).

to stage 1 no peak shift nor broadening was observed. These results show convincingly that for higher stages the transition proceeds through stage disordering in contrast with the previous results. Moreover the level of disorder increases with stage both in the stability and transition regions.

These findings are in agreement with other experimental observations by Fischer et al. [57] and Heiney et al. [5] about the existence of mixed stages and stage disorder in higher stage compounds. They are also consistent with the theory of Kirczenow [13] who predicted such disorder to be a consequence of finite intercalate domain sizes (see III.1). However the theory predicts continuous staging transitions while the results of Misenheimer and Zabel favor first-order ones from the coexistence of reflections from stage n and stage n+1 units in the transition region. Actually because of sample imperfections, inhomogeneities and the very sluggish kinetics of these transitions the argument about coexistence is not so convincing [4,58].

Finally let us note that the absence of stage disorder reported in earlier studies is likely to be due to a lack of angular resolution especially with the energy-dispersive method used by Nishitani et al. [35]. The quality of the host material and thermal history may also play a role. In this respect and considering that previous experiments used HOPG samples it would be interesting to carry out similar high-resolution work on single crystals.

In-situ X-ray scattering studies also prove to be attractive with the elegant work of Nishitani et al. [36] on the stage 2 - stage 1 and melting transitions in K-GIC. Taking advantage of the high intensity of the synchrotron radiation beam the authors were able to measure the scattered intensity along the $(q_a,0,0)$ direction. T_K (P_K) was varied while T_G was held constant (390°C) corresponding to the line AB in the schematic phase diagram of Fig. 14 (inset).

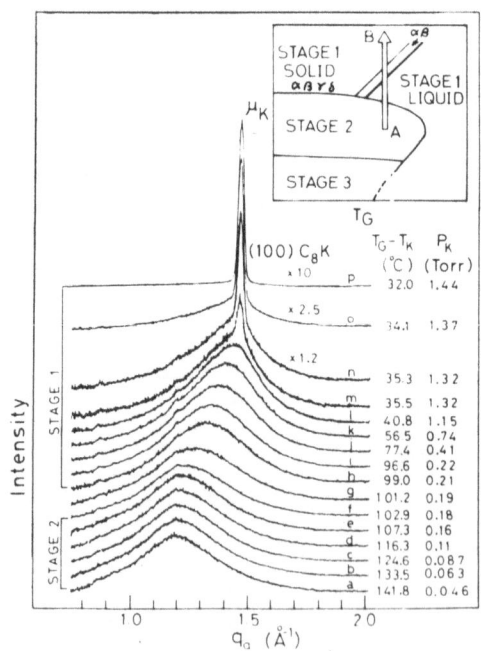

Fig.14

Evolution of X-ray diffraction spectrum along $(q_a,0,0)$ during the stage 2 - stage 1 and liquid-solid transitions in K-GIC (T_G is fixed at 390°C). Inset : schematic phase diagram, scan is along AB (from ref. 36).

At low T_K a broad peak centred at q_0 and characteristic of the liquid state is observed in the stage 2 and stage 1 phases. However q_0 increases gradually in both stages but abruptly at the staging transition. It saturates at $q_0 = 1.46$ $Å^{-1}$ as a sharp Bragg peak appears that corresponds to the (100) reflection of the stage 1 solid phase (i.e. the p (2×2) RO° structure of KC_8, see 1.4). Assuming a simple type of liquid scattering Nishitani et al. derived the in-plane K concentration from the wave vector q_0 of the liquid halo and they found that m (in $C_{mxn}K$) varies from 12.41 to 11.47 in stage 2, drops to m = 9.52 at the transition and decreases to m=8 in stage 1. This is in qualitative agreement with the observed variation of weight uptake [55].

Another important result of this study concerns the K interlayer ordering in the stage 1 solid. Nishitani et al find a narrow region (see Fig. 14, inset) where the K layer stacking is of the type $\alpha\beta$, between the liquid and the already known $\alpha\beta\gamma\delta$ stacking ragion. It is remarkable that such a succession of phases (i.e. disorder - $(2 \times 2)\alpha\beta$ - $(2 \times 2)\alpha\beta\gamma\delta$) was actually predicted to occur for some values of the interlayer interactions as a result of a lattice gas Landau-Lifshitz theory by Lee, Aoki and Kamimura [59].

3.2. High Pressure Phase Transitions

Hydrostatic pressure provides a very efficient means to induce stage transformations and in-plane structural changes in GICs as shown by several recent high-pressure X-ray and neutron diffraction experiments in alkali metal compounds [4]. Pionnering results were obtained by Clarke, Wada and Solin [38] using the diamond-anvil pressure cell technique and X-ray diffraction with single crystals and HOPG samples. For instance, in stage 2 KC_{24} (nominal composition) they observed a stage transformation to stage 3 extending in a broad pressure range from about P = 2.5 kbar to P= 6.5 Kbar. As the transition is reversible it was considered that no significant amount of K was expelled from the sample so that the overall K content remained constant. This implies an increase of the in-plane density which can be expressed as

$$KC_{12 \times 2} \rightarrow KC_{8 \times 3}$$

(in KC_{mxn} m and n represent the areal C/K ratio and the stage respectively).

At room temperature the K atoms in KC_{24} are known to be disordered and the KC_{12} areal composition is approximate (see 3.1 and 3.3.). Then it was inferred that the above densification corresponds to the ordering of the K atoms in the simple (2×2) structure realized in the stage 1 KC_8 compound. This was actually demonstrated by X-ray oscillation photographs as a function of pressure revealing the presence of characteristic (hk0) reflections from the (2×2) structure above P = 4 Kbar. Such correlated effects of in-plane densification and transition to higher stage were also observed in higher stage K - GICs [60] and for other alkali metal intercalants [61]. For example, in the case of stage 4 RbC_{48}, Wada [61] reported a transition to stage 6 which also agrees with a densification leading to an ordered (2×2) in-plane structure according to the relation : $RbC_{12x4} \rightarrow RbC_{8x6}$

A very recent extensive high pressure neutron diffraction study by Kim, Fischer, McWhan and Axe [40] on HOPG samples has provided a more detailed picture of the high-pressure effects and it is mostly coherent with previous data. From an Hendricks - Teller type of analysis of the (00ℓ) profiles the authors were able to identify stage disorder as an essential property of the pressure induced staging transitions. For stages higher than 1 the evolution of the structural parameters determined by Kim et al. (relative fraction of stage units, width of the (00ℓ) peaks, c lattice parameters) reveals a succession of continuous and discontinuous variations which reflects combined intralayer and staging effects. A clear understanding of these processes is still lacking because the in-plane structure could not be probed. For instance the transition from stage 2 to stage 3 was shown not to be continuous as first seen by Clarke et al [38] but two anomalies at P = 3.5 and 6.5 Kbar were detected (Fig. 15). They coincide with previously found resistivity anomalies. These anomalies could be related to different steps in the transition from the disordered K layer of KC_{12x2} to the ordered (2×2) structure of KC_{8x3} and/or to the stabilization of intercalate layer stacking correlations. Clearly a high-pressure study on single crystals should help by allowing study of the intercalate structure simultaneously.

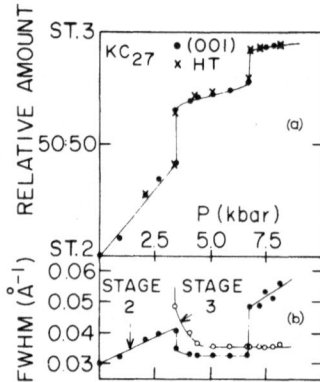

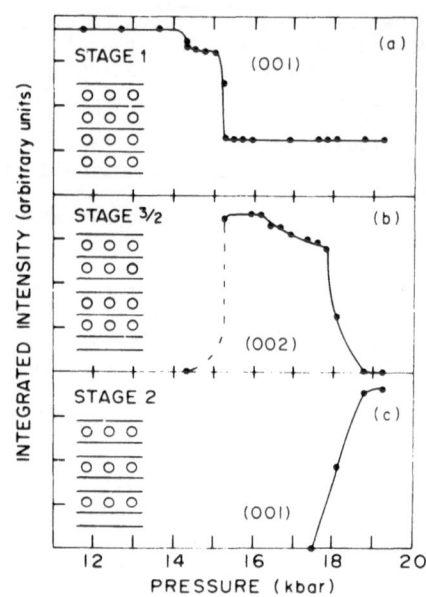

Fig. 15. Pressure dependence of the relative stage 3 - stage 2 fractions and (001) FWHM's in KC_{27} showing 2 transitions at P $\sim$ 3.5 and 6.5 kbar (from ref. 40).

Fig. 16 Integrated intensity vs pressure for the phases observed in KC_8. At 19 Kbar stage 3/2 disappears while stage 2 appears but stage 1 is unaffected (from ref. 39).

The behavior of stage 1 KC_8 deserves special attention. Previous X-ray work by Wada et al. [60] revealed no effect up to T = 12 Kbar while a resistivity anomaly was observed at about P = 15 Kbar [42].Then a neutron scattering study by Fuerst et al. [39] revealed very unusual staging effects which are now described. These authors showed that there are actually 2 phase transitions at approximately P = 15 and 19 Kbar. Starting from stage 1 at low pressure and scanning along the (00ℓ) direction they observed a new phase identified to be stage 3/2 that is a ..CK CCK.. carbon-potassium layer sequence. This phase coexists with stage 1 until P $\sim$ 19 Kbars where it disappears while a stage 2 phase emerges. This is shown in Fig. 16. It was suggested that these effects involve a densification with respect to the KC_8 ambient pressure limit to a $(\sqrt{3} \times \sqrt{3})KC_6$ structure. This hypothesis could not be checked however with neutron scattering on HOPG. A single crystal X-ray study using a diamond-anvil cell was carried out by Bloch et al. [41]. Diffraction photographs actually confirmed the existence of the $(\sqrt{3} \times \sqrt{3})$ structure from the observation of its characteristic (10ℓ) reflections. They also showed that the (2 x 2) and $(\sqrt{3} \times \sqrt{3})$ structures coexist in an intermediate pressure range but it is found to be much broader (10 - 23 Kbar) than the stability region of stage 3/2 (15 - 19 Kbar). Of interest is the report of very slow kinetics, large hysteresis of the transitions, stage disorder and graphite layer stacking faults. The connection of stage with in-plane structure through the staging transitions was suggested by Fuerst et al. [39] to follow the sequence :

$$6KC_{8x1} \rightarrow 4KC_{6x1} + K_2 C_{8x3} \rightarrow 4KC_{6x1} + 2KC_{6x2}$$

It assumes that the stage 3/2 phase has the (2 x 2) structure and this seems to be consistent with a comparison of observed vs calculated (00ℓ) intensities through the phase transitions by Kim et al. [40].

These studies should develop using higher pressures combined with low-temperature and they should extend rapidly to molecular intercalants as already initiated by Houser, Homma and Clarke [76] in graphite-$SbCl_5$ and graphite-HNO_3.

3.3. Structure and Low-Temperature Behavior of Stage 2 Heavy Alkali Metal GICs

For clarity and briefness I restrict myself to stage 2 GICs of the heavy alkali metals (AM) although they have close relations with higher stage compounds.

At room temperature the AM layers are uncorrelated while the graphite stacking has the sequence AB/BC/CA/ as first shown by Parry, Nixon and coworkers [62]. The ordering of the AM atoms is therefore two-dimensional (2D) and it is described as a liquidlike state. From experiments in HOPG it was found that the diffuse scattering associated with the liquid corresponds to an AM spacing of about 6 Å with a coordination number of 6 [63] . However single-crystal photographs in CsC_{24} by Parry [64] and recently by Rousseaux et al. in KC_{24} [47] and RbC_{24} [65] revealed complex anisotropic effects which are illustrated in Fig. 17, for KC_{24}. The liquid-like halo of diffuse scattering centered at the (000) reciprocal lattice (r.1) point shows a six-fold intensity modulation whose maxima are along the $[10.0]_G$ r.1. directions. Furthermore the halos are repeated by translations about the $[10.0]_G$ graphite reflections. Similar patterns are observed in RbC_{24} and CsC_{24} except that the intensity maxima are oriented along $[11.0]_G$ instead of $[10.0]_G$. These observations indicate that the liquid state exhibits pronounced orientational correlations and that the graphite-AM interaction is strong. The most reasonable cause of these effects is that the graphite host potential modulates the AM liquid state. Actually, this was very recently shown to be coherent with a theoretical treatment by Reiter and Moss [66] of a 2D liquid modulated by a periodic host. According to this theory the replication of the liquid halos about the $[10.0]_G$ r.1 points and their anisotropy appear as higher order effects of the potential. On the other hand the AM contribution to the graphite reflections is shown to be a first-order effect and a method is given for extracting the Fourier coefficients of this potential from the intensity of the graphite reflections [66] .

Dynamical aspects of the modulated liquid have been identified by inelastic and quasi-elastic neutron scattering [67]. It is found that the Rb atoms in RbC_{24} exhibit both solidlike and liquidlike diffusional behaviors at room temperature with the former increasing as T is reduced while the latter dominates at high temperature [67].

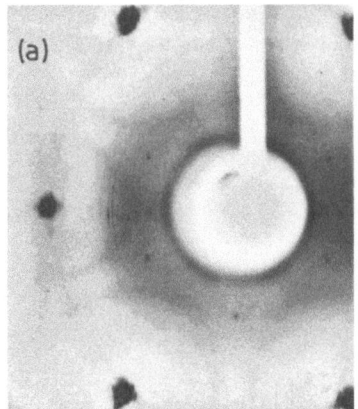
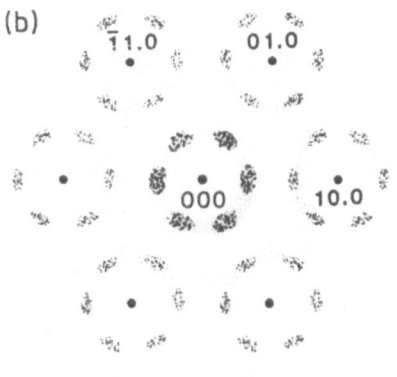

Fig. 17. (a) Fixed-crystal fixed-film X-ray photograph (cylindrical film) of KC_{24} at T ∿ 300 K showing diffuse halos with maxima along $[10.0]_G$ (from ref. 47) (b) schematic representation of the principal effects in (a) (from ref. 66).

Upon cooling the AM atoms order in the layers (via one or two transitions) and the liquid scattering features are replaced by Bragg reflections which, for the most part, emerge from the diffuse halos. For instance in KC_{24}, the K atoms order at $T_u = 124$ K into a triangular lattice, incommensurate with the graphite, whose fundamental r.l. vectors Q are rotated by 7.5° from the graphite [10.0] directions. This rotation angle θ is an important parameter whose value depends on the alkali metal and on the stage (e.g. 11° for RbC_{24} [65] and 14.5° for CsC_{24} [68]. It was found to be related to the length of the fundamental r.l. vectors through the trigonometric relation

$$\cos(30° - \theta) = (1 + 2z^2) / 2\sqrt{3}\,z \quad \text{where } z = \frac{|Q|}{|G|}$$

and G is the graphite r.l. vector. Furthermore the diffraction patterns contain many extra reflections which can be indexed as satellites due to an intermodulation of the AM and carbon lattices. Several models of the in-plane arrangement of the AM atoms have been developed to account for these effects. Following Clarke et al. [69] the most recent ones [70,71,72] are based on domain structures where the commensurate domains are made of the p $(\sqrt{7} \times \sqrt{7})$ R 19.11° structure. There are 7 equivalent sites leading to 7 different types of domains. In the domain structure these are arranged in an ordered honeycomb pattern (see Fig. 4, ref. [70]) which can be considered equivalently as a large superstructure. Recently calculated X-ray intensities based on the discommensuration-domain model by Suzuki [72] were compared with experimental data in KC_{24} [68] and CsC_{36} [73] with a better agreement for CsC_{36}. The corresponding domain pattern is shown in Fig. 18. Note that in this model the domain size and the orientation of the domain walls Ω are simply related to $|Q|$ and θ [72].

The validity of the domain model has been discussed recently [74] and it appears that while it provides a good description of the ordering in the case of Cs where the $(\sqrt{7} \times \sqrt{7})$ domains are well defined and large enough, it is less satisfactory as one goes to Rb and especially K because the domain sizes which are involved become very small. Therefore an alternative approach by Di Cenzo [75] is worth con-

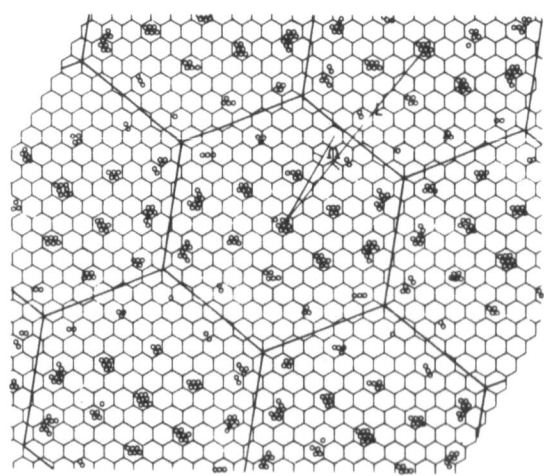

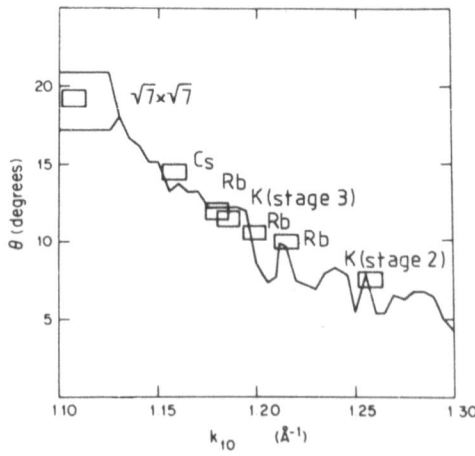

Fig. 18. Discommensuration-domain pattern for CsC_{36}. The small and large honeycomb arrays represent the graphite layer and domain walls respectively. Small circles symbolize the probability of Cs occupancy (from ref. 72).

Fig. 19. Optimized θ value (see text) as a function of $|Q|$ (noted k_{10}) for AM-GICs. Rectangles correspond to experimental values (from ref. 75).

sidering. It assumes that the AM atoms form an incommensurate close-packed array defined by Q and rotated by θ with respect to graphite. Then the AM atoms are allowed to relax to the nearest graphite hexagon, leading to a registered structure. This was done systematically as a function of $|Q|$ (Fig. 19), searching for the maximum θ value for which the minimum separation between AM atoms is $2 a_G$ (looking for θ max means that $\sqrt{7} a_G$ distances are favored since θ = 19.11° for the $\sqrt{7} \times \sqrt{7}$ structure). The agreement with the observed values of θ for several compounds is satisfactory. Actually more accurate intensity data would be needed to test these different models and it might eventually appear that the structure of stage 2 and higher stage K, Rb and Cs GICs cannot be conveniently described within a unique framework.

This is supported in some respects by the behavior of KC_{24} where a second transition at $T_\ell \simeq 95 K$ induces a dramatic change of the diffraction pattern [47]. The intricate r.l. pattern (Fig. 20) has been unravelled recently (several domains due to equivalent orientations are present, see 1.4) and explained with a 2D oblique lattice with unit cell parameters a = 6.1 Å, b = 5.81 Å and γ = 124.14° [74]. This corresponds to a breaking of the six-fold symmetry of the higher temperature phase in contrast with the behaviors of RbC_{24} and CsC_{24} where the hexagonal symmetry is preserved down to T ≃ 10 K.

4. STRUCTURE AND PHASE TRANSITIONS OF ACCEPTOR COMPOUNDS

Looking back to the review papers of 1981-82 [2,3] it is clear that few detailed structural studies of acceptor GICs had been done until the beginning of the eighties. This is certainly related to the complexity of the structural phenomena due to the molecular nature of the acceptor intercalants. In the past 3 or 4 years several important and unexpected results appeared and the development of these studies is now ensured. In this section I have chosen to summarize some results obtained with the bromine and nitric acid intercalants. However other compounds like the transition metal chlorides ($FeCl_3$, $MnCl_2$, $NiCl_2$, $CoCl_2$) or the pentahalides of antimony and arsenic ($SbCl_5$, AsF_5) deserve attention. In particular the graphite $SbCl_5$ compound is being studied extensively as it shows interesting effects

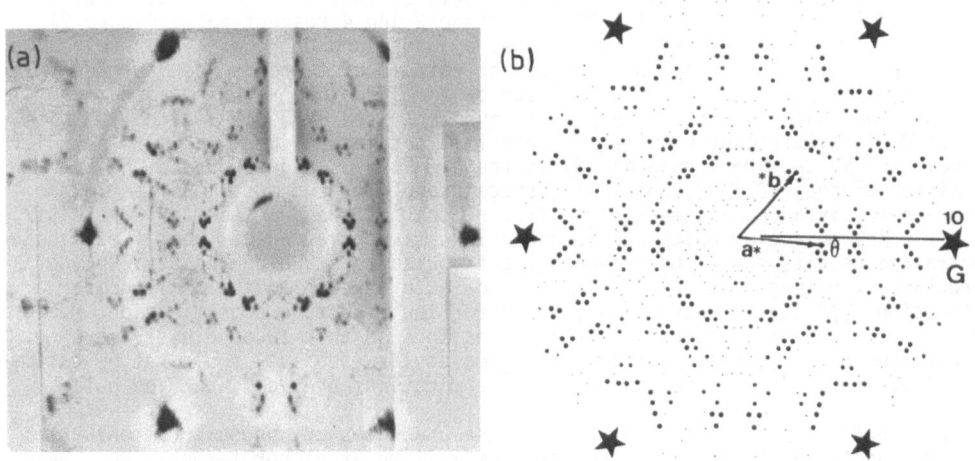

Fig. 20. (a) Fixed-crystal fixed-film X-ray photograph of KC_{24} at T ≃ 10 K (b) Simulated pattern obtained with an oblique reciprocal unit cell by superposing the symmetry-related domains and the graphite-induced modulation satellites (restricted to 1st order) (from ref. 65).

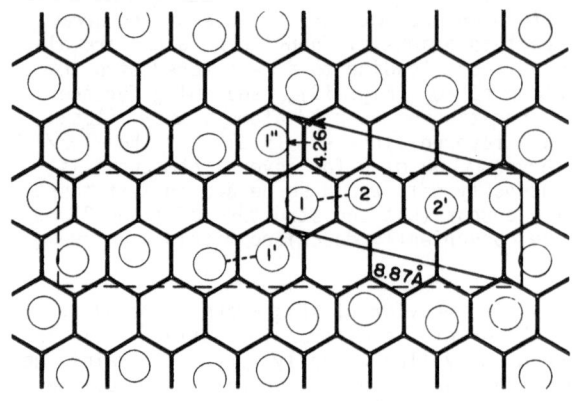

Fig. 21
Two-dimensional structure of the stage 2 $C_{3.5x2}$ Br compound at room temperature Br_1-Br_1, = 2.25 Å , Br_1-Br_2 = 2.38 Å, Br_1-Br_1" = 2.55 Å and Br_2-Br_2, = 2.86 Å. The broken line has been added to show the ($\sqrt{3}$ x 7) unit cell used by Erbil et al. (from ref. 25).

(existence of segregated species, $SbC\ell_5$, $SbC\ell_6^-$, $SbC\ell_3$, order-disorder transitions, glass phase due to electron-beam damage, pressure-induced molecular ordering [50, 76,77]).

4.1. Bromine GICs

Following the pioneering electron and X-ray structural work of Eeles and Turnbull in the mid-sixties [78] the bromine GICs have been studied extensively since 1980 and they prove to be excellent materials for the study of competing interactions in quasi-2D systems. There has been some confusion in the literature as regards the nature of the stable room-temperature phases of graphite-bromine. However it seems now established that several phases with different in-plane bromine densities can be stable for a specific stage (at least for low stages).

The highest in-plane bromine concentration of $C_{3.5}$ Br corresponds to a layer structure which has been determined from single-crystal X-ray diffraction by Ghosh and Chung [25] and Erbil et al. [26] for stage 2 and 4 respectively. Figure 21 shows the bromine atoms positions obtained by Ghosh and Chung within an oblique unit cell (space group p2). The largest interatomic distance is Br_2-Br_2' which is 2.86 Å and this lead the authors to describe the structure as being formed of poly-bromine chains built from Br_4 units and separated by 8.61 Å. Actually this structure is significantly different from that obtained by Erbil et al. (see Fig.6 in ref. 26) in C_{28} Br_2 (stage 4). First the symmetry is higher due to the existence of mirror planes ($c(\sqrt{3}\times 7)$ unit cell, space group cmm) and most importantly the interatomic distances are markedly different so that the bromine atoms are distributed in Br_2 molecular units and no chain-like structure can be identified. These contrasted results cannot be reasonably attributed to stage-dependent in-plane structures and they certainly reflect a poor accuracy due to the small number of reflections (18 and 9, respectively) used for the structure determination. Moreover both studies were restricted to (hk0) reflections so that z coordinates were not accessible and the likely tilt of the Br_2 molecules (see 1.4) could not be estimated. Clearly more accurate structure determinations should be carried out.

Upon heating this bromine-saturated phase undergoes an unusual commensurate incommensurate (C-I) transition followed by melting at higher temperature. These transitions were first revealed by Bardhan, Wu and Chung [79] and analyzed in great details by Kortan et al. [80] and Erbil et al. [26] in the case of C_{28} Br_2 (stage 4). The main findings of these high-resolution X-ray scattering experiments are now described. The (hk0) reciprocal plane is represented in Fig. 22a. In the third direction, c∗, the Bragg rods relative to the (hk0) points are modulated at room temperature but the modulation disappears by T = 325 K indicating uncorrelated bromine layers above this temperature. Then at T_c = 342.2 K the Bragg rods relative to the bromine structure shift along the k direction only, that is along the 7-fold direction as indicated by the arrows (Fig. 22a). This corresponds to a 1D incommensurate (I) modulation which was interpreted as a stripe-domain structure. It consists of commensurate (C) domains separated by domains walls running in one direction only, a (or $\sqrt{3}\ a_G$) in the present case. The theory of the stripe-domain phase was first formulated by Pokrovsky and Talapov [90] who predicted that the domain-wall density should exhibit a simple temperature behavior.

$$\rho \propto (T - T_c)^\beta \quad \text{with } \beta = 0.5$$

If Nb is the distance between the domain walls then $\rho = \frac{1}{Nb}$. Assuming a net displacement at the domain wall of $\frac{2b}{7}$ (or $2a_G$) the corresponding phase shift is $\frac{2b}{7} \times \frac{2\pi}{b} = \frac{4\pi}{7}$. The incommensurability ε, i.e. the peak shift wih respect to the nearest commensurate Bragg peak, is given by

$$\varepsilon \propto \frac{4\pi}{7Nb} = \frac{4\pi}{7}\rho$$

The shifts also depend on the peak indices as indicated in Fig. 22a with different symbols. The temperature dependence of ε for the 3 main peaks near the transition is shown in Fig. 22 b . The experimental points are very well fitted by a power-law $\varepsilon = \varepsilon_0 \left(\frac{T}{Tc} - 1\right)^\beta$ with $\beta = 0.50 \pm 0.02$ in agreement with the theory. At higher temperature the incommensurability saturates when the domain size decreases to about 130 Å because the repulsive interaction between domain walls limits their number.

Further high-resolution synchrotron X-ray experiments by Mochrie et al. [81]

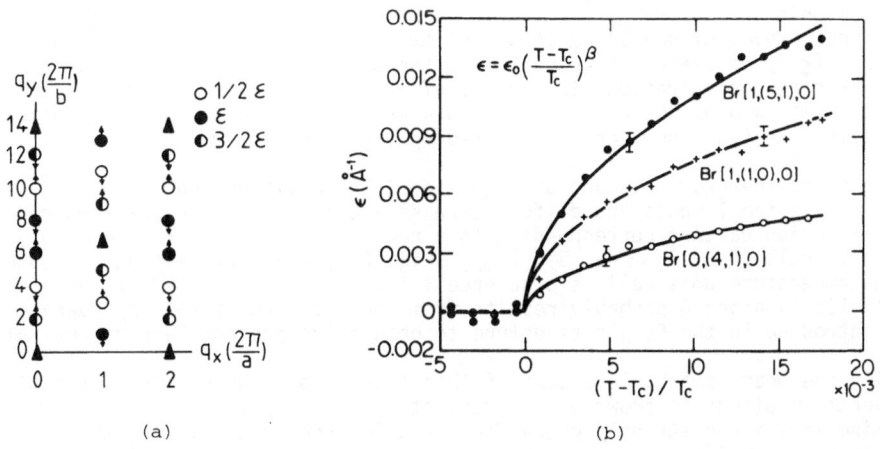

(a)   (b)

Fig. 22 a) Quadrant of the (hk0) reciprocal lattice of the bromine structure. Circles correspond to Bragg rods running along c∗ and triangles represent graphite + bromine peaks. The arrows show the principal peak shifts at the CI transition (from ref. 26).
b) Shifts of the principal peak positions as a function of reduced temperature. They correspond to 1/2 ε [0,(4,1),0], ε [1,(1,0),0] and 3/2 ε [1,(5,1),0] as indicated in (a). Solid lines are fit with power laws and β = 0.5 (from ref. 26).

on this remarkable C-I transition were devoted to an analysis of the intensity and profile of the satellites in order to test theoretical predictions that the order-parameter correlations should be of the form :

$$< \rho (\vec{r}) \rho (o) > \, \alpha \, r^{-n} \quad \text{(where r is the distance)}$$

and that $\eta_n = 2 \, n^2/p^2$ [82] (n is the order or harmonic number of the satellite and it is proportional to its displacement from the C position, p is the multiplicity of the C structure relative to the reference lattice, p = 7 here). Accordingly the structure factor i.e. the Fourier transform of the real space correlation function should be of the form $| \vec{q} |^{-2+n}$ ($\vec{q}$: reciprocal space wave vector). The measurements were performed at five successive harmonics of the I modulation that is reflections displaced by $1/2 \, \epsilon$, ϵ, $3/2 \epsilon$, 2ϵ and $5/2 \, \epsilon$ from the C positions. The analysis strongly supports the above predictions about the algebraic decay of the correlations and the temperature independence of η [81].

The stripe-domain phase is found to melt at 373.4 ± 0.1 K [26,79] leading to a liquid phase which shows interesting anisotropic effects like a retention of order in the $\sqrt{3}$ direction.

Turning now to a slightly lower in-plane bromine concentration of about C_4Br (as in stage 2 C_8Br) we obtain a compound characterized by a different room-temperature structure which is incommensurate with the graphite (in contrast with the above C phase of $C_{3.5}Br$) and was studied by several groups [49,79,83]. (hk0) diffraction photographs consist of both sharp spots and diffuse streaks. The spots can be accounted for using 1) a 2D oblique unit cell a = 4.36 Å, b = 8.63 Å , γ = 99.45° such as b is along the $\sqrt{3}$ direction, 2) graphite induced modulation satellites [34]. The position of the bromine atoms and the origin of the incommensurability have not been fully established (a 1D model was proposed however by Ghosh and Chung [49] but the structure is certainly closely related to the $C_{3.5}Br$ one with a rearrangement of the Br_2 molecular units due to the lack of one unit per two ($\sqrt{3} \times 7$) units cells as indicated by the C_4Br stoichiometry.)

Most interesting are the temperature-induced phase transitions of this compound. Upon heating an I-C transition was reported to occur at T_c = 319 K and it happens to lead to the $\sqrt{3} \times 7$ unit cell again [49] . However it is possible that this transition also involves a partial bromine de-intercalation and the in-plane concentration may decrease to about C_7Br. Actually this concentration is compatible with the ($\sqrt{3} \times 7$) unit cell as originally found by Eeles and Turnbull [78] . It corresponds to the removal of one Br_2 unit out of two in the $C_{3.5}Br$ structure considered above. The likely existence of these 2 structures with the same ($\sqrt{3} \times 7$) unit cell but with bromine concentrations in a ratio of 1 to 2 is the origin of some confusion in the literature. The need for more work on the phase diagram of graphite - bromine using in-situ experiments as in K-graphite is clear.

Upon cooling the C_8Br compound undergoes an I-C transition too. It occurs at T ≃ 275 K where the I spots and diffuse streaks are replaced by a well-ordered (hk0) diffraction pattern corresponding to a rectangular lattice (space group) with a large unit cell (A = 10 $\sqrt{3} a_G$, B = 7 a_G) (Fig. 23) [21]. Interestingly enough, this low-temperature unit cell is a supercell (ten-fold) of the ($\sqrt{3} \times 7$) unit cell. The multiplicity along A probably results from the ordering of the Br_2 "vacancies" one can introduce in the $C_{3.5}Br$ structure to obtain the present C_4Br concentration.

Finally the most remarkable aspect of this transition concerns the intercalate interlayer correlations as shown by Aberkane et al. [21,34]. While, at room temperature the bromine layers are strongly correlated as evidenced by the intensity modulation of diffuse rods running along $c*$, the correlations are lost in the low temperature phase as the modulations disappear (Fig. 23b). The establishment of in-plane order with a large unit-cell seems to inhibit correlations between the planes. Thus we have here an interesting example of a crossover from a high temperature 3D incommensurate phase to a low-temperature 2D commensurate phase.

The complex behavior of the graphite-bromine compounds with temperature originates essentially from the very different thermal contraction-expansion of the soft bromine and stiff graphite substructures.

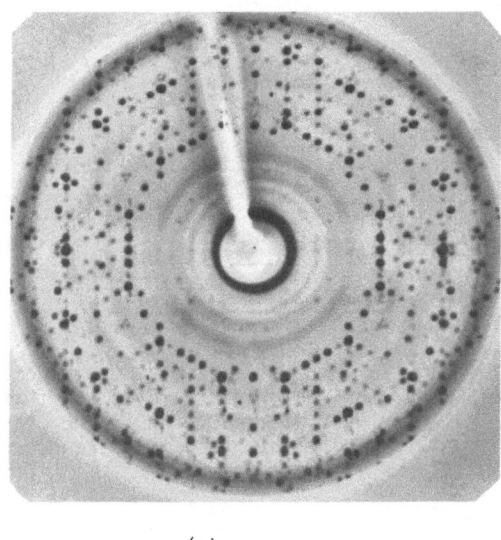

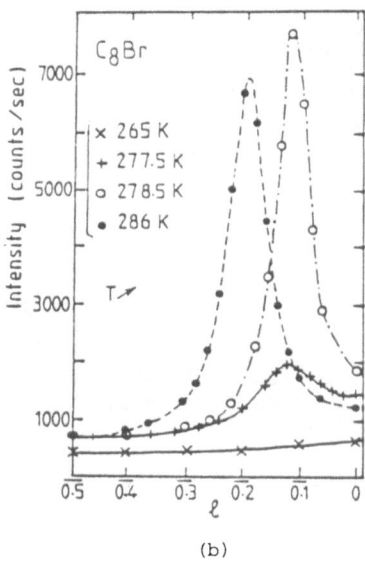

(a)   (b)

Fig. 23 a) (hk0) X-ray precession photograph of C_8Br at T ≃ 233 K. Three orientations of the domains are superposed (from ref. 84).
b) Diffractometer scans along a diffuse rod at q = 1.46 Å^{-1} showing the loss of
interlayer correlations at the I-C transition (T_c ≃ 275 K). (ℓ : c* unit)
(from ref. 34).

4.2. Nitric Acid GICs

Two varieties of HNO_3 GICs have been reported and they exhibit very different
structural properties. In anhydrous conditions one obtains α-C_{5n} HNO_3 which is characterized by a large intercalate sandwich distance d_s = 7.80Å [19] and for which
a number of different techniques have shown that the plane of the triangular NO_3^-
ions is almost perpendicular to the layer (a tilt of about 10° is often admitted)
[43,84,85,86] . The second variety β-C_{8n} HNO_3 is obtained by allowing the α variety
to sit in air for a few days which causes some de-intercalation and d_s drops to
6.55 Å . Thus the NO_3^- ions need to be almost parallel to the layers. The HNO_3 molecules are ordered at room temperature in the β phase [83 a] and their very complex
in-plane arrangement is probably stabilized by the presence of H_2O molecules as
suggested by Clinard et al. [87], leading to a structure similar to that of HNO_3,
H_2O.

However it is the α-C_{5n} HNO_3 compound which has received most attention in recent years because of the room temperature liquid HNO_3 layer and of the liquid -
solid transition at T ∿ 250 K, as first studied by Nixon et al. [19].I now describe
three peculiar structural aspects of α-$C_{5n}HNO_3$, the 2D anisotropic liquid at room
temperature, the liquid to incommensurate solid transition at T_m = 247 K and the
associated effects on the graphite layer stacking which were recently studied by
Samuelsen et al. [20] X-ray diffraction of the (hk0) plane at room temperature
shows a central halo of diffuse scattering at q ≃ 1. 77 Å^{-1} and no Bragg reflections [19,20,34] . This indicates that the disordered HNO_3 molecules have an average distance of very roughly $a_G\sqrt{3}$. A remarkable feature is that the intensity is
modulated in the halo with 6 broad maxima along the {110} graphite directions (see
Fig.1, ref. 34), showing orientational correlations between the molecules as a re-

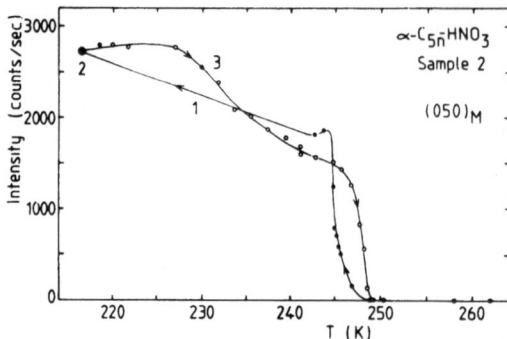

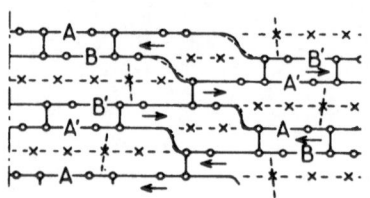

Fig. 24 Peak intensity vs temperature for a reflection from the HNO3 structure (from ref. 20).

Fig. 25 Schematic Daumas-Herold domain model for the graphite layer shift at T_m for a stage 2 α-C_{5n} HNO3. The possible supplementary tilt of the NO3 planes is indicated.

sult of the symmetry of the graphite potential. These correlations increase when the temperature is lowered. Furthermore the diffuse rods along $c\star$ relative to the halo show a monotonic intensity variation which is a signature of the 2D nature of the liquid. Recent quasi-elastic neutron scattering studies have shown that the diffusive motion of the molecules has both rotational and translational components [88].

At T_m = 247 K a liquid-solid transition showing hysteresis effects (Fig. 24) leads to a structure incommensurate with the graphite ($\vec{A}$ = (17 ± ε) $\vec{a}_G$, $\vec{B}$ = 2$\vec{a}_G$ + 9$\vec{b}_G$) and ε is found to be temperature dependent and decreases as T is reduced. In some freshly intercalated crystals it even ends up in a commensurate structure (ε = 0)[20]. However a different (hk0) pattern was recently obtained by Clarke et al. [86] in a stage 4 sample. It was interpreted as a coexistence of two registered structures.

Finally I mention that the liquid-solid transition has an interesting effect on the graphite layer stacking. In the liquid this stacking follows the common sequence A/AB/BA/A for stage 2 but at T_m the graphite layers are pulled out of A/A alignment as first noted by Parry [31] and confirmed recently [20], leading to a sequence of the type A/A'B'/BA/A' as shown in Fig. 25.

An attractive and simple model for the resulting shift of the graphite layers can be proposed if one remarks that the liquid-solid transition involves a shortening of the c axis by 0.04 - 0.05 A [89]. This may be due to a supplementary tilt of the HNO3 molecules on the order of 6° which could therefore be associated with a relative shift of the graphite bounding layers [20].

Acknowledgements

I wish to thank all the collaborators and colleagues who have contributed to my knowledge in this field. I owe thanks to the members of Professor Herold's group in Nancy, G. Furdin, H. Fuzellier, D. Guérard, M. Klatt, P. Lagrange and M. Lelaurain for the good quality of the intercalated crystals they provided. I acknowledge a fruitful interaction with F. Batallan, I. Rosenman and C. Simon. Finally, I am especially indebted to my close collaborators, F. Aberkane, F. Rousseaux and E. J. Samuelsen, who obtained some of the results presented in this chapter.

REFERENCES

1. A. Hérold in Physics and Chemistry of Materials with Layered Structures, F. Levy ed. Reidel, Dordrecht, (1979) p.323
2. M.S. Dresselhaus and G. Dresselhaus, Adv. Phys., 30, 139 (1981)
3. S.A. Solin, Adv. Chem. Phys. 49, 455 (1982)
4. R. Clarke and C. Uher, Adv. Phys. 33, 469 (1984)
5. P.A. Heiney, M.E. Huster, V.B. Cajipe and J.E. Fischer, Synth. Metals, 12, 21 (1985)
6. S. Hendricks and E. Teller, J. Chem. Phys. 10, 147 (1942)
7. J. Kakinoki and Y. Komura, J. Phys. Soc. Japan 7, 30 (1952)
8. N. Daumas and A. Hérold, C.R. Acad. Sci. Paris C 268, 373 (1969)
9. G. Timp and M.S. Dresselhaus, J. Phys. C 17, 2641 (1984)
10. S.A. Safran and D.R. Hamann, Phys. Rev. Lett. 42, 1410 (1979)
    S.A. Safran, Phys. Rev. Lett. 44, 937 (1980)
11. S. Ohnishi and S. Sugano, Solid State Commun. 36, 823 (1980)
12. G. Kirczenow, Phys. Rev. Lett. 49, 1853 (1982)
13. a) G. Kirczenow, Phys. Rev. Lett. 52, 437 (1984)
    b) G. Kirczenow, Phys. Rev. B 31, 5376 (1985)
14. D.E. Nixon and G.S. Parry, J. Phys. C 2, 1732 (1969)
15. D. Guérard, C. Zeller and A. Hérold, C.R. Acad. Sci. Paris C 283, 437 (1976)
16. S. Flandrois, J. Masson, J. Rouilloun, J. Gaultier and C. Hauw, Synth. Metals 3, 1 (1981)
    R.S. Markiewicz, J.S. Kasper and L.V. Interrante, Synth. Metals, 2, 363 (1980)
17. a) L. Pietronero and S. Strassler, Phys. Rev. Lett. 47, 593 (1981)
    b) M. Kerstesz, Mol. Cryst. Liq. Cryst. 126, 103 (1985)
18. N. Caswell, S.A. Solin, T.M. Hayes and S.J. Hunter, Physica 99B, 463 (1980)
19. D.E. Nixon, G.S. Parry and A.R. Ubbelohde, Proc. Roy. Soc. London A291 324 (1966)
20. E.J. Samuelsen, R. Moret, H. Fuzellier, M. Klatt, M. Lelaurain and A. Hérold Phys. Rev. B32, 417 (1985)
21. F. Aberkane, R. Moret and G. Furdin to be published
22. J.M. Cowley and J.A. Ibers, Acta Cryst. 9, 421 (1956)
23. D.D.L. Chung, G. Dresselhaus and M.S. Dresselhaus, Mat. Science and Engn. 31, 107 (1977)
24. W.T. Eeles and J.A. Turnbull, Proc. Roy. Soc. A283, 179 (1965)
25. D. Ghosh and D.D.L. Chung, Materials Res. Bull. 18, 1179 (1983)
26. A. Erbil, A.R. Kortan, R.J. Birgeneau and M.S. Dresselhaus, Phys. Rev. B28, 6329 (1983)
27. J.L. Feldman, E.F. Shelton, A.C. Ehrlich, D.D. Dominguez, W.T. Elam, S.B. Qadri and F.W. Lytle, Bull. Am. Phys. Soc. 28, 346 (1983)
28. D.P. DiVincenzo and E.J. Mele, Phys. Rev. B32, 2538 (1985)
29. D. Ghosh and D.D.L. Chung, Mat. Res. Bull. 18, 727 (1983)
30. M. Suzuki, Phys. Rev. B33, 1386 (1986)
31. G.S. Parry, Mater. Sci. Engn., 31, 99 (1977)
32. P. Lagrange, D. Guérard and A. Hérold, Ann. Chim. Fr, 3, 143 (1978)
33. J. Rossat-Mignod, D. Fruchart, M.J. Moran, J.W. Milliken and J.E. Fischer, Synth. Metals 2, 143 (1980)
34. F. Aberkane, F. Rousseaux, E.J. Samuelsen and R. Moret, Annales de Physique, Coll. N° 2, 11, 95 (1986)
35. R. Nishitani, Y. Uno and H. Suematsu, Phys. Rev. B27, 6572 (1983)
36. R. Nishitani, Y. Uno, H. Suematsu, Y. Fujii and T. Matsushita, Phys. Rev. Lett. 52, 1504 (1984)
37. M.E. Misenheimer and H. Zabel, Phys. Rev. Lett. 54 2521 (1985)
38. R. Clarke, N. Wada and S.A. Solin, Phys. Rev. Lett. 44, 1616 (1980)
39. C.D. Fuerst, J.E. Fischer, J.D. Axe, J.B. Hastings and D.B. McWhan, Phys. Rev. Lett. 50, 357 (1983)
40. H.J. Kim, J.E. Fischer, D.B. McWhan and J.D. Axe, Phys. Rev. B33, 1329 (1986)
41. J.M. Bloch, H. Katz, D. Moses, V.J. Cajipe and J.E. Fischer, Phys. Rev. B31, 6785 (1985)
42. S.Y. Leung, C. Underhill, G. Dresselhaus, T. Krapchev, R. Ogilvie and M.S. Dresselhaus, Phys. Lett. A76, 89 (1980)

43. P. Touzain, Synth. Metals 1, 3 (1979)
44. S.E. Hardcastle, M.E. Misenheimer and H. Zabel, Rev. Sci. Instrum. 54, 206 (1983)
45. a) F. Rousseaux, D. Tchoubar, C. Tchoubar, D. Guerard, P. Lagrange, A. Herold and R. Moret, Synth. Metals 7, 221 (1983)
    b) K. Oshima, S.C. Moss and R. Clarke, Synth. Metals (1985) 12, 125 (1985)
46. C. Thompson, S.C. Moss, G. Reiter and M.E. Misenheimer, Synth. Metals 12, 57 (1985)
47. F. Rousseaux, R. Moret, D. Guérard, P. Lagrange and M. Lelaurain, J. Physique Lett. 45, L1111 (1984)
48. R. Clarke, P. Hernandez, H. Homma and E. Montague, Synth. Metals 12, 27 (1985)
49. D. Ghosh and D.D.L. Chung, J. Phys. Lettres 44, L-761 (1983)
50. L. Salamanca-Riba, G. Roth, J.M. Gibson, A.R. Kortan, G. Dresselhaus and R.J. Birgeneau, Phys. Rev. B33, 2738 (1986)
51. E.L. Evans and J.M. Thomas, J. Sol. State Chem. 14, 99 (1975)
52. G. Timp and M.S. Dresselhaus, J. Phys. C 17, 2641 (1984)
53. D. Dorignac, M.J. Lahana, R. Jagut, B. Jouffrey, S. Flandrois and C. Hauw, Proc. Symp. MRS, 1982 Vol. 20 eds. M.S. Dresselhaus, G. Dresselhaus, J.E. Fischer and M.J. Moran, Elsevier, p.33
54. S.M. Heald and E.A. Stern Phys. Rev. B17, 4069 (1978)
55. A. Hérold, Physics of Intercalation Compounds, eds L. Pietronero and E. Tosatti, Springer Series in Solid State Sciences, 38, 7 (1981)
    B. Carton and A. Herold, Bull Soc. Chem , 4, 1337 (1972)
56. M.E. Misenheimer and H. Zabel, Phys. Rev. B27, 1443 (1983)
57. J.E. Fischer, C.D. Fuerst and K.C. Woo, Synth. Metals 7, 1 (1983)
58. S.A. Safran, Physics of Intercalation Compounds eds. L. Pietronero and E. Tosatti, Springer Series in Solid State Sciences, Vol 38, Springer, Berlin 1981, p. 43.
59. C.R. Lee, H. Aoki and H. Kamimura, J. Phys. Soc. Japan 49, 870 (1980)
60. N. Wada, R. Clarke and S.A. Solin, Synth. Metals 2, 27 (1980)
    N. Wada and S.A. Solin, Physica B 105, 268 (1981)
61. N. Wada, Phys. Rev. B24, 1065 (1981)
62. G.S. Parry and D.E. Nixon, Nature 216, 909 (1967)
    D.E. Nixon and G.S. Parry, J. Phys. D1, 291 (1968)
63. R. Clarke, N. Caswell, S.A. Solin and P.M. Horn, Phys. Rev. Lett 43, 2018 (1979)
64. G.S. Parry, Mat. Sci. Eng. 31, 99 (1977)
65. F. Rousseaux, R. Moret, D. Guérard, P. Lagrange and M. Lelaurain, Proceedings of the Int. Symp. on GICS, Tsukuba, Japan, 1985, Synth. Metals 12, 45 (1985)
66. G. Reiter and S.C. Moss, Phys. Rev. B33, 7209 (1986)
67. H. Zabel, A. Magerl, A.J. Dianoux and J.J. Rush, Phys. Rev. Lett 50, 2094 (1983)
    H. Zabel, M. Suzuki, D.A. Neumann, S.E. Hardcastle, A. Magerl and W.A. Kamitakara, Proceeding of the Int. Symp. on GICS, Tsukuba Japan, 1985, Synth. Metals 12, 105 (1985)
68. M. Mori, S.C. Moss, Y.M. Jan and H. Zabel, Phys. Rev. B25, 1287 (1982)
69. R. Clarke, J.N. Gray, H. Homma and M.J. Winokur, Phys. Rev. Lett. 47, 1407 (1981)
70. Y. Yamada and I. Naiki, J. Phys. Soc. Japan 51, 2174 (1982)
71. M. Suzuki and H. Suematsu, J. Phys. Soc. Japan 52, 2761 (1983)
72. M. Suzuki, Phys. Rev. B33, 1386 (1986)
73. M.J. Winokur and R. Clarke, Phys. Rev. Lett 54, 811 (1985)
74. F. Rousseaux, R. Moret, D. Guerard, P. Lagrange and M. Lelaurain, Annales de Physique, Coll. N° 2, 11, 85 (1986)
75. S.B. DiCenzo, Phys. Rev. B26, 5878 (1982)
76. B. Houser, H. Homma and R. Clarke, Phys. Rev. B30, 4802 (1984)
77. H. Homma and R. Clarke, Phys. Rev. B31, 5865 (1985)
    M. Suzuki, R. Inada, H. Ikeda, S. Tanuma, K. Suzuki and M. Ichihara, J. Phys. Soc. Japan 53, 3052 (1984)
78. W.T. Eeles and J.A. Turnbull, Proc. Roy. Soc. London A283, 179 (1965)
79. K.K. Bardhan, J.C. Wu and D.D.L. Chung, Synth. Metals 2, 109 (1980)
80. A.R. Kortan, A. Erbil, R.J. Birgeneau and M.S. Dresselhaus, Phys. Rev. Lett 49, 1427 (1982)
81. S.G.J. Mochrie, A.R. Kortan, R.J. Birgeneau and P.M. Horn, Phys. Rev. Lett. 53, 985 (1984)
82. H.J. Schulz , Phys. Rev. B22, 5274 (1980)

83.a) R. Moret, R. Comès, G. Furdin, H. Fuzellier and F. Rousseaux, Proceedings
   of Symposium on Intercalated graphite, Materials Research Society Symp.
   Proc. Ed. M.S. Dresselhaus, G. Dresselhaus, J.E. Fischer and M.J. Moran
   Vol. 20 (Elsevier, 1983) p. 27
   b)R. Moret, F. Rousseaux, G. Furdin and A. Hérold, Synth. Metals 7, 289,(1983)
84.H. Pinto, M. Melamud, O. Shalal, R. Moreh and H. Shaked, Physica 121B, 121
   (1983)
85.M.P. Conrad and H.L. Strauss, Phys. Rev. B 31, 6669 (1985)
86.R. Clarke, P. Hernandez, H. Homma and E. Montague, Proceed. Int. Symp. on GICS
   Tsukuba, Japan 1985, Synth. Metals 12, 27 (1985)
87.C. Clinard, D. Tchoubar, C. Tchoubar, F. Rousseaux and H. Fuzellier, Synth.
   Metals, 7, 333 (1983)
88.F. Batallan, I. Rosenman, A. Magerl and H. Fuzellier, Phys. Rev. B 32, 4810
   (1985)
89.M. Bottomley, G. J. Parry and A.R. Ubbelohde, Proc. Roy. Soc. London A. 279
   291 (1964)
90.V.L. Pokrovsky and A.L. Talapov, Phys. Rev. Lett. 42, 65 (1979)

MICROSCOPIC STUDIES OF INTERCALATED GRAPHITE

M.S. Dresselhaus and J.S. Speck

Massachusetts Institute of Technology
Cambridge, MA 02139, USA

INTRODUCTION

In recent years a variety of high resolution microscopic probes have become available, providing non–statistical information on many structural issues, including the c–axis structure, staging fidelity, the in–plane structure, novel surface features, intercalant island formation, domain structure and superlattice formation.[1] With regard to such issues as staging fidelity, the information provided by x–ray and microscopy probes is complementary,[2] whereas microscopy provides unique information about phenomena such as intercalant island morphology, the structure near surface steps and the nature of Daumas–Hérold domains. Microscopy is indeed a powerful tool for the study of charge density wave (CDW) phases in transition metal dichalcogenide systems and of phase transitions associated with such phases.[3] No clear evidence for CDW phases have yet been reported for GICs.

Microscopic probes often require very thin samples, the use of high vacuum conditions, and high voltage, high intensity electron beams. Sample thicknesses for transmission electron microscope (TEM) studies are typically 100Å to 1000Å, nevertheless containing many superlattice unit cells, in fact many more than are probed by surface sensitive techniques such as low energy electron energy loss spectroscopy (EELS), angle resolved photoemission spectroscopy (ARPES), and Auger electron spectroscopy (AES).[4] The preparation of thin GIC samples suitable for TEM studies is normally done by careful peeling of c–face sections using scotch tape, followed by removal of the sample from the tape to a TEM grid using an organic solvent such as ethanol. Intercalated fibers with diameters $\sim$ 5000 Å are suitable for TEM studies, without further thinning.

Experimentally, the applicability of thin foil TEM samples for the study of bulk properties has been established by the observation of similar in–plane and c–axis diffraction patterns by both x–rays (bulk) and electrons (thin foils) in prototype graphite intercalation compounds (GICs). On the other hand, degradation of the intercalation compounds by the electron beam has been observed, the extent of the degradation depending on the intercalant species and stage and on the electron beam voltage, current and observation time. Reliable TEM results have been obtained when attention has been given

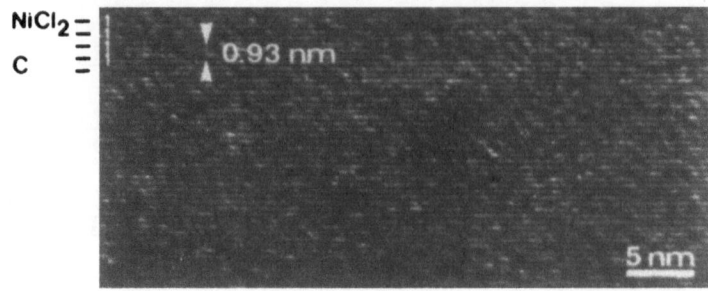

Figure 1: Lattice fringe pattern showing long range staging fidelity for a stage 1 GIC with a 9.3 Å C / NiCl$_2$ / C "sandwich" thickness, in very good agreement with that determined by x–ray studies.[7]

to minimizing beam damage effects. On the other hand, electron beam damage also gives rise to interesting phase changes in GICs, which have been studied in some detail for SbCl$_5$–GICs.[5]

In this review, we summarize the novel uses of microscopic techniques to reveal new or complementary information about c–axis structure, the fidelity of staging, in–plane structural order, island formation and evidence for Daumas–Hérold domains. The specific microscopic techniques include electron diffraction, bright field, dark field, lattice fringe imaging and image simulation, convergent beam analysis using the TEM;[6] microcompositional analysis and chemical mapping, microdiffraction and electron energy loss spectroscopy (EELS) using the scanning transmission electron microscope (STEM), microchemical mapping with the high resolution scanning ion microprobe (SIM), surface topology on an atomic level using the scanning tunneling microscope (STM), relative atomic registrations using Rutherford backscattering spectrometry (RBS)/channeling.

C–AXIS STRUCTURE

The formation of a superlattice along the c–axis[1] is the physical manifestation of the staging phenomenon, and every GIC sample is traditionally characterized for staging using (00ℓ) x–ray scans. More sophisticated x–ray characterization shows that staging infidelities or defects occur and these can be classified using the Hendricks–Teller theory according to whether mixed staging or random staging occur.[2] High resolution TEM studies provide a complementary means to view stage infidelities on a non–statistical, microscopic basis.

Information on the c–axis structure is obtained directly in graphite fibers or from bent–up portions of a thin sample foil where the electron beam of the TEM is directed along a basal plane direction. Interference of the (000) beam with one or more diffracted beams associated with intercalate superlattice reflections results in a lattice fringe pattern, such as shown in Fig. 1. In some cases, a high degree of staging fidelity is achieved, as for example in Fig. 1, where more than 100Å of essentially defect–free stage 1 fringes are shown for a NiCl$_2$–GIC sample both along the intercalate layers and perpendicular to them.[7] High resolution measurements of this kind clearly demonstrate

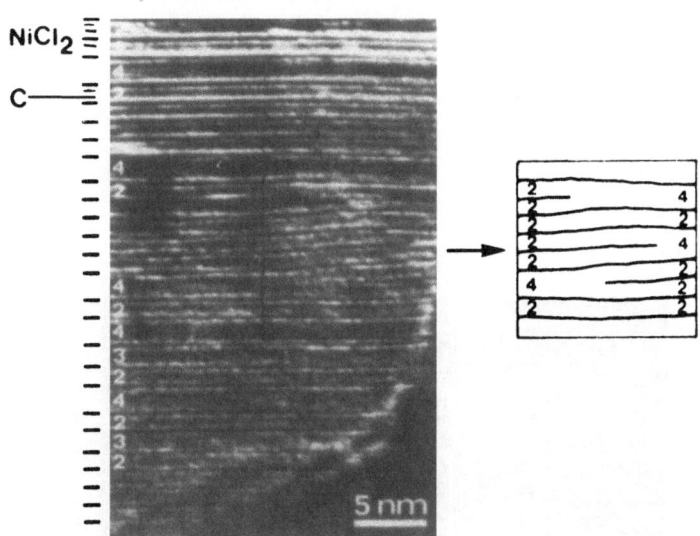

Figure 2: Lattice fringe TEM pattern from a nominally stage 2 $NiCl_2$–GIC showing mixed 2nd, 3rd and 4th stage sandwiches. The inset shows a region where edge dislocations of terminating $NiCl_2$ intercalate layers end, giving rise to interpenetrating regions of different stages.[7]

the long range nature of the staging phenomenon and the abrupt interfaces associated with intercalation superlattices.[1,2,8] In general, staging fidelity is observed to decrease with increasing stage, consistent with theories for staging.[2]

Similar measurements can be used to examine on a microscopic basis the types of staging defects that occur in GICs and their relative frequency. Such studies have been carried out on $NiCl_2$–GICs,[7] $FeCl_3$–GICs,[9] and KHg–GICs.[8] The use of c–axis lattice fringes to study staging imperfections is illustrated for a nominally stage 2 $NiCl_2$–GIC in Fig. 2. A schematic diagram for the interpretation of this photograph is also given in Fig. 2.[7] The photograph shows regions of stage 2 with admixtures of stage 3 and 4.

High resolution TEM has also been successfully applied to study microstructural defects such as dislocations and stacking faults in a number of acceptor and donor compounds.[8,10] The lattice image micrographs provide vivid evidence for isolated dislocations bounding the edge of an intercalate layer as is shown for a stage 1 $SbCl_5$–GIC in Fig. 3.[8] Also seen in this figure are some stage 2 sandwiches. Similar types of staging defects are seen in Fig. 3 for a stage 2 $SbCl_5$–GIC specimen. Another type of defect is an intercalate island trapped between two edge dislocations in desorbed GIC samples.[11] It is found that the Burger's vectors for the dislocations for Br_2–GICs and ICl–GICs are equivalent to the Burger's vector of the normal partial dislocations in hexagonal pristine graphite. This result is interpreted in terms of an interlayer shift of the graphite bounding layers to achieve identical crystallographic positions, and the resulting "A over A" stacking arrangement is denoted by A/A. In contrast, for $FeCl_3$–GICs where the intercalant structure is incommensurate with the graphite, the ABAB graphite stacking is preserved upon intercalation, in agreement with x–ray studies on these compounds. It is reasonable to expect that commensurate monolayer intercalate structures exhibit the symmetrical A/A stacking, while incommensurate structures have no symmetry reason

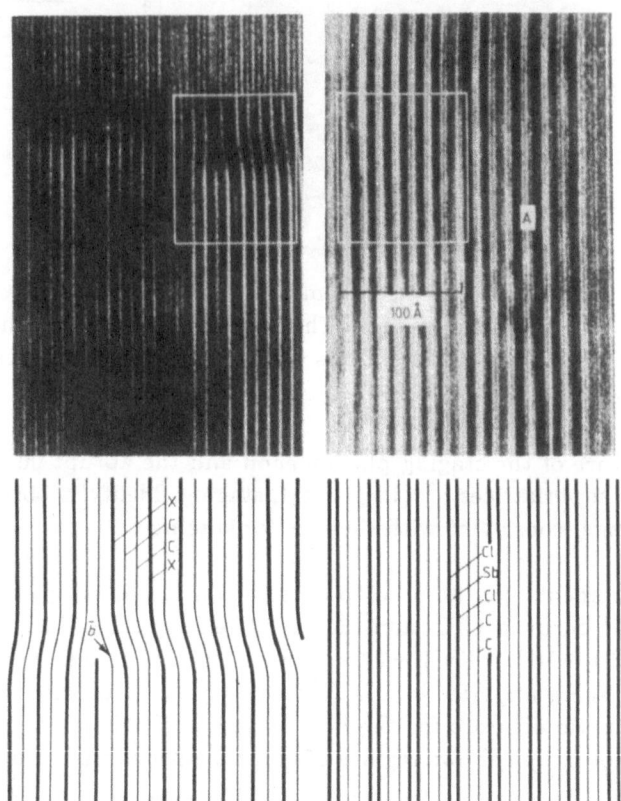

Figure 3: (left side) Lattice image of a stage 1 $SbCl_5$–GIC specimen showing inter-penetrating stages. The stage 2 region in this nominal stage 1 specimen terminates at an edge dislocation. (right side) Lattice image of a stage 2 $SbCl_5$–GIC foil showing a stage 2 region. The schematic diagrams below the respective lattice images refer to the regions highlighted in white on the positive prints.[8]

for disturbing the normal graphite stacking or causing shear between carbon layers.

IN–PLANE STRUCTURE

The layered nature of GICs make them especially amenable to the preparation of thin (100 Å to 1000 Å) samples, thereby facilitating the observation of $(hk0)$ diffraction patterns and the determination of in–plane crystal structure, microstructure and defects. In general, microscopy techniques are complementary to x–ray methods which can penetrate $\sim$ 1 mm while TEM electron diffraction patterns probe the sample structure on a 1 μm scale, and microdiffraction patterns can be obtained from samples of < 100Å in size using the STEM. Since the intercalate layer frequently consists of islands or domains with different in–plane crystal structures, the small spatial dimensions of electron microscopic probes offer an important advantage for structural studies of GICs. Because of the strong interaction of electron beams with the atoms of a crystal, multiple scattering events are frequent, so that it is difficult to carry out quantitative scattering intensity studies with electron beams.

Electron microscopy studies of the in–plane structure of acceptor compounds have generally been more successful than for donor compounds. In the case of acceptor GICs (such as with the intercalants Br_2, $SbCl_5$ and the transition metal chlorides), the observed electron diffraction patterns for the in–plane structure are in good agreement with in–plane x–ray studies, but electron diffraction has the advantage of examining structure within single domains. For example, electron diffraction patterns in a single structural domain of Br_2–GICs confirmed the commensurate rectangular $(\sqrt{3} \times 7)$ unit cell at room temperature, and confirmed the commensurate–incommensurate transition at $\approx$ 100°C, associated with the formation of a striped domain phase.[8] The observation of sharp electron diffraction patterns in Br_2–GICs was facilitated by the very large intercalate coherence distance (> 1μm in–plane) in these GICs.[12] Electron diffraction patterns for the simple commensurate in–plane superlattices have been observed,[2] as well as many complicated patterns which have not yet been fully interpreted.[11] For the acceptor compounds much activity and excitement in high resolution microscopy has focused on the island structure which is now under intense study for the transition metal chlorides.[13,14,15,16]

The in–plane structures for the transition metal chlorides are interesting in their own right. For these compounds, the in–plane structure is incommensurate with the graphite, and in many cases (though not all), the lattice constants are essentially identical with those of the parent bulk compound (such as for the intercalants $MnCl_2$, $FeCl_3$, $CoCl_2$ and $NiCl_2$); intercalants having modified in–plane structures include $CuCl_2$ and $MoCl_5$. Of particular interest is the orientational locking of the intercalant structure to that of the graphite host, so that a spot diffraction pattern is observed. Denoting the angle between the a–axis of the graphite and intercalant basis vectors by δ_{a-a}, it is found that δ_{a-a} is either close to 0° or to 30° for the transition metal chloride–GICs, depending on the magnitude of the lattice constant a for the intercalant.[17] For example, the spot diffraction pattern in Fig. 4a for a stage 1 $MnCl_2$–GIC clearly shows orientational locking for $\delta_{a-a} \approx$ 30°.[18] Typically, electron diffraction patterns for the transition metal chloride–GICs show a small arc of orientations (ranging from 0° to 4° in Fig. 4a) for the intercalant reflections around the 0° and 30° values, in contrast to the graphite spots

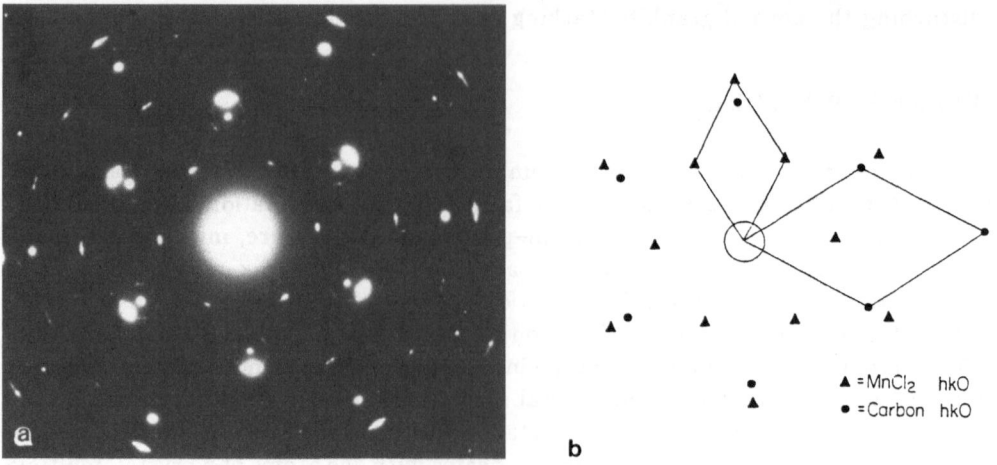

Figure 4: (a) Electron diffraction pattern for a stage 1 $MnCl_2$–GIC. (b) Schematic identification of graphite and intercalant spots.[18]

which are very sharp. Furthermore, the $(h0l)$ diffraction patterns show short range interlayer intercalate correlation at room temperature.[19] Also of interest is the difference in δ_{a-a} for stage 1 and stage 2 $CoCl_2$–GICs insofar as $\delta_{a-a} \sim 0°$ and 30°, respectively, for stage 1 and 2 compounds. The reverse angular values for the orientational locking are found for the stage 1 and 2 $CuCl_2$–GICs. The physical mechanisms responsible for the stage dependence of δ_{a-a} and for the small deviations of δ_{a-a} from the 0° and 30° values are under investigation.[18]

In–plane structural studies in donor compounds have been very interesting, but are not generally understood in relation to x–ray studies of the same materials. Of the various donor GICs that have been studied by microscopy techniques, the structure of KHg–GICs are perhaps the best understood.[8] X–ray measurements show that bulk KHg–GICs exhibit a (2×2) in–plane superlattice, which is also observed in electron diffraction patterns. The electron diffraction patterns furthermore shows a dominant $(\sqrt{3} \times \sqrt{3})R30°$ coexisting dense phase (see Fig. 5), with very long coherence length (1000Å) and very large defect–free regions of the sample.[8] No convincing explanation for the stabilization of the dense $(\sqrt{3} \times \sqrt{3})$ structure in these thin TEM foils has yet been given.

Even though the in–plane structure of alkali metal donor compounds has been studied most extensively by electron microscopy, the relation between these studies and the corresponding x–ray studies remains unexplained.[20,21] Nevertheless several important conclusions can be reached. Firstly, the same (2×2) in–plane superlattice structure obtained for the stage 1 alkali metal compounds C_8K, C_8Rb and C_8Cs from electron diffraction patterns is also found in the x–ray measurements; in all cases, long in–plane coherence distances are found. These observations establish that the TEM technique can be used to obtain structural information on bulk donor GICs.

However, for high stage $(n \geq 2)$ alkali metal GICs, though the electron microscopy data obtained by different groups have been generally consistent with each other,[20,21] the

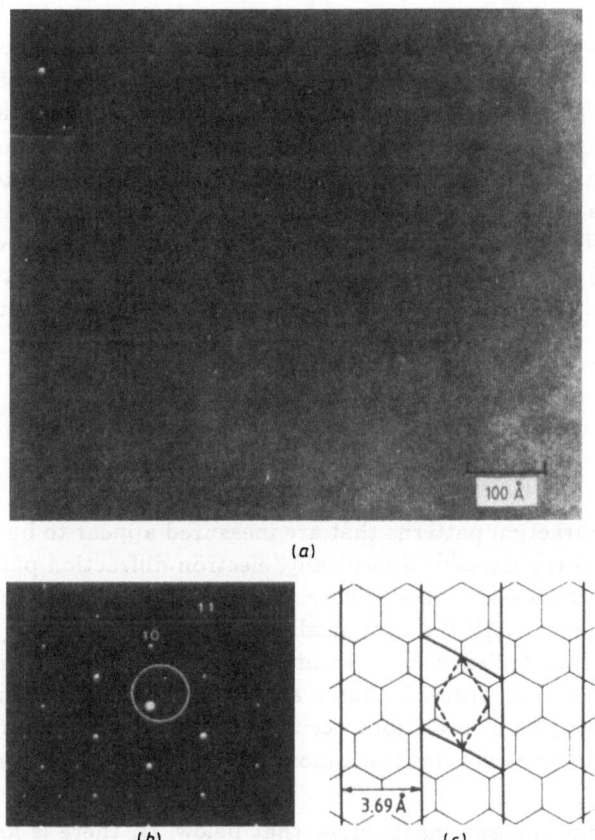

Figure 5: (a) A TEM image showing an in–plane $(\sqrt{3} \times \sqrt{3})R30°$ superlattice in a stage 1 KHg–GIC. (b) An electron diffraction pattern corresponding to the $(\sqrt{3} \times \sqrt{3})R30°$ structure. The ring defines the aperture used to generate the lattice image of (a). (c) A schematic representation of the lattice planes superimposed onto the graphite basal plane structure. The lattice–plane spacing of 3.69Å is measured from the optical diffractogram shown as an inset to (a).[8]

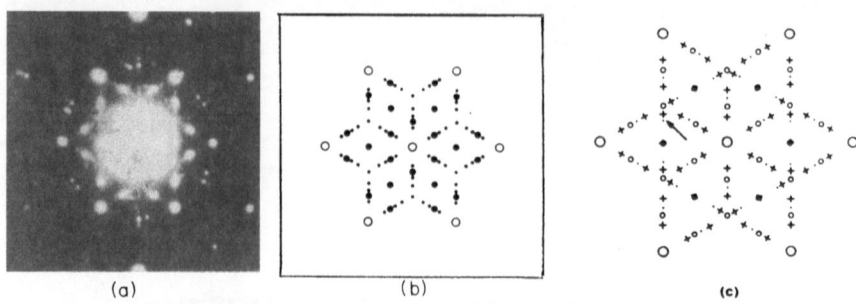

(a)   (b)   (c)

Figure 6: (a) High–temperature ($170 < T < 620$ K) in–plane electron–diffraction pattern for stage–2 Rb–GIC samples. Similar patterns are observed for samples with stages $n = 3$, 4 and 7. (b) A schematic representation of the diffraction pattern in (a). The open circles refer to the graphite pattern, the large solid circles to intense superlattice spots, and the small circles to weak superlattice spots.[20] (c) An interpretation of the diffraction patterns in (a) and (b) showing a superposition of diffraction spots from the $(\sqrt{3} \times \sqrt{7})\mathrm{R}(30°, 40.9°)$ (small dots), $(\sqrt{3} \times \sqrt{3})\mathrm{R}(30°, 60°)$ (small open circles), and $(\sqrt{3} \times 4)\mathrm{R}(30°, 60°)$ (crosses) superlattices, each with three equivalent orientations. The pair of spots indicated by the arrow and their equivalents are the only extra spots not observed in (b).[21]

detailed electron diffraction patterns that are measured appear to be incompatible with a huge quantity of x-ray data.[2,11] Specifically, electron diffraction patterns for K–GICs, Rb–GICs and Cs–GICs each show different diffraction patterns, but with relatively small stage dependences. All the thin alkali metal–GIC TEM foils (for $n \geq 2$) exhibit multiphase coexistence for each of these intercalate species. A number of interesting phase transitions between ordered phases are observed as a function of temperature. Though many of the phase transitions occur at the same temperature as are observed by x–rays, the in–plane structures are almost always different.[11,20,21]

Specifically, x–ray measurements show that below T_ℓ, there is long range interplanar and intraplanar ordering, while for temperatures in the range $T_\ell < T < T_u$, short range interplanar and long range intraplanar order occur. Above T_u, interlayer ordering disappears and only short range order (≤ 10 Å) remains in–plane. In contrast, electron diffraction results show sharp diffraction spots above T_u for all high stage alkali metal GICs. For example, for Rb–GICs, the in–plane structure implied by the electron diffraction pattern is $(\sqrt{7} \times \sqrt{7})\mathrm{R}19.1°$ below T_u,[20] and a novel pattern above T_u (see Fig. 6), which has been interpreted to contain a superposition of $(\sqrt{3} \times \sqrt{3})\mathrm{R}(30°, 60°)$, $(\sqrt{3} \times \sqrt{7})\mathrm{R}(30°, 40.9°)$, and $(\sqrt{3} \times 4)\mathrm{R}(30°, 60°)$ regions.[21] The corresponding in–plane ratio of carbon/intercalate atoms is 12:1, 10:1 and 16:1 which can be made consistent with the ratio of 14:1 for the $(\sqrt{7} \times \sqrt{7})\mathrm{R}(19.1°)$ structure by proper choice of the relative concentrations of the coexisting phases. Also of interest is the remarkable change in the intercalate microstructure at T_u shown by the dark field TEM patterns.[20]

Another example of multiphase coexistence that has been clearly documented is the

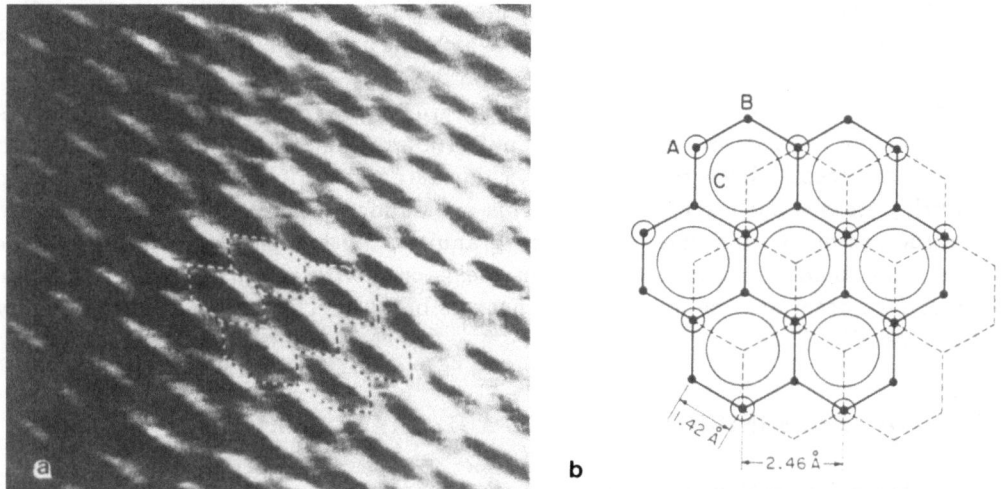

Figure 7: (a) A digitally filtered scanning tunneling microscope image for a 20Å × 20Å area of a graphite surface showing occupied A and B sites and a hollow at C, following the notation of (b). In (a), a hexagonal grid is superimposed to help visualization of the unit cell. A contour map (in the c–direction) corresponding to (a) was also measured.[24]

case of high stage Cs–GICs below T_ℓ (98K) where the electron diffraction patterns have been interpreted in terms of a superposition of commensurate (2×2), $(\sqrt{3} \times 2)$ and $(\sqrt{3} \times \sqrt{3})$ phases.[22] In contrast, x–ray diffraction results have been interpreted in terms of a modulated incommensurate triangular lattice, orientationally locked to the graphite lattice and coexisting with a (2×2) commensurate phase.[2] Others have modeled the x–ray results in terms of modulated discommensuration domains based on a commensurate $(\sqrt{7} \times \sqrt{7})$R19.1° structure.[2]

SCANNING TUNNELING MICROSCOPY OF GRAPHITE SURFACES

The scanning tunneling microscope (STM) provides a unique and powerful new tool for the direct determination of real space surface structure at the atomic level, including nonperiodic structures. In this microscope, a small metal tip is brought close enough to the surface to permit electron tunneling between the tip and the surface. The tip scans the surface in two–dimensions. By adjusting the height of the tip above the surface to maintain a constant tunneling current, it is possible to obtain a contour map of the surface.[23] Of all materials, graphite has yielded the best STM patterns to date, showing true atomic resolution (< 2 Å) due to the extreme rigidity and flatness of graphite surfaces.[24]

Amplitude traces of the tunneling current for the graphite surface are shown in Fig. 7, allowing identification of the A and B occupied surface sites and the C unoccupied site for the graphite honeycomb structure. The single atom resolution is attributed to an effective single atom tungsten tip.[23] In the bulk, sites A and B are coplanar while C lies 3.35 Å below (or above) the plane containing the B atom. Every atomic plane contains an A carbon atom while on sites B and C there are carbon atoms on alternating layers. The contour map provided by the STM indicates a 0.8 ±0.1 Å height difference between sites C and A at the surface; the STM traces indicate that at the surface the height of site A is ~ 0.2 Å above site B,[23] which has no neighbor in the lower plane.[24]

The STM offers great promise for a similar study on GICs, though such measurements have not yet been made. The first application of the STM to chemically reacted graphite has recently been reported for a chlorinated graphite surface in comparison to a freshly cleaved surface of graphite. This work shows a great deal of detail on the edge structure of pristine and chlorinated graphite. With regard to the basal plane surface, very smooth regions are shown in coexistence with rough regions. Oxidation of the surface decorates the line defects, uncovers basal interfaces and generates highly reactive sites.[25]

ELECTRON BEAM INDUCED DAMAGE AND PHASE CHANGES

The strong interaction between the electron beam in the electron microscope and the atoms of the lattice can result in lattice damage and phase changes in GICs. Thus electron microscopy studies in GICs are normally carried out using the lowest possible beam intensities when beam damage of the speciman is a possibility.

Electron beam damage phenomena have been studied quantitatively for GICs based on $SbCl_5$.[5] The damage process was examined as a function of electron dose and energy, while the annealing behavior was investigated as a function of the sample temperature. At room temperature $SbCl_5$–GICs exhibit a diffraction pattern (Fig. 8a), characteristic of the commensurate in–plane $(\sqrt{7} \times \sqrt{7})R19.1°$ superlattice structure. The main effect of the electron beam damage in the TEM is a phase change from a $(\sqrt{7} \times \sqrt{7})R19.1°$ commensurate structure (Fig. 8a) to a glass phase (Fig. 8b) above a damage threshold electron dose, depending on the electron beam energy. By showing that the lattice damage scales with the electronic stopping power, it was established that radiolysis is the dominant damage mechanism, producing atomic displacements by the creation of an excited electronic state.[5]

Defining the critical dose ϕ_c to induce a commensurate–glass phase change by the condition that the relative intensity R of the diffuse ring I_R to that of the sharp spot I_S on the electron diffraction pattern is given by $R = (I_S - I_R)/I_S \equiv 0.15$, an Arrhenius plot of ϕ_c vs $1/T$ yields activation energies at low temperatures and high temperatures (see Fig. 8c). From analysis of these data, a low activation energy of $E_R \sim 0.02$eV is found for the electron beam damage. Also a low activation energy $E_{r_2} = 0.03$eV is found for fast recombination at high temperature, but a much higher activation energy $E_{r_1} = 0.13$eV for slow recombination at low temperature. Thus low temperatures favor the observation of a glass phase.[5] It is likely that a similar phase change can be induced by the electron beam for other GICs which disproportionate. The electron beam in a TEM thus provides a convenient method for studying a commensurate–glass phase change under controlled conditions.

A specific model for the electron irradiation process has been proposed, consistent with the disproportionation of $SbCl_5$ during intercalation to yield the species $SbCl_5$, $SbCl_6^-$, $SbCl_3$, $SbCl_4^-$ within the GICs. According to this model, an electron is incident upon a mixture of $SbCl_6^- + SbCl_3$ species, which is associated with the ordered

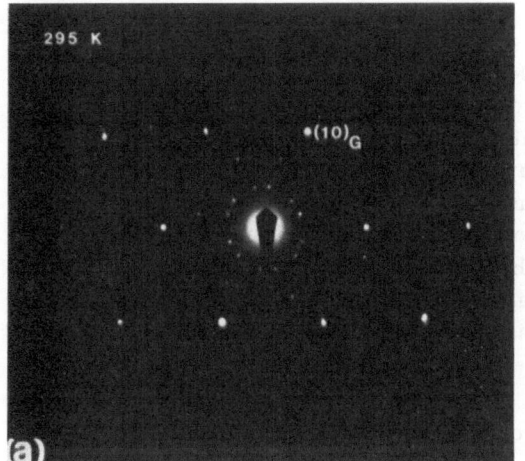

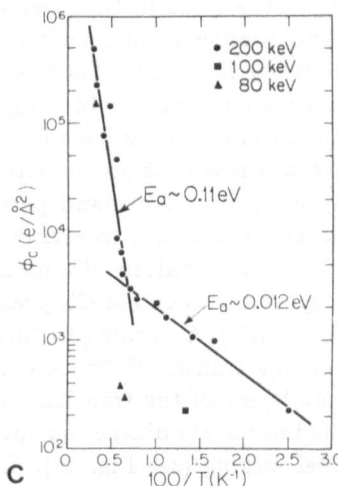

Figure 8: ($hk0$) electron diffraction patterns taken at (a) T = 295 K (crystalline phase) and (b) T = 50 K (glass phase) of a mixed–stage 2 and 3 vermicular graphite sample intercalated with $SbCl_5$. (c) Temperature dependence of the critical electron dose (ϕ_c) for the commensurate–glass phase change in $SbCl_5$–GICs for 200 ($\bullet$), 80 ($\blacktriangle$), and 100 ($\blacksquare$) keV electrons.

$(\sqrt{7} \times \sqrt{7})$R19.1° phase, converting it into a disordered phase of $SbCl_5 + SbCl_4^-$. This interpretation is consistent with image simulations of the lattice fringe images for various model structures.[5] This interpretation is however not consistent with STEM studies of the intercalate islands in $SbCl_5$–GICs discussed below.[26]

ISLAND FORMATION

It is believed that in many GICs the intercalant is arranged in an island morphology in the intercalate layer, especially in cases where the intercalant filling factor is significantly less than unity, or in cases where the intercalant disproportionates. The filling factor $\xi_f = W_a/W_x$ is defined as the ratio of the actual weight uptake W_a (e.g., measured by chemical means) to the weight uptake W_x implied by the diffraction pattern for the fully dense phase. If $\xi_f < 1$, a clustering of the occupied intercalated sites and of the empty sites leads to a minimization of the lattice strain, thereby favoring island formation and phase separation.[27] Although much of the previous modeling of the phase separation has been in terms of isolated islands, it is likely that in some GICs the intercalant forms a connected network of island–like structures, as was reported in the dark field images for the Rb–GICs and K–GICs for stages $n \geq 2$.[20]

The existence of intercalant island structures can have a profound effect on the properties of GICs, such as the low dimensional magnetic properties and charge transfer in acceptor GICs. With regard to low dimensional magnetic properties, magnetic vortices are expected in 2D–XY systems as the spins in the layer planes establish short range order.[28] The island size limits the number of magnetic vortices which interact coherently to form bound vortex pairs in the Kosterlitz–Thouless transition, thereby strongly affecting the magnetic properties near the vortex binding transition. With regard to charge transfer in acceptor GICs, several models have been proposed. One model proposed by Baron et al.[15] suggests that the islands are negatively charged by accumulation of an excess of anions (dangling bonds) on the island perimeter, so that electrons are transferred from the graphite to the intercalant to satisfy the valence requirements of the non–stoichiometric anions. Chemical analysis of a number of metal chloride acceptor compounds (MCl_n–GICs) prepared with excess Cl_2 pressure have indeed been found to exhibit excess chlorine (MCl_{n+x}–GIC). According to this model, knowledge of x and the island size determines the charge transfer.[13,14,15] A definitive determination of the island structure in the intercalate layers of the transition metal chloride–GICs is complicated by the *arc pattern* of the intercalate diffraction spots associated with the finite angular range of the orientational locking (see Fig. 4a). Interference between electron beams coming from islands with small differences in δ_{a-a} produce Moire patterns, which complicate observation of the intercalate microstructure.[17,19]

A clear demonstration of island morphology in acceptor GICs was demonstrated by Hwang et al.[26] for the case of $SbCl_5$–GICs using the STEM. Similar results were obtained by Salamanca–Riba et al.[5] using the TEM. Dark field images of a stage 4 $SbCl_5$–GIC[26] show high density islands, 500–1000Å in size, covering $(11 \pm 2)\%$ of the total imaged area (see Fig. 9a). Energy dispersive x–ray fluorescence spectra (see Fig. 9b) based on the Cl K lines and Sb L lines were interpreted to yield a chemical composition of $SbCl_{3.1\pm0.3}$ for the islands and $SbCl_{7.0\pm2.0}$ for the background regions. Microdiffraction studies[26] show that the ordered $(\sqrt{7} \times \sqrt{7})$R19.1° pattern is associated with the

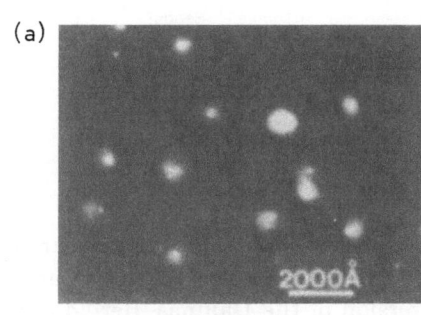

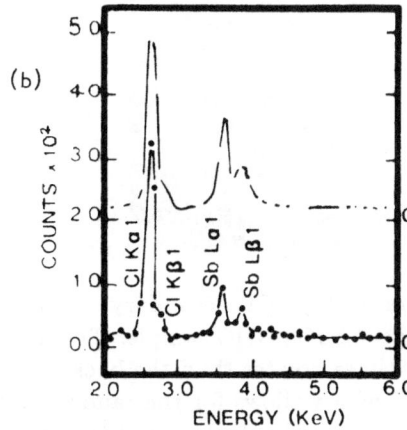

Figure 9: (a) Dark–field STEM image of a stage–4 SbCl₅–GIC sample. (b) Energy–dispersive x–ray fluorescence spectra of the island (open circles) and background (closed circles) regions.[26]

background, while the islands yield a diffuse ring pattern, consistent with a disordered structure. The islands are identified with an Sb–rich $SbCl_3 + SbCl_4^-$ region of average stoichiometry $SbCl_{3.2}$ and the background with an Sb–deficient $SbCl_5 + SbCl_6^-$ region of average stoichiometry $SbCl_{5.33}$ yielding a charge transfer of 0.31 per $SbCl_5$ molecule and a degree of disproportionation of 0.44. The spatial invariance of the 1s core level excitation observed in the energy loss spectra and the sharp Raman linewidths indicate that the island and background regions have the same Fermi energy shift with respect to the rigid graphite electronic bands.[26] The requirements of charge delocalization and uniform charge distribution for the interpretation of Fig. 9 yield a unique chemical composition $(SbCl_3)_{1/8}(SbCl_4^-)_{1/32}(SbCl_5)_{9/16}(SbCl_6^-)_{9/32}$ and the average stoichiometries of the islands and background given above. However, this model does not give a good fit to the lattice image simulation studies or an explanation for the formation of a glass phase by radiolysis in the TEM.[5] Further work is needed for a full elucidation of the real space structure of SbCl₅–GICs.

Using a high–resolution scanning–ion microprobe (SIM), secondary–electron images are obtained showing topographic and primary ion channeling contrast, revealing detailed information on the morphology of flat microcrystallites and low angle grain boundaries. When applied to a freshly cleaved SbCl₅–GIC sample, intercalate domains of $\sim 2000\text{Å}$ size are observed along crystal–surface defects and have been identified with Daumas–Hérold domains.[29] By the SIM technique, secondary ion mass spectroscopic mapping is done. When applied to the mapping of $^{35}Cl^-$ in the stage 2 and stage 4 SbCl₅–GICs, the chlorine distribution is found to be highly discrete, (see Fig. 10), forming beadlike lines with a high $^{35}Cl^-$ concentration relative to that of the background ($\sim 3\times$ for stage 2 and $\sim 10\times$ for stage 4). Sequential SIM mapping of the same SbCl₅–GIC surface indicates a loss of intercalant, since the sputtering yield of the intercalant is much higher than that of graphite. On sputtering, the line pattern diminishes before the bead–like domains do. This has been interpreted[29] to indicate that the bead–like domains are beneath the line pattern, suggesting that the beads represent the Daumas–

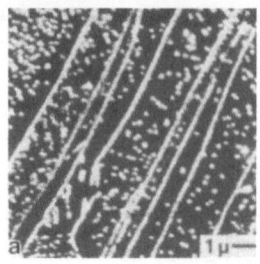 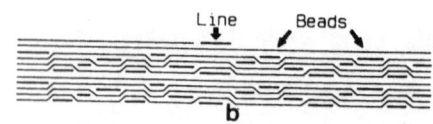

Figure 10: (a) $^{35}Cl^-$ secondary ion mass spectrometric map of a freshly cleaved surface of a stage–4 $SbCl_5$–GIC sample. The lines and bead–like domains have a Cl concentration ~ 10 times greater than the background. (b) A version of the Daumas–Hérold domain model which allows for the random nature of the domain size and lateral distribution. It also shows the depletion of the acceptor intercalant near a surface step.[29]

Hérold domains, one graphite monolayer underneath the surface. The lines are then interpreted as an accumulation of $SbCl_5$ on surface steps as illustrated in Fig. 10b. The strong sample dependence observed for the line patterns and the nearly parallel lines are typical of surface steps resulting from cleaving. The depletion of the bead density on one side of the lines suggests a diffusion mechanism for the intercalant and an accumulation of the intercalant into solid lines along the crystal defects. Further detailed work using SIM and other complementary techniques will be needed to elucidate the properties of Daumas–Hérold domain boundaries.

RUTHERFORD BACKSCATTERING/CHANNELING STUDIES

Unique information is obtained from application of the Rutherford backscattering spectrometry (RBS)–channeling technique to GICs, providing information on the long range ordering of the graphite skeleton and on the preferred intercalate sites relative to this skeleton, at fixed temperatures and during phase transitions. In Rutherford backscattering spectrometry, a beam of monoenergetic and collimated light particles (e.g., H^+ or He^+) impinges normal to the sample surface and the number and energy of particles scattered backwards by a small angle θ are monitored. Analysis of the energy distribution of the backscattered ions provides detailed stoichiometric information as a function of depth from the surface. If the probing beam is aligned nearly parallel to a close packed row of atoms in a crystalline target (channeling), structural information as a function of depth can be obtained. For GICs, channeling experiments can give information on a variety of structural properties: the stacking of the graphite planes, the commensurability of the intercalant layer and, when studied as a function of temperature, atomic rearrangement transitions can be monitored.

Stoichiometric and structural changes in stage 4 K–GICs have been studied as a function of temperature, in the range 100 K < T < 300 K, using the RBS/channeling technique, with a probing beam of 500 KeV protons.[30] Since the steering of the channeled particles is established by the host crystal lattice, the observation of the channeling effect in K–GICs indicates that the stacking between graphite layers is preserved above and below the intercalate layer, indicating well–established 3–dimensional crystalline order of the graphite skeleton in these samples. Furthermore, the similar value of the axial RBS half width at half maximum for the carbon signal, in both graphite and the K–GIC, indicates that the host graphite lattice is not greatly changed by intercalation. The increase in the minimum RBS yield of the C signal from the K–GIC with respect to that from pristine graphite can be attributed to the reduced crystalline order of the near–surface region of the intercalated sample. For a stage 4 K–GIC sample at room temperature, about 30% of the potassium atoms are found to be located above or below the center of the hexagon of carbon atoms of the adjacent graphite plane. However at 100 K, the K atoms are randomly located with respect to the rows of carbon atoms, although the crystalline structure of the graphite host is not changed. The transition between the "ordered" high temperature arrangement and the "disordered" low temperature arrangement occurs at about 110 K upon cooling and at about 213 K upon heating and is accompanied by a change in the areal density of K atoms.[30] The application of the RBS/channeling technique to other GICs could yield very interesting structural information, complementary to the type of information provided by scattering techniques.

ACKNOWLEDGMENTS

The authors wish to thank Dr. G. Dresselhaus for many important discussions. This work was supported by AFOSR contract #F49620–83–C–0011.

REFERENCES

1. M.S. Dresselhaus (this volume) p. 1.

2. R. Moret (this volume) p. 185.

3. W.Y. Liang (this volume) p. 31.

4. G. Dresselhaus and M. Lagües (this volume) p. 271.

5. L. Salamanca–Riba, G. Roth, J.M. Gibson, G. Dresselhaus and R.J. Birgeneau, *Phys. Rev.* **B33**, 2738 (1986); L. Salamanca–Riba, Ph.D. thesis MIT (1985), unpublished.

6. J.C. Spence, "Experimental High Resolution Electron Microscopy", (Oxford, Clarendon, 1981).

7. D. Dorignac, M.J. Lahana, R. Jagut, B. Jouffrey, S. Flandrois and C. Hauw, *Mat. Res. Soc. Symp.* **20** (1983), edited by M.S. Dresselhaus, G. Dresselhaus, J.E. Fischer and M.J. Moran, (Elsevier Science, North Holland, Amsterdam, 1983), p. 33.

8. G. Timp and M.S. Dresselhaus, *J. Phys.* C **17**, 2641 (1984).

9. J.M. Thomas, G.R. Millward, N.C. Davies and E.L. Evans, *J. Chem. Soc.* Dalton Trans., 2443 (1976): J.M. Thomas, G.R. Millward, R. Schlögl and H.P. Boehm, *Mat. Res. Bull.*, **15**, 671 (1980).

10. M. Heerschap, P. Delavignette and S. Amelinckx, *Carbon*, **1**, 235 (1964); M.M. Heerschap and P. Delavignette, *Carbon*, **5**, 383 (1967).

11. M.S. Dresselhaus and G. Dresselhaus, *Adv. Phys.* **30**, 139 (1981).

12. S.G.J Mochrie, A.R. Kortan, R.J. Birgeneau and P.M. Horn, *Phys. Rev. Lett.* <u>53</u>, 985 (1984).

13. S. Flandrois, J.M. Masson, J.C. Rouillon, J. Gaultier and C. Hauw, *Synthetic Metals* **3**, 1 (1981).

14. S. Flandrois, A.W. Hewat, C. Hauw and R.H. Bragg, *Synthetic Metals* **7**, 305 (1983).

15. F. Baron, S. Flandrois, C. Hauw and J. Gaultier, *Solid State Commun.* **42**, 759 (1982).

16. M. Matsuura, Y. Murakami, K. Takeda, H. Ikeda, and M. Suzuki, *Synthetic Metals* **12**, 427 (1985).

17. P. Behrens and W. Metz (this volume) p. 229.

18. J.S. Speck, (unpublished).

19. J. Speck, X. Hao, and M.S. Dresselhaus (this volume) p. 233.

20. N. Kambe, G. Dresselhaus and M.S. Dresselhaus, *Phys. Rev.* **B21**, 3491 (1980); A.N. Berker, N. Kambe, G. Dresselhaus and M.S. Dresselhaus, *Phys. Rev. Lett.* **45**, 1452 (1980).

21. D.M. Hwang, *Phys. Rev.* **B27**, 1119 (1983).

22. D. Hwang, N.W. Parker, M. Utlaut and A.V. Crewe, *Phys. Rev.* **B27**, 1458 (1983).

23. G. Binnig, H. Rohrer, C. Gerber, and E. Weibel, *Phys. Rev. Lett.* **49**, 57 (1982); G. Binnig, H. Fuchs, Ch. Gerber, H. Rohrer, E. Stoll, and E. Tosatti, *Europhys. Lett.* **1**, 31 (1986).

24. S.-I. Park and C.F. Quate, *Appl. Phys. Lett.* **48**, 112 (1986).

25. R. Wiesendanger, R. Schlögl, and H.-J. Güntherodt, Extended abstracts for Carbon '86, Baden–Baden Conference, p. 207 (1986).

26. D.M. Hwang, X.W. Qian and S.A. Solin, *Phys. Rev. Lett.* **53**, 1473 (1984).

27. S.A. Solin (this volume) p. 173.

28. G. Dresselhaus and M.S. Dresselhaus (this volume) p. 407.

29. R. Levi–Setti, G. Crow, Y.L. Wang, N.W. Parker, R. Mittleman and D.M. Hwang, *Phys. Rev. Lett.* **54**, 2615 (1985).

30. G. Braunstein, B. Elman, J. Steinbeck, M.S. Dresselhaus, T. Venkatesan and B. Wilkins, *Extended Abstracts for the Symposium on Graphite Intercalation Compounds*, Materials Research Society Proceedings 1984, edited by P.C. Eklund, M.S. Dresselhaus and G. Dresselhaus, p. 168.

ORIENTATIONAL LOCKING OF INCOMMENSURATE LATTICES

IN MENTAL CHLORIDE GICs

P. Behrens and W. Metz

Institute of Physical Chemistry
  of the University of Hamburg
Bundesstrasse 45, D-2000 Hamburg 13, FRG

Introduction

In the study of the structures of GIC the orientational locking of incommensurate lattices is a challenging problem: Although the incommensurability of the graphite and the intercalate in-plane lattice indicates
that host-guest interactions are weaker than the interactions between the
intercalate molecules, most of the incommensurate intercalate lattices
exhibit a definite rotational orientation δ with regard to the $\vec{a}$ axis of
the hexagonal in-plane lattice of graphite [1]. In order to look for the
reasons of this astonishingly behaviour we turned to the study of metal
di- and trichloride GICs with simple in-plane structures of the intercalate. These structures are shown in Fig. 1. They consist of a double
layer of close-packed chlorine atoms. Metal atoms occupy all or two thirds
of all octahedral sites of this double layer, leading to stoichiometries
$MeCl_2$ and $MeCl_3$, resp., for the metal chloride layer. The advantages of
this group of GICs are:

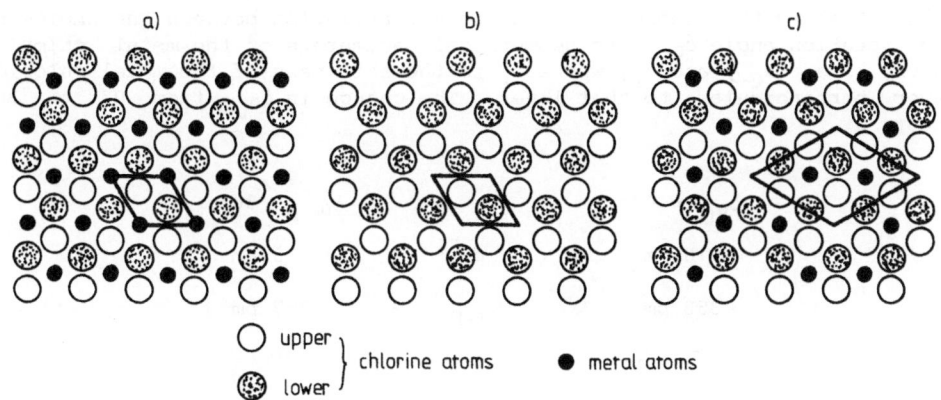

a)   b)   c)

○ upper ⎫
       ⎬ chlorine atoms   ● metal atoms
⊛ lower ⎭

Fig. 1: In-plane structures and unit cells of

a) $MeCl_2$ layer   b) close-packing   c) $MeCl_3$ layer
                           of chlorine atoms

for the structure type considered here (see text).

- simple structure of the intercalate

- a large number of metal chloride GICs possess this structure type

- only special values of the rotation angle δ_a between the $\vec{a}$ axes of the hexagonal in-plane lattices of graphite and of the intercalate are observed ($\delta_a = 0°$ or $30°$; an exception is the so-called "hexagonal phases" of $MoCl_x$-graphite [2,3])

We looked for relations between δ_a and characteristic geometrical parameters of the intercalate. To be able to compare $MeCl_2$- and $MeCl_3$-GICs we refer these parameters to the unit cell of the sublattice of the close-packed chlorine atoms (Fig. 1b) since this is the structural subunit of the intercalate which is in contact with the graphene layers. It can easily be seeen that this sublattice unit cell is identical to that of a $MeCl_2$ layer so that

$$a_{C-P} = a_{MeCl_2}$$

$$\delta_{a_{C-P}} = \delta_{a_{MeCl_2}}$$

where a is the length of the unit cell vector of the hexagonal cell and the index C-P refers to the unit cell of the close-packed chlorine atoms. The unit cell of a $MeCl_3$ layer is rotated by $30°$ with regard to this unit cell so that

$$\delta_{a_{C-P}} = \delta_{a_{MeCl_3}} \pm 30°$$

while a_{C-P} is calculated according to

$$a_{C-P} = a_{MeCl_3} / \sqrt{3}$$

in this case.

Results

Table 1 shows that there exists a close correlation between the choice of the rotation angle $\delta_{a_{C-P}}$ and geometrical parameters of the metal chloride layer, i.e. a_{C-P} and d_I, where d_I is the thickness of the metal chloride layer perpendicular to that layer. The change in $\delta_{a_{C-P}}$ from $0°$ to $30°$ occurs at

$\delta_{a_{C-P}} = 0°$				$\delta_{a_{C-P}} = 30°$
610 pm	<	d_I	<	613 pm
353 pm	<	a_{C-P}	<	357 pm

The only deviation from these relations is observed in the case of $AlCl_3$-GIC, where $a_{C-P} = 346$ pm suggests a value of $\delta_{a_{C-P}}$ of $0°$, while $\delta_{a_{C-P}} = 30°$ is observed. This may be explained by a considerable distortion of the close-packing of the chlorine atoms as it is found in pure $AlCl_3$. X-ray diffraction data extracted from monochromatic Laue photographs can neither confirm nor rule out such a distortion [14,15].

Table 1 : Values of d_I, δa_{c-p}, a_{c-p} for metal di- and trichloride GIC with the structure type shown in Fig. 1.

	NiCl$_2$	FeCl$_3$	CrCl$_3$	CoCl$_2$	MnCl$_2$	AlCl$_3$ a)	FeCl$_2$ b)	ZnCl$_2$ c)	CdCl$_2$ d)
d_I [pm]	595	603	610	615	613 616	617	616	618	' 616
δa_{c-p} [°]	0	0	0	30	30	30	30	30	30
a_{c-p} [pm]	348	350 353	344	357	369	346	359	362	(385)
Ref.	[4]	[5, 6, 7]	[8]	[9, 10]	[11, 12]	[13, 14, 15]	[16]	[15]	[15, 17]

Remarks: a) The structure type considered here is observed only in compounds prepared under special conditions [14, 15].

b) prepared by direct intercalation [16], not by reduction of FeCl$_3$-GIC (in this case $\delta a_{c-p} = 0°$ [18]).

c) The zinc atoms also might, to a small proportion, occupy tetrahedral sites [15].

d) a_{c-p} is taken from pure CdCl$_2$.

Discussion

Like in alkali-graphites [1], geometrical factors seem to have an important influence on the choice of the rotation angle in metal chloride GICs. Nevertheless, explanations given for the donor compounds are probably not applicable in the case of the compounds dealt with here. This can be concluded from the fact, that these theories predict a wide variety of different orientation angles, while in the case of MeCl$_2$- and MeCl$_3$-GIC only special values of δa are observed. This discrepancy might be due to the fact that the metal chloride layers are much more rigid than it is assumed for the alkali layers in the models described in [1].

References

[1] R. Moret, this issue, p. 185, and references cited therein
[2] A. Herold, in Intercalated Layered Materials, p. 321; Ed. F. Levy; Reidel, Dordrecht, Holland (1979)
[3] A.W.S. Johnson, Acta Cryst. 23, 770 (1967)
[4] S. Flandrois, J.M. Masson, J.C. Rouillon, J. Gaultier, C. Hauw Synth. Met. 3, 1 (1981)
[5] J.M. Cowley, J.A. Ibers, Acta Cryst. 9, 421 (1956)
[6] W. Metz, E.J. Schulze, Z. Krist. 142, 409 (1975)
[7] F. Rousseaux, R. Vangelisti, A. Plançon, D. Tchoubar Rev. Chim. Miner. 19, 572 (1982)
[8] R. Vangelisti, A. Herold, Carbon 14, 33 (1976)
[9] R. Heinrich, Thesis, Hamburg (1978)
[10] M. Elahy, M. Shayegan, K.Y. Szeto, G. Dresselhaus Synth. Met. 8, 35 (1983)

[11] F. Baron, S. Flandrois, C. Hauw, J. Gaultier
Solid State Commun. 42, 759 (1982)

[12] T. Dziemanowicz, W.C. Forsman, R. Vangelisti, A. Herold
Carbon 22, 53 (1984)

[13] W. Rüdorff, R. Zeller, Z. Anorg. Allg. Chem. 279, 182 (1955)

[14] P. Behrens, U. Wiegand, W. Metz, Proceedings of the 4th International
Carbon Conference Carbon '86, Baden-Baden (1986), p. 502

[15] P. Behrens, W. Metz, to be published

[16] G. Schoppen, Thesis, Hamburg (1978)

[17] W. Rüdorff, E. Stumpp, W. Spriessler, F.W. Siecke
Angew. Chem. 75, 130 (1963)

[18] K. Ohhashi, J. Tsukijawa, Journ. Phys. Soc. Japan 37, 63 (1974)

STRUCTURAL CORRELATIONS IN TRANSITION METAL GICs

J.S. Speck, X. Hao, and M.S. Dresselhaus

Massachusetts Institute of Technology
Cambridge, MA 02139, USA

We present here a discussion on the use of scattering measurements to provide information on structural correlations with particular reference to intercalate reflections in GICs. The real space assembly of atoms, known as the specimen function $R(\vec{r})$, which is the target for a scattering experiment (using x–rays, neutrons, or electrons), can be described by the product of a pair correlation function and a shape function,[1] written as $R(\vec{r}) = \langle \rho(\vec{r})\rho(0) \rangle S(\vec{r})$. For crystalline specimens, the pair correlation function can be replaced by the convolution of $F(\vec{r})$ with $G(\vec{r})$ so that the specimen function[2] is given by $R(\vec{r}) = [F(\vec{r}) * G(\vec{r})]S(\vec{r})$ where $F(\vec{r})$ is the scattering density within a unit cell, $G(\vec{r})$ is the sum over δ functions on sites of the real lattice, and $S(\vec{r})$ is the crystal shape function, defined as $S(\vec{r}) = 1$ inside the specimen and $S(\vec{r}) = 0$ outside, and "$*$" denotes the convolution of the indicated functions. Taking the Fourier transform of the specimen function, we obtain for 3–dimensional crystals[2] the relation $\bar{R}(\vec{q}) = [\bar{F}(\vec{q})\bar{G}(\vec{q})] * \bar{S}(\vec{q})$ where $\bar{F}(\vec{q})$ is the structure amplitude and $\bar{G}(\vec{q})$ and $\bar{S}(\vec{q})$ are respectively the Fourier transforms of $G(\vec{r})$ and $S(\vec{r})$. However, what we measure is $|\bar{R}(\vec{q})|^2$ which to a good approximation[1] is given by $|\bar{R}(\vec{q})|^2 = |\bar{F}(\vec{q})\bar{G}(\vec{q})|^2 * |\bar{S}(\vec{q})|^2$. Since the Fourier transform 'inverts' dimensions, large crystals give very narrow linewidths and small crystals give very broad linewidths. Hence, the FWHM linewidth in a given direction in $\vec{k}$–space for scattering from aggregates of crystallites is the reciprocal of the average crystal dimension in that $\vec{k}$ direction. If the linewidth of a reflection is governed by the specimen and not by the resolution function of the experimental set–up, then careful linewidth analysis on several reflections can provide a wealth of information on crystallite sizes and correlation lengths ξ in defective samples.

The situation is more complex for stacks of incommensurate crystals, such as are found in the incommensurate transition metal GICs. The results are described below for $(00l)$, (hkl), and $(hk0)$ interpenetrating incommensurate crystals which may have crystallographic correlations between different layers of a given species.

• For $(00l)$ reflections, scattering occurs from scattering density projected onto $\vec{c}^*$, the reciprocal lattice vector normal to the basal planes. The effective linewidth of the $(00l)$ reflection in the $\vec{c}^*$–direction is $1/d_{zf}$ where d_{zf} is the spacing in the c–direction between staging defects, so that d_{zf} measures the staging fidelity. Linewidths $\perp$ to the c^* direction are strongly dependent on the geometry and continuity of the intercalant layer

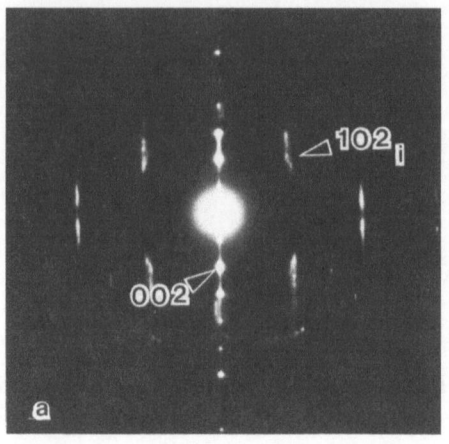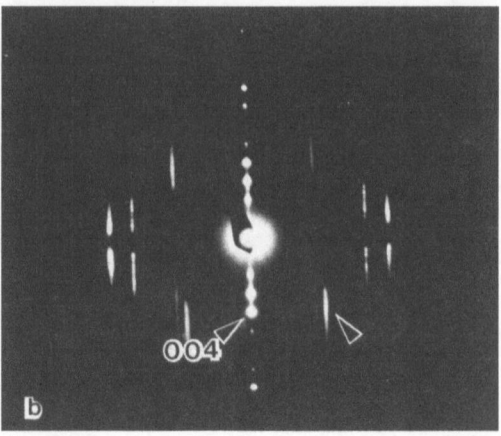

Figure 1: (a) Selected area electron diffraction pattern from a stage 1 $MnCl_2$–GIC sample showing (hkl) reflections (e.g., $(102)_i$) due to the intercalant. (b) Selected area electron diffraction pattern from a stage 2 $CoCl_2$–GIC sample showing no intensity modulation in the $\vec{c}^*$–direction for intercalant reflections.

planes. For circular intercalant islands, linewidths for scans taken $\perp$ to the c^* direction represent the average in–plane distance of the interplanar overlap of islands.

• For (hkl) reflections, coherent scattering occurs for correlated scattering density projected onto a direction normal to (hkl). The complete 3–dimensional (3D) lineshape is given by the Fourier transform of the correlated volume. That is, the linewidth of an (hkl) reflection in a given direction is proportional to $1/\xi$ in that direction.

• For $(hk0)$ reflections, coherent scattering occurs for both regions with 2D planar correlations and regions with 3D correlations. This leads to two shape transforms, so that the linewidth measured in the $\vec{c}^*$–direction is the sum of these two contributions.

An application of the results discussed above is in structural studies of transition metal GICs. The selected area electron diffraction pattern shown in Fig. 1a for a stage 1 $MnCl_2$–GIC clearly shows the presence of (hkl) reflections. Note that the (hkl) reflections are diffuse and streaked, whereas the $(00l)$ reflections for the same sample are sharp and discrete. The results indicate that interlayer correlations do occur, but over a much shorter distance than the average spacing between staging defects. It is anticipated that correlations between intercalant layers will be lost in higher stage $(n \geq 2)$ compounds. For example, the selected area electron diffraction pattern, shown in Fig. 1b for a stage 2 $CoCl_2$–GIC sample, exhibits very little intensity modulation in the $\vec{c}^*$–direction at the positions where $(hkl)_i$ reflections for the intercalant are expected. Hence, there is essentially no correlation between intercalant layers in this case.

The authors wish to thank Drs. J.M. Gibson and G. Dresselhaus for many useful discussions. This work was supported by AFOSR contract #F49620–83–C–0011.

REFERENCES

1. J. Cowley, *Diffraction Physics*, (North-Holland, New York) 1984.

2. L. Reimer, *Transmission Electron Microscopy*, (Springer–Verlag) 1984.

ELECTRONIC STRUCTURE AND OPTICAL PROPERTIES

C. Rigaux

Groupe de Physique des Solides
   de l'Ecole Normale Supérieure
24 rue Lhomond
75231 Paris Cedex 05, France

1. ELECTRONIC STRUCTURE

1.1 Introduction

Graphite exhibits a lamellar structure (Fig.1). The lattice consists of hexagonal carbon monolayers separated by an interplanar spacing (d=3.35Å) much larger than the in-plane nearest neighbour distance (b=1.42Å). The weak interlayer interactions allow the intercalation of a wide diversity of atomic and molecular species between the carbon layers. Graphite Intercalation Compounds (GIC) are characterized by the existence of stage ordering: a periodic sequence of n graphite layers and an intercalated layer, with n defining the stage of the compound [1-11].

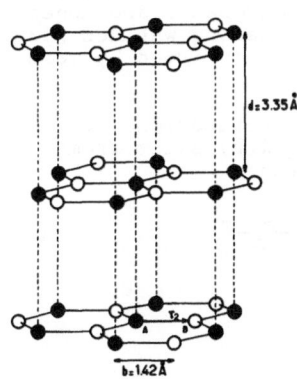

Fig. 1. Graphite lattice.

The graphite sublattice is essentially preserved after intercalation, the main effect consists of an increase in the c-axis spacing I_c between carbon layers bounding the intercalants: I_c varies from 3.71Å (LiC$_6$) to about 10Å in some molecular compounds [1,2]. While stage ordering is common to many GIC, the in-plane structure of the intercalant sublattice is quite varied, including both solid (commensurate or incommensurate with the graphite host) and liquid phase [1,3]. Few GIC are fully ordered at room temperature with commensurate intercalant and carbon sublattices (e.g. first stage of Alkali metal compounds MC$_6$, MC$_8$), whereas intercalated layers in most molecular compounds are disordered or in the liquid phase.

GIC display quite unique electronic properties. Most of them are related to the layered impurity (donor or acceptor) distribution, and to the excess charge transferred from intercalants to the graphite layers. The metallic behavior is a general feature of the entire graphite family: the intercalation is as a rule accompanied by a large increase in the in-plane conductivity and by the occurrence of a metallic reflectance. Fundamental interest of GIC is related to the variable dimensionality of this class of materials: by a proper choice of stage and intercalants, the anisotropy of conductivity can be varied over several orders of magnitude (from 10 to 10^6) [2,4,5]. The anisotropy of conductivity σ_a/σ_c is reduced in first stage donor compounds while it considerably increases for Graphite Acceptor Compounds (GAC), exceeding 10^6 in the most conducting materials. These features underline the two-dimensional character of GAC, in contrast to first stage Alkali metal GIC which display a three-dimensional anisotropic character, as shown by the appreciable c-axis conductivity (larger than in graphite) and the observation of plasma edges for $\vec{\varepsilon}/\!/\vec{c}$ [12].

Recent efforts have been directed towards fundamental understanding of the electronic properties of GIC. Transport measurements, spectroscopic techniques (optical reflectance, X-ray absorption, photoemission, electron loss spectroscopy), Fermi surface properties, nuclear and electronic magnetic resonance, magnetic susceptibility and specific heat measurements have been used to study the electronic structure of GIC [2,4,5-11]. A coherent interpretation of a great variety of accumulated experimental data requires the knowledge of electronic states in these materials. Rigorous band structure calculations have been performed for GIC of highest symmetry, e.g. first stage Alkali metal compounds. For more complex structures or for GIC whose lattice does not possess the translational symmetry, phenomenological models based on different assumptions have been elaborated. The comparison with experimental data constitutes a critical test for any proposed band structure model.

In this lecture, the essential characteristics of the electronic structure in GIC are reviewed. A detailed description of the electronic states in two-dimensional systems is presented, which offers a coherent interpretation of optical properties in GAC.

1.2 Electronic structure of graphite

In 2D graphite the bands are classified as σ (even) and π (odd) according to the symmetry of the wavefunctions with respect to reflection in the layer plane. Graphite has four valence electrons per carbon atom: three of these (2s, $2p_x$, $2p_y$) form covalent in-plane σ bonds, the fourth electron in the $2p_z$ state gives rise to the conduction and valence π bands.

The 2D hexagonal lattice of a single graphite layer is represented in Fig. 2; the unit cell contains two atoms A and B defined by $\vec{\tau}_1$ and $\vec{\tau}_2$ vectors. The primitve translation vectors are $\vec{a}_1$ and $\vec{a}_2$, $|\vec{a}_1|=|\vec{a}_2|=3^{\frac{1}{2}}b$, where b=$\overline{AB}$=1.42Å is the nearest neighbour distance. The two-dimensional Brillouin zone is represented in Fig. 3. At the points U, U', related by the time reversal operation, the valence and conduction π bands are degenerate by symmetry. These features were first established by WALLACE [13] from a simple tight binding method and have been later confirmed by more realistic band structure calculations [14,15].

In the tight binding approximation, the electron wavefunctions are chosen as linear combinations of the two functions $u_{i,\vec{\varkappa}}(\vec{r})$ built from atomic $2p_z$ orbitals $\Phi_z(\vec{r})$ centered at the A and B sites:

$$u_{i,\vec{\varkappa}}(\vec{r})=(\Omega/\Sigma)^{\frac{1}{4}} \sum_{\vec{\rho}_m} \exp[i\vec{\varkappa}.(\vec{\rho}_m+\vec{\tau}_i)]\ \Phi_z(\vec{r}-\vec{\rho}_m-\vec{\tau}_i) \qquad (1)$$
$$(i=1,2)$$

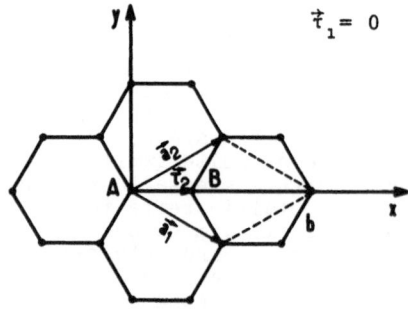

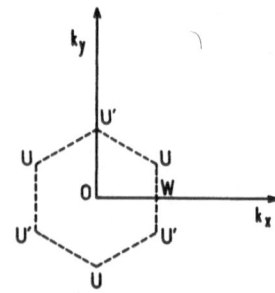

Fig. 2. Unit cell of a single graphite layer.

Fig. 3. 2D Brillouin zone.

$\vec{\varkappa}$ is the 2D wavevector with the origin at the O point, $\vec{\rho}_m = m_1\vec{a}_1 + m_2\vec{a}_2$ is the lattice vector, Ω and Σ are the areas of the unit cell and of the crystal respectively. The diagonalization of the one electron Hamiltonian in the basis (1) leads to the following secular equation:

$$\begin{vmatrix} -E & -\gamma_0 g(\vec{\varkappa}) \\ -\gamma_0 g^*(\vec{\varkappa}) & -E \end{vmatrix} = 0 \tag{2}$$

The resonance integral γ_0 is defined as

$$\gamma_0 = -\int \Phi_z(\vec{r}) \left[V(\vec{r}) - V_{at}(\vec{r}) \right] \Phi_z(\vec{r} - \vec{\tau}_2) \, d^3r \tag{3}$$

where $V(\vec{r})$ and $V_{at}(\vec{r})$ are the crystal and the atomic potentials respectively.

$$g(\vec{\varkappa}) = \exp(i\vec{\varkappa} \cdot \vec{\tau}_2) + \exp(i\vec{\varkappa} \cdot \overrightarrow{D_3\tau_2}) + \exp(i\vec{\varkappa} \cdot \overrightarrow{D_3^{-1}\tau_2}) \tag{4}$$

where D_3 is the operator of the $2\pi/3$ rotation around the c-axis. In writing (2), the overlap integrals of the pz functions centered on different atoms are neglected.

Energies of the conduction and valence π bands are given by the simple expression

$$E_{\substack{C\\V}}(\vec{\varkappa}) = \pm\, \gamma_0 |g(\vec{\varkappa})| . \tag{5}$$

In the vicinity of the U, U' points, $g(\vec{\varkappa}) \simeq \frac{3}{2} b(k_x - ik_y)$, and the band energies are isotropic functions of the wavevector $\vec{k} = \vec{\varkappa} - \vec{\varkappa}_u$:

$$E_{\substack{C\\V}}(k) = \pm\, \frac{3}{2}\, \gamma_0 bk \tag{6}$$

The degenerate π bands are of particular interest since in 2D graphite the Fermi level lies at the degeneracy points U, U'.

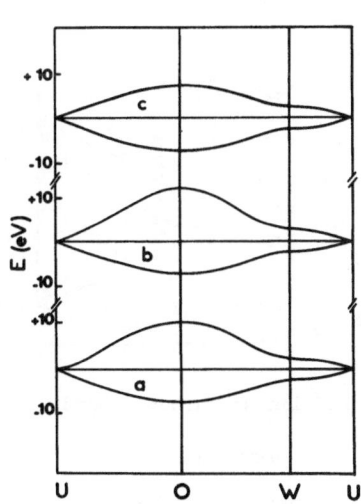

Fig. 4. Energy band structure of 2D graphite along the principal directions of the Brillouin zone. (a) Ref. [14], (b) Ref. [15], (c) Ref. [16].

The dispersion relations (5) for principal directions in the Brillouin zone, represented in Fig.4, reproduce fairly well the shape of the π bands. Each band has a saddle point at W and an extremum at O. The results of the simple tight binding method [16] are compared in Fig.4 with those obtained from more elaborated band structure calculations [14,15]. At the U point, the energy gap separating π and σ bands exceeds 10 eV [14].

In a 3D graphite lattice, the AB stacking sequence (Fig.1) of the hexagonal layers gives rise to four carbon atoms per unit cell. The 3D band structure model was established by SLONCZEWSKI and WEISS [17], and McCLURE [18] by applying a perturbation technique to the 2D π bands near the contact point, using a $\vec{K}.\vec{p}$ expansion in the plane and a Fourier expansion along the edges HK of the Brillouin zone. The band structure of 3D graphite is adequately described in terms of seven parameters. The key feature of the electronic structure of 3D graphite is the existence of a small valence-conduction overlap (36 meV) resulting entirely from weak interactions between every second layer. While 2D graphite is a zero gap semiconductor, 3D graphite is a true semimetal.

The comparison between different models for 3D band structure of graphite can be found in Refs. [19,20,21].

1.3.1 First principle band structure calculations have been performed for low stage Alkali metal compounds which are fully ordered at room temperature with commensurate intercalant and carbon sublattices. The intercalation compound of highest symmetry is the first stage of Li-graphite, LiC$_6$, with a p($\sqrt{3}$x$\sqrt{3}$)R 30° in-plane superlattice. The unit cell of LiC$_6$ is hexagonal and contains one Li and six C atoms, assuming αAαAα interlayer graphite (A) and intercalate (α) stacking order. The band structure of LiC$_6$ was calculated by HOLTZWARTH et al.[22] using an ionic potential C$_6^-$Li$^+$. The energy bands along the high symmetry directions of the hexagonal Brillouin zone are shown in Fig.5. The occupied bands are essentially those of carbon with A-A layer stacking. An s-like band derived from Li (2s) is completely empty, lying well above (at least 1.7eV) the Fermi energy. This band is weakly hybridized with the graphite π bands above the Fermi energy. The Li is fully ionized with a charge transfer of one electron per Li atom. The complex Fermi surfaces (Fig.6) consist of a distorted cylinder (upper band) and a 3D open surface (lower band).

KC$_8$ has a lower symmetry with 2K and 16C per unit cell. Self-consistent band structure calculations for KC$_8$ (with a p(2x2)R 0° superlattice

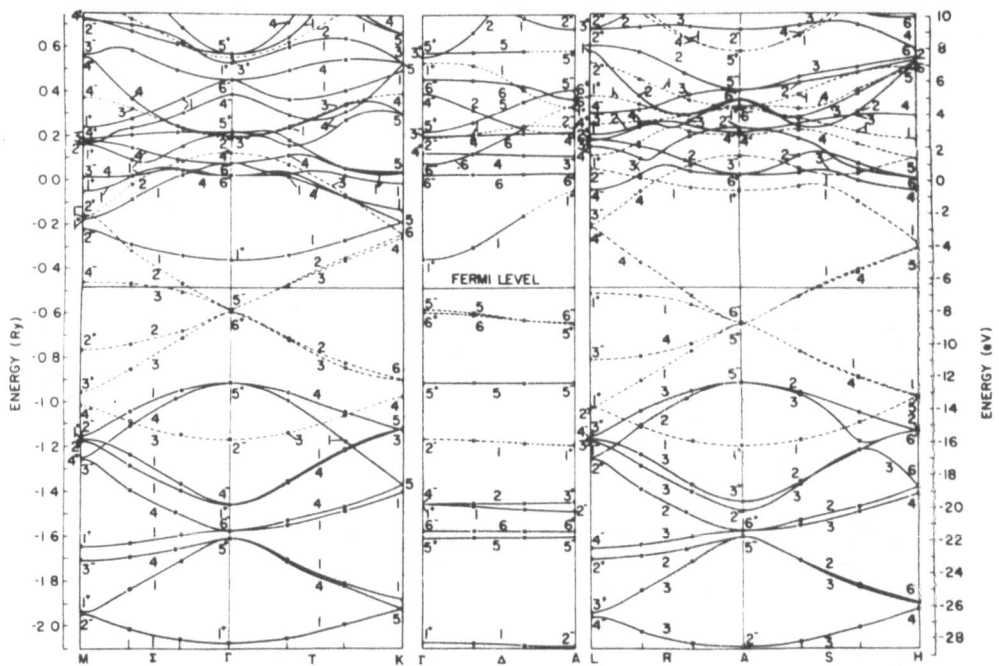

Fig. 5. Band structure of LiC$_6$ (from [22]). Dashed lines: carbon π bands.

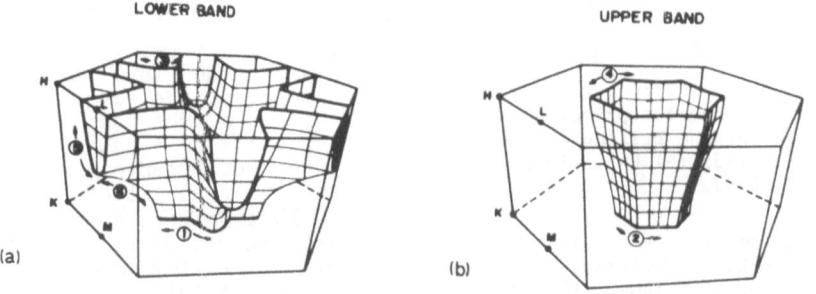

Fig. 6. Conduction band Fermi surfaces of LiC$_6$ (from [22]).

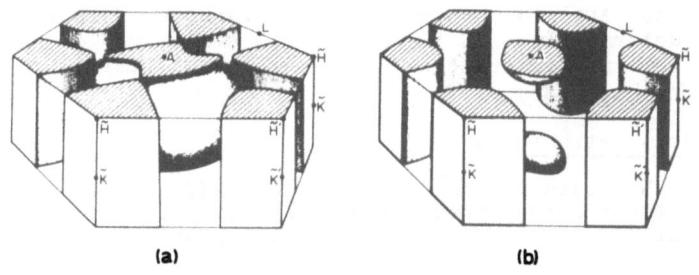

(a)   (b)

Fig. 7. Conduction band Fermi surfaces of KC_8 (from [24]).

and $\alpha\beta\gamma\delta$ intercalate interlayer stacking) were carried out by OHNO et al. [23] using pseudo-potential methods. The band structure is very similar to that calculated previously by INOSHITA et al. [24] from a tight binding method. The characteristics of this structure are reproduced by superposing the K(4s) bands on the graphite bands. Important hybridization effects occur near the Fermi energy. The charge transfer from K to C layers is found to be f=0.6. The Fermi surfaces are shown in Fig. 7: in the central part of the Brillouin zone, the conduction bands originating from the K(4s)-electron have predominantly isotropic character whereas at the zone edge, the Fermi surface is cylindrical reflecting the two-dimensional character of the graphite bands. Different conclusions were obtained by DIVICENZO and RABII [25], TARTAR and RABII [26], using a modified KKR method. These authors predict complete charge transfer from K(4s) to graphite π bands with an empty alkali-derived band lying 1.8 eV above the Fermi energy.

For LiC_6 and KC_8, the dispersion of the occupied π bands are in good agreement with the 2D graphite π bands folded into the smaller Brillouin zone of the intercalation compound [22] (Fig.8). The Brillouin zone of LiC_6 is 1/3 the area of the Brillouin zone of 2D graphite, the K point of 2D graphite mapping into the Γ point for LiC_6 structure and the M point (graphite) mapping into the M point for LiC_6. In the LiC_6 Brillouin zone, the two π bands (degenerate at the K point in the 2D structure of graphite) are translated into four bands centered at Γ point (Fig.8). The Brillouin zone of KC_8 is one fourth the area of that of 2D graphite, the M point maps into the zone center of KC_8 resulting in the folded π bands reported in Fig.8. The K point of the graphite is found at the K point of KC_8. The 2D Fermi surfaces are centered at K, in contrast to that centered at Γ point for LiC_6.

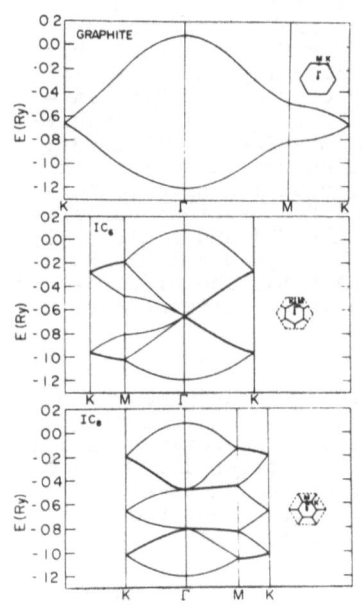

Fig. 8. π bands of 2D graphite folded into the Brillouin zone of the intercalation compounds. (a) 2D graphite, (b) MC_6, (c) MC_8 (after [22]).

1.3.2 The prospects of performing rigorous band structure calculations in GIC are very limited (even if the perfect 3D crystal order were established) owing to the large number of atoms in the elementary unit cell. DRESSELHAUS and LEUNG [27] have therefore developed a phenomenological model based on the symmetry properties of the superlattice structure of the intercalation compounds imposed on the graphite structure. The in-plane intercalate ordering gives rise to in-plane zone folding and the staging phenomenon introduces a c-axis superlattice structure which is incorporated into the calculation by a k_z-axis zone folding. The important step in this scheme is the construction of a layer representation which allows one to account for the change of interlayer interactions in GIC as compared to the case of pure graphite. This model was applied to the determination of the electronic structure of several GIC of stages 1-8. However, in practice, this method was limited to systems with translational symmetry.

1.4 Two dimensional structures

The crystallographic data indicate that in most GIC, over a broad temperature range, the rigorous translational symmetry is absent even if the stage is well defined. Typically, the intercalants form 2D ordered lattices commensurate with the graphite lattice, but the stacking sequence of these lattices remains irregular [28]. In other cases, the intercalant layers do not even exhibit the long range 2D order [29]. Thus, disregarding the in-plane structure of intercalant layers, phenomenological models based on charge transfer concept have been elaborated to describe the electronic states in GIC [16,30,33,35]. An intercalation compound of stage n is treated as a set of equivalent independent systems of n charged interacting graphite layers limited by two intercalant layers. A definite charge transfer from C atoms to intercalated species is assumed giving rise to mobile carriers in the graphite systems and to a rigid charge distribution in the intercalated layers. The charge flow across the intercalant layers is excluded. Neglecting the in-plane spatial distribution of intercalants, the intercalated layers are treated as uniformly charged sheets. Consequently they do not introduce potential energy variations within the graphite systems. According to this model, the properties of nth stage GIC reflect the properties of n interacting charged graphite layers. This 2D model is well adapted for GAC due to the large interlayer spacing I_c between the graphite layers bounding the intercalants, and to the highly anisotropic character of these compounds.

The electrostatic effects resulting from the excess charge accumulated on graphite layers have to be accounted for. For stages 1 and 2, these effects are not important as all graphite layers are equivalent and the same excess charge is accumulated on each of them. However for stage $n \geqslant 3$, the graphite layers are no longer equivalent and the electrostatic forces affect the equilibrium excess charge distribution which becomes highly inhomogeneous. The first quantitative analysis of the c-axis charge distribution was carried out by PIETRONERO et al. [31], by applying the Thomas-Fermi approximation and using locally the 3D rigid band model of graphite. In fact, in a system of n charged graphite layers, the band structure is principally different from that of a graphite crystal of macroscopic dimension: there exist 2n two-dimensional bands, each being specifically affected by the potential energy variations in the system. The band structure and the charge distribution have to be determined by a self-consistent procedure. Such self-consistent calculation have been performed for stage 3 of GAC by BLINOWSKI and RIGAUX [32], and for higher stages (n ≤ 8) of donor compounds by SAFRAN and HAMANN [33], SHIMAMURA and MORITA [34], OHNO and KAMIMURA [35]. The electrostatic energy is shown to play a significant role in stabilizing pure stage configurations [33].

Electronic states in GAC

For a single subsystem, the essential features of the π bands near the U point are obtained from the simplest tight binding method including only intra and inter layer nearest neighbour interactions between carbon atoms [16,32]. Following the procedure described in Section 1.2, the wave functions are approximated by linear combinations of the 2n tight binding functions $u_{j,\vec{\varkappa}}(\vec{r})$ built from $2p_z$ atomic orbitals $\Phi_z(\vec{r})$ centered at the non-equivalent sites A_i, B_i of consecutive layers (i=1,...n).

$$u_{j,\vec{\varkappa}}(\vec{r}) = (\Omega/\Sigma)^{\frac{1}{2}} \sum_{\vec{\rho}_m} \exp[i\vec{\varkappa}.(\vec{\rho}_m+\vec{\tau}_j)] \Phi_z(\vec{r}-\vec{\rho}_m-\vec{\tau}_j) \tag{7}$$

The positions of the A_i, B_i atoms in the layer i are defined by the vectors $\vec{\tau}_j$ (j=2i-1 and 2i for A_i and B_i sites respectively) (Fig. 9). $\vec{\rho}_m$ is the vector of the Bravais lattice in layer 1. $\Omega = \frac{3}{2} 3^{\frac{1}{2}}b^2$ is the area of the 2D elementary cell represented in Fig. 2, Σ is the crystal area and $\vec{\varkappa}$ is the 2D wave vector with the origin at the zone center (Fig. 3).

1.4.1 First and second stages

In the basis (7), the Hamiltonians for one and two layers are represented by the following matrices:

$$\begin{array}{c} \\ A \\ B \end{array}\begin{array}{cc} A & B \\ \begin{bmatrix} 0 & -x \\ -x^* & 0 \end{bmatrix} \end{array} \quad \text{one layer} \tag{8}$$

$$\begin{array}{c} \\ A_1 \\ B_1 \\ A_2 \\ B_2 \end{array}\begin{array}{cccc} A_1 & B_1 & A_2 & B_2 \\ \begin{bmatrix} 0 & -x & \gamma_1 & 0 \\ -x^* & 0 & 0 & 0 \\ \gamma_1 & 0 & 0 & -x^* \\ 0 & 0 & -x & 0 \end{bmatrix} \end{array} \quad \text{two layers} \tag{9}$$

where $x=\gamma_0 g(\vec{\varkappa})$ with $g(\vec{\varkappa})$ defined in (4). γ_0 and γ_1 are the dominant intralayer (A_i-B_i) and interlayer (A_i-A_{i+1}) resonance integrals. γ_0 is defined in (3).

$$\gamma_1 = \int \Phi_z(\vec{r})[V(\vec{r})-V_{at}(\vec{r})]\Phi_z(\vec{r}-\vec{\tau}_3)d^3r$$

In writing (8) and (9) several approximations are made: i) all overlap integrals for atomic pairs other than nearest neighbours are neglected, ii) the possible mixing of π and σ bands is disregarded, iii) the potential energy difference between non-equivalent atomic sites A_i,B_i of a given layer is neglected.

The band structure of the first stage is identical to that of the 2D graphite described in Section 2. The dispersion relations for first and second stages are:

First stage

$$E_{\substack{c \\ v}}(\vec{\varkappa}) = \pm|x| \tag{10}$$

Second stage

$$E_{c_1} = -E_{v_1} = [(\gamma_1^2+4|x|^2)^{\frac{1}{2}}-\gamma_1]/2 \equiv E_1$$
$$E_{c_2} = -E_{v_2} = [(\gamma_1^2+4|x|^2)^{\frac{1}{2}}+\gamma_1]/2 \equiv E_2 \tag{11}$$

The upper and lower signs in (10) correspond to conduction and valence bands respectively. Near the Brillouin zone corners, $x \simeq \frac{3}{2}\gamma_0 b(k_x-ik_y)$ where $\vec{k}=\vec{\varkappa}-\vec{\varkappa}_u$. Fig.10 shows the dispersion relations for stages 1 and 2

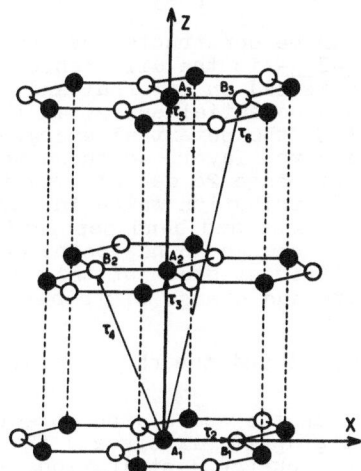

Fig. 9. Disposition of the atoms in the third stage.

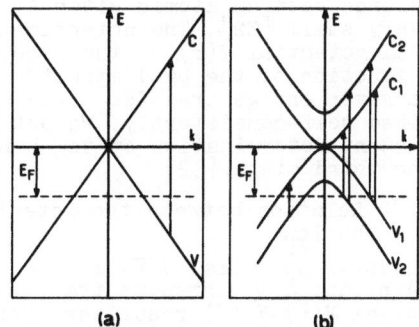

(a)   (b)

Fig. 10. Energy band structure near the U point. (a) first stage. (b) second stage.

respectively, in the vicinity of the U point. The first stage compound is a zero gap semiconductor with linear in k dispersion relations close to the band degeneracy points: $E_C = \pm \frac{3}{2} \gamma_0 bk$. For **stage 2**, the interlayer interactions (A_1-A_2) partially remove the band degeneracy and the band structure consists of two parallel valence (v_1,v_2) and conduction (c_1,c_2) bands separated by **an energy distance equal to** γ_1.

The hole Fermi energy is directly related to the hole densities η:

$$E_{F_1} = \frac{3}{2} \gamma_0 b (\pi\eta)^{\frac{1}{2}} \text{ for stage 1} \tag{12}$$

$$E_{F_2} = \frac{3}{2} \gamma_0 b (\pi\eta/2)^{\frac{1}{2}} \text{ for stage 2} \tag{13}$$

An important characteristic of GIC is the charge transfer coefficient f which is defined as the number of transferred electrons per intercalated molecule. For compounds of stage n=1,2 corresponding to the formula $(C_1)_n A$, the hole Fermi energy is related to the charge transfer f by

$$E_{F,n} = \gamma_0 (3^{\frac{1}{2}} \pi f / nl)^{\frac{1}{2}} \tag{14}$$

f, which depends on the intercalated species, has to be determined from the comparison with experimental data. The analysis of the optical properties has provided the determination of the charge transfer coefficients for stages 1 and 2 of several types of GAC.

1.4.2 Stage 3

1.4.2.1 Model Hamiltonian with electrostatic effects

For stage n $\geqslant$ 3, the electrostatic effects resulting from the non-equivalence of the charged graphite layers have to be accounted for. On the basis of the six Bloch functions $u_{j,\vec{\varkappa}}(\vec{r})$ (j=1,...6) (7), the approximated Hamiltonian for π bands in a system of three layers is represented by the following matrix :

$$H = \begin{array}{c} \\ \\ \end{array} \begin{array}{cccccc} A_1 & B_1 & A_2 & B_2 & A_3 & B_3 \\ \end{array}$$

$$H = \begin{bmatrix} \delta & -x & \gamma_1 & 0 & 0 & 0 \\ -x^* & \delta & 0 & 0 & 0 & 0 \\ \gamma_1 & 0 & -\delta & -x^* & \gamma_1 & 0 \\ 0 & 0 & -x & -\delta & 0 & 0 \\ 0 & 0 & \gamma_1 & 0 & \delta & -x \\ 0 & 0 & 0 & 0 & -x^* & \delta \end{bmatrix} \begin{array}{c} A_1 \\ B_1 \\ A_2 \\ B_2 \\ A_3 \\ B_3 \end{array} \tag{15}$$

The corresponding matrix for higher stages can be constructed by analogy. Near the U point, $x = \frac{3}{2} \gamma_0 b(k_x - ik_y)$; $2\delta = E_{ext} - E_{int}$ is the difference between the potential energy of an electron localized on external and internal layers ; $E_i = \int [\Phi_z(\vec{r} - \vec{\tau}_i)]^2 V(\vec{r}) d^3 r$ is the potential energy at the atomic sites of the layer i. We neglect in (15) the potential energy difference between atomic sites A_i ; B_i of a given layer, as these terms are very small [32]. The potential energy difference 2δ depends via the crystal potential $V(\vec{r})$ on the excess charge distribution which in turn is a function of the band structure, wavefunctions and band population. For a given charge transfer coefficient f, the energy 2δ has to be determined self-consistently. We outline the different steps of the self-consistent determination of the band structure and charge distribution in the third stage.[32]

1.4.2.2 Relation between the potential energy 2δ and the charge distribution

The crystal potential $V(\vec{r})$ may be considered as the sum of the potential $V_0(\vec{r})$ produced by three neutral, independent graphite layers and the potential $V(\vec{r}) - V_0(\vec{r})$ resulting from the excess charge distribution. The contribution from V_0 to 2δ results from the fact that the internal layer has neighbors on both sides whereas the external layer has neighbors on one side only. This contribution was roughly identified with one half of the average energy shift of π bands in 3D graphite [17] with respect to π bands in a simple layer.

The contribution to 2δ originating from the excess charge distribution was calculated by using a standard multipole expansion method. The excess charge distribution around each atom is assumed to have the same form as the $\Phi_z(\vec{r})$ atomic orbitals

$$\Phi_z(\vec{r}) = (\alpha^5/\pi)^{1/2} z\ e^{-\alpha z} \tag{16}$$

$$\alpha = 1.59/a_B \tag{17}$$

where a_B is the Bohr radius. We attribute to each atom of the layer i the charge e_i, the dipole moment $\vec{p}_i//z//c$ and a quadrupole moment tensor $\hat{Q}_i$ (the number of different elements is restricted by the symmetry of the system). Direct computer summation of the electric field contributions from lattice point charges proves that the electric field acting on a given layer due to excess charge and dipoles in other layers is practically uniform and depends only on the surface charge densities [32]. Consequently the dipole moments are the same for all atoms in a layer and their magnitude depends on the field according to the macroscopic formula

$$p = (\varepsilon_c - 1)(4\pi N \varepsilon_c)^{-1}\ E_{ext} \tag{18}$$

relating the atomic dipole moment p in 3D graphite to the external uniform field E_{ext} parallel to the c-axis. N is the number of atoms per unit volume and ε_c is the dielectric constant along the c-axis. For the stage 3, the surface density σ_{d1} of dipole moments induced on external layers is then

$$\sigma_{d1} = d\ \frac{(\varepsilon_c - 1)}{2\varepsilon_c}\ (\sigma_{c1} + \sigma_{c2}) \tag{19}$$

where σ_{c1} and σ_{c2} are the surface charge densities in external and internal layers respectively. Direct computer summations of the contributions to E_i from consecutive layers lead to the following expression for 2δ :

$$(2\delta)_{V-V_0} = -e(e_1 - e_2)(J - \frac{3.32}{b} - \frac{3.29\ q}{b^3}) + 2\pi\sigma_{c2} de - 2\pi|\sigma_{d1}|e\ . \tag{20}$$

$-ee_i J$ represents the electrostatic energy interaction of an electron in the state $\Phi_z(\vec{r} - \vec{\tau}_j)$ with the excess charge e_i at the atom at $\vec{\tau}_j$. $-eq$ is the zz component of the quadrupole moment tensor for the state $\Phi_z(\vec{r})$. The numerical factors in (20) result from the geometry of the graphite lattice. If we represent $\Phi_z(\vec{r})$ by Slater orbitals [17], $J = 0.622/a_B$ and $q = 2.373 a_B^3$ [32]. Substituting (19) in (20) and taking $\varepsilon_c = 2.34$ [36] for the c-axis dielectric constant, the energy 2δ can be expressed in terms of the total charge accumulated in a graphite system and to its fraction z localized on the internal layer :

$$2\delta = 0.1 + \frac{f}{1} (57.6\ z - 2.7)\ [eV] \tag{21}$$

where f is the charge transfer coefficient per intercalated molecule in the compound $(C_1)_n A$. For uniform charge distribution $z = 1/3$ and for typical charge transfer ($f/1 \simeq 0.03$), (21) gives $2\delta = 0.6$ eV. The potential energy difference 2δ can therefore be larger than $\gamma_1 = 0.377$ eV [38] and should not be disregarded in any band structure calculation.

Equation (21) forms a basis for a self-consistent determination of the band structure and charge distribution, for a given value of the charge transfer $f/1$.

1.4.2.3 Band structure

Near the U point, the dispersion relation for the three valence and conduction bands are

$$E_{1_{\substack{c\\v}}} = \delta \pm |x|$$

$$E_{2_{\substack{c\\v}}} = \pm\{\delta^2 + \gamma_1^2 + |x|^2 - [\gamma_1^4 + (4\delta^2 + 2\gamma_1^2)|x|^2]^{1/2}\}^{1/2} \tag{22}$$

$$E_{3_{\substack{c\\v}}} = \pm\{\delta^2 + \gamma_1^2 + |x|^2 + [\gamma_1^4 + (4\delta^2 + 2\gamma_1^2)|x|^2]^{1/2}\}^{1/2}$$

The upper and lower signs apply for conduction(c) and valence (v) bands respectively. The wavefunctions associated with each band are linear com-

243

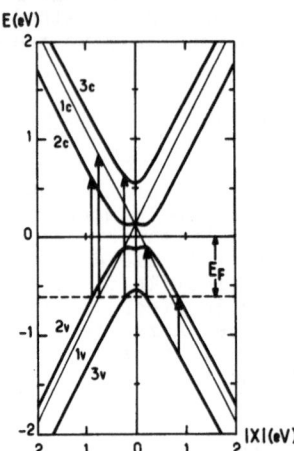

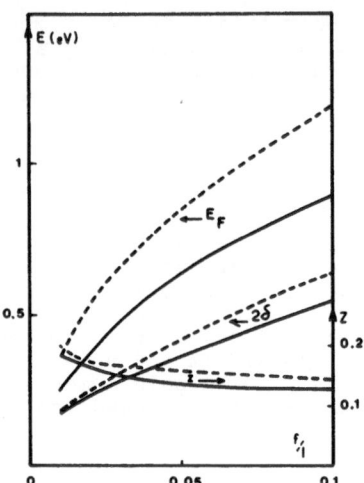

Fig. 11. Energy band structure near the U point in the third stage. $|x| = \frac{3}{2}\gamma_0 bk$, $\delta = \gamma_1/3$ (from [32]).

Fig. 12. E_F, 2δ (left scale) and z (right scale) vs f/1 for third stage. Solid lines: $\gamma_0 = 2.4$ eV, broken lines: $\gamma_0 = 3.12$ eV (from [32]).

binations of the basis Bloch function $u_{j\vec{\varkappa}}(\vec{r})$

$$\Psi = \sum_{j=1}^{6} a_j \, u_{j\vec{\varkappa}}(\vec{r}) \tag{25}$$

For the linear bands $E_{1\varsigma}$, the electrons are localized entirely on the external layers whatever the value of δ ($a_3 = a_4 = 0$). For all other bands, electrons are localized on both external and internal layers. Figure 11 illustrates the band structure for $\delta = \gamma_1/3$. The valence band 2V acquires a camelback structure at low k values, crosses the band 1V at $|x| = 2\delta$ and becomes practically linear for larger k. At low k values, the wavefunctions for the band 2V are practically localized on the internal layer and at high k values the contributions from the basis functions localized on external layers 1 and 3 dominate.

1.4.2.4 Fermi energy, charge distribution and potential energy difference

The hole Fermi energy E_F is determined from the dispersion relations (22) for different charge transfers 0.01<f/1<0.1, and for an arbitrarily chosen value of δ. This allows calculation of the band population and the fraction z of charge localized on the internal layer by summing up charge densities for all occupied band states and subtracting the charge densities for three neutral noninteracting graphite layers. Then (21) was used to correct the input value of δ and the calculations were repeated until self-consistency was achieved. The results of self-consistent calculations for E_F, z and 2δ are reported in Fig. 12 for charge transfer per C atom ranging from 0.01 to 0.1. Solid and broken lines correspond to $\gamma_0 = 2.4$ eV [16] and $\gamma_0 = 3.12$ eV [37] respectively.

For charge transfer coefficients determined from optical experiments in stage 2 GAC listed in Table II, the values of 2δ in stage 3 GAC vary from 0.24-0.28 eV for Br_2-graphite, 0.25-0.32 eV for $SbCl_5$-graphite, and 0.34-0.42 eV for AsF_5-graphite (the lower and upper limits corresponding to the different values of γ_0 considered here). For these compounds typically 13 to 16% of excess charge is expected to be localized on the central layer.

1.4.2.5 Comparison between different calculations

It is interesting to compare the results presented in Fig. 12 with those obtained from other determinations. The c-axis charge distribution was first calculated by PIETRONERO et al. [31] and later extended by SAFRAN and HAMANN [33] by using the semiclassical Thomas-Fermi approximation

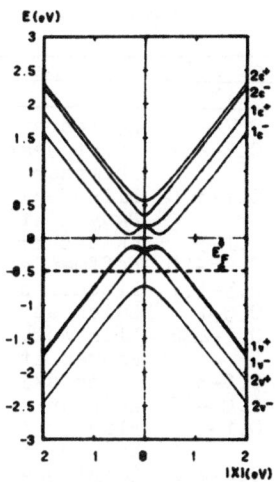

Fig. 13. Energy band structure near
the U point in stage 4.
$|x| = \frac{3}{2} \gamma_0 bk$, $\delta = \gamma_1/2$ (from
[32]).

which neglects the discrete nature of graphite layers. SAFRAN and HAMANN introduce a self-consistently determined layer potential term to the full tight binding Hamiltonian for π band electrons in a system of n graphite layers which are treated as uniformly charged sheets. SHIMAMURA and MORITA [34] took into account the polarization effects of the graphite π bands induced by the intercalant potential. More extensive self-consistent calculations were performed by OHNO and KAMIMURA [35] by including the whole energy bands of carbon ($1s, \sigma, \pi$) within the framework of the local density functional formalism. All these calculations are based on a system of n graphite layers bound by two intercalant layers treated as uniformly charged sheets.

The potential energy difference 2δ and the fraction of charge localized on internal layers z obtained from [33] and [32] for stage 3 GIC are in good agreement, e.g. for $f/1=12$, SAFRAN and HAMANN obtained $2\delta=0.597$ eV, $z=0.124$ ($\varepsilon_c=2$); $2\delta=0.494$ eV, $z=0.153$ ($\varepsilon_c=3$). BLINOWSKI and RIGAUX [32] data lead to $2\delta=0.494$ eV, $z=0.128$ ($\gamma_0=2.4$ eV); $2\delta=0.580$ eV, $z=0.150$ ($\gamma_0=3.12$ eV). The self-consistent calculations performed by OHNO and KAMIMURA [35] which included the whole graphite energy bands lead to an extremely inhomogenous charge distribution compared with the theoretical results obtained by considering π bands only.

1.4.3 Stage 4

The initial 8x8 secular equation for band energies in a system of four charged graphite layers can be factorized into the product of two 4x4 secular equations of the form [32]

$$
\begin{vmatrix}
\delta-E & -|x| & \gamma_1 & 0 \\
-|x| & \delta-E & 0 & 0 \\
\gamma_1 & 0 & -\delta-E\pm\gamma_1 & -|x| \\
0 & 0 & -|x| & -\delta-E
\end{vmatrix} = 0
\tag{24}
$$

Both equations (24) are biquadratic with respect to $|x|$, and the $|x|$ values for fixed energy E are obtained analytically. The resulting band structure is shown in Fig. 13 for $\delta=\gamma_1/2$. The labels + and - in (24) are used to distinguish the solutions of the two 4x4 equations. For typical values of the charge transfer considered above, the lower valence band $2V^-$ is completely filled for any value of δ and the hole Fermi energy shown schematically in Fig. 13 is smaller than in the third stage.

Table I

Intercalant	Stage	$\hbar\omega$(eV)	Reflectivity experiments in GAC
AsF$_5$	1,2,3,4	0.7-2	Hanlon et al. [38]
	1,2	0.5-2.2	Saint Jean et al. [39]
SbF$_5$	1,2,3,4		Thompson et al. [40]
SbCl$_5$	2	0.1-3	Blinowski et al. [16,42]
			Nguyen et al. [41]
	1,2,3,4,5	0.1-3	Eklund et al. [44]
	1,2,3,4	0.08-10	Hoffman et al. [45,46]
HNO$_3$	1,2,3	0.3-1.7	Fischer et al. [47]
	2-9	0.5-1	Shieh et al. [48]
Br$_2$	2,3,4	0.1-2.2	Blinowski et al. [16,42,43]
			Nguyen et al. [41]
ICl	2	0.1-3	Blinowski et al. [16]
	1	0.08-10	Hoffman et al. [46]
AlCl$_3$	1,2,3	0.1-3	Blinowski et al. [42,43]
			Nguyen et al. [41]
H$_2$SO$_4$	1,2,3	0.2-2.2	Saint Jean et al. [49,50]
			Zhang et al. [51]

For finite δ, the valence bands 1V$^+$ and 1V$^-$ exhibit a camelback shape for small k. For large k, the distance between both bands decreases but their dispersion relations can be reasonably approximated by δ-|x|. The analysis of the wavefunctions shows that carriers in all bands are partially localized on both external and internal layers.

For δ<γ_1/2, the bands 1V$^-$ and 1V$^+$ are degenerate at k=0, whereas for δ>γ_1/2, a finite gap between both bands appears and 1V$^+$ becomes degenerate at k=0 with the band 2V$^+$.

2. OPTICAL PROPERTIES OF GAC

2.1 Optical reflectance experiments

Optical properties of GAC were investigated mainly by reflectance experiments carried out in the IR and visible region. Most measurements were made at near normal incidence with unpolarized radiation propagating along the c-axis. Reflectivity spectra were reported for different stages of Acceptor Compounds with various intercalants. Experimental studies are listed in Table I. For each intercalated graphite system, the compounds were prepared from HOPG and the stage of the samples was carefully controlled by (001) x-ray diffractograms. For the most reactive compounds, the reflectance measurements were carried out on encapsulated samples which were either in the presence of intercalant vapor

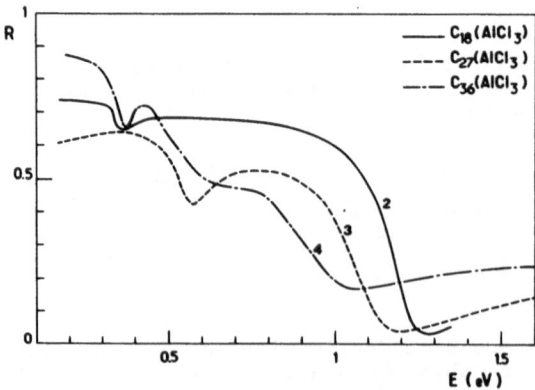

Fig. 14. Reflectance spectra of stages
2, 3, 4 of AlCl$_3$-graphite
(after [41]).

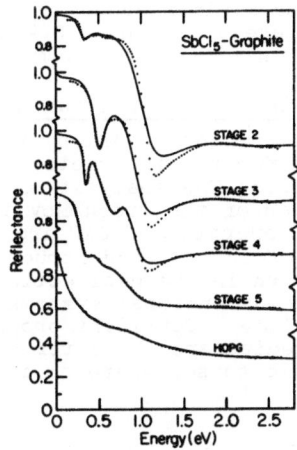

Fig. 15. Reflectance spectra of stages
2-5 of SbCl$_5$-graphite (after
[44]).

or in an inert atmosphere. In the case of H$_2$SO$_4$-graphite intercalation
compounds, in situ reflectance measurements were performed during the
electrochemical synthesis [49,50,51]. This technique provides interesting possibilities of studying the evolution of the optical reflectance
with increasing charge density (electrochemically controlled).

GIC exhibit a metallic reflectance: sharp plasma edges with well pronounced reflectance minima are observed in the near IR and visible region.
This behavior drastically contrasts with the reflectivity spectrum of
pristine graphite which presents a monotonic decrease between 0.3 and
1.8 eV (Fig. 16). Typical reflectivity spectra of acceptor GIC are illustrated in Figs. 14, 15, 18, 19. The position of the plasma edges depends on the intercalated species: typically, for stage 2, the energy of
reflectance minimum varies from 1 eV (for Br$_2$-graphite) up to 1.43 eV
(for AsF$_5$-graphite). For any intercalated graphite system, the plasma
edges are shifted towards lower energies with increasing stage number.
These features are illustrated in Figs. 14, 15, for AlCl$_3$ and SbCl$_5$-
graphite respectively. We use the notation I_c to denote the distance
between two graphite bounding layers containing the intercalant, often
denoted by d_s.

In the low energy region, the reflectivity spectra of GAC exhibit stage-
characteristic structures which are identified with intervalence band transitions. The position of these structures depends slightly upon the inter-

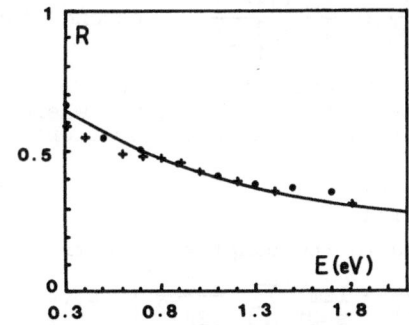

Fig. 16. Reflectivity spectrum in pure graphite.
Solid line: calculated from (35), experimental data: ● [2], + [52].

calated species: a reflectance minimum occurs at $\hbar\omega=0.37$ eV for stage 2, between 0.5 and 0.6 eV for stage 3, and near 0.4 and 0.7 eV in stage 4 GAC. Such structures do not exist in first stage compounds.

2.2 Frequency dependence of the dielectric function

Optical properties of solids are determined by the complex dielectric function $\varepsilon(\omega)$ which accounts for the interaction of electrons with the electromagnetic radiation field. The analysis of reflectance data in GIC requires the determination of the frequency dependence of the dielectric function. Optical properties of GAC are quantitatively interpreted within the framework of the 2D band structure model [16]. The compound of stage n of unit length in the c-direction consists of $[I_c+(n-1)d]^{-1}$ independent and identical two-dimensional systems of n charged graphite layers parallel to the x-y plane. For light propagating along the c-axis, optical properties of this medium are characterized by the component $\varepsilon_\perp=\varepsilon_{xx}=\varepsilon_{yy}$ of the dielectric tensor where $\varepsilon_\perp$ includes both intraband and interband contributions.

2.2.1 Intraband contribution

The intraband contribution due to free holes occupying the states of the valence bands is

$$\varepsilon_{intra} = - \sum_j \omega_j^2/\omega^2 \tag{25}$$

where ω_j is the plasma frequency for the jth band. The general expression for the plasma frequency is:

$$\omega_j^2 = 4\pi \, N_j \, e^2 \, \hbar^{-2} \left\langle \partial^2 E_j(\vec{K})/\partial k_x^2 \right\rangle \tag{26}$$

$$\left\langle \partial^2 E_j(\vec{K})/\partial k_x^2 \right\rangle = \frac{1}{\eta_j} \frac{4}{4\pi^2} \int_{BZ} \frac{\partial^2 E}{\partial k_x^2} f[E_j(\vec{K})] \, d^2k \tag{27}$$

N_j is the hole density per unit volume which is related to the areal hole density η_j by

$$N_j = \eta_j/[I_c+(n-1)d] \tag{28}$$

In (27), $f[E_j(\vec{K})]$ is the Fermi function and the factor 4 accounts for the spin degeneracy and the existence of two valleys near the U and U' points. For degenerate carriers $(kT \ll E_F)$

$$\left\langle \partial^2 E_j(\vec{K})/\partial k_x^2 \right\rangle = \frac{1}{\eta_j} \frac{4}{4\pi^2} \int_{k<k_F} [\partial^2 E_j(\vec{K})/\partial k_x^2] \, d^2k \tag{29}$$

Using the dispersion relations (10) and (11) for the valence bands in first and second stages respectively, the intraband contribution (25) is given by the following expressions

$$\varepsilon_{intra} = - 4e^2 E_F/I_c \hbar^2 \omega^2 \qquad \text{first stage} \tag{30}$$

$$\varepsilon_{intra} = - \frac{8e^2 E_F}{(I_c+d)\hbar^2\omega^2} \frac{E_F^2-(\gamma_1^2/2)}{E_F^2-(\gamma_1^2/4)} \qquad \text{second stage } (E_F>\gamma_1) \tag{31}$$

The expressions (30) and (31) exhibit an unusual concentration dependence : contrary to the case of parabolic bands where $\omega_j^2 \sim N_j$, here the intraband term is proportional to the hole Fermi energy E_F and therefore to $N_j^{1/2}$.

2.2.2 Interband contribution

The interband contribution to the complex dielectric function is

$$\varepsilon_{inter}-1 = - \frac{4\pi e^2}{m^2\omega[I_c+(n-1)d]} \lim_{s\to 0^+} \sum_{j,j'} \frac{2}{(2\pi)^2} \int_{BZ} f(E_j(\vec{K}))[1-f(E_{j'}(\vec{K}))]$$

$$\times [|P_{j\to j'}|^2/\omega_{jj'k}] [(\hbar\omega_{jj'k}-\hbar\omega-is)^{-1}-(\hbar\omega_{jj'k}+\hbar\omega+is)^{-1}] \tag{32}$$

where $\hbar\omega_{jj'k}=E_j(\overline{k})-E_{j'}(\overline{k})$ and $|P_{j\to j'}|^2=\frac{1}{2}[|\langle j'k|P_x|jk\rangle|^2+|\langle j'k|P_y|jk\rangle|^2]$

We consider successively the case of pure graphite, first and second stage compounds.

2.2.2.1 Pure graphite

It has been shown that the 2D model provides a correct description of the near IR and visible properties of graphite. If we treat graphite as a system of independent layers, ε_{inter} is given by (32) where n=1 and I_c=d. The matrix element $P_{v\to c}$ is obtained by identifying the $\overline{k}.\overline{\pi}$ expression derived by the SW model [17] with the result of the tight binding calculations in the vicinity of the U point:

$$|P_{v\to c}|^2 = \frac{9}{8}\gamma_0^2 \, b^2 \,\, (m^2/\hbar^2) \tag{33}$$

In the range of validity of the linear dispersion relation (10) for initial and final states (which corresponds to the photon energy region $\hbar\omega \leqslant \gamma_0$), the expression (32) leads to the imaginary part of ε_{inter}

$$\text{Im } \varepsilon_{inter} = \pi e^2/d\hbar\omega \tag{34}$$

The frequency dependence of the real and imaginary parts established experimentally by TAFT and PHILIPP [53] shows that Re ε_{inter} is nearly constant (ε_0=4.4) for photon energies between 0.8 and 3.5 eV. If we approximate the dielectric function by

$$\varepsilon_\perp^o(\omega) = \varepsilon_o+(i\pi e^2/d\hbar\omega) \tag{35}$$

the calculated reflectivity spectrum reported in Fig.16 is in excellent agreement with the experimental data.

2.2.2.2 First stage

Important contributions to the dielectric function originate from strong valence to conduction transitions. These transitions start at $\hbar\omega=2E_F$. For $kT \ll E_F$, the interband contribution can be expressed as

$$\varepsilon_{inter} - 1 = \frac{d}{I_c} [\varepsilon_\perp^o(\omega)-1] - \Delta\varepsilon$$

$\Delta\varepsilon$ is the contribution from the valence to conduction transitions which are excluded due to the depopulation of the valence level:

$$\Delta\varepsilon = \lim_{s\to 0^+} \frac{e^2}{I_c\hbar\omega} \int_0^\infty \left(\frac{1}{y-\hbar\omega-is} - \frac{1}{y+\hbar\omega+is}\right) \frac{dy}{1+\exp[(y-2E_F)/2k_BT]}$$

The real and imaginary parts of the total dielectric function, including intraband and interband contributions are

$$\text{Re } \varepsilon_\perp(\omega) = 1 + \frac{d}{I_c}(\varepsilon_0-1) + \frac{e^2}{2I_c\hbar\omega} F_1(2E_F) - \frac{4e^2E_F}{I_c\hbar^2(\omega^2+1/\tau^2)} \tag{36a}$$

$$\text{Im } \varepsilon_\perp(\omega) = \frac{\pi e^2}{I_c\hbar\omega} F_2(2E_F) + \frac{4e^2E_F}{I_c\hbar^2\omega\tau(\omega^2+1/\tau^2)} \tag{36b}$$

where

$$F_1(2E_F) = \ln \{[(\hbar\omega+2E_F)^2+4k_B^2T^2]/[(\hbar\omega-2E_F)^2+4k_B^2T^2]\}$$

$$F_2(2E_F) = 1 + \frac{1}{\pi} \text{arc ctg}(\frac{\hbar\omega+2E_F}{2k_BT}) - \frac{1}{\pi} \text{arc ctg}(\frac{\hbar\omega-2E_F}{2k_BT}).$$

For T→0 the function $F_2(x)$ tends to the standard step function $\theta(\hbar\omega-x)$. The last term in (36) represents the intraband contribution. $\hbar/\tau$ is a damping constant introduced phenomenologically to account for carrier scattering.

2.2.2.3 Second stage

Several types of interband transitions are allowed: i) the intervalence transitions ($v_2 \to v_1$) between the parallel bands are expected at $\hbar\omega = \gamma_1$, ii) the valence to conduction transitions ($v_2 \to c_2$ and $v_1 \to c_1$) between symmetrical bands, the threshold of these transitions is at $\hbar\omega = 2E_F$. Weaker transitions $v_1 \to c_2$ and $v_2 \to c_1$ are expected, the thresholds of these transitions are at $\hbar\omega = 2E_F \pm \gamma_1$.

For $kT \ll E_F$ and $\gamma_1 \ll 2E_F$, the real and imaginary parts of $\varepsilon(\omega)$ for second stage, including both intraband and interband contributions, are given by the following expressions:

$$
\begin{aligned}
\text{Re } \varepsilon_\perp(\omega) = 1 \; &+ \; \frac{2d(\varepsilon_0 - 1)}{I_c + d} \; + \; \frac{e^2}{(I_c + d)\hbar\omega} \left(1 - \frac{\gamma_1^2}{\hbar^2\omega^2}\right) F_1(2E_F) \\
&+ \frac{e^2 \gamma_1^2}{2(I_c + d)\hbar^3\omega^3} \left[F_1(2E_F - \gamma_1) + (F_1(2E_F + \gamma_1) \right] \\
&+ \frac{e^2 \gamma_1^2}{(I_c + d)\hbar\omega E_F} \left(\frac{1}{\gamma_1 - \hbar\omega} - \frac{1}{\gamma_1 + \hbar\omega} \right) - \frac{8e^2 E_F}{(I_c + d)\hbar^2(\omega^2 + 1/\tau^2)} \left(1 - \frac{\gamma_1^2}{4E_F^2}\right) \quad (37a)
\end{aligned}
$$

$$
\begin{aligned}
\text{Im } \varepsilon_\perp(\omega) = \; &\frac{2\pi e^2}{(I_c + d)\hbar\omega}\left(1 - \frac{\gamma_1^2}{\hbar^2\omega^2}\right) F_2(2E_F) + \frac{\gamma_1^2 e^2 \pi}{(I_c + d)\hbar^3\omega^3}\left[F_2(2E_F - \gamma_1) + F_2(2E_F + \gamma_1) \right] \\
&+ \frac{e^2 \gamma_1^2 \pi}{(I_c + d)\hbar\omega E_F}\, \delta(\hbar\omega - \gamma_1) + \frac{8e^2 E_F}{(I_c + d)\omega\tau\hbar^2(\omega^2 + 1/\tau^2)}\left(1 - \frac{\gamma_1^2}{4E_F^2}\right) \quad (37b)
\end{aligned}
$$

Figure 17 shows the frequency dependence of Re $\varepsilon_{inter}(\omega)$ and Im $\varepsilon_{inter}(\omega)$ for stage 2. Each threshold of interband transitions at $\hbar\omega = 2E_F, 2E_F \pm \gamma_1$ is characterized by a peak of Re $\varepsilon_{inter}(\omega)$ and a step of Im $\varepsilon_{inter}(\omega)$, with the most pronounced features for $v_2 \to c_2$ and $v_1 \to c_1$ transitions. Both real and imaginary parts of $\varepsilon_{inter}(\omega)$ are singular at $\hbar\omega = \gamma_1$, corresponding to the intervalence transition.

2.3 Comparison between theory and experiments

Optical reflectance data obtained for various GAC in the spectral region 0.1-3 eV were quantitatively interpreted within the framework of the 2D band structure model.

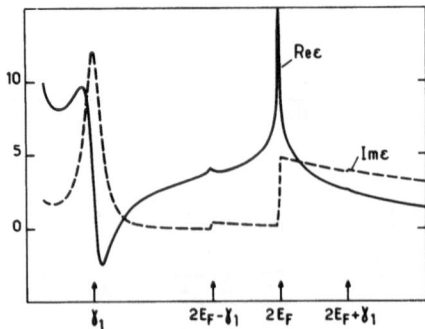

Fig. 17. Frequency dependence of the real and imaginary parts of ε_{inter} for stage 2 (after [42]).

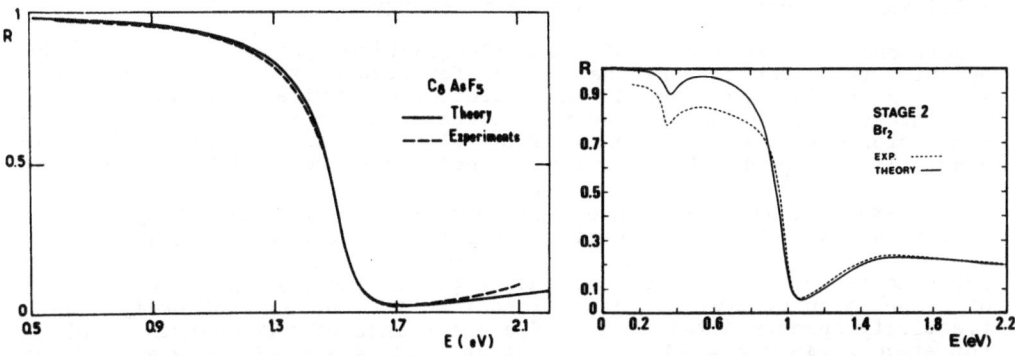

Fig. 18. Comparison between calculated (solid line) and experimental (dotted line) reflectance spectra for C_8AsF_5 (after [39]). The theoretical fit corresponds to $E_F=1.28eV$, $\tau=5\times10^{-14}s$, $T_{eff}=350K$.

Fig. 19. Comparison between calculated (solid line) and experimental (dotted line) reflectance spectra for $C_{16}Br_2$ (after [42]). The theoretical fit corresponds to $E_F=0.70eV$, $\tau=3\times10^{-14}s$, $T_{eff}=450K$.

2.3.1 Determination of the Fermi energy from the plasma reflectance analysis

Using the expressions (36) and (37) for $\varepsilon(\omega)$, the reflectance $R=|[(\varepsilon(\omega))^{1/2}-1]/[(\varepsilon(\omega))^{1/2}+1]|^2$ was calculated and compared with the experimental data for first and second stage GAC. The dielectric function (independent of γ_0) depends mainly on E_F which enters in both intra and interband contributions and which is treated as an adjustable parameter. Below the threshold of interband transitions, $\varepsilon_\perp(\omega)$ is dominated by ε_{intra} whereas at photon energies $\hbar\omega \geqslant 2E_F$ the important interband contributions to Im $\varepsilon_\perp(\omega)$ drastically influence the frequency dependence of $R(\omega)$ which presents a slow decrease in the high field region similarly to the case of pure graphite. The position of the reflectance minimum depends drastically on E_F which is accurately determined (ΔE_F < 10 meV) from the best fits of $R(\omega)$ to the experimental spectra. The comparison between theory and experiments for GAC of stages 1 and 2 is illustrated in Figs.18,19. The hole Fermi energy E_F and the damping constant τ obtained from the quantitative analysis are reported in Table II. In calculating $R(\omega)$, an effective temperature T_{eff} larger than the experimental temperature was introduced in expressions (36) and (37) to account phenomenologically for the effect of scattering on interband transitions.

Table II

GAC	E_F(eV)	$\tau(10^{-14}s)$	f	f/l	Ref.
C_8AsF_5	1.28	5	0.35	0.044	(a)
$C_{12-14}SbCl_5$	1.28	3	0.43	0.033	(b)
$C_{16}Br_2$	0.70	3	0.18	0.023	(c)
$C_{16}ICl$	0.75	1	0.21	0.026	(c)
$C_{24}SbCl_5$	0.75	1	0.32	0.026	(c)
$C_{24-28}SbCl_5$	0.79	..	0.33	0.026	(b)
$C_{18}AlCl_3$	0.90	5	0.34	0.038	(c)
$C_{16}AsF_5$	1.02	5	0.39	0.048	(a)

(a) [39]   (b) [45]   (c) [16,42,43]

2.3.2 Relation between charge and Fermi energy

Electrochemical synthesis of the H_2SO_4-graphite system offers a unique opportunity to control the charge transfer between graphite layers and acceptor species and to study the evolution of optical properties during the intercalation process. As the charge accumulated in pure stage configuration may be varied within certain limits, this system provides the possibility to probe the relationship between charge and Fermi energy.

In situ optical reflectance measurements were carried out by SAINT JEAN et al.[49,50] in the spectral region 0.2-2.2 eV, on H_2SO_4-graphite system during the electrochemical synthesis. At all times during the charging cycle, the accumulated charge per C atom,$1/x$,in the compounds of composition $C_x{}^+SO_4H^-$ is determined from the current intensity. Reflectivity spectra obtained at different points of the charging cycle for stage 2 ($48 \leqslant x \leqslant 58$) and for stage 1 ($22 < x \leqslant 28$) show a shift of the plasma edges towards higher energies with increasing charge per C atom in pure stage compounds. Hole Fermi energies were determined from the theoretical fits of the calculated reflectance $R(\omega)$ to the experimental spectra. Thus independent determination of E_F (optically) and $1/x$ (electrochemically) enabled us to prove the validity of the 2D band structure model. Within this model the charge per C atom and the hole Fermi energy are related by

$$1/x = E_F^2/3^{1/2}\pi\gamma_o^2 \qquad (38)$$

Experimental values (Ref.[49,50]) of x and E_F determined at numerous points of the charging cycle show that xE_F^2 is independent of x: for stage 1, $xE_F^2=37.4\pm0.9$ (eV)^2; for stage 2, $xE_F^2=42.7\pm1.2$ (eV)^2. This important result confirms the validity of the model and provides a direct determination of the intralayer resonance integral γ_0 in 2D GIC. For stage 1, SAINT JEAN et al.[49,50] obtained $\gamma_0=2.62\pm0.03$ eV; this value is in excellent agreement with the theoretical band structure calculation of PAINTER and ELLIS for a single graphite layer [14]. For stage 2, a larger value was found, $\gamma_0=2.81\pm0.04$ eV, which may result from the fact that more distant neighbour resonance integrals (γ_3,γ_4) were neglected in the simplified model.

2.3.3 Charge transfer coefficients

The charge transfer coefficients for different GAC corresponding to the formula $(C_l)_n$ A (n=1,2) are determined from the hole Fermi energies using the relation (14). Numerical values of f, reported in Table II, are obtained from the above values of γ_0. In Ref.[16] and [43], charge transfer coefficients were established for the lower and upper limits of γ_0 i.e. $\gamma_0=2.4eV$ which corresponds to the 2D band structure of graphite [4], and $\gamma_0=3.12eV$ [37] established for 3D graphite in the complete SW model. The charge transfer coefficients deduced by HOFFMAN et al.[45] were calculated for $\gamma_0=3eV$.

2.3.4 Intervalence band transitions

The observation of stage characteristic structures in the IR reflectivity spectra of GAC provides evidence of the intervalence transitions (IVT) predicted by the 2D band structure model. Compounds of stage 2 exhibit a reflectance minimum systematically observed at $\hbar\omega=0.375\pm0.005$ eV whatever the intercalants: $AlCl_3$ [41], Br_2 [41], $SbCl_5$ [41,45], SO_4H_2 [50]. These features are illustrated in Fig.20 for several compounds. The position of the reflectivity structure corresponds to the energy of the IVT allowed for unpolarized radiation propagating along the c-axis. These transitions take place at $\hbar\omega=\gamma_1$ between the occupied states of v_2 and the empty states of v_1 (Fig.10). As the dielectric function is singular at $\hbar\omega=\gamma_1$, a Lorentzian function with an appropriate line width was introduced in ε_{inter} to account phenomenologically for the effect of scattering on IVT. Fig.20 shows that the calculated reflectance reproduces quite well the position and the lineshape of the peak observed at 0.375eV in all investigated GAC.

In third stage GAC an intense reflectivity structure is observed in the spectral region 0.50-0.57 eV with the position of the reflectance minimum depending on the intercalated species (Figs.14,15,21). The peak position appears at 0.50eV for Br_2 [41], 0.56eV for $SbCl_5$ [44], 0.52-0.55 eV for SO_4H_2 [49,50] and at 0.57eV for $AlCl_3$ [41].

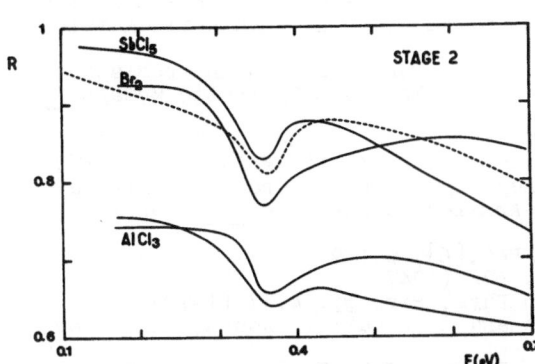
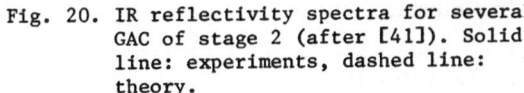

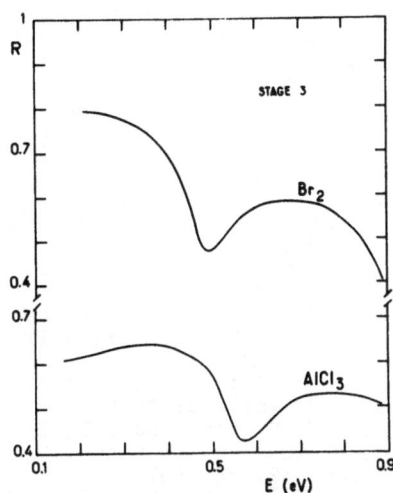

Fig. 20. IR reflectivity spectra for several GAC of stage 2 (after [41]). Solid line: experiments, dashed line: theory.

Fig. 21. IR reflectivity spectra for stage 3 of AlCl$_3$ and Br$_2$-graphite (after [41]).

The probabilities for interband transitions induced by unpolarized radiation propagating along the c-axis were calculated from the wavefunction (23). The IVT 3v→2v are allowed for k=0 and their probabilities decrease rapidly to zero with increasing k (Fig.11). These transitions are predicted at $\hbar\omega=\sqrt{2}\gamma_1=0.53$eV in the case $\delta=0$. Several interband transitions between the valence and conduction bands are also allowed but the onsets of these transitions are located at higher energies. For $\delta\neq0$, the 3v and 2v valence bands are no longer parallel and the 3v→2v transitions should give rise to an intense absorption in the energy region $(2\gamma_1^2+\delta^2)^{\frac{1}{2}}-\delta < \hbar\omega < (2\gamma_1^2+4\delta^2)^{\frac{1}{2}}$. The absorption increases rapidly in the vicinity of the upper limit, i.e. for transitions between almost parallel bands and an asymmetrical peak is expected slightly below $(2\gamma_1^2+4\delta^2)^{\frac{1}{2}}$. The δ-values were estimated from the self-consistent procedure for the charge transfer coefficients listed in Table II. The expected energies of the 3v→2v transitions are very close to the positions of the strong reflectance structures. The dependence of the IVT upon δ implies a positive energy shift of the peak with increasing charge transfer, in qualitative agreement with the experimental features.

Compounds of stage 4 exhibit two reflectance structures which have been observed for SbCl$_5$ [44], Br$_2$ [41] and AlCl$_3$ [41] graphite. The lowest energy peak occurs at 0.37eV and the second structure appears near 0.74eV. This latter structure was particularly well resolved in the spectrum of SbCl$_5$-graphite (Fig.15).

Referring to the band scheme reported in Fig.13 the allowed IVT for stage 4 are 2v⁻→1v⁺ and 2v⁺→1v⁻ at the energies $\hbar\omega=(\sqrt{5}\pm1)\gamma_1/2$ for $\delta=0$. For $\delta\neq0$, all interband transitions are allowed for any k. One can still however expect two strong transitions between nearly parallel valence bands with the final states above the Fermi energy. The value $\delta=\gamma_1/2$ accounts for the experimental reflectivity peak positions observed in several stage 4 GAC at 0.37 and 0.74 eV.

References

1. A. Herold: In Physics and Chemistry of Materials with Layered Structures, ed. by F.A. Levy, D. Reidel (Dordrecht 1979), Vol.6 (Intercalated Layered Materials), p.323
2. J.E. Fischer: In Physics and Chemistry of Materials with Layered Structures, ed. by F.A. Levy, D. Reidel (Dordrecht 1979, Vol.6 (Intercalated Layered Materials), p.383
3. S.A. Solin: In Advances in Chemical Physics
4. P. Pfluger and H.H. Gunterodt: In Advances in Solid State Physics, Vol.21 (Vieweg, Braunsweig 1981)

5. M.S. Dresselhaus and G. Dresselhaus: In Adv. Phys. 30, 139 (1981).
6. Proceedings Franco-American Conference on Intercalation Compounds of Graphite (La Napoule, France), Mat. Sci. Eng. 31 (1977)
7. Proceedings of the International Conference on Layered Materials and Intercalated (Nijmegen, The Netherlands), Physica 99B (1980)
8. Proceedings of the Second International Conference on Intercalation Compounds of Graphite (Provincetown, Mass., USA), Synthetic Metals 2/3 (1980/1981)
9. Proceedings of the Third International Conference on Intercalation Compounds of Graphite (Pont-à-Mousson, France), Synthetic Metals 8 (1983)
10. Proceedings of the Fourth International Conference on Intercalation Compounds of Graphite (Tsukuba, Japan), Synthetic Metals 12 (1986)
11. Proceedings of the French-Japanese Symposium on Graphite Intercalation Compounds (Paris, France), Annales de Physique N°2 (Suppl.2) 11 (1986)
12. M. Zanini and J.E. Fischer: In Ref.[6], p.169
13. P.R. Wallace: In Phys. Rev. 71, 622 (1947)
14. G.S. Painter and D.E. Ellis: In Phys. Rev. B1, 4747 (1970)
15. M. Tsukada, K. Nakao, Y. Uemura and S. Nagai: In Journ. Phys. Soc. Jap. 32, 54 (1972)
16. J. Blinowski, Nguyen Hy Hau, C. Rigaux, J.P. Vieren, R. Le Toullec, G. Furdin, A. Herold, J. Melin: In Journal Physique (Paris) 41, 47 (1981)
17. J.C. Slonczewski and P.R. Weiss: In Phys. Rev. 109, 272 (1958)
18. J.W. McClure: In Phys. Rev. 108, 612 (1957)
19. A. Zunger: In Phys. Rev. B17, 626 (1978)
20. R.C. Tatar and S. Rabii: In Phys. Rev. B25, 4126 (1982)
21. I.L. Spain: In Proceedings of the International Conference on Semimetals and Narrowgap Semiconductors (Pergamon Press, New York 1973), p.177, and in Chemistry and Physics of Carbon, ed. by P.L. Walker (New York:Marcel Dekker), Vol.8, p.105
22. N.A.W. Holtzwarth, L.A. Girifalco and S. Rabii: In Phys. Rev. B18, 5190 (1978) and in Phys. Rev. B18, 5206 (1978)
23. T. Ohno, K. Nakao, H. Kamimura: In J. Phys. Soc. Jap. 47, 1125 (1979), and H. Kamimura: In Ref.[11], p.39
24. T. Inoshita, K. Nakao and H. Kamimura: In J. Phys. Soc. Jap. 43, 2137 (1977)
25. D.P.DiVicenzo and S. Rabii: In Ref.[7], p.406, and in Phys. Rev. B25, 4110 (1982)
26. T.C. Tatar: In Ph. D. Thesis (University of Pennsylvania, USA,1985)
27. G. Dresselhauss and S.Y. Leung: In Phys. Rev. B24, 3490 (1981)
28. W.D. Ellenson, D. Semmingson and J.E. Fischer: In Mat. Sci. Eng. 31 (1977)
29. G.S. Parry and D.E. Nixon: In J. Phys.C 2, 2156 (1969)
30. N.A.W. Holtzwarth: In Phys. Rev. B21, 3665 (1980)
31. L. Pietronero, S. Strassler, H.K. Zeller and M.J. Rice: In Phys. Rev. Lett. 41, 763 (1978), and in Sol. St. Comm. 30, 399 (1979)
32. J. Blinowski and C. Rigaux: In Journ. Physique (Paris) 41, 667 (1980), and in Ref.[8], p.297
33. S.A. Safran and D.R. Hamann: In Phys. Rev. B22, 3490 (1981), and in Phys. Rev. B23, 565 (1981), and in Ref.[8], p.1
34. S. Shimamura and A. Morita: In J. Phys. Soc. Jap. 51, 502 (1982)
35. T. Ohno and H. Kamimura: In J. Phys. Soc. Jap. 52, 223 (1983), and in Ref.[11], p.179
36. D.L. Greenaway, G. Harbeke, F. Bassani and E. Tosatti: In Phys. Rev. 178, 1340 (1969)
37. M.S. Dresselhaus, G. Dresselhaus and J.E. Fischer: In Phys. Rev. B15, 3180 (1977)
38. L.R. Hanlon, E.R. Falardeau, D. Guerard, J.E. Fischer: In Mat. Sci. Eng. 31, 161 (1977), and in Sol. St. Comm. 24, 377 (1977)
39. M. Saint Jean, Nguyen Hy Hau, C. Rigaux, G. Furdin: In Sol. St. Comm. 46, 55 (1983)
40. T.E. Thompson, E.R. Falardeau and L.R. Hanlon: In Carbon 15, 39 (1977)
41. Nguyen Hy Hau, J. Blinowski, C. Rigaux, R. Le Toullec, G. Furdin, A. Herold, R. Vangelisti: In Synth. Metals 3, 99 (1981)
42. J. Blinowski, Nguyen Hy Hau, C. Rigaux, J.P. Vieren: In J. Phys. Soc. Jap. 49, Suppl.A, 915 (1980)(15th Intern. Conf. Phys. Semicond.)

43. C. Rigaux and J. Blinowski: In Physics of Narrow Gap Semiconductors (4th Intern. Conf.), p.352, Lecture Notes in Physics N°152 (Springer, Berlin 1982)
44. P.C. Eklund, D.S. Smith and V.R.K. Murthy: In Synth. Metals 3, 111 (1981)
45. D.M.Hoffman, R.E. Heinz, G.L. Doll and P.C. Eklund: In Phys. Rev. B32, 1278 (1985)
46. D.M. Hoffman, P.C. Eklund, R.E. Heinz, P. Hawrylak and K.R. Subbaswamy: In Phys. Rev. B31, 3973 (1985)
47. J.E. Fischer, T.E. Thompson, G.M.T. Foley, D. Guerard, M. Hoke and F.L. Lederman: In Phys. Rev. Lett. 37, 769 (1976)
48. C.C. Shieh, R.L. Schmidt and J.E. Fischer: In Phys. Rev. B20, 3351 (1979)
49. M. Saint Jean, M. Menant, Nguyen Hy Hau, C. Rigaux, A. Metrot: In Synth. Metals 8, 189 (1983)
50. M. Saint Jean: In Thesis (Paris 1983)
51. J.M. Zhang, D.M. Hoffman and P.C. Eklund: In Phys. Rev. B (in press)
52. E.A. Taft and H.R. Philipp: In Phys. Rev. 138, A197 (1965)
53. F. Bassani, G. Pastori-Parravicini: In Nuovo Cimento 50B, 95 (1967)

OPTICAL PROPERTIES OF DONOR-TYPE GRAPHITE INTERCALATION COMPOUNDS

P.C. Eklund, M.H. Yang, and G.L. Doll

University of Kentucky
Lexington, KY 40506

I. Introduction

Pristine graphite is a highly anisotropic semimetal exhibiting weakly overlapping conduction and valence bands with approximate mirror symmetry. These π bands exhibit small basal plane electronic masses. In the lowest order of approximation (rigid band model [1]) the electrical properties of donor- and acceptor-type graphite intercalation compounds (GICs) [2,3] are determined by shifting the Fermi level up (donors) or down (acceptors) in the rigid π band, consistent with the charge transferred between the carbon and intercalate layers. The rigid band model, however, does not address the effects which might arise from the new c-axis periodicity (stage index), in-plane zone folding [4] of the π band associated with the intercalate layer superlattice (if one exists), or the intercalate-derived states themselves. Considerable progress has been made in our ability to account for these effects, and this has come from the interplay of energy band theory and many diverse experimental probes.

In this paper we will show how optical reflectivity results have contributed to our knowledge of the electronic structure. We focus here on the donor GICs. The optical properties of acceptor GICs are addressed in a separate paper in this volume [5]. Two-dimensional (2D) band models have been proposed by Blinowski et al. [6] and Ohno et al. [7] for GIC systems in which significant hybridization between carbon and intercalate states does not occur. 2D models have been used recently by Doll et al. and Zhang et al. to interpret the optical properties of the low stage index ternary donor GICs graphite-KH_x [8] and stage 1 $KC_{24}(NH_3)_x$ [9], respectively. Yang and Eklund [10] have recently observed low energy interband absorption structure in stage 3,4 graphite-K and interpreted their results in terms of a 2D band model due to Saito [11]. Three dimensional (3D) energy band calculations have been carried out for two stage 1 donor compounds: KC_8 by DiVincenzo and Rabii [12] and LiC_6 by Holzwarth et al. [13]. Dielectric functions have been calculated by Chen and Rabii for the LiC_6 band structure [14] and compared to optical data by Fischer et al. [15]. Recent optical results on the new ternary compounds graphite-$CsBi_x$ have been reported by Yang et al. [16]. Other donor compound spectra published prior to 1985 are discussed briefly in ref. 15.

II. Kramers-Krönig (K-K) Analysis of the Reflectance, and the Separation of the Free-Carrier and Interband Contributions to the Dielectric Function

The complex dielectric function $\varepsilon(\omega)$ of a metal can be written as the sum of three terms

$$\varepsilon(\omega) = \varepsilon_{free}(\omega) + \varepsilon_{inter}(\omega) + \varepsilon_{phonon}(\omega) \qquad (1)$$

$\varepsilon_{inter}(\omega)$ is due to electronic transitions between bands, and $\varepsilon_{phonon}(\omega)$ is the contribution from IR-active phonons. IR phonons in GICs are considered in another paper in this volume [17]. $\varepsilon_{free}(\omega)$ is the contribution from free carriers which is written in the Drude approximation as

$$\varepsilon_{free}(\omega) = 1 - \frac{\omega_p^2}{\omega(\omega+i/\tau)} \qquad (2)$$

where $\omega_p = [4\pi Ne^2/m_{opt}]^{1/2}$ is the plasma frequency, m_{opt} is the optical effective mass, N is the carrier density and τ is the average free carrier lifetime.

The interband term $\varepsilon_{inter}(\omega)$ is given by [18a]

$$\varepsilon_{inter}(\omega) = \frac{4\pi^2 e^2\hbar^2}{m^2\omega^2}\left(\frac{1}{4\pi^2}\right) \int\limits_{S} dS \frac{|\vec{n}\cdot M_{s's}(\vec{k})|}{|\nabla_k\{E_{s'}(\vec{k})\}|} \qquad (3)$$

This term describes the contribution to $\varepsilon(\omega)$ from vertical transitions ($\Delta\vec{k}=0$) between electronic states in different bands (s,s') with the same wavevector ($\vec{k}$). The integration is over the k-space surface S ($\vec{n} \perp S$) defined by $E_{s'}(\vec{k})-E_s(\vec{k})=\hbar\omega$, which reflects the conservation of energy in the photon absorption process. The wavevector of the photon is negligible, and thus the wavevectors of the electron in the initial and final states are set equal. $M_{s's}(\vec{k})$ is the electric dipole matrix element. The joint density of states (JDOS) is defined by [18a]

$$JDOS(\omega) = \frac{1}{4\pi} \int\limits_{S} \frac{dS}{|\nabla_k\{E_{s'}(\vec{k})-E_s(\vec{k})\}|} \qquad (4)$$

Thus if the wavevector dependence of $M_{s's}$ is ignored, $\varepsilon_{inter} \propto JDOS/\omega^2$. From eq. (4) we see that the JDOS is large when the bands s and s' are approximately parallel over a large fraction of the area S.

Experimental values of the dielectric function $\varepsilon(\omega)$ are obtained from the Reflectance $R(\omega)$ via the following expressions [18b]:

$$r(\omega) = \frac{E_{ref}}{E_{inc}} = \rho(\omega)e^{i\theta(\omega)} \qquad (5a)$$

$$R(\omega) = |r(\omega)|^2 = \rho^2(\omega) \qquad (5b)$$

At near-normal incidence we have [18b]

$$r(\omega) = \frac{1 - \sqrt{\varepsilon}}{1 + \sqrt{\varepsilon}} \qquad (5c)$$

where $r(\omega)$ is known as the reflectivity coefficient, and E_{inc} and E_{ref}

are the incident and reflected electric fields, respectively. The phase shift is not measured, but is calculated from the Kramers-Kronig integral [19]

$$\theta(\omega) = - \frac{\omega}{\pi} P \int_0^\infty \frac{\ln\left[R(\omega')/R(\omega)\right]}{\omega'^2 - \omega^2} d\omega' \qquad (6)$$

in which P denotes the principal value of the integral. The calculation of θ (eq. 6) requires a knowledge of $R(\omega)$ over an infinite range of frequency. The data must therefore be properly extended to higher and lower frequency. In Fig. 1 we show the reflectance data of Fischer et al. [15] for KC_8, LiC_6, $KHgC_4$ and graphite in the range 0-40 eV. The spectra pertain to the basal plane ($\vec{E} \perp \vec{c}$) components of the dielectric tensor. They are obtained from c-face reflectance data below ~12 eV and electron energy loss (EELS) data ($\delta\vec{k} \perp \vec{c}$) in the range 10-40 eV, where $\delta\vec{k}$ is the momentum transfer to the electron beam. By analogy with pristine graphite [20], the peaks near 5 and 15 eV are assigned [15], respectively, to transitions between graphitic π and σ bands. The regions of π and σ interband absorption are well-separated, and lie respectively below and above the deep notch in R at ~8 eV. No significant structure is expected above 40 eV and thus the data are extended to higher energy in the standard way [18]: up to some frequency ω'', a functional form ω^{-s} is assumed, where s,ω'' are parameters adjusted to prevent negative phase shifts, and for $\omega>\omega''$, the relation ω^{-4} is assumed, characteristic of free electrons. Below the lowest experimental energy (0.5 eV, which in this case is the transmission limit of quartz) the data are extended using a Drude model. The results of the phase shift calculations, as outlined above, determine ε_1 and ε_2 and are shown in Fig. 2 for several donor compounds.

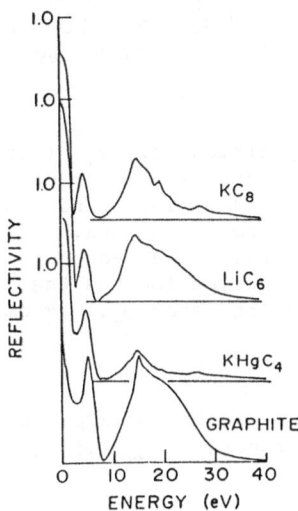

Fig. 1

Reflectance data from ref. 15 for KC_8, LiC_6, $KHgC_4$, and graphite. Spectra were obtained by c-face reflectivity (E<12 eV) and by EELS data (10< E< 40 eV).

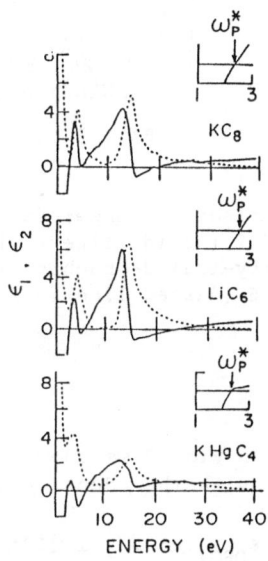

Fig. 2

K-K analysis of spectra in Fig. 1 showing ε_1 (solid line) and ε_2 (dashed line). Insets show the first zero crossing of $\varepsilon_1(\omega)$ at the screened plasma frequency ω_p^*. From ref. [15].

At low energy (ω), ε_1 and ε_2 are seen to turn sharply negative and positive, respectively, due to the strong contribution from the metallic free carrier term ε_{free}. The insets to Fig. 2 show the lowest energy zero-crossing of ε_1 at ω_p^*, which is referred to as the screened plasma frequency. The zero-crossing of $\varepsilon_{1,free}$ yields the "unscreened" plasma frequency ω_p. ω_p^* can be greater or less than ω_p depending on the nature of the interband contribution (ε_{inter}).

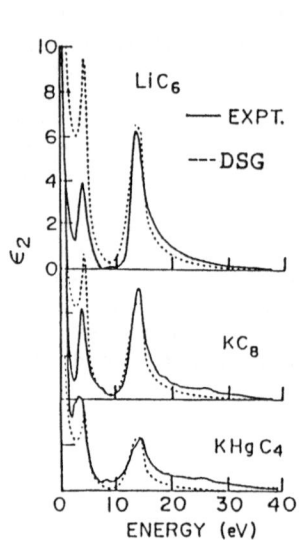

Fig. 3

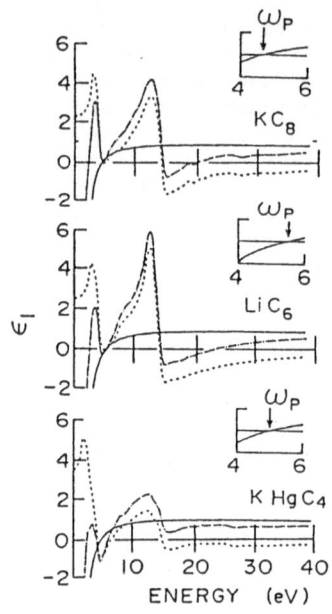

Fig. 4

Comparison of observed ε_2 from Fig. 2 (solid line) with the density-scaled-graphite (DSG) of eq. 6 (dashed line). From ref. [15].

Decomposition of ε_1 (dash-dot line) into free-carrier (solid line) and interband (dashed line) contributions. Insets show the zero-crossing of $\varepsilon_{1,free}$ which locates the "bare" plasma frequency ω_p. From ref. [15].

In Fig. 3 we show the comparison [15] of ε_2 found for KC_8, LiC_6 and $KHgC_4$ with the density-scaled graphite (DSG) function $\varepsilon_{DSG}(\omega)$ given by

$$\varepsilon_{DSG,}(\omega) = 1 + \frac{n \cdot 3.35}{I_c} \left(\varepsilon_G(\omega) - 1 \right) \tag{7}$$

where $\varepsilon_G(\omega)$ is the dielectric function obtained for graphite, n is the stage index ($n=1$ in these cases) and I_c is the distance between successive intercalate layers. The thickness of a single carbon layer is taken as 3.35 Å [2,3].

The agreement between ε_2 and $\varepsilon_{2,DSG}$ is seen to be good in the region $\omega > 8$ eV for LiC_6 and KC_8, where the spectrum is dominated by inter-σ-band transitions. The agreement above ~20 eV is observed to be not as good for $KHgC_4$. Below 8 eV the differences in ε_2 and $\varepsilon_{2,DSG}$ are due to: (1) the shift of E_F in the π-bands of the GIC, (2) the effects of hybridization on the carbon π-bands and (3) intercalate-derived states near E_F. These are precisely the interesting topics which low-energy optical studies would like to address.

The method used by Fischer et al. [15] to separate free and bound (or interband) contributions to $\varepsilon(\omega)$ involves a graphical procedure due to Ritsko [21] in which the leading edge of the ~4 eV peak in $\varepsilon_2(\omega)$ is extrapolated to zero to determine a threshold ω_T for interband absorption. Using the general K-K relation, they calculate

$$\varepsilon_{1,inter}(\omega) \cong \frac{2}{\pi} \int_{\omega_T}^{\infty} \frac{\omega'\varepsilon_2(\omega')}{\omega'^2 - \omega^2} d\omega' \qquad (8)$$

where for $\omega > \omega_T$ they have set $\varepsilon_{2,inter} \approx \varepsilon_2(\omega)$ in the integrand. Ignoring the small contribution from ε_{phonon}, eq. 1 is then used to determine $\varepsilon_{1,free}$ and $\varepsilon_{2,free}$. The results obtained for $\varepsilon_1(\omega)$ by this method of decomposition are shown in Fig. 4. A value for ω_p is then obtained from the zero-crossing of $\varepsilon_{1,free}(\omega)$ which is shown in the insets to the figure. ω_p is the "bare" or unscreened plasma frequency.

The overall agreement between ε_2 and $\varepsilon_{2,DSG}$ observed in the 8-25 eV region indicates that density-scaled graphite (DSG) should provide a reliable high energy extension for $R(\omega)$ in GICs. This is an important observation, because it suggests that quantitative information regarding the low energy dielectric function can be obtained from spectra which end at ~6 eV, which is near the practical upper limit for spectrometers which operate in air. We demonstrate that this extension is accurate by comparing the results obtained for KC_8 by Fischer et al. in the range 0-40 eV to those obtained by Yang and Eklund [10] in the 0.5-6 eV range using the DSG data extension. In Fig. 5 we show on a $\log(\omega)$ scale the KC_8 results of Yang and Eklund [10] in the region 0.1-25 eV: the data are the dots, the heavy line is the spectrum of density-scaled graphite (DSG) calculated according to eq. 8 using $I_C = 5.35$ Å and n=1 (KC_8). The thin line is the low energy Drude extension (eq. 2) plus a core constant ε_{core} [18], and the dashed line is the vacuum UV extension, which is DSG in the region 8-25 eV and $\sim\omega^{-4}$ thereafter.

Yang and Eklund achieve the decomposition of $\varepsilon_{free}(\omega)$ and $\varepsilon_{inter}(\omega)$ in the conventional way [20] by fitting ε_2 at low energy to the Drude model (eq. 2). The decomposition is then carried out according to eq. 1, ignoring the phonon term. The fit is shown in Fig. 6, where we plot $1/\omega\varepsilon_2$ vs. ω^2, which yields a straight line for the Drude contribution. The intercept and slope are sensitive to the values of ω_p and τ. The Drude parameters obtained from the analysis of Fig. 6 are $\omega_p = 4.6$ eV and $\omega_p\tau = 14.5$. The value of ω_p so obtained is in excellent agreement with that obtained by Fischer et al. [15] (4.65 eV). In Fig. 7 we compare the data of Fischer et al. and Yang et al. The data are seen to be in good agreement, and would superimpose if a scale factor of 1.05 were used. The solid and dashed lines representing the free-carrier reflectance are both calculated in the Drude approximation using $\omega_p = 4.65$ eV and a core constant $\varepsilon_\infty = 3.92$ [22] --the two curves differ in the choice of $\omega_p\tau$: $\omega_p\tau = 14.5$ (solid) and $\omega_p\tau = 141$ (dashed). The value 14.5 is consistent with the fit to ε_2 (Fig. 6) and, as can be seen in Fig. 7, fits $R(\omega)$ as well. The value $\omega_p\tau = 141$ is obtained by Fischer et al. [15] from their fit to $\varepsilon_{1,free}(\omega)$ [23]. The larger value of $\omega_p\tau$ however, is seen to result in a poor fit to $R(\omega)$.

261

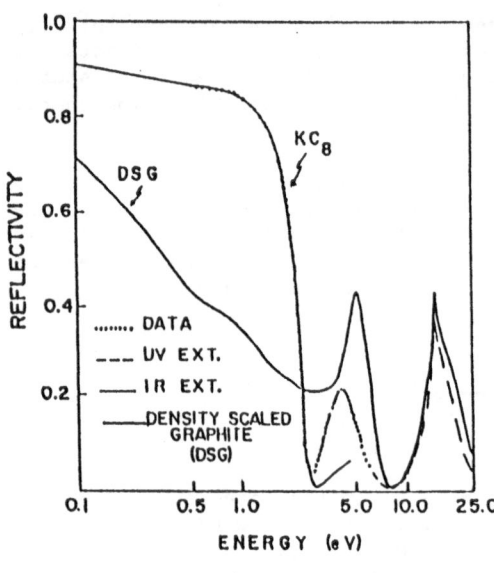

Fig. 5

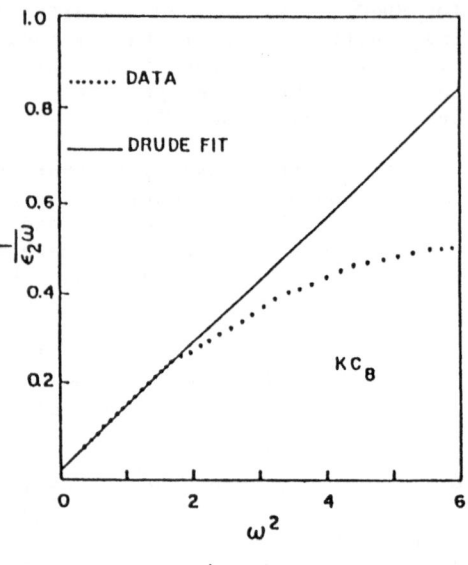

Fig. 6

Reflectivity of KC_8 vs. $\log(\omega)$ from ref. 10 in the region .1-25 eV. Data are the dots, the heavy line denotes the spectrum of DSG from eq. 7, the thin line represents the low energy Drude extension (eq. 2), and the dashed line is the vacuum UV extension which is DSG for $8 < \omega < 25$ eV and ω^{-4} for $\omega > 25$ eV.

Conventional decomposition of $\varepsilon_{free}(\omega)$ and $\varepsilon_{inter}(\omega)$ obtained from Fig. 5 for KC_8. Data (dots) are plotted as $1/\omega\varepsilon_2$ vs. ω^2, and compared to a Drude fit (solid) line). Plotted in this manner, the Drude contribution is linear with the slope and intercept dependent upon ω_p and τ. From ref. 10.

The Drude analysis shown in Fig. 6 is generally applicable to all GICs if the $R(\omega)$ data is taken to low enough energy. In Fig. 7 we show similar Drude analysis for KC_{24} and the hydrides $KH_{.8}C_4$ and $KH_{.8}C_8$ [8].

The optical conductivity ($\sigma_{opt} = \omega_p^2\tau/4\pi$) is often compared to the dc conductivity obtained from transport data. The value $\omega_p\tau=141$ brings σ_{opt} into reasonable agreement with $\sigma_{dc}=1.1\times10^5$ (Ohm-cm)$^{-1}$ [24], but the value $\omega_p\tau=14.5$, which fits the optical data, results in a value $\sigma_{opt}\sim\sigma_{dc}/9$. The fact that the experimental and theoretical values of ω_p are in good agreement suggests that $\tau_{opt}<<\tau_{dc}$. This condition is also found in other donor-GICs [8,9,10,16] and remains unexplained.

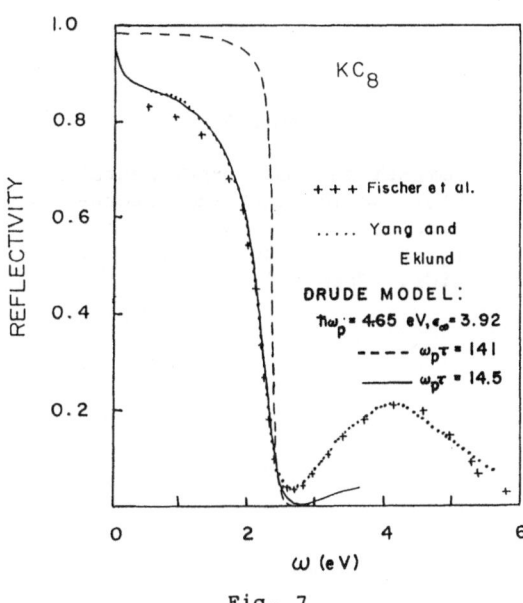

Fig. 7

Fig. 8

Reflectivity spectra for KC_8 measured by Yang and Eklund (···) and Fischer et al. (+++) are plotted together in the energy region 0-6 eV. For ω_p=4.65 eV and ε_∞=3.92, the Drude fit is plotted for different choices of $\omega_p\tau$: $\omega_p\tau$=141 [15] (dashed line) and $\omega_p\tau$=14.5 [10] (solid line).

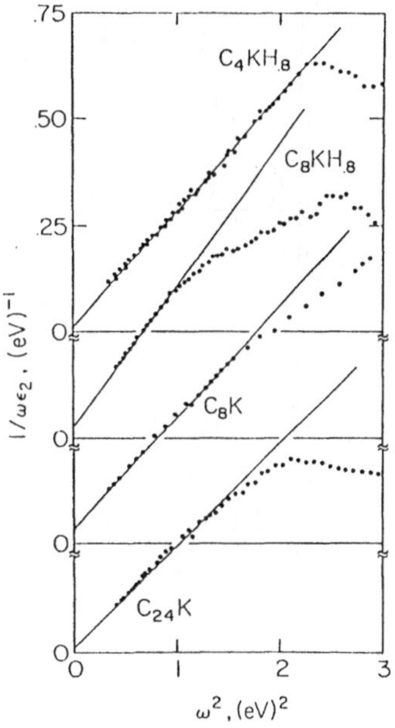

$(\varepsilon_2\omega)^{-1}$ vs. ω^2 for $KH_{.8}C_4$, $KH_{.8}C_8$, KC_8, and KC_{24}. From ref. 8.

III. Recent Results

LiC$_6$

In Fig. 9, we show a comparison of experimental (solid line) [15] and theoretical (dashed line) values for the ($\vec{E} \perp \vec{c}$) dielectric functions for LiC_6. The theoretical results are due to Chen and Rabii [14], and are based on an ab-initio self-consistent-field band structure. Overall, theory and experiment are found to be in quite good agreement. Good agreement is obtained for the interband contribution, except near the region of transparency (~10 eV). The agreement between the calculated and experimentally determined plasma frequencies (6.8 eV: theory and 5.5 eV: expt.) is not as good as found for KC_8. Theoretical and experimental results were compared for the component ($\vec{E} \parallel \vec{c}$). The experiments were carried out in the range 0.5-3 eV on sputter-etched a-faces, which are quite difficult to prepare. The experimental and theoretical positions for the midpoint of the Drude edge were observed to be 1.5 and 1.2 eV, respectively.

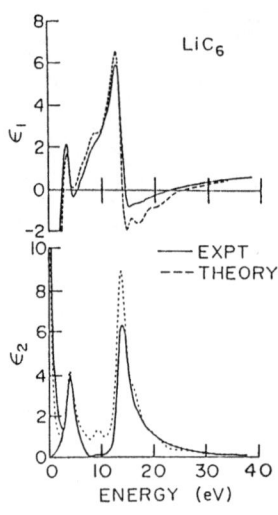

Fig. 9

Comparison of experimental [15] (solid line) and theoretical [14] (dashed line) ε_1 and ε_2 with $\vec{E} \perp \vec{c}$ for LiC_6. From ref. 15.

Stage 2-4 Graphite-K

The preceeding discussion has emphasized the low energy contribution from ε_{free}. Low energy interband transitions between π conduction bands (C→C) in stage n>2 donor compounds are also possible. Similar low energy interband transition have been studied in the acceptor compounds where the transitions are between π valence band (V→V) [5]. Recently, prominent low energy ($\omega \approx 1 eV$) structure in the reflectance of stage 3 and 4 graphite-K has been reported by Yang and Eklund [10]. They have identified this structure with inter-conduction-band (C→C) transitions between carbon p_z or π bands near the K-point. Their reflectivity data for the stage 1-4 compounds are shown in Fig. 10. Results of their K-K analyses [10] for stage 2-4 are shown in Fig. 11. Strong peaks in ε_2 are observed below 1 eV for stages 3,4 but not in stage 2. They interpret the optical data in terms of a phenomenological 2D band structure due to

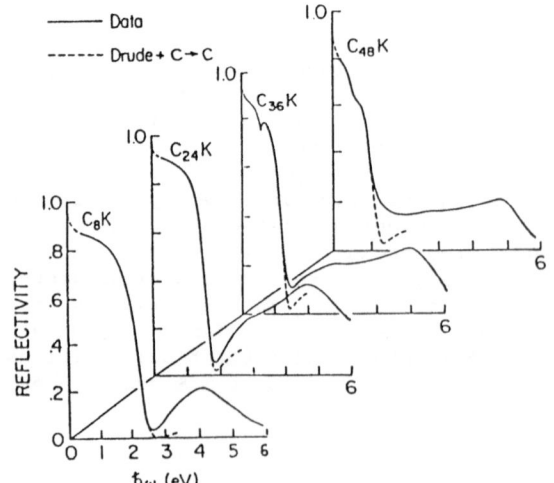

Fig. 10

Reflectance spectra for stages 1-4 graphite-K compounds. Data are represented by solid lines, while dashed lines are the results of Drude fits to the data of KC_8 and KC_{24}. For KC_{36} and KC_{48} the dashed lines are the calculated reflectance due to free-carrier and low energy interband absorption (C→C). The C→C contributions are modelled as Lorentz oscillators. From ref. 10.

Saito [11]. Saito has adjusted the band parameters in his tight-binding model to bring the energy bands into agreement with the self-consistent results of Ohno and Kamimura [25]. In Fig's 12a-c we show the energy bands calculated by Yang and Eklund [10] using Saito's model. The results are plotted in the vicinity of the K point and they indicate that electrons are located in trigonal cylinders at the six corners of the hexagonal Brillouin zone. The 2D Brillouin zone is shown as an inset to Fig. 12a. We note that there are n conduction- and n valence-bands for a stage n compound. The asymmetry between the (K-Γ) and the (K-M) dispersion is evident in the figure. E_F is chosen consistent with complete charge transfer from the K(4s) band. (C→C) transitions observed in the experiments are indicated by the vertical arrows. In stage 2 there is a considerable directional dependence in the conduction band splitting near E_F, which is not as pronounced in stage 3 and 4. This directional dependence broadens the associated peak in the joint density of states making a C→C feature difficult to detect, in agreement with experiment. Optical matrix elements have been calculated between the various conduction and valence bands by Blinowski and Rigaux [26]. In stages 3 and 4 there exists near degeneracies between some of the conduction bands, consistent with a number of features observed in the respective low energy region. Lorentz oscillator fits [10] to the low energy stage 3 and 4 structure in $\varepsilon_2(\omega)$ reveal that stage 3 exhibits a single peak at 0.59 eV and stage 4 exhibits an unresolved doublet (.60 eV and .75 eV). Saito's band parameters lead to the following values for the average band spittings which would give rise to optical structure at .96 eV (stage 3) and .77 eV, 1.08 eV (stage 4). The agreement between theory and experiment is reasonably good, and small adjustments in the band parameters can be made to bring theory and experiment into better agreement.

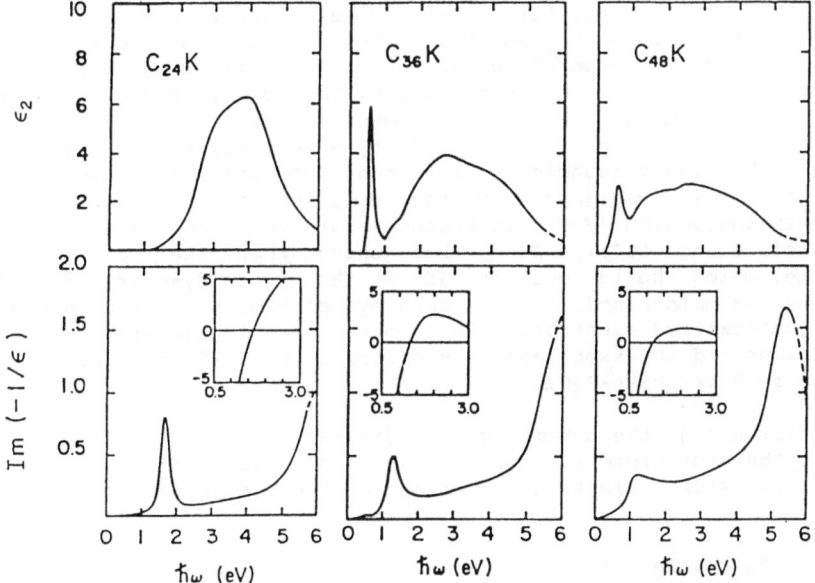

Fig. 11 a-c. $\varepsilon_{2,inter}(\omega)$ and Im(1/ε) obtained from the K-K analyses of the data in Fig. 9 for KC_{24}, KC_{36}, and KC_{48}. Insets show the lowest energy zero-crossing of ε_1 which locates the screened plasma frequency (ω_p^*). A high energy extension based on density-scaled graphite (DSG) was used.

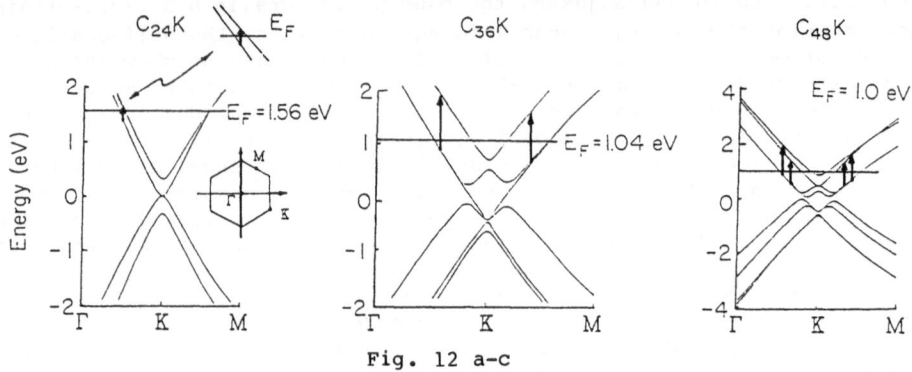

Fig. 12 a-c

The carbon p_z or π energy bands calculated by Yang and Eklund [10] using the tight-binding model of Saito [11]. Low energy interband transitions are indicated by arrows in the figures. The inset in Fig. 12a shows the 2D Brillouin zone of graphite, as well an an enlargement of the region near E_F along Γ-K. Note that only two resolvable transition energies are expected from the stage 3,4 band structures because of the near degeneracies between some of the bands.

Stage 1 and 2 Graphite-KH$_x$

Optical results for the stage 1 and 2 KH$_x$-GICs have been obtained recently by Doll et al. [8]. The samples were prepared by the direct action of KH powder with graphite. They have interpreted their data in terms of a model incorporating 2D π and 3D K(4s) bands, and find that the K(4s) band is empty, or nearly empty. The schematic density of states model they used for graphite-KH$_x$ is shown in Fig. 13, and the K-K results for $\varepsilon_2(\omega)$ for the stage 1 and 2 ternary and binary K-GICs are shown in Fig.'s 14a,b. The threshold for interband absorption is indicated as E_T, which they have assigned to the midpoint in the threshold for valence-to-conduction (V→C) interband absorption, consistent with the theory of Shung [27]. Shung has shown that the (V→C) interband threshold, which should occur at ~2E_F if the mirror symmetry of the band structure is maintained, is broadened by scattering (electron-electron) of the photoexcited electron. Shung obtained good agreement between his calculations and the experimentally observed [28] V→C thresholds in stage 1 and stage 2 graphite-SbCℓ_5.

Referring to the schematic DOS diagram of Fig. 13, it is evident that for the stoichiometry $C_{4n}KH_x$ [29], the fractions f_Y (Y=K,H,C) per atom of the K(4s) electron residing in the respective (Y) bands are related by

$$f_K + f_H + 4nf_C = 1 \qquad (9)$$

where n = 1, 2 is the stage index of the hydride. If E_F falls within the K(4s) band, then $f_K > 0$. Assuming (1) the H(1s) band is full (i.e. $f_H = 1$) and (2) the stoichiometry of Guérard et al. [29] (i.e. x=.8), they arrive at the following expression for the square of the plasma

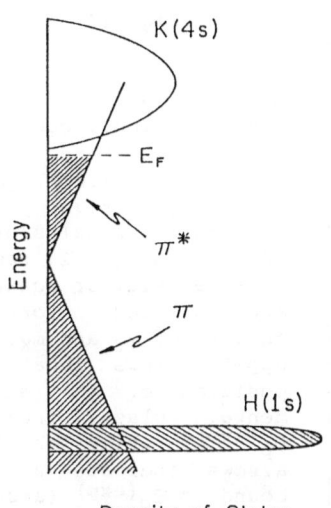

K(4s)

E_F

π^*

π

H(1s)

Energy

Density of States

Fig. 13

Schematic density of states model for KH_xC_{4n} (n=1,2). Electrons are emptied from the K(4s) band into low-lying H(1s) states.

frequency for the stage 1 hydride

$$\omega_p^2 (eV^2) = 7.87\gamma_0(.2-f_K)^{1/2} + 15.4(f_K/m_K) \qquad \text{(stage 1)} \qquad (10)$$

where m_K is the optical mass of the K(4s) electrons. The first and second terms in eq. 10 are the contributions from the π electrons in a 2D linear band model [6,30] and 3D K(4s) electrons, respectively. γ_0 is the next-neighbor overlap integral [6], which has been found [8,28,31] to have a value of ~3 eV, as in pristine graphite. If $f_K=0$, we note that the K(4s) band is empty, and if $f_K=.2$, then the carbon π band is empty. An expression similar to eq. 10 was obtained [8] for stage 2. In Fig. 15 we show the results of Doll et al. [8] for ω_p^2 vs. f_K for the stage 1 and 2 hydride. Two curves are drawn for each stage corresponding to values of

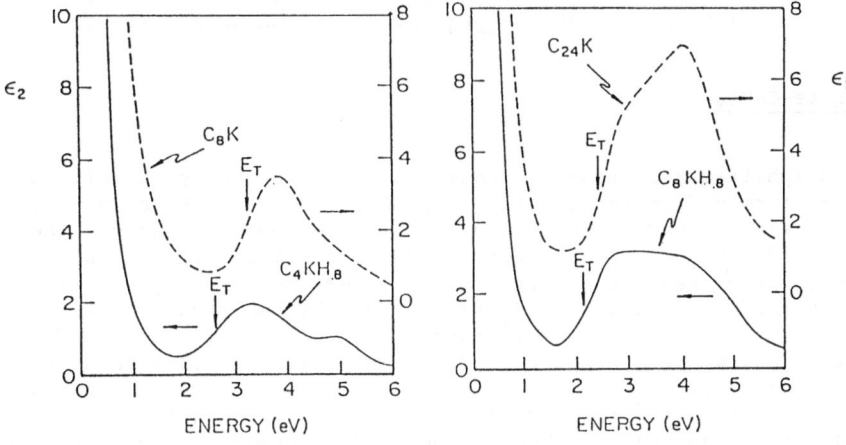

Fig. 14 a,b

$\varepsilon_2(\omega)$ obtained from the reflectivity spectra [8] of stage 1 (a) and stage 2 (b) K-and KH-GIC's.

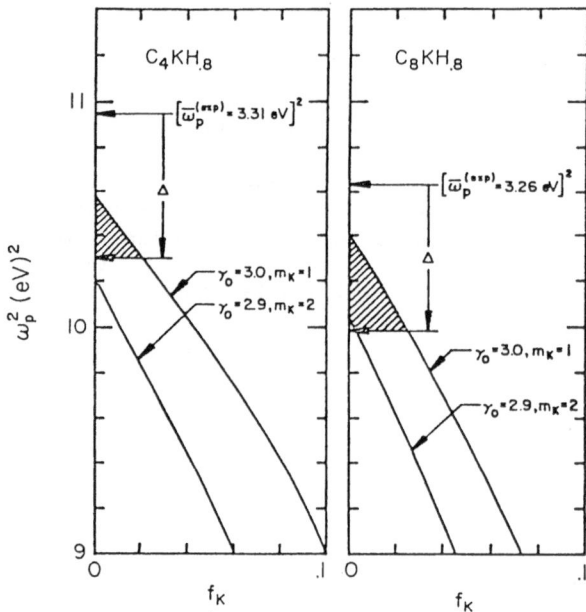

Fig. 15

ω_p^2 vs. f_K, where f_K is the fractional occupation of the K(4s) band (eq. 7) for $C_4KH_{.8}$ (left panel) and $C_8KH_{.8}$ (right panel). The stage 1 results of eq. 10 and stage 2 results from a similar equation, are plotted for two choices of γ_0 and m_K. The upper arrows locate the position of the experimental plasma frequency $\bar{\omega}_p(exp)$, while the lower arrows indicate a lower bound to $\bar{\omega}_p(exp)$ (assuming an experimental uncertainty of $\pm$ 0.1 eV). Shaded regions therefore denote the range of values for f_K allowed within experimental error.

the parameters γ_0 and m_K indicated. The shaded regions in the figure are to be identified with the range of values (ω_p, f_K) allowed, if the estimated experimental error ($\pm$ 0.1 eV) in ω_p is introduced. The analysis therefore indicates that the K(4s) band is empty, or nearly empty, in both stage 1 and stage 2.

From a different perspective, they also have considered the following assumptions: (1) the K(4s) band is empty and (2) the π-band dispersion is 2D and γ_0=2.9 eV. This set of assumptions results in values of x=0.8 for both stage 1 and 2, in excellent agreement with the reported stoichiometry [29] $KH_{.8}C_4$ (stage 1) and $KH_{.8}C_8$ (stage 2).

Graphite-CsBi$_x$

Graphite-CsBi$_x$ is an example of a new ternary alkali metal GIC system formed from the reaction of MBi$_x$ alloys (M=K,Rb,Cs) [32]. The stage 1 and 2 compounds can form in the α- or β-phase, where the β-phase has a ~1 Å thicker CsBi$_x$ intercalate layer [16]. Yang et al. [16] have recently reported the results of an (00ℓ) x-ray study which indicates that the Cs/Bi ratio (x) is different for the two phases: x=.5 (α-phase) and x=1 (β-phase). The reflectance spectra (0–6 eV) for the stage 1 and 2 CsBi$_x$ GICs are shown in Fig. 16. Strong interband absorption indentified [16] with intercalate states appears as peaks just above the Drude minimum in the reflectance. In Fig. 16 we show the K-K results for $\varepsilon_2(\omega)$ for these GICs. The dashed and dash-dot lines refer to the interband and free carrier contribution, respectively. The location of the intercalate states peak is at ~1.2 eV for the stage 1 and 2 β-phase compounds, and upshifts to ~1.6 eV for the α-phase stage 1 compound. The threshold for interband absorption does not involve the π states whose threshold has been estimated by Yang et al. to correspond to

Fig. 16

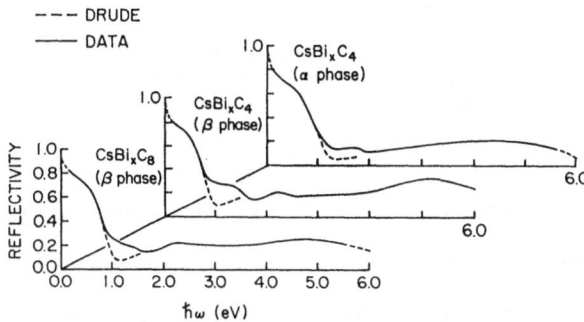

Reflectivity spectra for stage 1 (α and β phase) CsBi$_x$C$_4$ and stage 2 (β phase) CsBi$_x$C$_8$ [16]. Data are displayed as solid lines, while dashed lines denote Drude fits to the data. Deviations of the data from Drude minimum are due to strong inter-band absorption associated with intercalate states.

the dotted lines in the figure. By taking the midpoint of the V→C π band threshold in $\varepsilon_2(\omega)$ as a measure of ~$2E_F$, they arrive at estimates for the Fermi Energy: 1 eV (stage 1 α-phase), 0.9 eV (stage 1 β-phase) and 0.9 eV (stage 2 β-phase). Strong interband absorption due to HgK-derived states in KHgC$_4$ has also been reported by Preil et al. [33]. Similar to the case of the CsBi$_x$ GICs, Preil et al. report a threshold for interband absorption at 1.3 eV associated with the intercalate states (KH$_g$) and a π-π* threshold at 2.5 eV. However, the threshold energies in the paper of Preil et al. refer to the energy at which ε_2 extrapolates to zero. The plasma frequencies for the CsBi$_x$ GICs are considerably lower than those reported for the KHg GICs: 2.5 eV (stage 1 CsBi$_x$ α- and β-phase); 2.6 eV (stage 2 CsBi$_x$ β-phase) [16]; 5.0 eV [15], 5.1 eV [16] (stage 1 KHg); 4.7 eV (stage 2 KHg) [16]. These plasma frequencies show clearly that these two ternary GIC systems are quite different. As Yang et al. have pointed out [16], substantially different contributions to ω_p from the KHg-and CsBi$_x$ layers could account for the difference. Band structure calculations are needed in both these ternary GIC systems for more progress to be made in understanding their electronic properties.

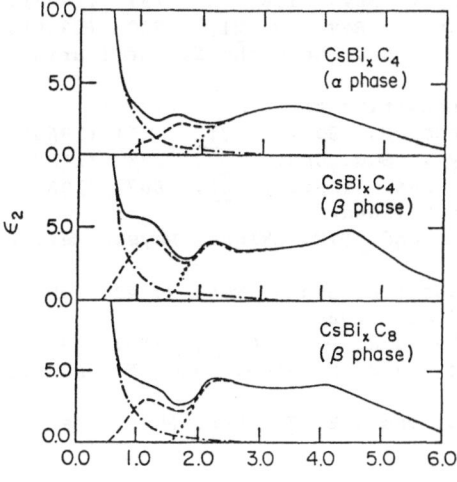

Fig. 17

K-K results for ε_2 obtained from Fig. 16 for the α and β phases of CsBi$_x$C$_4$ and CsBi$_x$C$_8$. The solid, dashed and dash-dot lines refer to the data, interband, and free-carrier contributions, respectively. Dotted lines represent interband contributions arising from the graphitic π bands.

Acknowledgement

This work is supported in part by DOE # DE-FG05-84ER45151. The authors are grateful for past collaborations in this area with D.S. Smith, R.E. Heinz, D.M. Hoffman, and J.M. Zhang.

References

1. M.S. Dresselhaus, G. Dresselhaus, and J.E. Fischer, Phys. Rev. B 15, 3180 (1977).
2. M.S. Dresselhaus and G. Dresselhaus, Advances in Physics, 30, 139 (1981).
3. S.A. Solin, Advances in Chemical Physics 49, 455 (1982).
4. S.Y. Leung and G. Dresselhaus, Phys. Rev. B 24, 3490 (1981).
5. C. Rigaux, "Optical Properties of Acceptor-Type Graphite Intercalation Compounds", in this volume, p. 237.
6. J. Blinowski, N.H. Hau, C. Rigaux, J.P. Vieren, R. LeToulec, G. Furdin, H. Herold, and J. Melin, J. Phys.(Paris) 41, 47 (1980).
7. T. Ohno, N. Shima, and H. Kamimura, Solid State Commun. 44, 761 (1982).
8. G.L. Doll, M.H. Yang, and P.C. Eklund, Bull. Am. Phys. Soc. 31, 687 (1986). (submitted Phys. Rev. B).
9. J.M. Zhang, P.C. Eklund, S.W. Qian, and S.A. Solin, Bull. Am. Phys. Soc. 31, 687 (1986).
10. M.H. Yang and P.C. Eklund, Bull. Am. Phys. Soc. 31, 689 (1986).
11. R. Saito, Ph.D. Thesis, University of Tokyo, Tokyo, Japan (1984), (unpublished).
12. D.P. DiVincenzo and S. Rabii, Phys. Rev. B 25, 4110 (1982).
13. N.A.W. Holzwarth, D.P.DiVincenzo, R.C. Tatar, and S. Rabii, Intl. J. Quantum Chem. 23, 1223 (1983).
14. N. Chen and S. Rabii, Phys. Rev. B 31, 4784 (1985).
15. J.E. Fischer, J.M. Bloch, C.C. Shieh, M.E. Preil, and K. Jelley, Phys. Rev. B 31, 4773 (1985).
16. M.H. Yang, R.E. Heinz, and P.C. Eklund, Bull. Am. Phys. Soc. 31, 689 (1986).
17. P.C. Eklund, "Optical Spectroscopy of the Lattice Modes in GICs", in this volume, p. 323.
18a. J.N. Hodgson, Optical Absorption and Dispersion in Solids (Chapman & Hall, London, 1970), pp. 41-56.
18b. F.J. Wooten, Optical Properties of Solids (Academic Press, New York, 1972), p. 90.
19. Ibid, p. 248.
20. E.A. Taft and H.R. Phillipp, Phys. Rev. 138, A 197 (1965).
21. J.J. Ritsko and E.J. Mele, Phys. Rev. B 21, 730 (1980).
22. ε_∞ is defined here as $\varepsilon_\infty = 1 + \varepsilon_{inter}(\omega=0)$, where the factor 1 arises from the Drude term.
23. Note: For $\omega_p\tau \gg 1$, $\varepsilon_{1,free}$ is not sensitive to τ.
24. J.J. Murray and A.R. Ubbelohde, Proc. R. Soc. A 312, 371 (1969).
25. T. Ohno and H. Kamimura, J. Phys. Soc. Jpn., 52, 223 (1983).
26. J. Blinowski and C. Rigaux, J. Phys. (Paris) 41, 667 (1980).
27. K.W.-K. Shung, Phys. Rev. B (to be published).
28. D.M. Hoffman, R.E. Heinz, G.L. Doll, and P.C. Eklund, Phys. Rev. B 32, 1278 (1985).
29. D. Guérard, C. Takoudjou, and F. Rousseaux, Synth. Met. 7, 43 (1983).
30. N.A.W. Holzwarth, Phys. Rev. B 21, 3665 (1980).
31. H. Zaleski, P.K. Ummat, and W.R. Datars, J. Phys. C 17, 3167 (1984).
32. P. Lagrange, A. Bendriss-Rerhrhaye, J.F. Mareche, and E. McRae, Synth. Met 12, 201 (1985).
33. M.E. Preil, L.A. Grunes, J.J. Ritsko, and J.E. Fischer, Phys. Rev. B 30, 5852 (1984).

ELECTRON SPECTROSCOPIES

G. Dresselhaus[†] and M. Laguës[‡]

[†]Massachusetts Institute of Technology, Cambridge, MA 02139, USA
[‡]Ecole Supérieure de Physique et de Chimie Industrielles
Paris CEDEX 05, FRANCE

INTRODUCTION

The use of electron spectroscopies to investigate the electronic structure of solids is generally applicable to study the microscopic excitations of intercalation compounds for all ranges of energy. Electron spectroscopies in graphite intercalation compounds (GICs) are the focus of this chapter. The common characteristic of electron spectroscopies is that they require the spectral analysis of an electron beam emitted from a solid. Similarly, photon spectroscopies require the measurement of the directions, polarizations and energies of photons transmitted through or reflected from solids.[1,2] Electron spectroscopies often involve both photons and electrons (e.g., photoemission or Auger electron spectroscopy) or can be restricted solely to electrons (e.g., characteristic energy loss spectra). In this chapter, we will also briefly discuss positron annihilation and μ meson spin rotation (μ–SR) which though bulk measurements give information closely related to that obtained by Auger or other surface spectroscopy techniques. Since the application of electron spectroscopies to the study of the electronic structure gives information about both surface and bulk states, it is important to use diverse experimental techniques to obtain a consistent model for the bulk electronic structure of a given GIC.

The relative ease in building electron and photon sources with energies ranging from a few milli–electron–volts (meV) to many million–electron–volts (MeV) allows one to measure core levels as well as some details about the valence and conduction bands in solids. Generally the resolution available in electron spectrometers is somewhat lower than that obtainable by optical means, so that optical experiments are somewhat favored for accuracy. However recent improvements in the resolution of electron spectrometers as well as greatly enhanced intensity for UV and x–ray light sources, made possible by synchrotron radiation, have greatly increased the available experimental information. In particular, Angle Resolved Photo Emission Spectroscopy (ARPES) is rapidly becoming the method of choice for directly measuring the $E(\vec{k})$ relations along specific symmetry axes in crystalline solids.[3,4] The application of these techniques to intercalation compounds has given some of the most accurate determinations of the electronic structure and has stimulated sophisticated band calculations for this class of materials.

The basic electron spectroscopy experiment consists of an energetic photon (or electron) with energy $\hbar\omega$ incident on a crystalline solid (such as a GIC) with an energy sufficiently large to emit an electron. The solid will be left in an excited state which is described in the one–electron approximation by a hole with energy $E(\vec{k})$ in an otherwise filled valence band. We write this excitation symbolically as

$$\hbar\omega + \mathrm{GIC} \rightarrow \mathrm{GIC}^* + e^- \tag{1}$$

in which the emitted electron is external to the GIC and the excited state of the solid is written as GIC*. If the energy of the photon is high enough so that the state GIC* represents a hole in the deep core state, then a subsequent Auger process can occur in which

$$\mathrm{GIC}^* \rightarrow \mathrm{GIC}'^* + e^- \tag{2}$$

and the emitted electron energy $E(e^-)$ is characteristic of the energy difference between the two pertinent core states given by

$$E(e^-) = E(\mathrm{GIC}'^*) - E(\mathrm{GIC}^*). \tag{3}$$

If one of the states GIC'* is a valence state, then the Auger line shape gives information about the density of states in the valence band. Since the atomic core states are at known energies, the small shifts in the energies of these core levels (as for example observed in the x–ray photoemission spectroscopy (XPS) experiment) give unambiguous information about the local bonding environment of the various atomic species in the solid. The energy resolution of electron spectroscopies (except for Auger spectroscopy) depends firstly on the excitation energy spread, and secondly on the electron analyzer resolution. In principle, one can achieve energy resolutions as good as 3 meV (High Resolution Electron Energy Loss Spectroscopy or HREELS). With regard to Auger spectroscopy, the intrinsic width of peaks is ≥ 1 eV, thus offering little hope for measuring energies, as for example chemical shifts, with an accuracy much better than 1 eV. The basic characteristics of the electron spectroscopies are the same, whether the initial excitation occurs via an incident energetic photon or electron.

SURFACE SENSITIVITY

One of the primary considerations in the interpretation of any electron spectroscopy experiment involves the high surface sensitivity of the technique, resulting from the strong interaction between the electrons. In metals this extreme surface sensitivity is a direct consequence of the plasmon energy loss of the electron before its escape from the surface. In semiconductors one can also excite electron–hole pairs at energies above the band gap so that the attenuation length will vary from one material to another in the low energy region. An electron with kinetic energy above the plasma energy (~ 5 eV in metals and ~ 15 eV in insulators) experiences energy loss via plasmon creation. Electrons that lose as much as one plasmon quantum of energy (typically ~ 10 to 15 eV) are effectively removed from contributing to the "full energy" peak or structure. Thus, the mean attenuation length shown in Fig. 1 exhibits a marked decrease above the plasmon energy, and the effective sampling depth ξ_s for electrons contributing to the full–energy peaks shows a very broad minimum ($\xi_s \sim 5$Å) for the electron kinetic energy range near 100 eV. Most of the electron spectroscopy experiments (including x–ray photoelectron spectroscopy XPS) on GICs correspond to electron energies near the minimum escape depth. The increase in attenuation length at high energies shown in Fig. 1 results from the decreased probability for plasmon creation. The main loss mechanism at high energies is associated with nuclear events.

The "escape depth curve" gives directly the depth from which the emitted electron provides first order information about the sample. For example, photoemission using a He I UV source at 21.22 eV would involve background (multiple scattering) contributions from emitted electrons coming from depths > 10 Å, and possible structure from emitted electrons excited within the first 10 Å of the sample surface. The measured signal will of course involve electron events occurring over a distribution of depths, determined by the penetration depth of the incident particle or radiation. Though a UV or X–ray photon may have a penetration depth of $\sim 10^{-2}$cm and photon–electron current may be excited throughout the sample, only those events occurring within ~ 10 Å of the surface result from a single primary event and give direct information about $E(\vec{k})$. This surface sensitivity has for example, been exploited in the experimental work on surface adsorbed gas phases on graphite. On the other hand, experiments purported to give information about the bulk properties of an intercalation compound must be carefully examined to identify features associated with surface phenomena.

The "escape depth curve" shown in Fig. 1 is plotted for electrons emitted normal to the surface and thus represents surface sensitivity for normal electrons. However, if electrons are accepted at an angle θ from the normal to the sample plane, then the surface sensitivity is enhanced by a factor of $\cos^{-1}\theta$ giving up to an order of magnitude increase in the surface sensitivity. Both the intrinsic surface sensitivity and its tunability, by varying energy and angle, offer great opportunities for applying photoemission to the study of surface phenomena. Likewise, the energy and angular dependence of the electron spectra can be used to distinguish qualitatively the surface–related features from features associated with the "bulk".

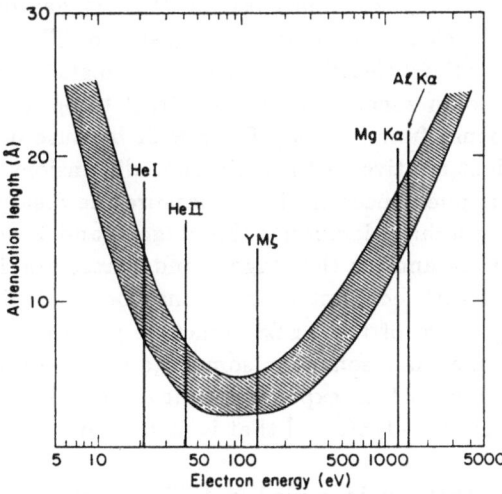

Figure 1: The universal "escape depth" curve of electron attenuation length in various heavy metals, drawn as a band that encompasses experimental data for most materials, including graphite and its intercalation compounds. The photon energies of several laboratory photon sources are shown.[3,4]

The surface sensitivity in semimetals and semiconductors is similar to that in metals, while in ordinary molecular solids, the surface sensitivity appears to be much lower, presumably because the plasmon loss mechanism is weaker. Since GICs are usually good metals with well defined plasma frequencies in the 1–2 eV range for polarization in the plane, high surface sensitivity is expected. In contrast, the extreme anisotropy of acceptor GICs gives a molecular solid behavior for polarization normal to the plane, so that normal polarization phenomena are more easily probed using electron spectroscopies on a c–face (i.e., electrons propagating along the c–axis) rather than with optical spectroscopies on an a–face.

Most of the applications of electron spectroscopies to GICs discussed in these lectures are attempts to obtain information about their electronic structures. Early attempts to deduce the bulk electronic structure of GICs were greatly impeded by surface sensitive phenomena, often coupled with instability effects characteristic of intercalation compounds. Even those compounds which are relatively stable in the bulk, have been shown to exhibit different surface stoichiometries. However, these surface problems are well known in surface science and some of the difficulties can be avoided by *in situ* sample cleaving or can be corrected for by proper analysis of the data. However, surfaces have their own intrinsic electronic structure, which is interesting in its own right; electron spectroscopies provide an excellent technique for probing surface states.

From this discussion, we conclude that the detected electrons in principle couple to the bulk electronic structure. However if they leave the sample surface with energies < 200 eV, then bulk information is firstly mixed with information from the surface and secondly the bulk information is integrated over a large range of wave vectors Δk_z due to the small thickness that is probed.

ELECTRON ENERGY LOSS SPECTROSCOPY (EELS)

Low energy electron energy loss spectroscopy (EELS) has provided an important technique for studying the electronic structure of graphite. This inelastic electron scattering technique measures the dielectric response function $\varepsilon(\omega, q)$ for a range of $\vec{q}$ vectors which can span the Brillouin zone. This is in contrast to optical measurements which yield the dielectric response function only for $q \approx 0$, because of the small magnitude of the wave vector of light relative to Brillouin zone dimensions. Electron energy loss spectroscopy has been applied successfully to measure the dielectric response function $\varepsilon(\omega, q)$ for intercalated graphite, including the stage 1 and 2 donor compounds C_8K and $C_{24}K$, for stage 1 C_6Li and for the stage 1 and 2 acceptor $FeCl_3$–GICs. Electron loss structure associated with both interband transitions and plasma resonances have been identified. Though most of the EELS structure pertains to graphite–derived interband transitions and plasma resonances, some intercalant–derived structure has also been identified. In the actual EELS experiments, it is the electron energy loss function (imaginary part of $-1/\varepsilon$) $Im[-1/\varepsilon(\omega, q)]$ that is measured.

The various features that are identified in EELS spectra up to 40 eV include the free carrier intraband plasma resonance (in the 1–3 eV range), the π–electron interband plasma resonance (in the 6–7.5 eV range), the all–valence electron plasmon (in the 25–26 eV range), the M point $\pi - \pi^*$ interband transitions (in the 4–5 eV range), the $\sigma - \sigma^*$ interband transitions (in the 13–15 eV range) and other weaker interband transitions which have been identified with zone folding effects. Energy loss peaks have also been identified with interband transition between intercalate levels, based on the similarity of the energy loss spectra to those from the pristine intercalant. Attempts have also been made to determine the charge transfer from analysis of the EELS plasma resonance lineshape.

An example of EELS spectra showing these plasma resonances and interband transitions for the graphite–derived energy bands is shown in Fig. 2 for C_6Li in the energy range $0 < \hbar\omega < 40$ eV and for very small momentum transfers near $q = 0.1$ Å^{-1}.[5] In first stage C_6Li, the energy loss peak at 2.85 eV associated with the free carrier plasma oscillations is in good agreement with the energy of the plasma edge in the optical reflectivity.[5,1,2] The peak in the dielectric function associated with the π–electron plasma resonance is found in pristine graphite at 7.5 eV, but downshifted in C_6Li to 6.3 eV, similar to that for C_8K,[6] while the all–valence electron plasma resonance is found at 25.6 eV. The plasma dispersion relation for the π–electrons in the alkali metal intercalation compounds,

$$E_p(q) = E_p(0) + z\hbar^2 q^2/m, \qquad (4)$$

is essentially the same as in graphite ($z = 0.42$), as has been established by detailed studies in C_8K.[6] With regard to interband transitions, the dominant interband M–point π–band transition at 4.9 eV in graphite is downshifted to 4.35 eV in C_6Li. In the EELS experiments, the contributions to the interband transition peak involve a significant region of the Brillouin zone, so that an identification of peaks in $1/\varepsilon$ with interband transitions at specific points in the Brillouin zone is only approximate.

The interband transition between graphite σ–bands near 13 eV is found at 13.3 eV in C_6Li, and is somewhat broadened (see Fig. 2). The weak structure at 9.7 eV in the spectra of Fig. 2 is identified with several closely spaced interband transitions predicted

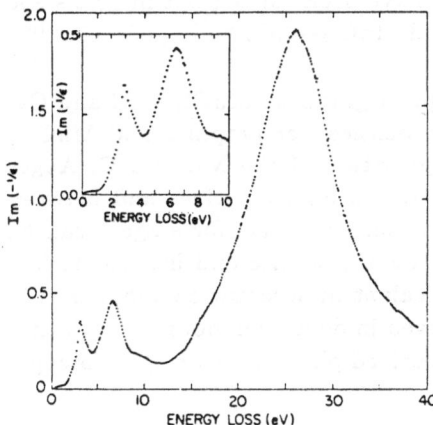
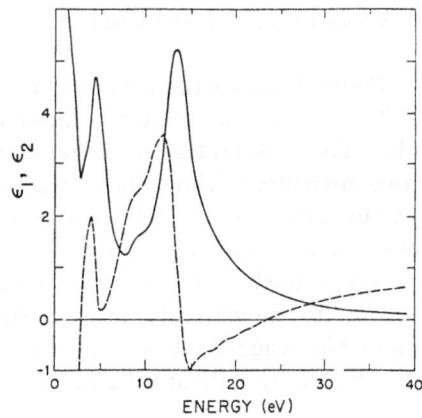

Figure 2: The electron energy loss function $Im[-1/\varepsilon(\omega, q)]$ for C_6Li obtained using the EELS technique for $q \sim 0$. The Kramers–Kronig analysis of these data has been used to obtain $\varepsilon_1(\omega)$ (dashed) and $\varepsilon_2(\omega)$ (solid).[5]

by theory.[5] Additional structure in this energy range in the EELS spectra of C_8K has been identified with zone folding effects;[6] the corresponding structure is absent for C_6Li because of the different zone folding of the K point to the Γ point for the $(\sqrt{3} \times \sqrt{3})R30°$ in–plane superlattice for C_6Li compared with the folding of the M point to the Γ point for the $(2 \times 2)R0°$ in–plane superlattice for C_8K. The recent EELS determination of $Im[-1/\varepsilon(\omega,0)]$ in C_6Li has been compared with optical reflectivity determinations of the complex dielectric constant[7] and has been used in a Kramers–Kronig analysis of the reflectivity data to obtain $\varepsilon(\omega,0)$.

For the case of the $FeCl_3$–GICs, electron loss peaks at 2.75, 3.9, 8.0 and 9.0 eV have been identified with transitions between $FeCl_3$ levels, since similar peaks occur in pristine $FeCl_3$ at the same energies.[8] Similarly, electron loss peaks have been identified with a transition from the Li $1s$ level to the Fermi level (57.1 eV) and the Li $1s - 2p$ transition at 63.0 eV.[5] By studying transitions with significant electron momentum transfer, Fermi level shifts of 0.9, 0.7 and ~ 1 eV relative to the graphite Fermi level were reported for the stage 1 and 2 $FeCl_3$–GICs and for C_8K, respectively.

The main conclusions reached from EELS experiments is that for the electronic transitions which are accurately determined by EELS the graphite π and σ bands in the intercalation compounds are well described by a rigid band model. As a form of spectroscopy, EELS provides invaluable information on the energy of the graphitic and intercalate levels in the GICs.

AUGER ELECTRON SPECTROSCOPY (AES)

A study of the Auger spectrum in GICs provides information about both the intercalant and the carbon bonding. The Auger process is initiated by an incident photon (or electron) which creates a hole in the $1s$ carbon K level (or intercalate core level). The carbon KVV signal corresponds to the recombination of this $1s$ carbon K level hole with a valence electron and the simultaneous emission of an electron from a valence state. Studies of the carbon KVV Auger lineshape have been carried out in first stage C_8Rb and C_8Cs by Oelhafen et al.[9] and in C_8Cs and $MnCl_2$–GICs by Laguës et al.[10]

Figure 3 shows the carbon KVV Auger lineshape of graphite, $MnCl_2$–GIC and Cs–GIC, where we see that the lineshapes are nearly identical for graphite and $MnCl_2$–GIC. This observation and the quantitative interpretation of the Mn and Cl Auger peak intensities[10] show that the intercalate concentration on the surface and between the top two graphite planes is very low, so that the surface layers for stage 1 can be approximated by $GGIGIG\ldots$, where G and I represent graphite and intercalate layers, respectively. The surface depletion of the intercalant in acceptor compounds is in contrast to the intercalant surface enrichment observed in donor compounds,[11] which is implied by Auger spectroscopy, by ARPES (angle resolved photoemission spectroscopy) studies and by XPS studies on C_8K.[13]

Laguës et al.[10] have suggested that the surface intercalant depletion in acceptor–GICs and the surface intercalant enrichment in donor GICs, arises from the electrostatic screening associated with a surface state. The electrostatic energy associated with this screening ($\simeq 1$ eV per intercalant atom) is larger than the typical chemical potential difference between the various GIC stages. The intercalant concentration is thus expected to change at least in the uppermost intercalated layer, in order to reach a free–energy minimum. This description applies not only to GICs but to any compound. However binding energies and diffusion activation energies are usually much too large so that the equilibrium situation at the surface is not very different from the bulk. The occupation of a surface state level results in a local electron enrichment. This picture is consistent with dangling π bonds at the surface and the large value of the work function. Thus in the case of GICs, this favors a surface enrichment of positive ions (donors) and depletion of negative ions (acceptors).

We now apply these concepts to discuss the carbon KVV lineshape and its relation to the valence density of states in donor GICs. Referring to Fig. 3, we see that the Auger spectrum for the Cs–GIC exhibits a modified line shape relative to graphite, with an additional sharp structure not present in the graphite spectrum, as well as several weaker structures. This sharp feature arises from the increased occupation of the graphite π^* band conduction states as the Fermi level rises due to donor intercalation. An explanation for the lineshape of the sharp feature is provided by the Lander model for the Auger lineshape. According to the Lander model,[14] the Auger intensity $I(E)$ is given by the sum over all possible transitions leading to the final state energy E. Lineshape effects are particularly important for KVV transitions where the density of states (DOS) of the valence band $n(E)$ enters the calculation. Neglecting the energy dependence of the matrix elements, the lineshape $I(E)$ is simply the self–convolution of

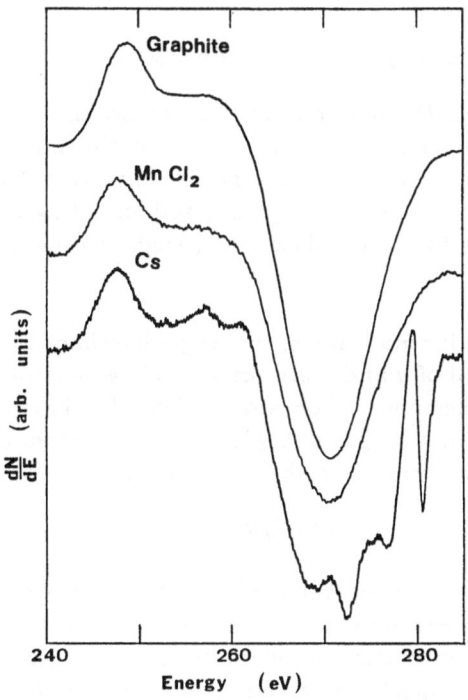

Figure 3: The Auger spectra of pristine graphite, stage 1 MnCl$_2$–GICs and stage 1 Cs–GICs.[10]

$n(E)$ given by

$$I(E) \propto \int n(x)n(E-x)dx \equiv n*n. \qquad (5)$$

This simple description assumes that the Auger transition does not involve localized states and thus is a true band structure effect. Large screening effects of the initial K state hole or final valence state hole would modify the lineshape. The energy dependence of the transition matrix element would also modify the lineshape. For example, in a tight binding description, we write

$$n(E) = n_s(E) + n_p(E) \qquad (6)$$

where $n_s(E)$ and $n_p(E)$ respectively represent the density of states associated with s and p bands. In this description, n_s*n_s, n_s*n_p and n_p*n_p terms occur in the self convolution of Eq. (5) and should have different matrix elements. Finally, the Lander model which works well for pristine graphite, assumes direct emission of the Auger electron to the vacuum. Each Auger electron reaches the surface with a definite probability of experiencing an energy loss, which can be determined by an EELS experiment, and must be included in a detailed analysis of the lineshape.

The application of the Lander model to GICs makes use of the well established graphitic nature of the graphite derived bands in GICs and the model is able to explain the observed lineshape. The density of states $n(E)$ is written as a sum of the graphite DOS $g(E)$ and of the sharp feature in the DOS $p(E)$ induced by intercalation. We note that $p(E)$ also corresponds to a sharp peak in the UPS (ultraviolet photoelectron spectroscopy) spectra at the Fermi level.[10] Following the Lander model, we write the convolution for the GICs as

$$I_{GIC}(E) \propto g*g + 2g*p + p*p \qquad (7)$$

where $I_G(E) \propto g*g$ is the KVV lineshape of pure graphite. Thus, it is possible to simplify the analysis of $I_{GIC}(E)$ by subtracting the experimental Auger spectrum of graphite $I_G(E)$. Since $I_{GIC}(E) \approx I_G(E)$ in the energy range $E < 260$ eV, the contribution of the sharp feature $p(E)$ can be found by subtraction, noting that the second term $2g*p$ in Eq. (7) corresponds to the graphite DOS, slightly broadened by convolution with $p(E)$.

According to the quantitative modeling of the Auger lineshape by Laguës et al.,[10] the relative weights of the integral of g and p are respectively 4 electrons and f_C electrons, where f_C is the charge transfer per carbon atom. Thus, the integral in Eq. (7) has a weight of 16 for the $g*g$ term, a weight of $8f_C$ for the $2g*p$ term and a weight of f_C^2 for the $p*p$ term. After normalization to the graphite peak weight $g*g$, the relation

$$T^2 = 4S \qquad (8)$$

is obtained where T and S are respectively the normalized integrals of $2g*p$ and $p*p$. Thus, the charge transfer per carbon atom f_C can be obtained either by convolution of g with p to obtain T, or the convolution of p with itself to obtain S. A detailed analysis shows that the graphite Auger spectrum can be subtracted from the corresponding GIC spectrum to obtain the density of states g and p appropriate for the GIC. The functions g and p thus obtained successfully reproduce the features in the Cs–GIC Auger lineshape with regard to the energies of the various peaks in the spectrum. The model explains the magnitude of T, but fails to explain the magnitude of S or of the sharp intercalation–induced peak near E_F, which is 5 to 10 times larger than calculated on the basis of complete charge transfer. An attempt to modify the Lander model[15] by taking into account the local perturbation of the band structure due to the K hole, enhances the magnitude of the expected sharp peak, but is inconsistent with the experimentally observed lineshape. A possible explanation of this discrepancy is that the Auger feature near E_F is related to a different electronic level which should not be convoluted with the valence band in Lander's model. Though this level could be a bulk state, it is probably a surface state (or a surface–driven charge density wave) which is observed at the Fermi level by ARPES[16] and described below.

It is clear that a more detailed study of the band structure by a complementary technique is needed to explain the lineshape and intensities of the Auger spectra for the alkali metal–GICs.

ANGLE RESOLVED PHOTOEMISSION SPECTROSCOPY (ARPES)

In general, photoemission, both in the x–ray (XPS) and ultraviolet (UPS) regimes involves measurement of the energy distribution of the emitted electrons from a photoexcited surface. The spectra obtained from the angle integrated experiments give a direct measure of the density of states of the occupied levels in a solid. Much of the early work on photoemission from GICs was done with this type of experiment and the graphitic nature of the electronic structure of GICs was clearly established. With the development of more intense light sources in the soft x–ray and ultraviolet range it became possible to carry out *angle resolved* photoemission experiments, in which the energy and wave vector $E_{vac}(\vec{k}^{ext})$ and $\vec{k}^{ext} = (k_\parallel, k_\perp)$ are simultaneously measured for each emitted electron. From these two measurements it is possible to deduce the energy and exact position in the Brillouin zone for the electron which was initially photoexcited, and thereby to map out the $E(\vec{k})$ relation for the occupied levels in the solid.

The theory for ARPES involves an initial interband optical transition from a state $E_i(\vec{k}_i)$ to an excited state $E_f(\vec{k}_f)$ followed by an emission of the electron into the vacuum state with energy $E_{vac}(\vec{k}^{ext})$. The theory for the interband optical process in GICs is covered elsewhere in this volume.[1,2] Because of the small wavevector of light, one assumes that direct transitions dominate the photoexcitation process (i.e., $E_f(\vec{k}_f) = E_i(\vec{k}_i) + \hbar\omega$ and $\vec{k}_f = \vec{k}_i$). The emission process has been modeled by a nearly free Bloch electron moving in a "muffin tin" potential, coupled through a surface barrier to a vacuum state. This emission model is not sensitive to the Bloch state model and suggests that a number of different wavevectors normal to the surface can couple to the vacuum state. However, the conservation of in–plane wavevector is rigorous for a perfectly specular surface. Thus the boundary conditions for ARPES experiments assume specular surfaces marking the boundary between the Bloch function solution appropriate to a perfect crystal and a plane wave solution for the vacuum state. Any surface imperfection gives rise to a broadening of the spectral features[17] while localized surface states give rise to distinct features in the spectrum. The ARPES technique is particularly attractive for GICs because the electronic states have only a small k_z dispersion and the 2D description (requiring no model for the excited electronic state) holds very well, at least for the graphite–derived levels. This interpretation of the ARPES experiment relates to measurements of the bulk dispersion relations $E_i(\vec{k}_i)$. The information provided by the ARPES measurement however comes from only a small depth (~ 5 Å), so that the surface affects the electronic levels both directly and indirectly. The sudden discontinuity of the crystal potential at the surface directly gives rise to surface states. The intercalation–induced changes in binding, coordination number and binding energies all modify the electronic structure locally and indirectly, through modification of the chemical composition, and the crystal structure near the surface. In GICs we expect both surface related effects to be important.

In our discussion of Auger electron spectroscopy, evidence was presented favoring an enriched intercalate concentration near the surface for donor GICs and a depleted intercalate concentration for acceptor GICs. Since the emitted electron originates from approximately the same depth for both Auger and ARPES techniques, acceptor GICs are expected to exhibit ARPES spectra very similar to those for pristine graphite; this may account for the absence of ARPES studies for acceptor GICs in the literature. To increase the sensitivity of the ARPES technique to the bulk electronic structure, it may be possible to carry out ARPES measurements on acceptor GICs at lower incident electron energies where the penetration depth may be significantly larger than at 50 eV (see Fig. 1), where most of the ARPES measurements have thus far been made.

Angle–resolved photoemission studies have been reported for both graphite and graphite donor intercalation compounds. Early work was done by Eberhardt et al.[18] to yield an accurate determination of the energies of the Γ–point valence bands for C_6Li showing general agreement with band calculations of Holzwarth et al.[19] More detailed results for $E(\vec{k})$ in Cs–GICs have recently been reported[11] and typical energy distribution curves (EDCs) are shown in Fig. 4 for C_8Cs as a function of polar angle $\vec{k}^{ext}$. The main peak which is easily resolved up to 30° corresponds to the graphite π bands. The spectra shown in Fig. 4 correspond to the C_8Cs ordering which can be observed using low energy electron diffraction (LEED) on a freshly cleaved surface of Cs–GICs. A second surface phase was observed in different spatial positions of the cleaved surfaces, corresponding to the C_6Cs ($\sqrt{3} \times \sqrt{3}$)R30° superlattice ordering; the ARPES spectra for this surface phase are not shown here.[20]

From the peaks in Fig. 4, the $E(\vec{k})$ relations for C_8Cs are deduced along high symmetry axes, and are shown in Fig. 5. Here the lines connect the experimental points. The shape of the π bands (upper heavy solid lines) is qualitatively the same as for graphite, but the band nearest the Fermi level has a lineshape different from that expected for the π^* conduction band. A gap is observed in Fig. 5 between the graphite π band and the very flat band near the Fermi level. This gap is either due to the bulk band–structure, or more likely to a surface effect. Very similar ARPES results have been reported for C_8K by Takahashi, et al.[21]

The experimental zone–folded π or σ bands for C_8Cs show excellent agreement with the calculations for C_8K of Tatar and Rabii,[22] and by Ohno et al.,[23] indicating that the π bands in C_8Cs present a non–rigid deformation of the graphite bands. The simple behavior of the EDCs around $\theta = 0°$ (Fig. 4) suggest that if a nearly free electron band is present at the center of the Brillouin zone, it must either be nearly empty, or its cross section should be extremely small. This conclusion holds for either of the two surface phases though not necessarily for the bulk.

This difference in composition between the uppermost layer and the bulk is a crucial point in the analysis of all the electron spectroscopy results. It is useful to think about this first intercalated layer, as a surface phase in equilibrium with the bulk compound. In this limit, the concept of stage has to be replaced by an appropriate analysis of the surface phase. The ARPES measurements also support the surface enrichment of the alkali metal in the case of the Cs–GIC in three different ways: 1) the large Fermi level shift of about 2 eV, suggests a large electron transfer, 2) the Fermi surface dimensions are too big for stage 2 or higher, and 3) the symmetry of the observed bands agree with either the C_8Cs or C_6Cs surface composition for the first one to three graphite layers, but not with the $C_{14}Cs$ structure typical of stage 2 Cs–GICs.

There is at present no definitive interpretation for the flat band near E_F. Since this level relates most closely to the charge transfer between the intercalate and graphite–derived bands, the ARPES experiment in its present form does not yield a definitive

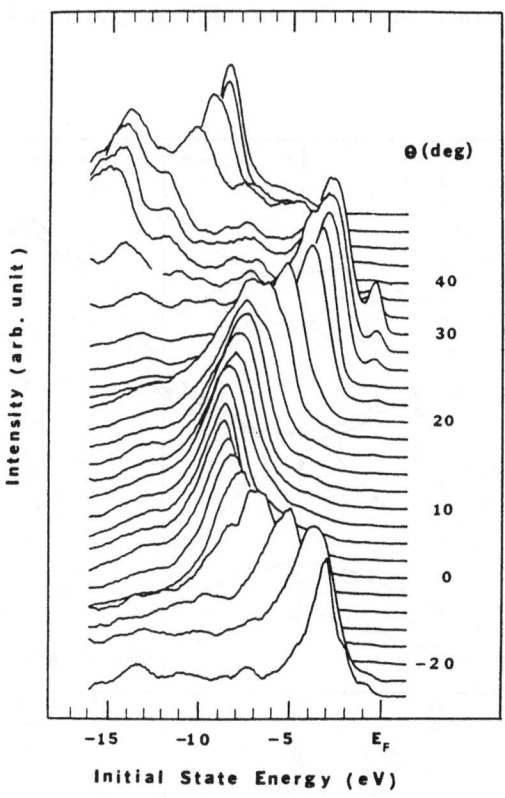

Figure 4: The ARPES energy distribution curves from a sample with a surface composition C_8Cs, at a photon energy of 50 eV. The polar angle is varied from $-26°$ to $+56°$. For $-30° < \theta < 30°$, the main peak is identified with the graphite π–bands. For $\theta > 30°$, the peak near -10 eV corresponds to the graphite σ–bands. The peaks occurring at about -15 eV for $\theta \geq 40°$ may be associated with the two $5p$ Cs levels.[20]

value for the charge transfer. If it were possible to carry out the ARPES measurements with photons down to an energy of ~ 5 eV, then a more definitive separation between surface states and bulk states might be possible. In this low photon energy regime, the electron escape depth for the electron may also be large enough to obtain definitive spectra for higher stage donor compounds. Because of the intercalant enrichment near the surface, the surface–related phenomena affect the higher stage compounds even more critically than for stage 1. For this reason it will be important to carry out ARPES measurements as a function of incident photon energy, and to obtain detailed spectra in a photon energy regime where the escape depth is significantly larger than I_c.

X–RAY PHOTOEMISSION SPECTROSCOPY (XPS)

When the energy of the electron in a photoemission experiment is increased to probe deeper into the bulk, then it is no longer possible to carry out accurate angle resolved measurements since all electrons excited from the first Brillouin zone are then emitted nearly normal to the surface. However a signal related to the density of occupied states can still be obtained. Electrons with a kinetic energy of 1 keV have a mean free path in solids of ~ 15 Å. Thus the electrons detected in the typical XPS measurement originate in the first few atomic layers of the sample. For GICs photoemission data are always taken under ultra–high vacuum conditions, with samples usually cleaved *in situ*. To

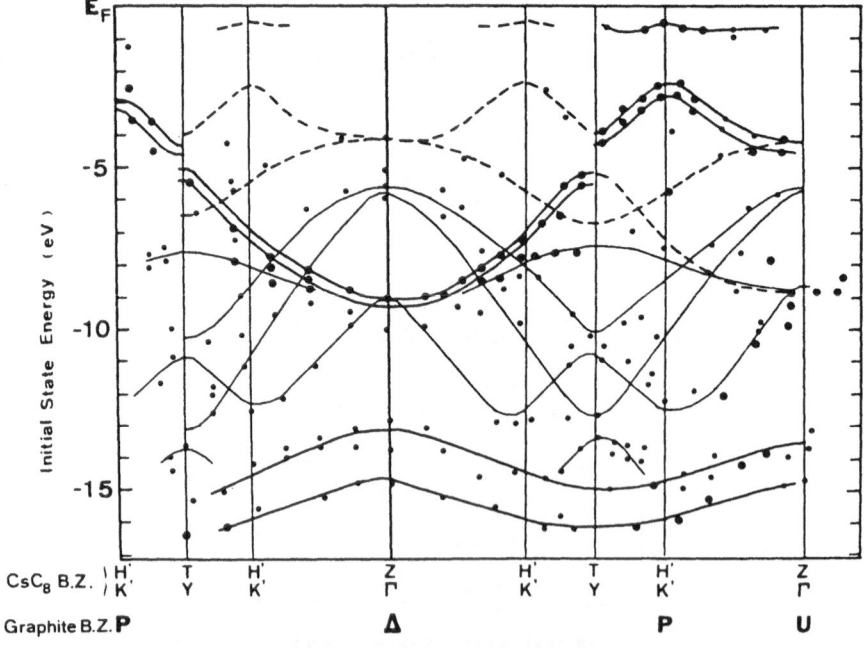

Figure 5: The electronic energy bands for a sample with a surface composition C_8Cs. The $E(\vec{k})$ relations are determined by ARPES measurements at a photon energy of 50 eV.[20] The solid curves are associated with the extended zone graphitic bands and the dashed lines correspond to zone–folded graphite bands. The heavy lines near -15 eV correspond to intercalate–derived energy levels.

verify that the data are representative of the bulk, it is useful to compare spectra taken at different emission angles.

Since the XPS cross–section is strongly dependent on energy, it is possible to greatly reduce the XPS intensity for s orbitals relative to the p orbitals, by increasing the photon energy.[12] For the case of graphite intercalation compounds, the use of the XPS technique thus makes it possible to distinguish the s intercalate electron contribution near the Fermi level from that from the graphite π bands. An example of the usefulness of valence band XPS in this context is the recent XPS study by DiCenzo[13] for C_8K which is shown in Fig. 6. The π orbitals of graphite have essentially zero cross–section at the photon energies of this experiment, and thus any intensity observed near the Fermi energy is due to electrons occupying states derived from K $4s$ orbitals. By comparison of the photoelectron yield near E_F with that of the K $3p$ peak (at about 20 eV) and noting the relative $3p$ and $4s$ cross sections, an estimate of the number of $4s$ electrons per K atom is obtained. After correction for surface related effects it is concluded that less than 0.05 valence electrons remain on each K atom. On the other hand, charge density associated with the interlayer hybridized intercalate state may not contribute strongly to the XPS spectra.

The XPS technique also gives information on the structure through study of the dependence of the binding energy on the chemical environment of the atom. This is illustrated in Fig. 7, where the C $1s$ spectrum of a liquid nitrogen cooled stage 2 K–GIC is presented.[13] In the stage 2 compound the carbon atoms occupy a variety of sites relative to the K atoms. If the K atom is in registry with the graphite, then about half the carbon atoms are immediately adjacent (closest nearest neighbor interlayer distance) to the K atoms. These atoms are in an unique environment, whereas the other half of the carbons can occupy a variety of sites, each with its own unique binding energy. If the compound were to form an unique space group, then all sites could be specified explic-

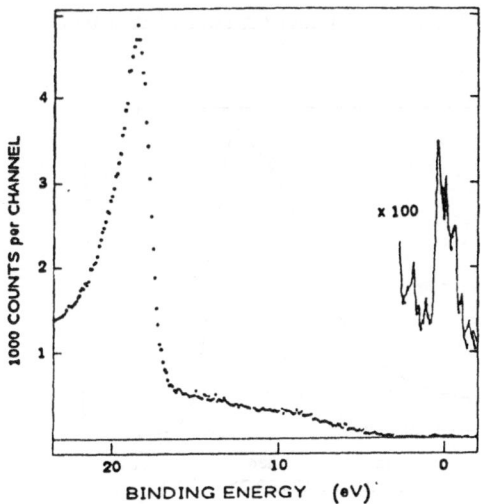

Figure 6: The x–ray photoemission spectrum (Al Kα, 1486 eV) of C_8K. A small peak at the Fermi level (0 binding energy) is shown on an expanded scale and is identified with $4s$ electrons on potassium atoms. This K $4s$ signal is primarily of surface origin.[13]

itly. However, x–ray data indicate that these materials are disordered (perhaps forming a lattice gas). Nevertheless the C $1s$ XPS spectrum has a reproducible structure. Approximately half of the intensity is identified with a peak at the lowest energy, in Fig. 7, corresponding to the 6 nearest neighbor carbons, and having the largest charge transfer. In addition to these unique sites, four other distinctly different sites are identified, so that a five peak fit provides a good model for the spectrum. This result offers strong support for the notion of registry of K atoms in higher stage K–GICs. The excellent agreement between the measurements and the model calculation suggests that the K is in registry with the graphite in this temperature range and is consistent with a lattice gas–type model.

CORE LEVEL SPECTROSCOPIES

Core level shifts give valuable information on the electronic environment of atoms in solids and can be probed by a number of techniques such as electron energy loss spectroscopy, Auger spectroscopy and X–ray (or ultraviolet) photoemission spectroscopy, XPS (or UPS) also called ESCA (Electron Spectroscopy for Chemical Analysis). The most common use of XPS is in core level spectroscopy which is illustrated above for the carbon K $1s$ level edge shown in Fig. 7 for stage 2 K–GICs. Since the XPS spectrum of each element is unique because of its binding energy and cross section, the XPS technique can be used for the identification of chemical species. The charge state of the atom is inferred from the shift in the core electron binding energy, which can vary by several eV depending on the chemical environment of the atom. Various calibrations of XPS shift vs. charge state have been made. Although large shifts can clearly signal a change in the charge state of the atom, core level spectroscopy is often complicated by our inability to distinguish between the initial and final state contributions to the shift.

An example of the use of EELS for core level spectroscopy is the case of FeCl$_3$–GICs which was studied by Mele and Ritsko[24] with regard to the broad structure near 287 eV associated with transitions from the $1s$ carbon core states to the empty carbon $2p$ state

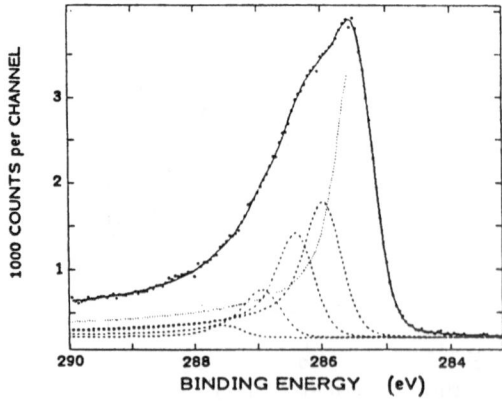

Figure 7: The x–ray photoemission spectrum of stage 2 K–GICs showing the C $1s$ spectrum. The dashed curves represent the result of a five–component least–squares fit.[13]

at the Fermi level. Their results on a stage 1 $FeCl_3$–GIC sample show this electron loss peak to shift because of the lowering of the Fermi level due to intercalation, yielding a shift in Fermi energy of 0.9 eV relative to graphite.

POSITRON AND MUON DECAYS

Positron annihilation and muon decay are two techniques for studying bulk solids which involve the radioactive decay of elementary particles and have been used as microscopic probes of the electronic structure of GICs. These techniques involve positrons (e^+) and muons ($\mu^\pm$) and exploit the fact that these particles rapidly achieve thermal equilibrium in the solid and interact strongly with the valence and conduction electrons before radioactive decay. These two solid state experiments are most sensitive to the momentum distribution (positrons) and the magnetic distribution (muons) in the material.

In both cases the experiment monitors the solid state behavior by measurement of the angular distribution of the radioactive decay products which are gamma rays for positrons and energetic electrons (positrons) for muons. Both e^+ and μ^+ are known to bind electrons in vacuum to form positronium and muonium. The properties of these hydrogenic species are well known and their observation in a solid state experiment is only possible if a void occurs which is large enough to accommodate the hydrogenic entity (about 1 Å) and is thus a clear indication of voids in the material. The most important solid state information is obtained when no voids are present in the material, and the particle then thermalizes by interacting with the elementary excitations of the solid. In this case, an analysis of the decay process contains microscopic information about the dispersion relations for the elementary excitations in the solid.

Positron Annihilation

Positron annihilation experiments performed in GICs have shown very different behavior for positrons in donor and acceptor GICs. In acceptor GICs, positrons are attracted by the anions of the intercalant and positron–anion bound states are formed. The angular correlation distribution (ACD) obtained in acceptor–GICs can readily be compared with calculation of free positronium–halides.[25,26] In contrast, positrons in donor GICs are repelled from the positively charged intercalant and are annihilated by the delocalized Bloch states. The annihilation process is dominated by Bloch electrons in the interstitial regions between the carbon layers. Consequently, the positron characteristics are strongly related to the electron momentum distribution, i.e., the band structure in the vicinity of the Fermi level in donor GICs.

The annihilation of a positron at rest by an electron at rest gives 2 equal energy (0.5 MeV) γ rays at 180° with respect to one another. If either the electrons or the positrons have a distribution of momenta before annihilation, then the angular correlation distribution (ACD) of the γ rays gives an experimental measure of the electron momentum distribution.

Positron annihilation experiments in GICs have been carried out by Cartier et al.[25] and by Berko[26] in alkali metal GICs. In these experiments the positron is thermalized and after thermalization, the positron decay is monitored by means of γ ray counters which monitor the 2 γ decays. Whereas the angular correlation distribution (ACD) for C_6Li is found to be quite similar to graphite (see Fig. 8), indicating no occupation of a free electron type band, the curves for the K, Rb, and Cs intercalants show a sharp feature near $p_z = 0$ indicating a free electron component of the in–plane electron distribution, and suggesting occupation of a free electron or s–like alkali metal band. It has been verified[26] that this feature is not associated with voids and is a true measure of an electronic effect. Since positron annihilation is a bulk effect, the surface phase proposed by Lagües[10] does not apply to the positron annihilation experiment.

From the high intensity of the narrow component of the ACD in Fig. 8 (especially for C_8K), it is concluded that the related electrons occupy the interstitial region between the graphite layers, which is preferentially sampled by the positrons. The shape of the narrow component for $p_z \perp c$ accentuates the free–electron character of these electrons parallel to the graphite planes, as emphasized in the interpretation of the experiments.[25] The 2D character of the free electron contribution is confirmed by the significant broadening of the ACDs in the orientation $p_z \parallel c$, as expected for a cylindrical Fermi surface. While positron annihilation provides a powerful tool to probe the free charge in the interstitial region between graphite planes, it does not clearly distinguish between charge distribution near the alkali metal atoms (s band character) or in the interstitial region away from the alkali metal atoms (interlayer states). Subsequent to the experiments,[25] calculations of the charge distribution in the interstitial region has emphasized these distinctions.[27]

μ Spin Rotation by Polarized Muon Beams

Muons are produced by accelerators from the decay of pions, which in turn are formed when a high–energy proton beam strikes a pion production target. The reaction for muon formation is then

$$\pi^+(\pi^-) \rightarrow \mu^+(\mu^-) + \nu_\mu(\nu_\mu) \tag{9}$$

where pions decay with a lifetime of 26 ns into a charged muon and a neutral neutrino. In the rest frame of the pion, the muon is 100% spin–polarized either along (μ^-) or opposite (μ^+) to its momentum direction. In metals, even ferromagnets, the muon spin does not precess during the rapid deceleration associated with thermalization so that the implanted muon remains spin polarized.

Finally the polarization of the muon is detected by means of the parity–violating weak decay interaction

$$\mu^+(\mu^-) \rightarrow e^+(e^-) + \overline{\nu}_\mu(\overline{\nu}_\mu) + \nu_e(\nu_e) \tag{10}$$

where the positron (electron) is detected. Measurements of the angular distribution of the $e^\pm$ enable one to determine the polarization direction of the muon prior to the decay.

Positive and negative muons behave differently after thermalization in a solid. Negative muons are trapped in tightly bound atomic states near nuclei (note the Pauli principle does not apply to exclude the added muon from the lowest $1s$ orbit). In

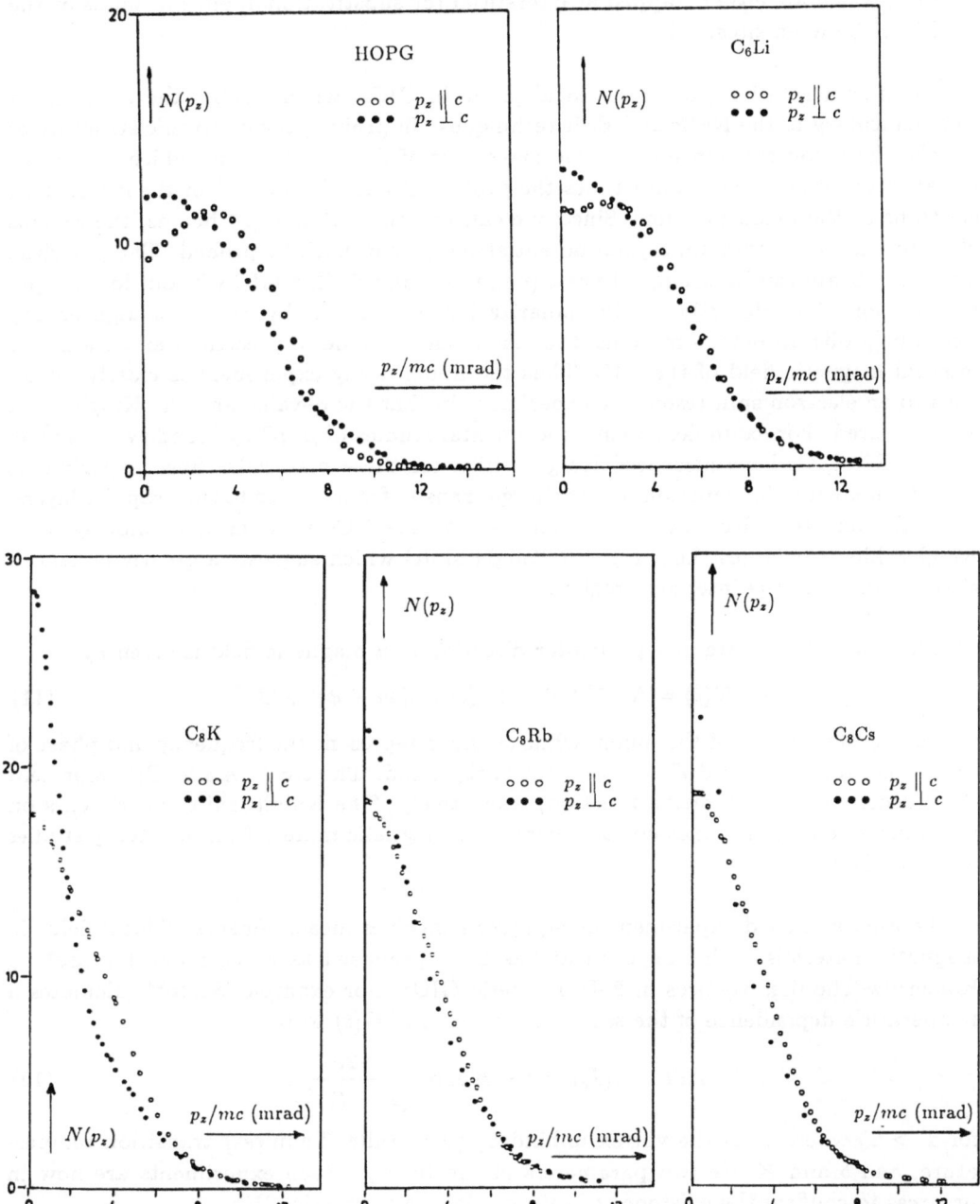

Figure 8: The number of 2γ coincidences *vs.* angle or angular correlation distributions (ACDs) for positron annihilation in HOPG and stage 1 alkali metal–graphite compounds. The results for K, Rb and Cs show a sharp central component for $p_z \perp c$–axis consistent with an in–plane delocalized electron.[25]

graphite, the μ^- binds to a carbon atom to form $\mu^-{:}C$ which behaves chemically like a substitutional boron atom. The vast majority of solid state experiments are carried out with μ^+, which are either trapped at interstitial (or substitutional) sites in solids or the μ^+ diffuse between sites.

Muon spin rotation provides a local probe in GICs, which probes the bulk and is complementary to the NMR and ESR techniques. In probing the electronic structure of metals, the muon resonance technique makes use of the Knight shift, which is proportional to the spin susceptibility times the probability amplitude $|\psi|^2$ of the conduction electrons at the muon position. Since one can use both μ^+ and μ^- beams, the regions of minimum and maximum crystal potential energy can both be probed. The polarized $\mu^\pm$ meson beam can be brought to rest (i.e., thermalized) in a GIC without loss of spin polarization. The detection of the polarization direction is by means of high energy electron (positron) detectors. Thus the muon can be made to precess in an external or internal magnetic field of the GIC. Thus the muon decay experiment is closely analogous to an electron spin resonance experiment in that the g–value and the Knight shift are measured. For example, recent experimental studies of $\mu^-{:}C$ by Kondow et al.[28] on stage 2 Rb–GICs have measured Knight–shifts for these materials. Such experiments provide important information on the charge transfer from the Rb to the graphite layers. The experiments of Kondow et al.[28] on stage 2 Rb–GICs show an anomalously large Knight shift (the largest known μ–SR Knight shift) which suggests a greatly enhanced charge density in the interlayer region.

The muon decay rate in a particular direction in a magnetic field is given by

$$N(t) = N_0 e^{-t/\tau}[1 + AG(t)\cos(\omega t + \phi)] + B \tag{11}$$

where τ is the lifetime of the muon, while ω and ϕ represent the frequency and phase of the muon precession, and B is a constant background. The terms A and $G(t)$ represent the anisotropy and relaxation function, respectively. The Knight shift experiments on non–magnetic materials measure ω, whereas for magnetic materials, muon decay studies focus on $G(t)$.

Another standard experiment using muons involves measurements of local fields in magnetic materials. This experiment has been proposed as a technique for probing Kosterlitz–Thouless vortices in 2–D magnetic GICs. For example, Szeto[29] calculates a temperature dependence of the second moment Δ of $G(t)$ to be

$$\Delta(T)/\Delta(T_c) = 1 - K\exp\left(\frac{-2b}{(T-T_c)^{\frac{1}{2}}}\right) \tag{12}$$

for $T > T_c$ where T_c is the vortex unbinding (Kosterlitz–Thouless) transition temperature, and b and K are two parameters of the theory. Such experiments are now in progress to confirm the existence of spin vortices in magnetic GICs.

Thus μ–spin rotation experiments give all the information about local fields that one can obtain from NMR. They have the great advantage over NMR that all the experiments can be carried out in all materials and one does not have to depend on the existence of a suitable nucleus. The single important disadvantage is that only a few sites in the world have muon beams available.

ACKNOWLEDGMENTS

The authors wish to thank Professors M.S. Dresselhaus and A.P. Legrand for many important discussions. The work at MIT was supported by AFOSR contract #F49620–83–C–0011. ML was supported by CNRS.

REFERENCES

1. C. Rigaux, (this volume) p. 235.

2. P.C. Eklund, (this volume) p. 257.

3. Topics in Applied Physics, Photoemission in Solids I & II, edited by L. Ley and M. Cardona, (Springer–Verlag, Berlin, Heidelberg, New York, 1979) Vol. **26** & **27**.

4. E.W. Plummer and W. Eberhardt, *Advances in Chem. Phys.* **49**, 533 (1982).

5. L.A. Grunes, I.P. Gates, J.J. Ritsko, E.J. Mele, D.P. DiVincenzo, M.E. Preil, and J.E. Fischer, *Phys. Rev.* **B28**, 6681 (1983).

6. D.M. Hwang, N.W. Parker, M. Utlaut, and A.V. Crewe, *Phys. Rev.* **B27**, 1458 (1983).

7. J.E. Fischer, J.M. Bloch, C.C. Shieh, M.E. Preil, and K. Jelly, *Phys. Rev.* **B31**, 4773 (1985).

8. J.J. Ritsko and E.J. Mele, *Phys. Rev.* **B21**, 730 (1980).

9. P. Oelhafen, P. Pfluger, and H.J. Güntherodt, *Solid State Commun.* **32**, 885 (1979).

10. M. Laguës, D. Marchand, C. Frétigny, and A.P. Legrand, *Solid State Commun.* **49**, 739 (1984).

11. M. Laguës, D. Marchand, and C. Frétigny, *Annales de Physique* **11**, 49 (1986).

12. J.H. Scofield, *J. Elec. Spec. and Related Phenomena* **8**, 129 (1976).

13. S.B. DiCenzo, *Synthetic Metals* **12**, 251 (1985).

14. J.J. Lander, *Phys. Rev.* **91**, 1382 (1953).

15. B.I. Dunlap, D.E. Ramaker, and J.S. Murday, *Phys. Rev.* **B25**, 6439 (1982).

16. C. Frétigny, D. Marchand, and M. Laguës, *Phys. Rev.* **B32**, 8462 (1985).

17. S.D. Kevan, *Phys. Rev.* **B33**, 4364 (1986).

18. W. Eberhardt, I.T. McGovern, E.W. Plummer, and J.E. Fischer, *Phys. Rev. Lett.* **44**, 200 (1980).

19. N.A.W. Holzwarth, S.G. Louie, and S. Rabii, *Phys. Rev.* **B29**, 1013 (1983).

20. D. Marchand, C. Frétigny, M. Laguës, F. Batallan, Ch. Simon, I. Rosenman, and R. Pinchaux, *Phys. Rev.* **B30**, 4788 (1984).

21. T. Takahashi, H. Tokailin, T. Sagawa, and H. Suematsu, *Synthetic Metals* **12**, 239 (1985).

22. R.C. Tatar and S.Rabii, *Phys. Rev.* **B25**, 4126 (1982).

23. T. Ohno, K. Nakao, and H. Kamimura, *J. Phys. Soc. Japan* **47**, 1125 (1979).

24. E.J. Mele and J.J. Ritsko, *Phys. Rev. Letters* **43**, 68 (1979); J.J. Ritsko, E.J. Mele, and I.P. Gates, *Phys. Rev.* **B24**, 6114 (1981).

25. E. Cartier, F. Heinrich, U.M. Gubler, P. Pfluger, V. Geiser and H.–J. Güntherodt, *Synthetic Metals* **8**, 119 (1983).

26. S. Berko, private communication.

27. H. Kamimura and S. Rabii, private communication.

28. T. Kondow, Y. Kuno, R. Kadono, J. Imazato, K. Nagamine, and T. Yamazaki, *Physics Letters* **110A**, 319 (1985).

29. K.Y. Szeto, private communication.

TERNARY GRAPHITE INTERCALATION COMPOUNDS

S.A. Solin

Department of Physics and Astronomy
Michigan State University
East Lansing, MI   48824-1116

INTRODUCTION

Ternary graphite intercalation compounds (T GIC's) are materials which contain two distinct intercalant species in the host galleries. The two species may be donor ions such as potassium and rubidium [1], acceptor molecules such as ferric chloride and aluminum chloride [2], or combinations of donor and acceptor species [3]. They may also be microscopically distinct chemical and/or structural phases of the same species [4]. Although T GIC's were first synthesized several decades ago [5], interest in them on the part of physicists, chemists, and physical chemists has only recently been aroused. This awakening is a result of the general realization that T GIC's offer exciting new opportunities for the exploration of physical phenomena in lower dimensions and for practical applications as well.

The field of T GIC's is already too vast to adequately cover in the space of this paper and is the current subject of an extended review article [6]. Therefore, this brief treatise will be limited to a classification of the variety of the distinct structural forms which T GIC's can adopt and to a discussion of some of the novel physical phenomena which are a consequence of those structural forms.

Classification of Ternary Graphite Intercalation Compounds

The three major structural classifications of T GIC's are shown schematically in Fig. 1. Consider first the homogeneous form. The two intercalating species in this form are present in the same gallery, but the two-dimensional structure of the guest layer is disordered. When this disorder is chemical and the hexagonal sites of the graphite layers are randomly occupied by species A or B (either of which could in principle be a vacancy) the label "lattice gas" applies. Such a structure may also be called a 2D solid-solution [7]. In contrast, the 2D structure of the intercalated layers may also be that of a binary (2 component) liquid [8]. The particular planar structure adopted by the guest species will depend on the relative magnitudes of the guest-guest and guest-host interactions.

When the two intercalating species are confined to different galleries in the host structure, or when they form separate layers within the same gallery, the resultant compound is referred to as a heterogeneous T GIC.

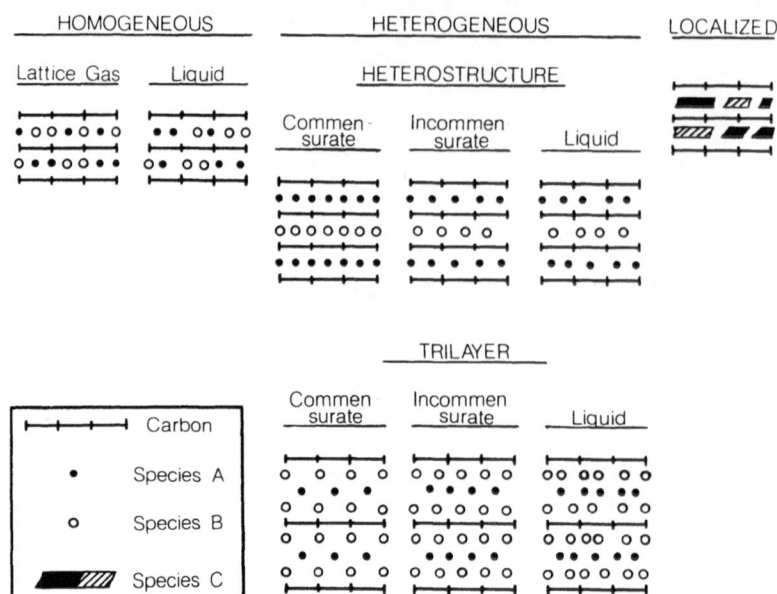

Fig. 1. Classification diagrams for ternary graphite intercalation compounds. Species A, B, and C each represent atoms, ions, or molecules.

This form, which includes both heterostructure and trilayer arrangements, can be further subdivided into three categories labeled respectively as commensurate when the A and B layers are both epitaxial to the bounding carbon layers. The intercalant layers are incommensurate when they "crystallize" into 2D structures, the lattice parameters of which are not a rational fraction of the graphite in-plane lattice constant. And they are liquid when their planar positions do not exhibit long-range order. [Here "liquid" is a generalized nondynamical structural term which also encompasses the frozen disordered form or "glass".]

The third category of T GIC depicted in Fig. 1 is the localized T GIC which as the name and diagram imply is a manifestation of the very high spatial resolution with which physical phenomena can be probed using modern techniques such as analytical electron microscopy [9]. In a localized T GIC, what appears and is often presumed in macroscopic measurements to be a single phase of one species of intercalant may, on microscopic examination, be revealed to be phase-separated structurally and/or chemically (e.g. valence alternation) into two distinct "species". Since the proper interpretation of macroscopically measured physical properties may depend critically on the microstructure of the GIC, the identification of localized T GIC's is an important step in understanding some "binary" compounds.

PHYSICAL PHENOMENA AND STRUCTURE

We now consider some of the unusual physical phenomena exhibited by T GIC's. Where appropriate, the structural origin of these phenomena will be elucidated. Note, however, that in addition to the myriad of new structural variations realizable with T GIC's they offer two distinct additional research advantages over the more commonly studied binary GIC's. First, the dominance of the guest-host interaction over the guest-guest interaction which is apparently endemic to binary GIC's [10,11] does not obtain for some of their ternary counterparts. The T GIC's therefore exhibit some unusual physical properties not found in binary GIC's. Second, the unique staging phenomenon [12], which is the noteworthy virtue of binary GIC's, is also a vice because it severly limits the composition range over which a given physical property can be studied at constant stage. This composition "handle" which is one of the most useful variables at our disposal in solid state physics is often retrieved without the loss of staging when we study T GIC's.

Homogeneous T GIC's

The Lattice Gas Form. One of the best and most heavily studied examples of the lattice gas homogeneous T GIC is the dual alkali stage-1 system $K_{1-x}Rb_xC_8$ [13] which can be synthesized with compositions in the entire range $0 \leq x \leq 1$. The structure of these T GIC's is very similar to that of KC_8, but the K and Rb ions (apparently) randomly occupy the hexagonal sites of the (2 x 2)R 0° in-plane superlattice. To date, there is no direct evidence of either long-range or short-range in-plane chemical order even though the compound $K_{.33}Rb_{.67}C_8$, which may be expected to

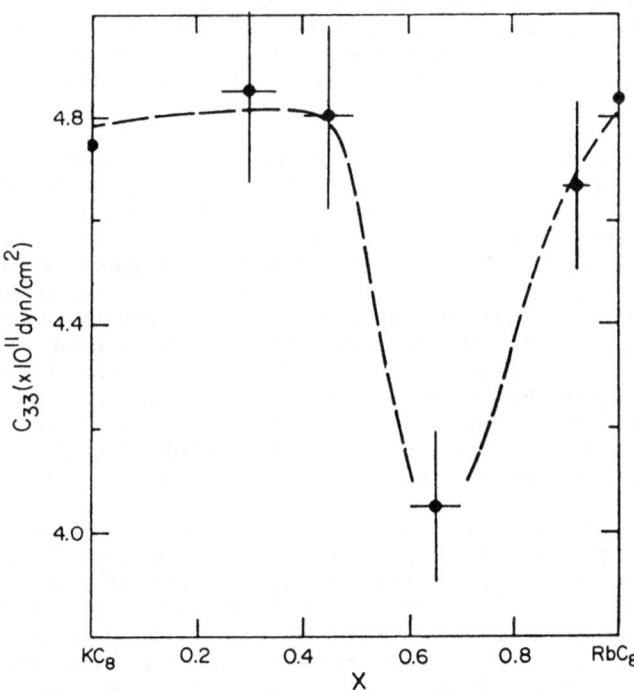

Fig. 2. The composition dependence of the elastic constant C_{33} (solid circles) (from ref. 16) of $K_{1-x}Rb_xC_8$ (the dashed line is a guide to the eye).

exhibit a $(2\sqrt{3} \times 2\sqrt{3})R$ 30° superlattice structure, has been studied with X-rays from room temperature down to liquid nitrogen temperature [14]. Nevertheless, the potassium-rubidium-graphite system exhibits some very striking anomalies in several of its physical properties at the composition corresponding to x = 2/3 [15-17]. An example of such an anomaly is given in Fig. 2 which shows the composition dependence of the C_{33} elastic stiffness constant of $K_{1-x}Rb_xC_8$ [16].

In addition to the dip in the elastic stiffness constant at x = 0.67, the magnetic susceptibility [15] exhibits a pronounced spike and the Raman active M-point phonon frequency [17,18] exhibits a composition-dependent dip at that value of x. Two semiquantitative explanations for the composition anomalies in $K_{1-x}Rb_xC_8$ have been advanced to date. Solin et al. have suggested that the anomalies are a consequence of short-range chemical order associated with the formation of KRb_2 clusters which are statistically numerous at the composition x = 2/3 [17]. But such clusters should be and have not yet been revealed in the above-mentioned X-ray studies [14]. Akera and Kamimura [19] have attributed the x = 2/3 anomalies to unusual features of the electronic band structure of the T GIC. In particular, they associate the softening of the C_{33} elastic constant with a van Hove singularity in the density of states which occurs just at the Fermi level for x ~ 2/3 [19].

The Liquid Form. Although Setton and coworkers [20] and others [21] have investigated the properties of several alkali-molecular T GIC's such as potassium-benzene-graphite which may be expected to exhibit a liquid-like intercalant structure, the only T GIC with a confirmed liquid structure is the alkali-ammonia system [8,22] $K(NH_3)_xC_{24}$, $0 \leq x \leq 1$. This system is perhaps the most heavily investigated of all T GIC's and has been probed by elastic X-ray [22] and neutron [23] diffraction, inelastic and quasielastic neutron scattering [24], C^{13} and proton NMR [25], Raman scattering [26], optical reflectivity [27], and resistivity measurement techniques [28]. In addition, the key features of the ternary Gibbs phase diagram of $K(NH_3)_xC_{24}$ have been established from absorption isotherm measurements [29].

The keen interest in the potassium-ammonia T GIC stems from the recent discovery [8] that the $K-NH_3$ layers in graphite are the 2D structural analog of the bulk 3D metal-ammonia solutions which are famous for the metal-insulator transition which they exhibit [30]. Verification of this similarity is given in the theoretical and experimental X-ray diffraction data of Fig. 3. The solid line in that figure is a one-parameter fit to the in-plane diffuse scattering (dotted line) from the uncorrelated [31] potassium-ammonia layers intercalated into the HOPG host material. The theoretical fit was calculated using the Monte Carlo computer-generated structural model for the 2D liquid shown in the inset of Fig. 3. As can be seen from that inset, the model contains two types of ammonia molecules, those which are symmetrically 4-fold coordinated to a potassium ion and those which are considered dynamically uncoordinated spacers.

The mismatch between experiment and theory in Fig. 3 is a manifestation of the symmetry constraint imposed on the $K-NH_3$ clusters [31]. This mismatch can be significantly reduced if that constraint is relaxed, but only at the expense of the introduction of several additional fitting parameters, the incorporation of which would mask the essential feature of the liquid structure--the planar 4-fold coordination. That coordination is expected on physical grounds by analogy with the 3D potassium-ammonia solutions in which the potassium ions are octahedrally 6-fold coordinated to the surrounding ammonia molecules [30]. But such an

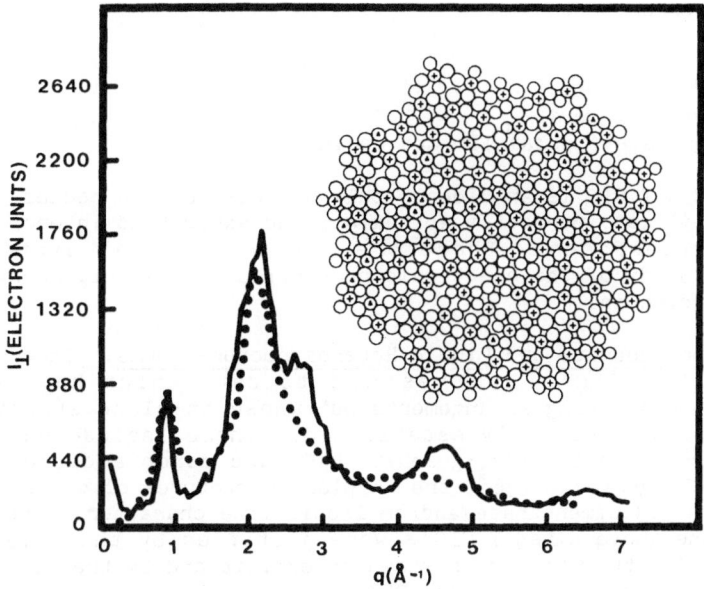

Fig. 3. The calculated (solid line) in-plane X-ray diffraction pattern of the K-NH$_3$ liquid in K(NH$_3$)$_{4.33}$C$_{24}$ scaled at q = 0.88Å to the measured room temperature diffraction pattern (dotted line) which was recorded using MoKα radiation. The calculated pattern was deduced from the 4-fold coordinated distribution shown in the inset which depicts K$^+$ ions (⊕) bound to 4 NH$_3$ molecules (○) and also shows unbound NH$_3$ molecules (◉).

octahedral cluster is too large to fit in the gallery height available in K(NH$_3$)$_x$C$_{24}$ [31]. However, if the opposing apical ammonia molecules are removed from the octahedral cluster, the resultant configuration is a planar 4-fold coordinated potassium-ammonia cluster exactly like those which dominate the liquid structure shown in Fig. 3.

Heterogeneous T GIC's

The Commensurate Heterostructure Form. The only commensurate heterostructure T GIC verified to date is the compound KCsC$_{16}$ which contains alternating intercalant layers of K and Cs which both form a (2 x 2)R 0° superlattice on the graphite hexagonal net. This compound was synthesized by Solin and coworkers [32] by exposing stage-2 CsC$_{24}$ to liquid potassium at a temperature just above its melting point.[24] The resultant compound exhibited long-range in-plane and c-axis correlations with a high degree of structural order [33] as evidenced in part by the sharp (00ℓ) X-ray diffraction pattern shown in Fig. 4. The heterostructure stacking arrangement and in-plane ordering are confirmed respectively by the excellent 2-parameter fit to the structure factor shown in Fig. 4 and by the reciprocal space diagram given in the inset of that figure. Moreover, the basal spacing deduced by indexing the reflections on a ...C K C Cs C K C Cs... stacking arrangement is d$_H$ = 11.27Å where, to within experimental error,

$$d_H = d_{1K} + d_{1Cs}. \qquad (1)$$

Here $d_{1K} = 5.35Å$ and $d_{1Cs} = 5.94Å$ are the basal spacings of the corresponding stage-1 potassium and cesium binary GIC's.

The KCsC$_{16}$ T GIC is an "ideal" bulk heterostructure because its layers are atomically flat, perfectly epitaxial, and exhibit no interdiffusion at the layer interfaces. Therefore, it is an ideal system for the experimental study of the electronic properties of a Kronig-Penny [34] one-dimensional potential.

The Incommensurate and Liquid Heterostructure Forms. There have been a number of T GIC's synthesized to date which have confirmed heterostructure stacking arrangements but whose in-plane structures have not yet been definitively established. These include dual acceptor compounds such as CoCl$_2$-FeCl$_3$-graphite [35] and donor acceptor compounds such as K-CoCl$_2$-graphite [36], the in-plane structures of which are likely to possess an incommensurate and/or liquid-like character. The efforts to synthesize the above cited T GIC's were justified by the novel magnetic properties which they have been shown to exhibit and by the rich variety of their magnetic phase transitions [35].

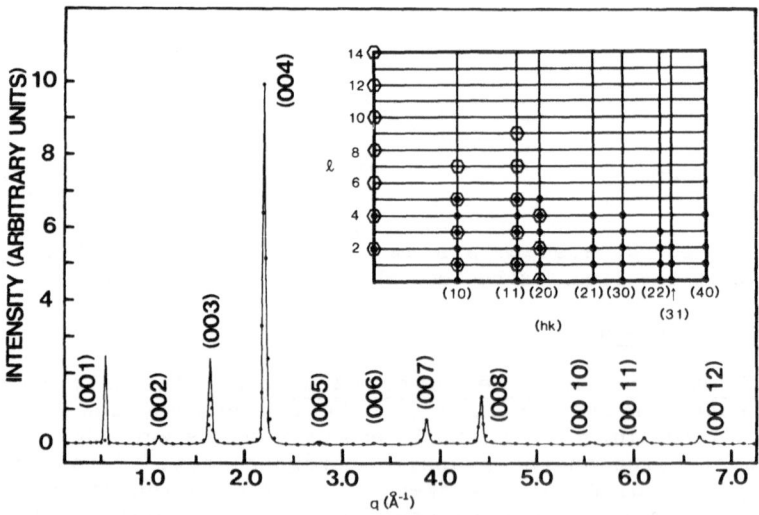

Fig. 4. The (00ℓ) X-ray diffraction pattern of the KCsC$_{16}$ heterostructure T GIC. Background scattering from the encapsulating glass envelope and peaks due to KC$_8$ have been removed for clarity. The dotted trace is the experimental result while the solid line is a theoretical fit to the data using the (004) reflection as the intensity normalizing peak. Inset: reciprocal space diagram of all of the (h k l) heterostructure diffraction peaks observed. (●) represents reflections observed while scanning along a constant level (h,k,l = const.), and (○) along a constant row (h = const., k = const., l).

The Trilayer Heterogeneous Forms. The trilayer heterogeneous forms are the second example of a class of T GIC's which have been very heavily studied. In these compounds a layer of, say, B species is sandwiched between two layers of a species and this triad is intercalated as a unit into the graphite galleries to form T GIC's of various stages. An example of one of the best characterized trilayer T GIC's is the stage-1 compound $KHgC_4$ [37] the in-plane and sandwich structures of which are shown in Fig. 5. Note that $KHgC_4$ is a commensurate trilayer T GIC because the intercalated species adopt an epitaxial arrangement with respect of the graphite host layers [37]. A similar structure is apparently formed by the newly synthesized stage-1 compound $CsBi_{0.55}C_4$ [38] which has aroused much interest because it exhibits a relatively high superconducting transition temperature of ~2K, it can be prepared in higher stages and it appears to be air-stable [38]. Another very heavily studied T GIC which exhibits a commensurate trilayer structure is K-H-graphite [7]. In contrast, the sodium-barium T GIC's form an incommensurate trilayer stage-2 structure in which a layer of barium is sandwiched between two layers of sodium in a crystalline arrangement with an in-plane lattice parameter that is not integrally related to the in-plane lattice parameter of graphite [7]. Finally, definitive examples of the trilayer liquid form have not yet been identified and this form is included in Fig. 1 for completeness only.

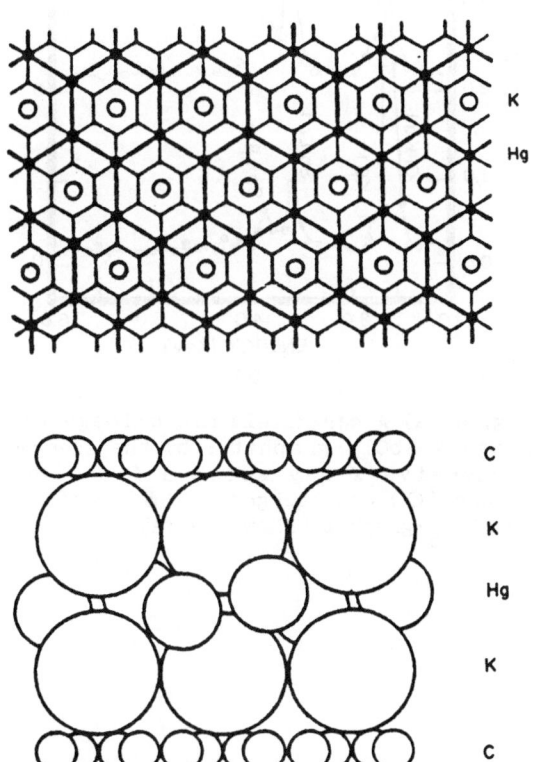

Fig. 5. The structure of the metallic layers in $KHgC_4$ (from ref. 37).

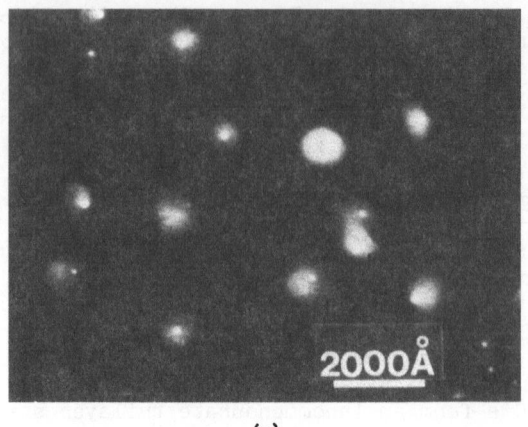

(a)

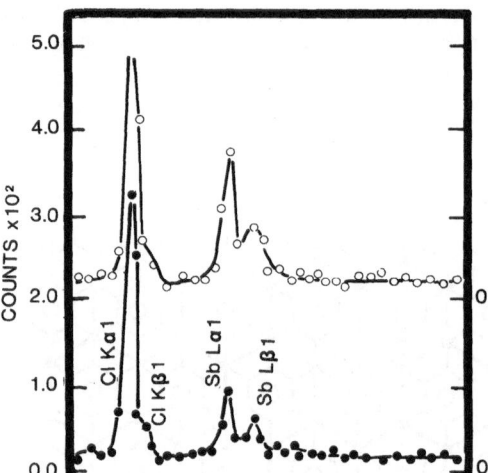

(b)

Fig. 6. a) A dark-field micro-image of stage-4 $SbCl_5$-graphite; b) the energy dispersive X-ray spectra from the island (O) and background (●) regions in $SbCl_5$-graphite shown in (a).

Localized T GIC's

The identification of the new class of localized T GIC's is so recent that there is to date but one definitive member of this class; namely, antimony pentachloride-graphite [4]. Fig. 6a shows a dark-field image of stage-4 $SbCl_5$ graphite taken with a field-emission scanning transmission electron microscope using a beam focal spot size of ~50A. Islands of enhanced density superposed on a continuous background sea can be clearly discerned in the figure. Figure 6b shows the energy dispersive X-ray spectra of the island and background regions and clearly shows that while the Cl content is essentially uniform across the exposed region, the Sb content in the island regions is roughly double that in the background [4,40]. The actual compositions of the island and background regions were found from the measurements shown in Fig. 6 to be $SbCl_{3.1\pm0.3}$ and $SbCl_{7.0\pm2.0}$, respectively. It is this result which characterizes $SbCl_5$-graphite as a localized T GIC for the chemical forms of antimony-chloride detected are quite distinct.

CONCLUDING REMARKS

The study of T GIC's is truly in its infancy. New ternary compounds are now being synthesized and the novel phenomena which they exhibit are the subjects of several ongoing investigations. Moreover, although I have focused here on stage-1 compounds, higher stage T GIC's [37] offer the challenging opportunity to couple one-dimensional c-axis effects with an enriched variety of 2D in-plane structural effects.

ACKNOWLEDGEMENTS

Beneficial interactions with D.M. Hwang, D.R. Stump, S.D. Mahanti, B.R. York, S.K. Hark, H. Zabel, P. Chow, Y.B. Fan, X.W. Qian, and Y.Y. Huang are gratefully acknowledged. This work was supported by the U.S. National Science Foundation under grants DMR 82-11554 and DMR 85-17223.

REFERENCES

1. P. Chow and H. Zabel: Syn. Metals 7, 243 (1983)
2. E. Stumpp: Physica B+C 105, 253 (1981)
3. M. Suzuki, P.C. Chow, and H. Zabel: Phys. Rev. B32, 6800 (1985)
4. D.M. Hwang, X.W. Qian, and S.A. Solin: Phys. Rev. Lett. 53, 1473 (1984)
5. W. Rudörff and E. Schultze: Angew. Chem. 66, 305 (1954)
6. S.A. Solin and H. Zabel: to be published
7. A. Hérold: in The Physics of Intercalation Compounds, ed. by L. Pietronero and E. Tosatti (Springer, Berlin, 1981), p. 7
8. X.W. Qian, D.R. Stump, B.R. York, and S.A. Solin: Phys. Rev. Lett. 54, 1271 (1985)
9. Introduction to Analytical Electron Microscopy, ed. by J.J. Hren, J.I. Goldstein and D.C. Joy (Plenum, New York, 1979); also D. Cherns, A. Howie and M.H. Jacobs: Z. Naturforsch 28A, 565 (1973), and P. Duncumb: Phil. Mag. 7, 2101 (1962)
10. H. Homma and R. Clarke: Phys. Rev. B31, 5865 (1985)
11. H. Homma and R. Clarke: Phys. Rev. Lett. 52, 629 (1984)
12. S.A. Solin: Adv. Chem. Phys. 49, 455 (1982)
13. See ref. 1 and references therein
14. H. Zabel: private communication
15. G. Furdin, B. Carbon, A. Hérold, and C. Zeller: C.R. Acad. Sc. Paris, Ser. B 280, 653 (1975)
16. D.A. Neumann, H. Zabel, J.J. Rush, and N. Berk: Phys. Rev. Lett. 53, 56 (1984)
17. S.A. Solin, P. Chow, and H. Zabel: Phys. Rev. Lett. 53, 1927 (1984)
18. See p. 315 of these proceedings
19. H. Akera and H. Kamimura: Syn. Metals 12, 275 (1985)
20. R. Setton, F. Beguin, J. Jegondez, and C. Mazieres: Rev. Chim. Miner. 19, 360 (1982)
21. F. Beguin, B. Gonzalez, J. Conard, and H. Estrade-Szwarckopf: Syn. Metals 12, 187 (1985)

22. B.R. York and S.A. Solin: Phys. Rev. B$\underline{31}$, 8206 (1985)
23. Y.B. Fan, S.A. Solin, D. Neumann, and H. Zabel: to be published
24. D. Neumann, H. Zabel, Y.B. Fan, and S.A. Solin: to be published
25. H.A. Resing, B.R. York, S.A. Solin, and R.M. Fronko: Proc. of the 17th Carbon Conference, Louisville (1985), p. 192
26. P. Vora, B.R. York, and S.A. Solin: Syn. Metals $\underline{7}$, 355 (1983)
27. D.M. Hoffman, A.M. Rao, G.L. Doll, P.C. Eklund, B.R. York, and S.A. Solin: Mats. Res. Society Bull., in press
28. Y.Y. Huang, D.R. Stump, S.A. Solin, and J. Heremans: to be published
29. S.A. Solin and Y.B. Fan: Proc. of the 17th Carbon Conference, Lexington (1985), p. 190
30. J.C. Thompson: Electrons in Liquid Ammonia (Clarendon, Oxford, 1976)
31. X.W. Qian, D.R. Stump, and S.A. Solin: Phys. Rev., in press
32. B.R. York, S.K. Hark, and S.A. Solin: Phys. Rev. Lett. $\underline{50}$, 1470 (1983)
33. S.K. Hark, B.R. York, and S.A. Solin: J. Chem. Phys. $\underline{82}$, 921 (1985)
34. G.H. Wannier: Elements of Solid State Theory (Cambridge, London, 1960) p. 135
35. M. Suzuki, I. Oguro, and Y. Jinzaki: J. Phys. C $\underline{17}$, L575 (1984)
36. See ref. 3 and references therein
37. A. Hérold, D. Billaud, D. Guérard, P. Lagrange, and M. El Makrini: Physica $\underline{105B}$, 253 (1981)
38. P. Lagrange, A. Bendriss-Rerhrhaye, J.F. Mareche, and E. McRae: Syn. Metals $\underline{12}$, 201 (1985)
39. M. Colin and A. Hérold: Bull. Soc. Chim. Fr. 1982 (1971)
40. D.M. Hwang, R. Levi-Setti, G. Crow, Y.L. Wang, N.W. Parker, R. Mittleman, X.W. Qian, and S.A. Solin: Syn. Metals $\underline{12}$, 73 (1985)

THE STRUCTURE OF TERNARY ALKALI

GRAPHITE INTERCALATION COMPOUNDS

P. C. Chow and H. Zabel

Loomis Lab of Physics
University of Illinois
Urbana, Il. 61801 USA

We have studied the structure of $Rb_xK_{1-x}C_8$ with x-rays. The major motivation of the work is to investigate the correlation in the positions of the alkali metal atoms in between the carbon layers. This is interesting because anomalies in the physical properties (magnetic susceptibility[1], Raman-active M-point phonon frequency[2], and elastic constant C_{33}[3]) have been measured for x=0.67. Two different explanations have been proposed to account for these anomalies. Solin[2] postulated that short range order of the alkali metal atoms can account for the Raman data. Akera[4] argues that the Van Hove singularity in the density of states at the Fermi level gives rise to the anomalies in the data. X-ray evidence for short range order will help distinguish between the two models.

By sequential intercalation of K and Rb into highly oriented pyrolytic graphite (HOPG), we have produced $Rb_xK_{1-x}C_8$, $0 < x < 1$. The composition of the samples was determined by comparing the measured ratio of (00L) intensities to the calculated intensities.[5] The d-spacing was found from Bragg's law. The resulting d-spacing versus composition curve (see Fig. 1) deviates from a linear relationship and indicates that the carbon layers are buckled in the ternary compound.

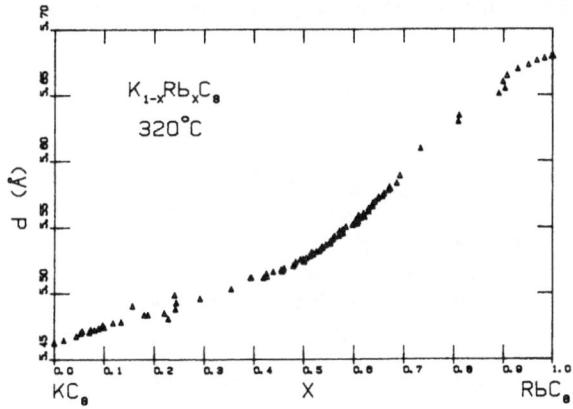

Figure 1. The d-spacing versus composition curve was found from an in-situ x-ray diffraction experiment.

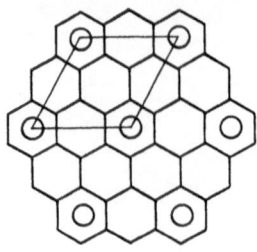

Figure 2. The in-plane 2x2 alkali
metal lattice. αβγδ is the stacking sequence for the alkali layers
in the ĉ-direction.

The in-plane structure of the alkali metal atoms in the ternary compound is of considerable interest. At room temperature, x-ray (HKL) scans
confirm that the overall alkali metal lattice is 2x2 with αβγδ stacking
(see Fig. 2), as one would expect, because both KC_8 and RbC_8 have this
same structure.

As the sample is cooled, there are at least three possibilities for the
alkali metal structure, depending on the relative magnitudes of the alkali/
alkali and alkali/graphite interactions. 1) The K and Rb may form an ordered
structure, as shown in figure 3 for x=0.33. In this case, x-ray scans should
show superstructure peaks, corresponding to the extra periodicity of the
alkali metal lattice. We have calculated the diffraction pattern of ordered
$Rb_{.33}K_{.66}C_8$ shown in Fig. 3. The intensities of the superstructure peaks
are orders of magnitude smaller than that of the overall 2x2 alkali lattice.
2) If the alkali atoms prefer like-neighbors, they could phase separate into
regions of K atoms and regions of Rb atoms on a 2x2 lattice. In this case,
we would see a superposition of two sets of (00L) peaks in the diffraction
pattern. 3) It is also possible that the difference in the K and Rb interactions is very small, requiring the sample to be cooled to a very low temperature. At such a low temperature, the diffusion of the alkali metal atoms
may be kineticlly hindered, so no correlations in positions can be detected.

We have completed scans in selected areas of reciprocal space for a
range of temperatures, looking for diffuse scattering, which would indicate
correlations in the alkali metal positions in $Rb_{.33}K_{.66}C_8$. To date, we have
no indication for either superstructure, phase separation, or short range
order.

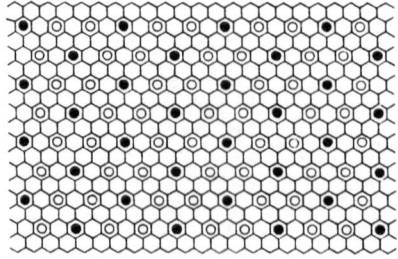

Figure 3. A possible ordered
superstructure for $Rb_{.33}K_{.67}C_8$
●...Rb, o...K

Because of the large background, due to the Rb fluorescence, it is difficult to see and diffuse scattering. There are several changes in our experimental technique that would improve our chances of finding evidence for correlations. Single crystals hosts, instead of HOPG, would concentrate the diffracted intensity in reciprocal space, making it easier to detect diffuse scattering. Laue photographs, instead of diffractometer scans, allow for quicker data collection. These experiments are in progress.

REFERENCES
1. G. Furdin, B. Carbon, A. Herold and C. Zeller: C. R. Acad. Sc. Paris, Ser. B. 280,653 (1975)
2. S. A. Solin, P. Chow, and H. Zabel, Phys. Rev. Lett. $\underline{53}$ 1927 (1984)
3. D. A. Neuman, H. Zabel, J. J. Rush, and N. Berker, Phys. Rev. Lett. $\underline{53}$ 56 (1984)
4. H. Akera and H. Kamimura, Syn. Metal $\underline{12}$ 275 (1985)
5. P. C. Chow and H. Zabel, Syn. Metals $\underline{7}$, 243 (1983)

IN-PLANE ELASTIC NEUTRON SCATTERING FROM THE TERNARY GRAPHITE INTERCALATION COMPOUND $K(ND_3)_{4.38}C_{24}$

Y.B. Fan and S.A. Solin

Department of Physics and Astronomy
Michigan State University
East Lansing, MI   48824-1116

D. Neumann* and H. Zabel

Department of Physics
University of Illinois
Urbana, IL   61801

J.J. Rush

Institute for Materials Science and Engineering
National Bureau of Standards
Gaithersburg, MD   20899

In-plane elastic neutron scattering studies of $K(ND_3)_xC_{24}$ were carried out to further test and refine the x-ray derived structural model [1] of the 2D K-NH$_3$ liquid intercalant layers. Additional structural information was obtained because the neutron and x-ray scattering processes emphasize the different pair correlation functions of the K-ND$_3$ disordered intercalant system.

In our experiment, 30 pieces of pyrolytic graphite of dimension 1.5 x 1.0 x 0.5 cm^3 were used to form binary stage-2 KC_{24} GIC's using the standard 2-bulb method. These peices were then stacked with their c-axes well aligned into an aluminum sample can in a glove box. The aluminum can was connected to a gas handling system for the ammoniation of KC_{24} [2] and the stage-1 $K(ND_3)_{4.3}C_{24}$ was obtained by exposure of the KC_{24} to the vapor of ND$_3$ at room temperature. All of our measurements were carried out at room temperature with incident neutrons monochromated to an energy of 35meV. The in-plane structure factor shown as a solid line in Fig. 1(b) results from removing the Bragg peaks generated by the graphite host and the sample can. The transmission of the sample was estimated to be in excess of 95% making absorption and multiple scattering corrections unnecessary. The Placzak corrections and the fast neutron background were also found to be negligible. Besides the in-plane diffuse scattering, c* scans along (2.0,0,ℓ) were taken to ascertain that no c-axis correlation existed between the intercalant layers.

In the computer generated 2D cluster model shown in Fig. 1(a), each atom or molecule was represented by a hard disc. 2165 "ammonia" disc and 500 "K" disc were used to form the computer-generated structure which

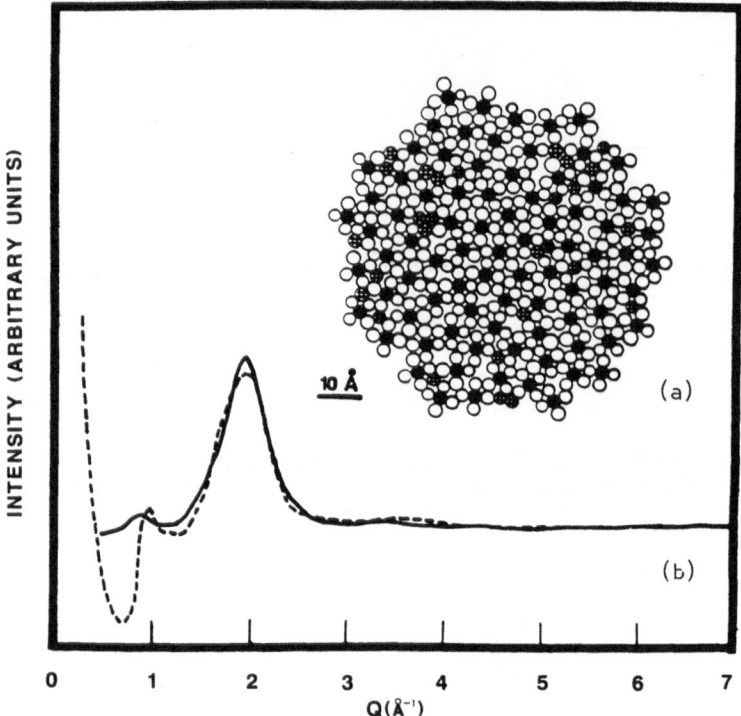

Fig. 1. (a) The computer-generated 4-fold coordinated cluster model of the structure of $K(ND_3)_{4.3}C_{24}$. The symbols have the following meaning ($\bullet$): K, $\bigcirc$: bound ND_3, $\oplus$: unbound (ND_3). (b) The calculated (-----) and measured (———) in-plane neutron scattering structure factor of $K(ND_3)_{4.3}C_{24}$. Bragg reflections from the graphite host and the aluminum sample can have been removed for clarity.

contains 500 4-fold coordinated symmetric $K-ND_3$ clusters and 165 unbound ammonia molecules (detailed information can be found elsewhere [1,2]).

The calculated neutron scattering structure factor for this model is shown as a dashed line in Fig. 1(b) and compared with the experimental data (solid line). The calculated result was deduced using three fitting parameters, the intramolecular radius of ND_3, a constant background and a scale factor. As can be seen from Fig. 1(b), the fit is excellent in the range $Q > 0.8$, but has serious discrepancies at low values of Q.

Considering the Bessel-function back-transforms (BFBT) of the data and calculation of Fig. 1(b), these transforms are shown in Fig. 2. Since the ammonia contribution dominates the neutron scattering, the BFBT of the experimental data (solid line) is approximately the ammonia radial density distribution. The dashed line in Fig. 2 is the corresponding function deduced from the cluster structure model. Clearly, the dashed line exhibits more sharp structure, the envelope of which is similar to the solid line. This indicates that in the real $K-ND_3$ 2D liquid the correlations between molecules are not as strong as indicated by the model. The strong correlation or the high frequency component is a characteristic of the symmetric cluster while the low frequency component (the envelope) results from the 4-fold clustering and is independent of the angular

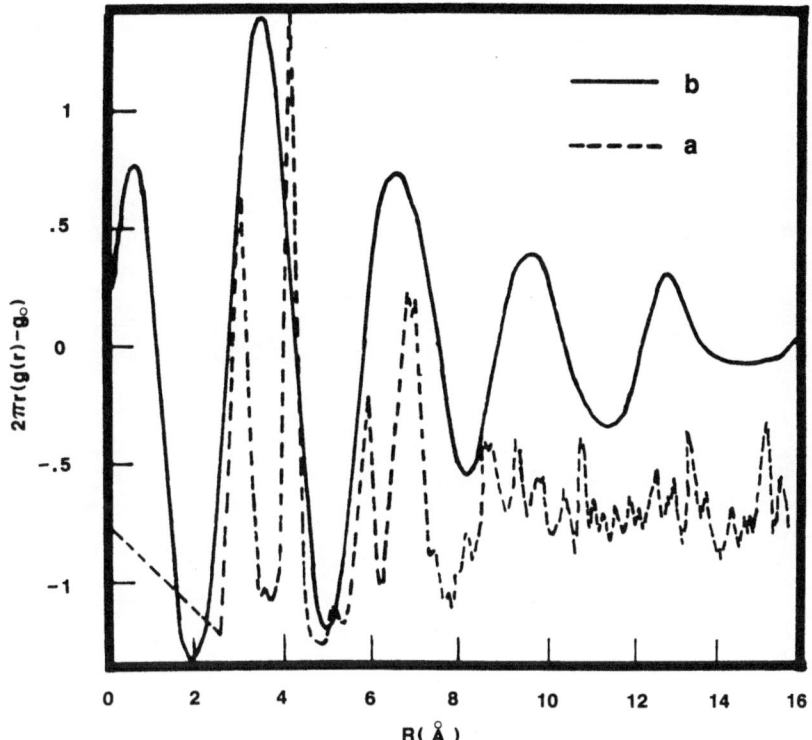

Fig. 2. The Bessel function back-transforms of the measured (b, ———) and
calculated (a, -----) structure factors shown in Fig. 2(b).

symmetry in the cluster. From the BFBT, it is easy to show that the high
frequency component contributes strongly to the low Q discrepancy while
very weakly to the other values of Q.

In conclusion, the 4-fold cluster model contains the essential
features of the structure of the 2D $K-ND_3$ liquid in GIC's, while the real
liquid exhibits a much lower degree of rotational symmetry of the $K-ND_3$
clusters.

ACKNOWLEDGEMENTS

We thank X.W. Qian, D.R. Stump, and S.D. Mahanti for many valuable
discussions. This work was supported by the U.S. National Science
Foundation under grants DMR 85-17223 and DMR 82-11554 (SAS) and grant
DMR 86-29733 (HZ).

*AT&T Bell Laboratories Scholar.

REFERENCES

1. X.W. Qian, D.R. Stump, and S.A. Solin, Phys. Rev. B33, 5756 (1986).
2. Y.B. Fan, S.A. Solin, D.A. Neumann, H. Zabel, and J.J. Rush, to be
   published.
3. R. Clarke, N. Caswell, and S.A. Solin, Phys. Rev. Lett. 42, 61 (1979).

ELASTIC NEUTRON SCATTERING STUDIES OF H_2 AND D_2 PHYSISORBED STAGE 2

GRAPHITE-POTASSIUM

G.L. Doll*, P.C. Eklund*, and G. Senatore+

*University of Kentucky, Lexington, Kentucky,
+Università de Trieste, Trieste, Italy

The stage 2 MC_{24} (M=K,Rb,Cs) compounds exhibit low temperature physisorption for a variety of gases (H_2,CH_4,N_2,etc.) [1]. Physisorption of gases such as these has also been observed in the Cs-intercalated layer-chalcogenide $Cs-ZrS_2$ [2]. A particularly striking feature of hydrogen physisorption occuring in the GIC systems is that hydrogen isotopes are physisorbed more readily than hydrogen itself [3,4,5]. In K-graphite, to accomodate the adsorbate species within K-intercalant layers, the interlayer separation between carbon bounding layers expands by $\sim.30\text{Å}$. This is in contrast to stage 2 Rb- and and Cs-graphite where no expansion is observed due to the larger diameter of the Cs^+ and Rb^+ ions.

We have performed in situ elastic neutron scattering measurements during the 77K physisorption of $H_2(D_2)$ to study the c-axis lattice expansion of $C_{24}K(G)_x$ ($G=H_2,D_2$) as a function of adsorbate concentration (x). Powdered graphite (Fischer Scientific) was used as the host to first make $KC_{24+\delta}$ by the two-temperature method [6]. δ was determined by measuring the weight uptake of K in a helium glovebox. The powder was then loaded into a thin-walled aluminum can heat sunk (T=77K) to the cold plate of the dewar shown in Fig. 9a in ref. 6. $H_2(D_2)$ was introduced to the sample via the gas-handling system also illustrated in this figure. Neutron scattering measurements were performed at the High Flux Isotope Reactor facility located at Oak Ridge National Laboratory (Oak Ridge, Tennessee USA) employing a triple-axis spectrometer in an elastic scattering configuration.

Values for the c-axis repeat distance (I_c), derived from a least square fit to the 00ℓ (ℓ=2-4) diffraction peaks were obtained for well-staged $C_{24}K$-powder samples. The expansion ΔI_c resulting from D_2 and H_2 adsorption is plotted vs. H_2 or D_2 concentration x, where x is the $H_2(D_2)$/K concentration ratio in Fig. 1. This work is the first to study the expansion as a function of x. Previous studies have reported values for the maximum expansion $\Delta I_c(x_{sat})$, where x_{sat} is the maximum $H_2(D_2)$ uptake. Differences between D_2 and H_2 expansion curves in the figure reveal a large isotope effect. The shaded and open circles refer to the D_2 and H_2 data, respectively.

The initial slope of the D_2 data is seen to be much steeper than the H_2 data. We have fit the experimental results to a simple two-parameter phenomenological function:

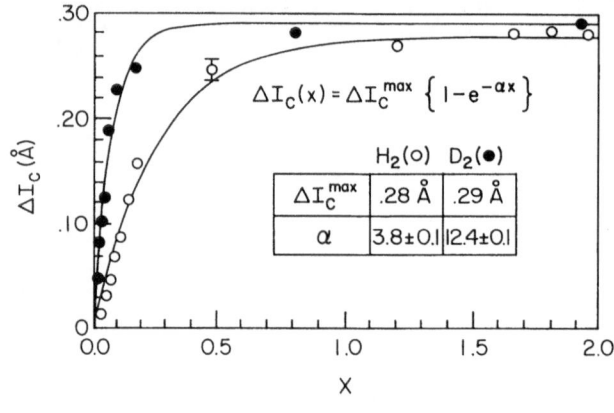

$$\Delta I_C(x) = \Delta I_C^{max} \left\{ 1 - e^{-\alpha x} \right\}$$

	$H_2(\circ)$	$D_2(\bullet)$
ΔI_C^{max}	.28 Å	.29 Å
α	3.8±0.1	12.4±0.1

Fig. 1

c-axis lattice expansion (ΔI_C) vs. adsorbate ($G = H_2, D_2$) concentration (x) in $C_{24.3}K(G)_x$-powder at T=77K. The solid lines are the result of fits of eq. (1) to the data. Parameters obtained from the fit are displayed in the inset. From ref. 7.

$$\Delta I_C(x) = \Delta I_C^{max} \left(1 - e^{-\alpha x}\right) \qquad (1)$$

These parameters ($\Delta I_C^{max}, \alpha$) were adjusted to yield the fits (solid lines) shown in the figure. The maximum expansion value ΔI_C^{max} for D_2 was found to slightly exceed that of H_2. Typical error bars are shown. The exponent α for D_2 is found to be ~3 times larger than that of H_2, consistent with the much steeper initial rise of the D_2 curve. As can be seen in the figure, H_2 and D_2 expansion data are fit quite well by the simple 2-parameter function. At this time we do not have a model which can produce such a function (eq. 1).

The data show that the majority of the expansion takes place for concentrations of x significantly less than the saturation value x_{sat} ~2. This suggests that a multi-site model for occupation of physisorbed H_2 and D_2 should be considered. For simplicity we propose the following 2-site model, with low and high energy sites S_1 and S_2, respectively. We propose that eq. (1) represents the contribution to the c-axis expansion from the low energy site S_1 and that the high energy site S_2 makes only a small contribution. ΔI_C is ~90% of full expansion at x ~.3 and ~.7 for D_2 and H_2 physisorption, respectively. We take this as qualitative evidence for ~25% of the sites to be associated with the low energy site (i.e. S_1 is nearly saturated at x ~.5). If the energy of sorption is dominated by the G-K interaction, then the low energy sites S_1 would occur at positions in the intercalate layer furthest removed from the K sites. This proposal is similar to that made in the study of H_2 physisorption in Cs-ZrS_2 [2].

The occupation probability for site S_1 is proposed to be isotope-dependent and responsible for the different values found for exponents (α). Low temperature (T~10K) studies of monolayers of gases physisorbed onto graphite substrates found that H_2 has a higher surface mobility than D_2 [8]. D_2 molecules become localized at lower concentrations than H_2 molecules and the ($\sqrt{3} \times \sqrt{3}$) adsorbate superlattice has been reported to appear at lower coverages for D_2 than H_2 physisorption. Adsorbate-adsorbate interaction at higher coverages eventually localizes H_2 molecules. The differences in mobility are associated with the differences in the zero-point internal vibration energy and its effect upon isosteric heats of sorption [3]. We suggest that the interactions governing the occupation of the low energy site S_1 are primarily responsible for the observed isotope effect in both the isotherms (Pressure vs. x at T=77K) and c-axis expansion of $\Delta I_c(x)$.

Acknowledgements

We are grateful to Dr. J.L. Zarestky (Ames Laboratory) and Dr. R.M. Nicklow (ORNL) for their assistance in this study. This work is supported in part by the U.S. Department of Energy (DE-FGO5-84-ER45151).

References

1. K. Watanabe, M. Soma, T. Onishi, and K. Tamaru, Nature 233, 160 (1971).
2. L. Christiany, H. Fuzellier, P. Lagrange and A. Herold, Physica 99B, 477 (1980).
3. K. Watanabe, T. Kondow, M. Soma, T. Onishi and K. Tamaru, Proc. Royal. Soc. Lond. A. 333, 51 (1973).
4. P. Lagrange, M. Portmann and A. Herold, C. R. Acad. Sc. Paris, 283, 511 (1976).
5. T. Terai and Y. Takohaski, Synth. Met. 7, 49 (1983).
6. P.C. Eklund, Synthesis of Graphite Intercalation Compounds, this volume, p. 163.
7. G.L. Doll, P.C. Eklund, and G. Senatore, in preparation.
8. M. Nielsen, J.P. McTague and W. Ellenson, J. Phys. Colloq. C4, 10 (1977).

PHONON PROPERTIES OF GRAPHITE INTERCALATION COMPOUNDS

S.A. Solin

Department of Physics and Astronomy
Michigan State University
East Lansing, MI   48824-1116

INTRODUCTION

Graphite intercalation compounds (GIC's) exhibit a rich and varied range of vibrational excitations or phonons which have been extensively studied during the past decade and heavily reviewed [1-3]. Several techniques have been employed to ascertain the phonon properties of GIC's. These include indirect phonon probes such as specific heat measurements [4] and electrical transport studies [5] as well as primary direct probes such as inelastic neutron diffraction [6], Raman scattering [2], optical reflection/absorption [7], and secondary direct probes such as acoustic attenuation measurements [8]. Given the space limitations for this paper, I feel that is appropriate to focus here on the primary direct phonon probes which are the main source of our current understanding of vibrations in GIC's. The purpose of this paper is to set the groundwork for an understanding of phonons in GIC's and apprise the reader of some of the latest exciting developments in that field.

HOST EXCITATIONS

Prerequisite to an understanding of the phonon modes of GIC's is a thorough awareness of the phonons in the host graphite itself. The phonon properties of graphite have long been the subject of experimental and theoretical interest. In addition to phonon studies of graphite employing all conceivable experimental techniques [9,10], there have been many theoretial studies of its phonon dispersion curves [11-15]. Some of these theoretical studies yielded mutually exclusive results, but nevertheless claimed to provide excellent fits to the experimental data. This anomalous situation resulted from the propensity of the early theoretical efforts to gauge their success by the quality of fit to the available low frequency neutron diffraction measurements of the phonon dispersion curves [16] and in more ambitious efforts to calculate the temperature dependent specific heat as well.

Two recently discovered key features of the phonon dispersion curves of graphite provided sufficient additional experimental constraints to distinguish valid theories from those which were inadequate. First, it was learned from second-order Raman scattering studies of HOPG that the highest single phonon energy was not at the Brillouin zone-center Γ point as is usual for most crystals, but was on the Σ line near the midpoint between Γ

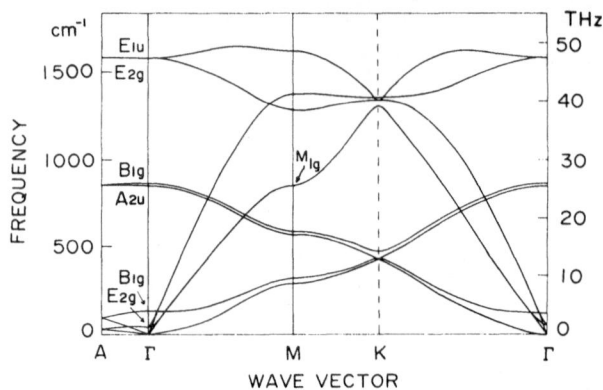

Fig. 1. The phonon dispersion curves of graphite in the [001], [100], and [110] directions (from ref. 14).

and M [17]. Thus, an essential feature of the theoretical dispersion curves must be a rise along the Σ line away from the Γ point. Second, careful infrared reflectivity measurements [18] fixed the zone center A_{2u} out-of-plane optic mode energy at $868 \pm 1 cm^{-1}$, whereas the calculated values of this mode energy had ranged from $600 cm^{-1}$ to $1400 cm^{-1}$ [11-15].

The first theory to properly account for these new experimental facts and in addition provide a good fit to the low energy neutron data was that of Maeda et al. [14] who used a Born-von Karman force constant model to calculate the phonon dispersion curves of graphite which are shown in Fig. 1. Although the A_{2u} mode energy was an input parameter to the calculation, the required and very pronounced rise along the highest Σ branch between Γ and M was a result. Notice also from Fig. 1 that the flatness of the phonon dispersion curves at the M point will give rise to a high vibrational density of states at that point as well as at the Γ point. This observation will be relevant to subsequent discussion.

The labels of the zone center Γ point phonons in Fig. 1 are based on a group theoretical analysis of the hexagonal structure of pristine graphite which crystallizes in the D_{6h}^4 space group and contains 4 atoms in its biplanar primitive cell. The 12 zone-center vibrations transform according to the following irreducible representations of the D_{6h} point group [19]:

$$\Gamma_{vib} = 2A_{2u} + 2B_{1g} + 2E_{1u} + 2E_{2g} \ . \tag{1}$$

Of these, three are acoustic modes ($A_{2u} + E_{1u}$), three are infrared active ($A_{2u} + E_{1u}$), four are Raman active ($2E_{2g}$) and two are silent ($2B_{1g}$). To date, all of the optically active Raman and IR modes including the heretofore elusive A_{2u} mode and the low frequency rigid-layer shear mode have been detected experimentally as shown in Fig. 2 which also gives the Eigen-vectors for those modes.

VIBRATIONS OF BINARY GRAPHITE INTERCALATION COMPOUNDS

Optical Spectra

Although IR measurements [7] of GIC's have yielded important contributions to our understanding of their zone-center vibrational

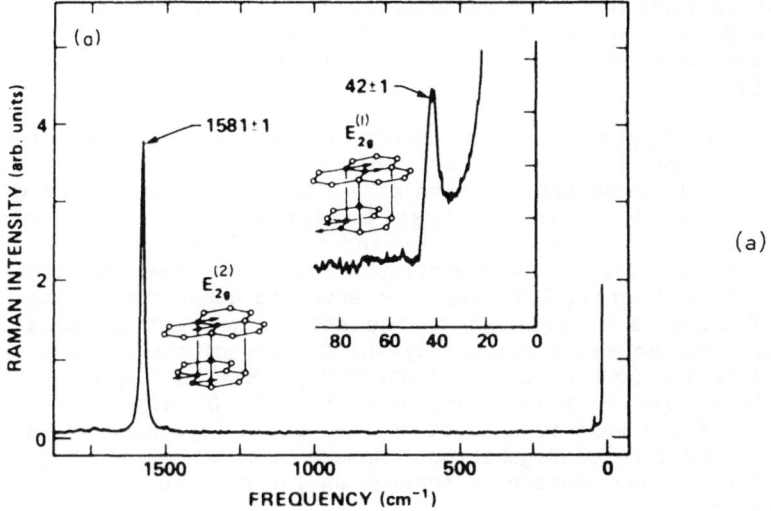

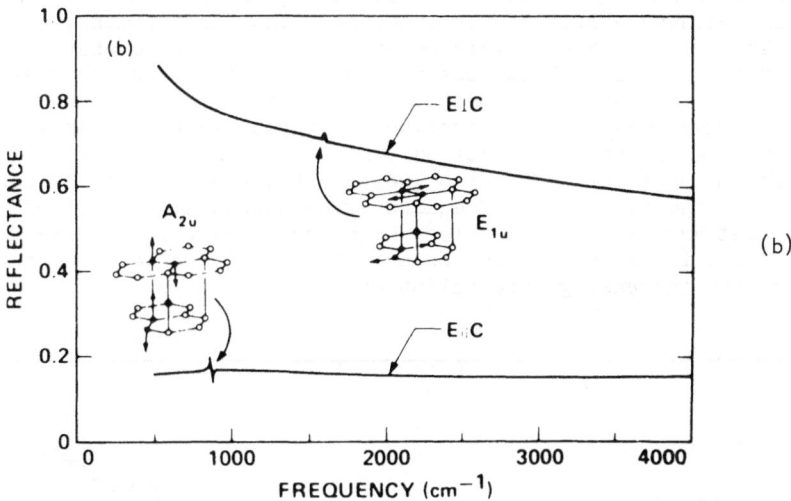

Fig. 2. a) The Raman spectra obtained from a cleaved surface of graphite, and the atomic displacements of the Raman active E_{2g} modes. b) The $E \perp C$ and $E \parallel C$ reflectance of graphite, and the atomic displacements of the IR active modes. The $E \perp C$ spectrum has been normalized for surface damage.

excitations, the primary source of such knowledge has been and remains Raman scattering. The Raman spectra of the stage-1 heavy alkali donor GIC's which are shown in Fig. 3 are unique to all GIC's and their interpretation has been the source of some controversy [20]. Recall that the stage-1 heavy alkali GIC's crystallize into a 3-dimensional (3D) superlattice structure at room temperature [21], whereas the structures of higher stage alkali and all other GIC's including stage-1 to stage-n acceptor compounds are much more complex and exhibit some marked measure of disorder [21].

The high frequency (Raman shift, $\tilde{\nu} > 300 \text{cm}^{-1}$) spectra of Fig. 3 are characterized by a continuum background on which are superposed an asymmetric Fano-like [22,23] band at ~ 1460cm^{-1} and a triplet of modes at ~560cm^{-1}. The former is universally agreed to result from a Fano interaction between the continuum and the 1581cm^{-1} E_{2g} mode of the host graphite layers. However, the continuum itself has been attributed to both multiphonon scattering [24] and to electronic Raman scattering [20]. The triplet of modes is attributed to the set of M-point phonons in Fig. 1 which have become Raman allowed by the folding of the M-point of pristine graphite into the zone-center Γ point of MC$_8$, M = K, Rb, Cs. This zone-folding is a result of the in-plane (2 X 2)R 0° superlattice structure [3,21] which the alkali ions adopt on the carbon layers as a consequence of their 3D crystalline configuration. Some components of the M-point triplet also experience Fano broadening through an interaction with the continuum background [1,2].

Notice too from Fig. 3 that the energies of the M-point and Fano-broadened E_{2g} phonons shift up with increasing intercalant mass, e.g. $M_{Cs} > M_{Rb} > M_K$, but $\tilde{\nu}_{Cs} > \tilde{\nu}_{Rb} > \tilde{\nu}_K$. This behavior is contrary to what one would expect from a traditional mode mass analysis [25] in which the heavier ion yields the lower frequency of vibration. This novel mass dependence of the vibrational frequency results from the fact that in the Eigen-vectors in question, the alkali ions are stationary so their mass does not enter into a mode mass calculation. But the guest-host interaction force constants which determine the vibrational frequencies do depend on ion size through the guest-host contact interaction which increases with size as does the ion mass. Thus, it is the force-constant change which drives up the phonon frequencies with increasing intercalant mass/size.

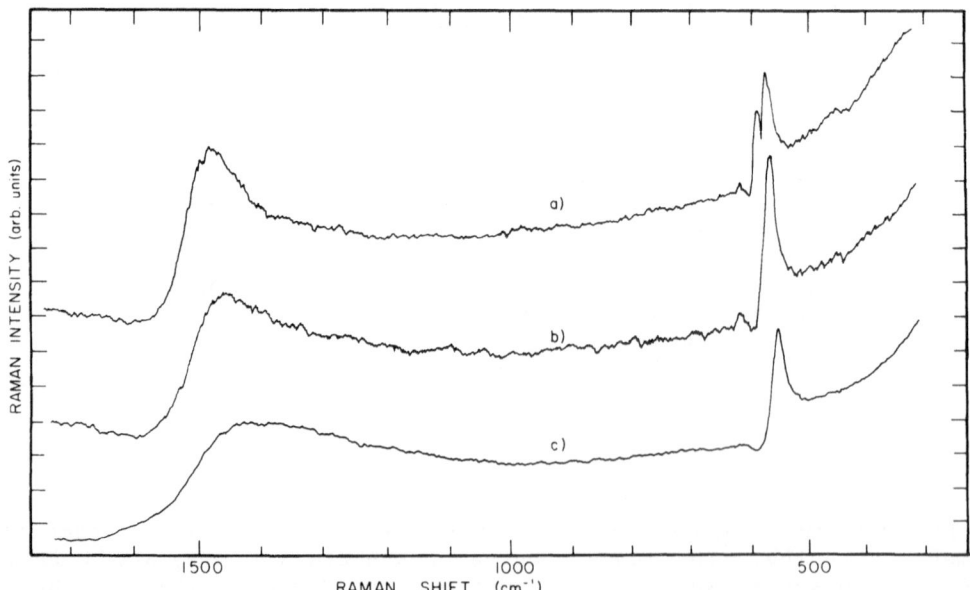

Fig. 3. The Raman spectra of stage-1 a) CsC$_8$; b) RbC$_8$; and c) KC$_8$ recorded with the incident and scattered radiation polarized parallel to one another and perpendicular to the scattering plane.

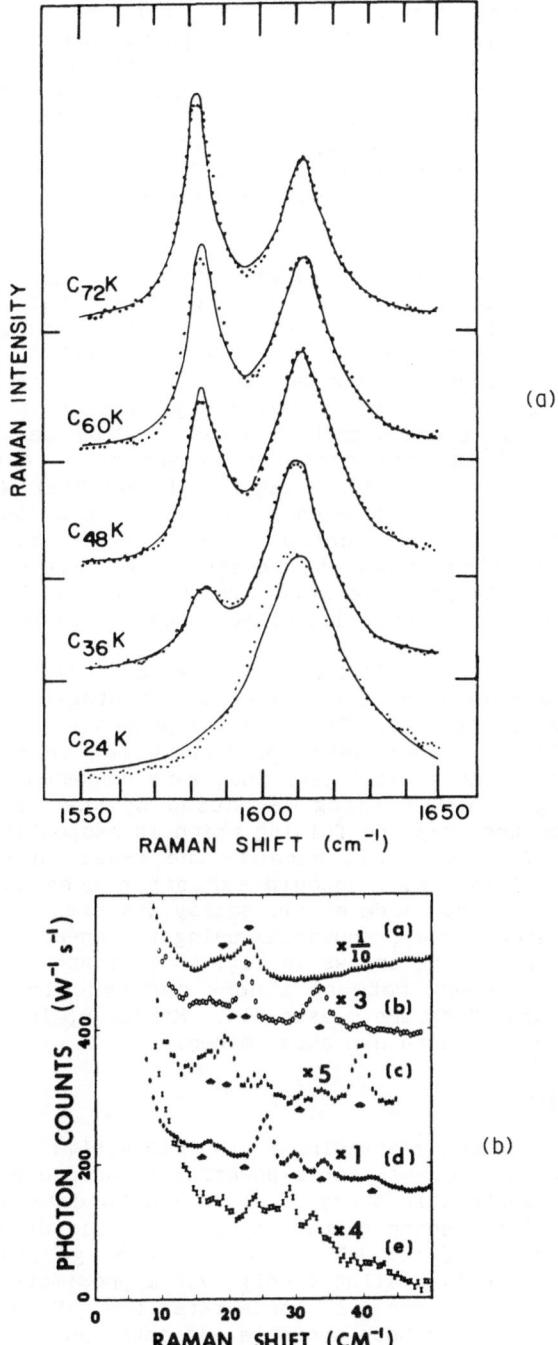

Fig. 4. a) The high-frequency Raman spectra of KC_{12n}, n = 2-6. The dotted and solid lines correspond respectively to the data and to a Lorentzian fit. b) The low-frequency Raman spectra of KC_{12n}, n = 2-6. The arrows indicate the calculated values of the phonon frequencies (see text) (from ref. 27).

Unlike the stage-1 alkali GIC spectra, the Raman spectra of all other GIC's exhibit, near 1580cm^{-1}, either a relatively sharp (but in some cases slightly Fano-broadened) singlet mode for stages 1 and 2 or a sharp doublet for stages higher than 2 [1]. One of the doublet pairs occurs at an energy very close to, but slightly upshifted from, that of the E_{2g} host layer phonon while the other may be significantly upshifted by 20-30cm^{-1}. Moreover, there is an intensity exchange between the doublet pair with increasing stage number, the lower energy member growing at the expense of the upper component. This behavior is shown in Fig. 4a for the potassium-graphite system KC_{12n}, n = 2-6 and has been explained by a nearest layer (NL) model [3,21].

In the NL model the energy upshift in the modes of stage n $\geq$ 2 GIC's is attributed to interlayer coupling between the intercalant and graphite layers [3]. In addition, with respect to the carbon layers in the system, two types of nearest layer environment will exist for n > 2, one labeled type B in which the carbon layer is bounded on one side by a graphite layer and on the other side by an intercalate layer, and another labeled type C in which the carbon layer is bounded on both sides by other carbon layers. To first order, each of these environments generates a characteristic carbon layer phonon frequency thus accounting for the doublet when n > 2. But stage-2 GIC's contain only type B environments which accounts for the singlet character of their Raman spectra. Since the number of type C layers grows with increasing stage at the expense of type B layers, the intensity exchange alluded to above results. The NL model also provides a quantitative description of the above described intensity exchange [26].

In a recent experimental tour de force, Wada et al. [27] succeeded in recording the very low frequency Raman spectra of stages 2-6 potassium graphite which are shown in Fig. 4b. The observed phonon modes in Fig. 4b were attributed by them to zone-center phonons which involve "rigid" displacements of both host and guest layers that were rendered Raman active by c-axis zone-folding. That folding is caused by the layer stacking structure in contrast to the in-plane folding which is responsible for the M-point modes of Fig. 2. Note that because the c-axis unit cell of a stage-n GIC contains n + 1 layers, it should exhibit n zone-center rigid layer shearing optical modes. Wada et al. employed a simple linear chain model to calculate the shear mode frequencies using the shear constants as adjustable parameters [27]. The arrows in Fig. 4b indicate the results of this calculation. The agreement between theory and experiment is quite reasonable for those modes which are measurable. Moreover, in no case does a stage-n compound exhibit more than n shear modes.

Neutron Scattering Studies

Although inelastic neutron scattering is the definitive probe for the study of phonon dispersion curves, this powerful technique has only been recently applied to the systematic study of GIC's and to date most of the emphasis has been on alkali donor compounds [6]. One disadvantage of the neutron scattering method vis-à-vis GIC's and pristine graphite is that useful neutron flux is readily available only over approximately the lower one-third of the vibrational spectrum of those materials. Nevertheless, in that energy range which includes both acoustic and optic phonon branches of interest, the neutron technique can yield superb results such as those shown in Fig. 5 for the (001) longitudinal modes of stages 1-3 potassium-graphite [28].

The layer breathing modes whose dispersion curves are depicted in Fig. 5 correspond to rigid layer c-axis displacements of the guest and host layers. They are thus longitudinal modes which propagate perpendicular to the basal planes of the GIC. The n + 1 layers of the c-axis cell (see

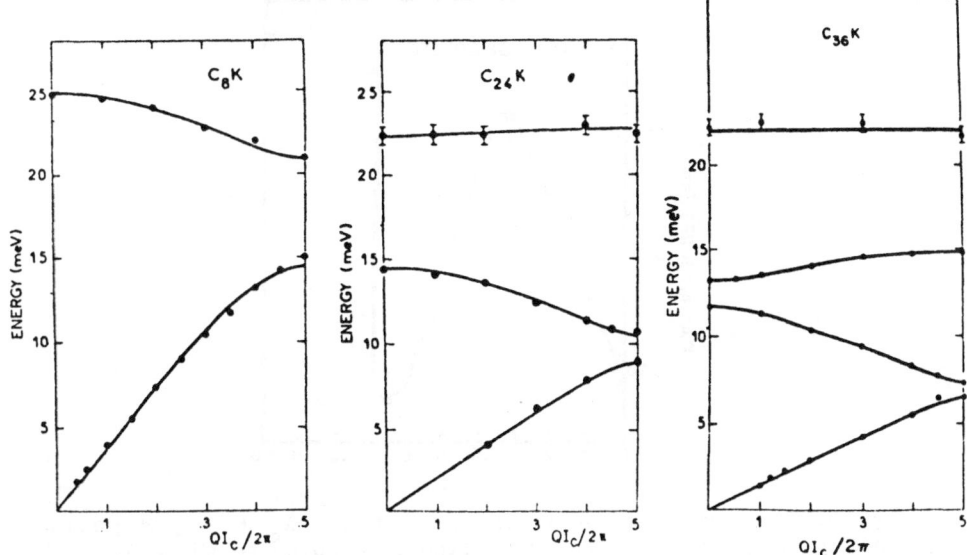

Fig. 5. The longitudinal [001] phonon dispersion curves of stages 1 to 3 potassium-graphite. The solid lines are best fits using one-dimensional mixed Born-von Karman and ion-shell models (from ref. 28).

discussion above) generate one acoustic branch and n interlayer optic branches in the low energy vibrational spectra. Thus the stage-1 KC_8 dispersion curves of Fig. 5 contain only 2 branches, those of stage-2 contain 3 branches, stage-3 yields 4 branches, etc. The solid line fits to the data of Fig. 5 were calculated using a mixture of one-dimensional Born-von Karman and ion-shell models [28]. Obviously, the agreement between theory and experiment is excellent.

PHONON ANOMALIES IN TERNARY GRAPHITE INTERCALATION COMPOUNDS

Until now we have discussed only binary GIC's. However, recent studies [29,30] of the dual alkali stage-1 ternary GIC's $M_{1-x}M'_xC_8$ where M and M' are heavy alkali metals have revealed some very unusual and exciting phonon properties which warrant brief discussion here. Consider the system $K_{1-x}Rb_xC_8$, $0 \leq x \leq 1$, the structure of which is known to be very similar to that of KC_8 except that x% of the potassium ions in (randomly selected?) hexagonal graphite interlayer sites are replaced by Rb ions. Thus, the in-plane intercalate structure is that of a (2 x 2)R 0° K-Rb lattice-gas. To date, despite several attempts [31,32], there is no direct evidence for either short-range or long-range K-Rb in-plane structural correlations.

The Raman spectra of $K_{1-x}Rb_xC_8$ are very similar to the spectra of the pure binary compounds shown in Fig. 1, but each of the detected phonons shows a small composition dependent frequency shift which is most pronounced for the relatively sharp M-point modes as shown in Fig. 6.

Notice the striking anomalous drop in the phonon frequency in Fig. 6 at a composition of x = 0.67. Anomalies in the composition dependence of the magnetic susceptibility [33] and the neutron scattering derived C_{33} elastic force constant [29] of $K_{1-x}Rb_xC_8$ have also been observed at the

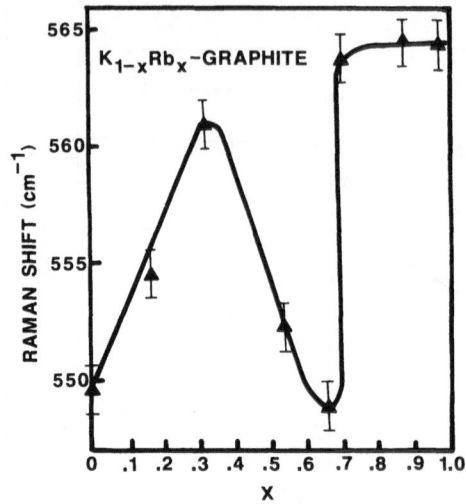

Fig. 6. The composition dependence of the A_g + B_{1g} M-point phonons in $K_{1-x}Rb_xC_8$. The solid line is a guide to the eye.

same value of x. The origin of these anomalies has not yet been definitively established, but it has been suggested that they may be associated with as yet undetected short-range K-Rb structural correlations [30] or with the novel electronic bandstructure of $K_{.33}Rb_{.67}C_8$ [34].

ACKNOWLEDGEMENTS

It is a pleasure to acknowledge useful discussions with H. Zabel, P. Chow, H. Kamimura, and S. Rabii. This work was supported by the U.S. National Science Foundation under grants DMR 85-17223 and DMR 82-11554.

REFERENCES

1. S.A. Solin: Physica 99B, 443 (1980)
2. S.A. Solin and N. Caswell: J. Raman Spectroscopy 10, 129 (1981)
3. M.S. Dresselhaus and G. Dresselhaus: Adv. Phys. 30, 139 (1981); Topics in Applied Physics, Vol. 51, "Light Scattering in Solids III", eds. M. Cardona and G. Guntherodt, (Springer, Berlin, 1982)
4. B.J.C. Van der Hoeven and P.H. Keesom: Phys. Rev. 130, 1318 (1963)
5. R.W. Lynch and H.G. Drickamer: J. Chem. Phys. 44, 181 (1966)
6. H. Zabel: Physica B, in press
7. C. Underhill, S.Y. Leung, G. Dresselhaus, and M.S. Dresselhaus: Solid State Comm. 27, 769 (1979)
8. D.M. Hwang: Proc. of the Mat. Res. Soc. Symposium on Graphite Intercalation Compounds, Boston, MA, November 1982 (North-Holland, New York, 1983), p. 295
9. See ref. 3 and references therein
10. R. Clarke and C. Uher: Adv. Phys. 33, 469 (1984)
11. A.A. Ahmadieh and H.A. Rafizadeh: Phys. Rev. B7, 4527 (1973)
12. K.K. Mani and R. Ramani: Phys. Status Solidi B61, 659 (1974)
13. A.P.P. Nicholson and D.J. Bacon: J. Phys. C10, 2295 (1977)
14. M. Maeda, Y. Kuramoto, and C. Horie: J. Phys. Soc. Japan 47, 337 (1979)
15. H.C. Gupta, J. Malhotra, N. Raui, and B.B. Tripathi: Solid State Comm. 57, 263 (1986)

16. R. Nicklow, N. Wakabayashi, and H.G. Smith: Phys. Rev. B5, 4951 (1971)
17. R.J. Nemanich and S.A. Solin: Solid State Comm. 23, 417 (1977)
18. R.J. Nemanich, G. Lucovsky, and S.A. Solin: Solid State Comm. 23, 117 (1977)
19. L.J. Brillson, E. Burstein, A.A. Maradudin, and T. Stark: in Physics of Semimetals and Narrow Gap Semiconductors, D.L. Carter and R.T. Bates, eds. (Pergamon, Oxford, 1971), p. 187
20. N. Caswell and S.A. Solin: Phys. Rev. B20, 2551 (1979)
21. S.A. Solin: Adv. Chem. Phys. 49, 455 (1982)
22. U. Fano: Phys. Rev. 124, 1866 (1961)
23. R.J. Nemanich, S.A. Solin, and D. Guérard: Phys. Rev. B16, 2965 (1977)
24. M.S. Dresselhaus and G. Dresselhaus: Mat. Sci. and Eng. 31, 141 (1977)
25. A.A. Maradudin, E.W. Montroll, G.H. Weiss, and I.P. Ipatova: Theory of Lattice Dynamics in Harmonic Approximation (Academic, New York, 1971)
26. S.A. Solin: Mat. Sci. Eng. 31, 153 (1977)
27. N. Wada, M.V. Klein, and H. Zabel: J. de Physique 42, C6-350 (1981)
28. H. Zabel and A. Magerl: Phys. Rev. B25, 2463 (1982)
29. D.A. Neumann, H. Zabel, J.J. Rush, and N. Berk: Phys. Rev. Lett. 53, 56 (1984)
30. S.A. Solin, P. Chow, and H. Zabel: Phys. Rev. Lett. 53, 1927 (1984)
31. R. Clarke: private communication
32. H. Zabel: private communication
33. G. Furdin, B. Carton, A. Herold, and C. Zeller: C.R. Acad. Sci. Paris, Ser. B 280, 653 (1975)
34. H. Akera and H. Kamimura: Syn. Met. 12, 275 (1985)

OPTICAL SPECTROSCOPY OF THE LATTICE MODES IN GRAPHITE INTERCALATION

COMPOUNDS

P.C. Eklund

University of Kentucky
Lexington, KY 40506

I. Introduction

The weak coupling between carbon and intercalate layers in graphite intercalation compounds (GIC) [1,2] results in a lattice mode picture consisting of perturbed graphitic modes, and new modes derived from the intercalate layer. Lattice modes in GICs have been studied using Raman scattering, IR reflection spectroscopy and inelastic neutron scattering. Neutron studies are typically limited to the frequency range $\omega < 400$ cm^{-1} and require much larger samples, such as can be prepared from polycrystalline pyrolytic graphite. The crystallites are preferentially aligned in this material, resulting in a well defined sample c-axis. Neutron scattering studies have therefore been largely concerned with the low frequency interlayer c-axis modes. The strong intralayer bonding between relatively light carbon atoms gives rise to high frequency optic branches which have been extensively studied using Raman and infrared (IR) spectroscopy. Most of these spectra are taken with radiation incident on the c-face (cleavage face). A limited number of optical experiments have been carried out [3] on a-faces, however. The a-face is important for studying interlayer modes which are excited when the electric vector $\vec{E}$ is parallel to the graphitic c-axis ($\vec{E} \parallel \vec{c}$).

In this paper we review the contributions of Raman and IR spectroscopies to our understanding of the microscopic properties of GICs. An extensive review of light scattering in GICs by Dresselhaus and Dresselhaus [4] has appeared recently. We dwell on a few of the developments in this area since the writing of that review.

II. Pristine Graphite

Extensive studies of the stage dependence of the graphitic modes have been carried out by optical and neutron scattering techniques [1,2,4]. To understand these results, we must first appreciate the phonon dispersion of the host material. In Fig. 1 we show the phonon dispersion of pristine graphite calculated by Al-Jishi and Dresselhaus [5]. The Brillouin zone appears to the left in Fig. 2. The phonon dispersion in Fig. 1 is in good agreement with Raman, neutron and IR data, and fits the M point frequencies somewhat better than a previous calculation by Maeda et al. [6]. Both of these lattice dynamics models have been extended to describe the stage dependence of the lattice modes in GICs [7-9].

Optical probes of the phonon dispersion necessarily involve excitations with nearly zero crystal momentum: crystal momentum must be conserved and the incident photon carries negligible momentum. Therefore, single phonon processes involve excitation of the zone-center modes at the Γ point. The zone-center (q=0) modes which are Raman active (even parity g modes) or IR active (odd parity u modes) which are shown in Fig. 3. The A_{2u} (868 cm^{-1}) mode is a nondegenerate, interlayer mode which has been observed on the a-face. The rest are doubly degenerate intralayer modes (only one of the degenerate modes is shown): E_{2g_1} (42 cm^{-1}), E_{2g_2} (1582 cm^{-1}) and E_{2u} (1588 cm^{-1}) [4]. The splitting between the high frequency IR and Raman modes (6 cm^{-1}) and the low frequency observed for the E_{2g_1} mode (42 cm^{-1}) are spectroscopic evidence for very weak coupling between next neighbor carbon layers. Two phonon optical excitations ($\omega_1+\omega_2,\vec{q}_1+\vec{q}_2$), where $\vec{q}_1+\vec{q}_2 \sim 0$ are also possible. They have been studied using Raman scattering by Nemanich and Solin [10] and Elman et al. [11]. These experiments are sensitive to the high frequency dispersion at the Σ and M points. The second order, or two phonon spectrum of pristine graphite (Fig. 4) exhibits three

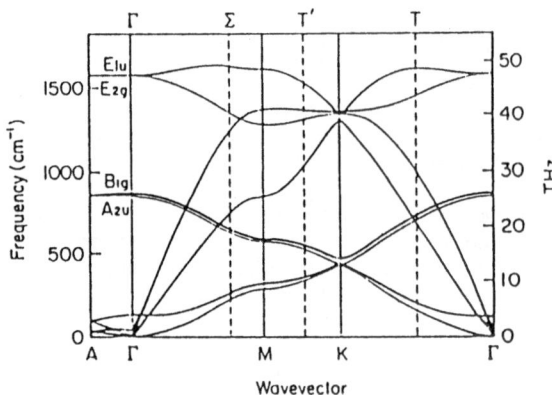

Fig. 1. The phonon dispersion relations of pristine graphite.

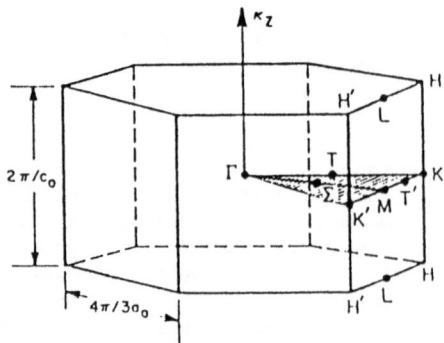

Fig. 2

The hexagonal Brillouin zone of pristine graphite.

324

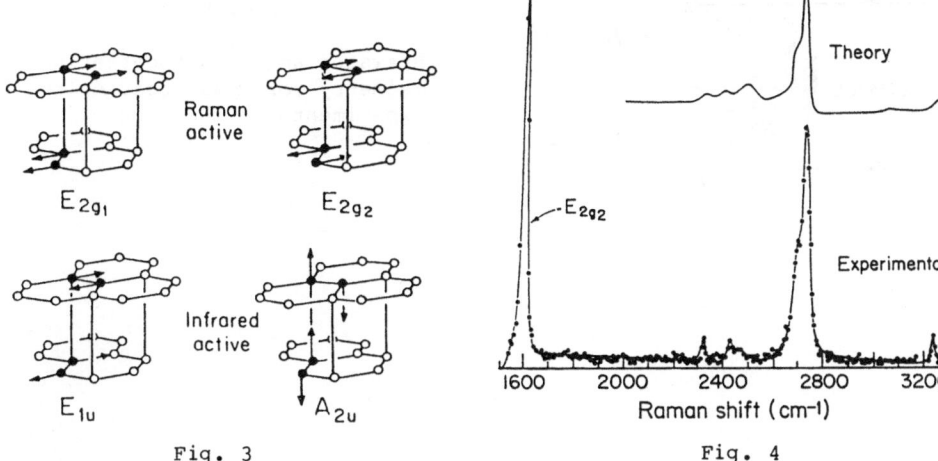

Fig. 3

Optically active lattice modes of pristine graphite.

Fig. 4

High frequency Raman spectrum of pristine graphite.

prominent features: one narrow symmetric feature at 3248 cm^{-1}, and two wider, asymmetric features at 2450 cm^{-1} and 2735 cm^{-1}. The peaks are associated with regions of low dispersion in Fig. 1, which provide a high single phonon density of states. On the basis of the lattice dynamics model of Al-Jishi and Dresselhaus [5], these frequencies have been identified (cf. theoretical curve in Fig. 4) with: (1) overtone (2ω) scattering from the highest frequency branch at the Σ point (2x1620 cm^{-1} ~ 3248 cm^{-1}), (2) overtone scattering from the zone edge LA modes at the M point (2x1360 cm^{-1} ~2735 cm^{-1}), and (3) combination scattering ($\omega_1+\omega_2$ ~ 2450 cm^{-1}) involving phonons near the Γ point (ω_1 ~ 840 cm^{-1}) and Σ point (ω_2~1620 cm^{-1}). The Γ point 868 cm^{-1} modes are also IR active and have been observed in reflectance studies by Nemanich et al. [12] on the a-face. All of the features seen in Fig. 4 have their counterpart in GICs, and have been studied as a function of stage index.

III. Experimental

Raman experiments on opaque materials are usually carried out in the backscattering geometry. A common scattering configuration used to study the intralayer modes in GICs (Fig. 5) is referred to as the "Brewster angle backscattering" geometry: (1) The $\vec{E}$ field is in the plane of incidence, (2) backscattered light is collected normal to the optical surface as shown, and (3) θ is chosen near the minimum in R(θ). The incident laser power must be kept low (<100 mW, or lower, depending on the particular GIC) to prevent laser damage to the sample. Focusing the incident laser beam to a stripe with a cylindrical lens, rather than a spot with a spherical lens, reduces the (power/unit area). Raman experiments are routinely carried out on cleaved c-faces because this surface is more specular than an a-face. Natural a-faces do not exist and must be prepared by careful polishing, followed by Ar-ion sputter etching [3] to remove both the a-face damage and the embedded c-face material produced while polishing.

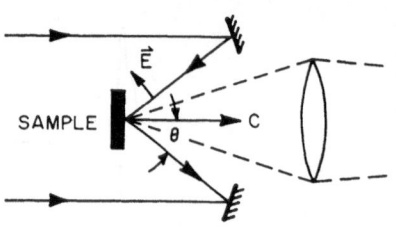

Fig. 5

Brewster angle backscattering geometry for Raman scattering experiments on the c-face.

Reflectance spectra from the c-face reveal the IR active E_{1u} mode and its GIC counterparts. These spectra have been obtained at near-normal incidence using Fourier transform spectrometers (FTIR). In a-face studies, the light is also incident at a near-normal angle, but polarized light is used to separate interlayer modes ($\vec{E} \parallel \vec{c}$) from intralayer modes ($\vec{E} \perp \vec{c}$) [12]. Extracting lineshape parameters for the IR modes requires a knowledge of the free carrier contribution to the spectra as well. The free carrier contribution can usually be obtained from the region $1 < \omega < 6$ eV [13].

The instability of GICs in air complicates matters somewhat, particularly in the case of IR spectroscopy. Suitable optical cells have been fabricated with glass or IR transmitting windows. Raman scattering experiments are routinely carried out with the sample sealed off in a rectangular cross section pyrex tube. Helium gas may be diffused later through the ampoule walls to provide better thermal contact with the sample. This can be accomplished at room temperature with an external He pressure of 1-2 atm. The reader is referred to further comments on sample handling in a separate contribution to this volume [14]. Finally, it is worth noting that the depth of the IR and Raman probe is ~1000 Å, which is considerably less than Mo or Cu x-rays (20-50 microns).

IV. Raman and IR Studies of Graphitic Modes in Donor and Acceptor GICs

In Figs. 6 and 7 we display the well known [4] Raman doublet for acceptor (graphite-FeCl$_3$ [15]) and donor (graphite-Rb [16]) GICs. The lower and upper frequency line have been identified with the interior (C_i) and bounding (C_b) carbon layers, respectively. A variety of studies have shown [1,2] that most of the charge transferred between carbon and intercalate layers resides in the C_b layers. Intercalate (I) and C_b layers are coupled by the attractive Coulomb force, whereas the C_i layers, which have two carbon next neighbor layers, find themselves in a more "graphitic" environment. The interior layer mode therefore exhibits the mode frequency closest to that of pristine graphite (1582 cm^{-1}). In agreement with the results of Figs. 6 and 7, lattice dynamical calculations in GICs by Leung et al. [8] find that for a stage $n < 2$ compound, a single bounding layer branch is split off from n-1 nondegenerate (yet spectroscopically unresolvable) interior layer branches. The splitting between the interior and bounding layer Γ point modes is observed [4] to be ~20 cm^{-1} and ~30 cm^{-1} in acceptor and donor GICs, respectively. The larger values found in the donor compounds are consistent with higher charge transfer, which strengthens the C_b-I interlayer coupling.

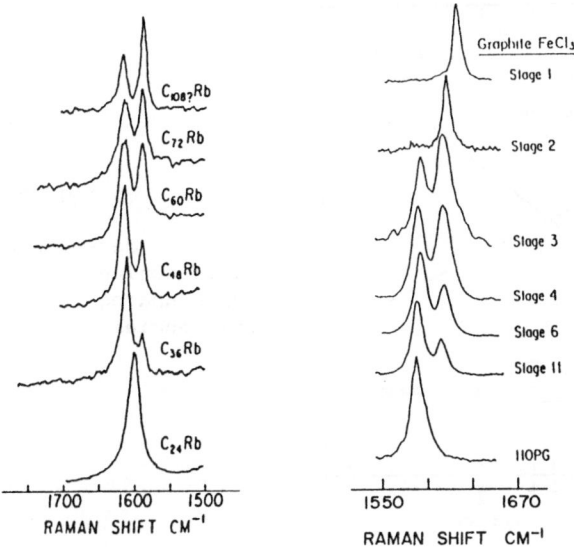

$C_{108?}Rb$

$C_{72}Rb$

$C_{60}Rb$

$C_{48}Rb$

$C_{36}Rb$

$C_{24}Rb$

Graphite FeCl₃

Stage 1

Stage 2

Stage 3

Stage 4

Stage 6

Stage 11

HOPG

1700   1600   1500
RAMAN SHIFT CM⁻¹

1550   1670
RAMAN SHIFT CM⁻¹

Figs. 6 & 7

Graphitic bounding and interior layer Raman modes in typical donor (Rb:Fig. 6) and acceptor (FeCℓ_3:Fig. 7) GICs. Figs. 6 and 7 are from refs. [15] and [16], respectively.

The stage 1 binary donor GICs (CsC$_8$, RbC$_8$, KC$_8$, and EuC$_6$ [17]) all exhibit an asymmetric Fano or Breit Wigner (FBW) lineshape which is discussed in the next section. Returning to the Lorentzian features shown in Figs. 6 and 7, we plot in Fig. 8 the peak frequency of the interior and bounding layer modes vs. reciprocal stage (1/n) for a large variety of donor (dashed lines) and acceptor (solid lines) compounds. The data for several GICs in Fig. 8 are taken from ref. [4] and more recent articles. The lines are the result of a least squares fit to the data. The stage 1 result for the donor LiC$_6$ is included because the FBW broadening is small or absent altogether, and the peak is nearly Lorentzian. The figure shows that donor modes downshift, and the acceptor modes upshift, with increasing (1/n). The (1/n) behavior of the bounding layer mode frequencies is consistent with the observed stage dependence of the contraction (acceptors) and expansion (donors) of the C-C bondlength [1,2,18]. A contracted C-C bond stiffens, and vice versa. Presumably, the C-C bond contraction (expansion) which is driven by charge transfer [18,19] in the C$_b$ layers, is transferred to the C$_i$ layers through elastic interactions, thereby transferring the strain. This has been proposed [4] to account for the similar slopes observed for the interior and bounding layer modes in the acceptor GICs (Fig. 8). The interior layer mode in the donor GICs is seen to be nearly independent of stage index, in contrast to that observed in the acceptor compounds. However, the donor data is largely due to work in Rb-graphite [16], and other donor GICs should be measured.

The charge transfer (f$_c$) dependence of the bounding layer modes in stage 1 and 2 graphite-H$_2$SO$_4$ has been studied recently by Eklund et al. [20]. f$_c$ is the number of holes in the π band per c atom. They report an upshift in the mode frequency which is linear in 1/f$_c$, for fixed stage index. These results are discussed in another contribution to this volume [21].

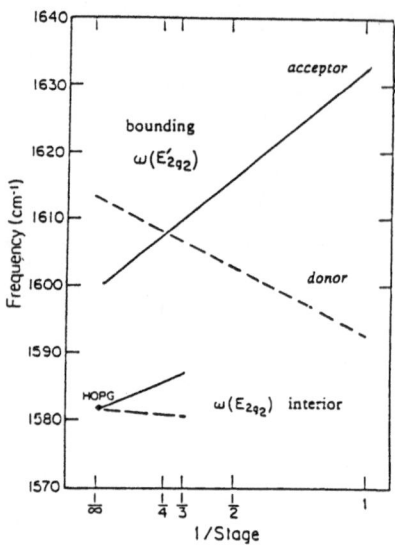

Fig. 8

(1/n) dependence of the Raman active interior (E_{2g_2}) and bounding (E'_{2g_2}) layer mode frequencies.

Even though the inversion symmetry of pristine graphite is lost upon intercalation, the GIC modes retain the "g" and "u" character of the host. That is, the loss of inversion symmetry does not render the modes both IR and Raman active. Group theory predicts that a single graphitic carbon layer will not exhibit any u (IR active) modes [22]. This is consistent with the fact that no E_{1u} derived modes have been observed in stage 1 donor or acceptor compounds [22,23]. The stage dependence of the IR modes (~ 1600 cm^{-1}) have been studied in graphite-Rb [22], graphite-AℓCℓ$_3$ [23] and graphite-FeCℓ$_3$ [24]. In Fig. 9 we show the reflectance spectra of Underhill et al. [24] for several stages of graphite-FeCℓ$_3$. The structures shown amount to $\sim 3\%$ modulation of the near normal reflectance, which at these energies is dominated by free carrier absorption. The dramatic change in line shape, from peaks at very high stage indices, to dips in the reflectance of lower stage compounds is due to the free carrier background. The dotted lines in the figure are the result of a line shape analysis in which the dielectric reponse of the GIC is written as the sum of free carrier, core and phonon contributions

$$\varepsilon(\omega) = \varepsilon_{core} + \varepsilon_{free}(\omega) + \varepsilon_{phonon}(\omega). \qquad (1)$$

The (core + free carrier) contribution is calculated in the Drude approximation [13], and the phonon term is written as the sum of Lorentzian oscillators. The reflectance (dotted line, Fig. 9) is then calculated from eq. (1) by adjusting the oscillator parameters. The lineshape analysis of the stage n>2 spectra results in two unresolved modes split by ~ 6 cm^{-1}, which have been identified [22-24] with the interior and bounding intralayer graphitic modes. The identification is based on the fact that the higher mode frequency and associated oscillator strength extrapolates in the high stage limit to values consistent with the E_{1u} IR mode in pristine graphite. The slope reported for the (1/n) dependence of these IR mode frequencies for graphite-FeCℓ$_3$ matched the slope observed for the corresponding Raman active modes, consistent with the strain model [4]. This simple (1/n) dependence was not observed [23] in graphite-AℓCℓ$_3$. In graphite-Rb, the IR modes were observed to soften with increasing (1/n), consistent with observations of the Raman active bounding and interior layer modes. In contrast to the stage 2 acceptor compounds, no E_{1u} mode was observed in stage 2 graphite-Rb.

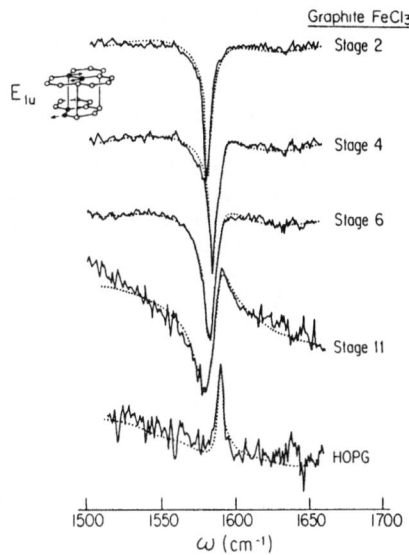

Graphite FeCl₃

E_{1u}

Stage 2

Stage 4

Stage 6

Stage 11

HOPG

1500   1550   1600   1650   1700

ω (cm⁻¹)

Fig. 9

Reflectance structure in graphite-FeCl$_3$ associated with IR lattice modes for various stages. The IR active E_{1u} mode is shown in the inset. From ref. [24].

Two phonon, or 2nd order Raman spectra have been reported by Giergiel and Eklund [25] in acceptor (SbCl$_5$, FeCl$_3$) and donor (Rb) GICs, and are found to be similar to that observed in pristine graphite (Fig. 4). The GIC results were therefore interpreted [25] in the context of the approximate symmetry of pristine graphite. In the acceptor compounds the overtone and combination scattering features were followed to the lowest stage index studied (stage 2), whereas in a prototypical donor compound, graphite-Rb, the 2nd order features were obseved for high stage compounds (stage 3 and 4) but not for stages 1 and 2. The absence of structure in the high frequency 2nd order spectra of stage 1,2 graphite-Rb suggests that the dispersion in the bounding layer modes of low stage (n<3) donor GICs is more significantly altered than in the case of acceptor GICs. This observation is consistent with the stronger C_b-I interlayer coupling, and the larger charge transfer in the donor compounds.

In Fig. 10 we plot <u>half</u> the observed frequency of the two phonon Raman features (dashed lines) vs. (1/n). For comparison, we also plot the 1st order Γ point E'_{2g_2} and E_{2g_2} modes (solid line). The overtone scattering associated with the M and Σ points is seen to have distinctly different (1/n) behavior for donors and acceptors. The highest phonon frequency in pristine graphite has been shown by Nemanich et al. [10] to be sensitive to second neighbor in-plane force constants, whereas the Γ point (~1582 cm⁻¹) frequency is sensitive to the next neighbor in-plane force constant. The calculations of Leung et al. [8] show the entire upper optic branch upshifting with decreasing stage index for acceptor GICs, in disagreement with the observed bahavior. The two phonon spectra discussed above present a quantitative test of the high frequency dispersion in GIC lattice dynamics models. The agreement between theory and experiment could undoubtedly be improved if some of the simplifying assumptions regarding the stage dependence of the force constants are reconsidered. However, the models and experiment can be brought into good overall agreement if sufficient experimental data exist to choose proper force constants. This point is well illustrated for the case of stage 2 graphite-Rb, where inelastic neutron, and high and low frequency optical data on various lattice modes are available.

329

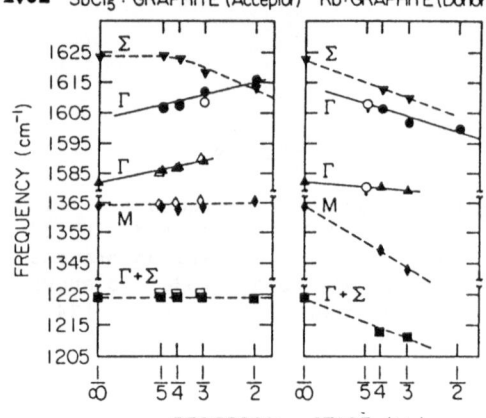

△◊○□ FeCl₃ : GRAPHITE (Acceptor)
▼▲◆●■ SbCl₅ : GRAPHITE (Acceptor)   Rb : GRAPHITE (Donor)

Fig. 10

(1/n) dependence of the 2nd order (dashed lines) and the 1st order (solid line) Raman scattering features. The features are plotted at half their frequencies. From ref. [25].

In Fig. 11 we show the results of lattice dynamics calculations for stage 2 RbC$_{24}$ by Al-Jishi and Dresselhaus which are found to be in good agreement with the experimental data [26]: (1) a Raman doublet at the low frequencies of 18,22 cm^{-1}, identified with transverse shear modes (2) a Raman singlet at 115 cm^{-1}, identified with an out-of-plane mode and in agreement with neutron data, (3) a Raman mode at 35 cm^{-1}, observed below the phase transition [27] at T=170 K, and identified [26] with zone-folded modes rendered Raman active by commensurate rubidium layer ordering at low temperatures, (4) a bounding layer Raman active mode at 1602 cm^{-1}, and (5) c-axis dispersion of the low frequency interlayer modes obtained from inelastic neutron scattering measurments [27]. The inset shows an enlargment of the low frequency dispersion along the c-axis. The open circles are obtained from neutron scattering, and the dark squares represent the low frequency Raman active modes.

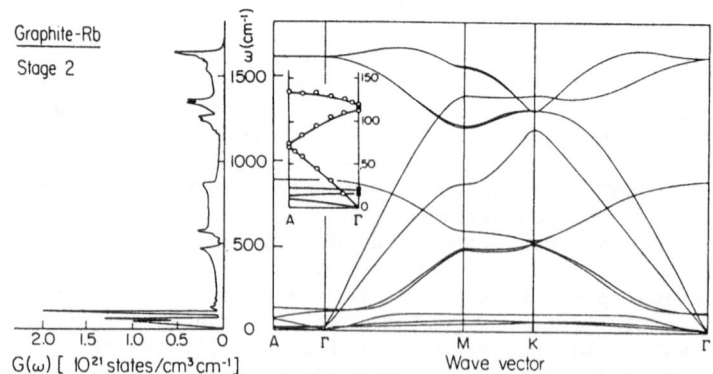

Fig. 11

Phonon dispersion curves and density of states for stage 2 graphite-Rb [26].

A Lorentzian line associated with a discrete Raman active mode at frequency ω_0 can be asymmetrically broadened and shifted through the coupling of the mode to a Raman active continuum. A lucid discussion of this Fano-Breit-Wigner (FBW) scattering has been given by Klein [28], and examples of this phenomenon have been well documented [4] in the stage 1 donor GICs. In Fig. 12, we display the spectra of Hwang and Guérard [17] for EuC_6, CsC_8 and LiC_6. Broad asymmetric features are evident for EuC_6 and CsC_8 in the ~1500 cm^{-1} region. LiC_6 exhibits a weakly asymmetric line near ~1600 cm^{-1}. Other binary donor GICs (RbC_8 and KC_8, but not the ternary GICs) exhibit similar spectra. The coupling of the discrete phonon to the continuum is small in LiC_6 [17,29] and the resulting FBW lineshape is therefore nearly Lorentzian. Eklund and Subbaswamy [30] have demonstrated that the 1510 cm^{-1} FBW peak in MC_8 (M=K,Rb,Cs) compounds could be understood as a downshifted, or "renormalized" graphite bounding layer mode (E'_{2g_2}), with an "uncoupled" frequency ω_0 ~1585 cm^{-1}. They showed that proper account of the frequency dependence of the continuum was necessary to account for the position (1510 cm^{-1}) of the FBW peak. Several origins of the continuum interfering with the ~1585 cm^{-1} discrete mode have been proposed: multiphonon [30,31], electronic [32] and disorder induced one phonon (stacking faults) [17]. The feature at ~1830 cm^{-1} in the EuC_6 spectrum (Fig. 12) has been identified [17] with the surface formation of the carbide EuC_2, while the structure at ~590 cm^{-1} in the CsC_8 spectrum (Fig. 12) has been identified with a triplet [32] or doublet [33] of zone-folded M point modes. EuC_6 and LiC_6 exhibit the $(\sqrt{3} \times \sqrt{3})R30°$ superlattice, whereas the MC_8 (M=K, Rb,Cs) compounds exhibit the (2×2) R0° structure. The M point modes are not zone-folded by the $(\sqrt{3} \times \sqrt{3})$ R30° structure [4], in agreement with the results in Fig. 12.

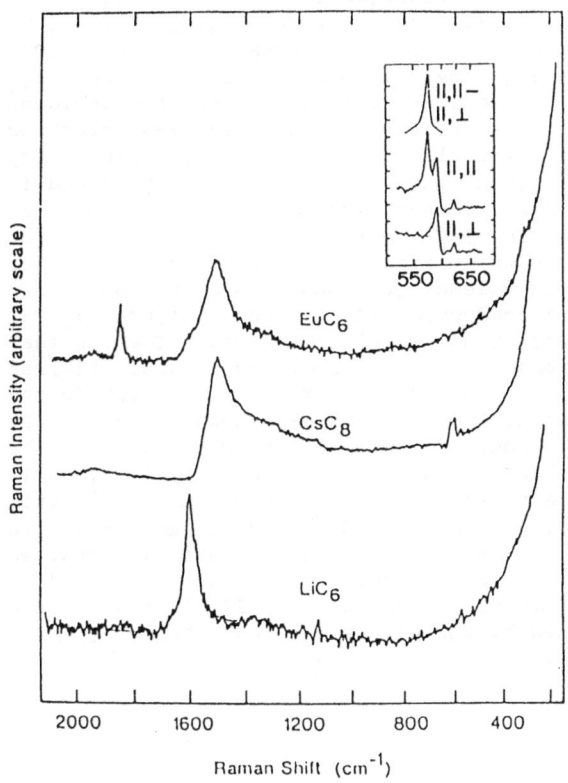

Fig. 12

Raman spectra of EuC_6, CsC_8 and LiC_6 [17]. The inset shows the frequency dependence of the renormalized frequency of the ~590 cm^{-1} zone-folded mode in CsC_8 [32].

The inset to Fig. 12 shows the data of Caswell and Solin [32] for CsC_8 in the 600 cm^{-1} region where three peaks (574,594 and 620 cm^{-1}) are evident. The 594 cm^{-1} component is also a FBW resonance which was studied [32] through the Cs layer melting transiton at T_M=608 K. The interfering continuum, in this case, was shown to be due to one phonon scattering. The renormalized frequency of the ~594 cm^{-1} peak was observed to suddenly soften by ~1 cm^{-1} at T_M. For T>T_M, the feature must therefore be assigned to disorder induced scattering. Presumably clusters of Cs atoms must exist above T_M with the same short range order as found in the ordered state.

The M point modes of the two-dimensional random alloys $K_{1-x}Rb_xC_8$ (0<x<1) were studied recently by Solin et al. [34]. The samples are prepared by sequential intercalation of rubidium and potassium to form stage 1 solid solution ternary GICs [35]. They observed a composition dependent softening of the M point mode in the vicinity of x=0.67. The magnetic susceptibility [35], and the interlayer force constants [36] both exhibit anomalous x dependence at x=0.67. The anomalies were attributed [34] to electronic band structure effects associated with the formulation of KRb_2 clusters.

A FBW resonance involving the E'_{2g_2} bounding layer mode in C_8AsF_5 has also been recently reported by Ohana et al. [37]. The coupling is observed to be weak compared to that observed in the stage 1 donor compounds, and the continuum in this case has been identified with electronic transitons. This is the only report of a FBW resonance in acceptor compounds, which generally exhibit symmetric peaks well fit by a Lorentzian lineshape.

VI. Resonant Raman Scattering in GICs

Resonant Raman scattering has been observed recently in stage 1 and 2 graphite-H_2SO_4 by Eklund et al. [38] and in stage 1 graphite-AsF_5 by Ohana et al. [39]. The phenomenon should be observable in most GICs. The cross section for scattering involving the emission of the E'_{2g_2} bounding layer phonon was found to exhibit a maximum when the incident laser photon energy matched the center of the threshold for π-π interband absorption. Graphite-H_2SO_4 is particularly attractive for this study because the position of this interband threshold can be easily moved in energy through control of the electrochemical reaction used to prepare these GICs. The π band structure [39] near the Fermi energy E_F for the stage 1 and 2 compounds, is shown in Fig. 13. The solid lines refer to the stage 1 conduction (C_1) and valence (V_1) bands, and the dashed and solid lines refer to the pair of conduction (C_1,C_2) and valence (V_1,V_2) bands derived from two-dimensional band models for GICs. The vertical arrows indicate the strong interband transitions between π bands [39] at the threshold for absorption. Because of the mirror symmetry in the bands, the threshold energy is at $\hbar\omega$~$2E_F$. Shung has shown recently [40] that the threshold is broadened about $2E_F$ due to the scattering of the photoexcited electron. The interband thresholds are indicated in Fig. 14 by the rise in the imaginary part $\varepsilon_2(\omega)$ (solid line) of the dielectric function in the range 1-2 eV. Below the threshold, $\varepsilon_2(\omega)$ rises with decreasing ω due to free carrier absorption. $\varepsilon_2(\omega)$ data is obtained from a Kramers-Krönig transformation of the reflectance data [41]. Data are shown for the $C_p^+(HSO_4)^-(H_2SO_4)_x$ compounds (p=28,21:stage 1 and p=48,60:stage 2); p is determined by the electrochemical charge Q which is measured during intercalation.

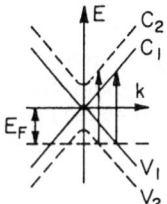

Fig. 13

The stage 1 and 2 band structure near E_F. Stage 1:dashed
bands, and stage 2:dashed and
solid bands.

The scattering cross section, or the number of scattered photons per
incident photon, is also plotted in the figure. The dashed lines are
guides for the eye. The cross section data were taken using various
lines from Krypton- and Argon-ion lasers. Optical reflectance and Raman
spectra were taken on different samples in different optical cells. For
each compound, the peak in the cross section is observed to nearly line
up with the midpoint of the interband threshold. As can be seen in the
C_{21} data (Fig. 14), the Raman cross section at the peak (on resonance) is
~30 times larger than measured in the low energy wing (off resonance).
The cross section data have been interpreted in terms of a microscopic
model for the resonant scattering process [38]. The data and calculations are in reasonably good agreement. The calculation includes
explicitly the wavevector dependence of both the electron-phonon and $\vec{A} \cdot \vec{p}$
matrix elements and the lifetime of the photoexcited electrons is used to
account for the width of the cross section resonance [38], broadening the
resonance in much the same way as the threshold for interband absorption
[40].

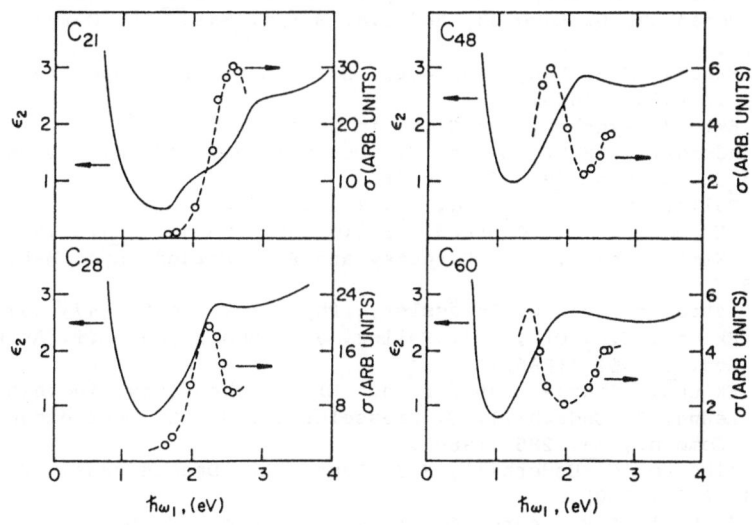

Fig. 14

Experimental data [38] for the imaginary part of the
dielectric function $\varepsilon_2(\omega)$ (solid lines) and the Raman
cross section σ (dashed lines) for the four compounds
$(C_p^+ HSO_4^- (H_2SO_4)_x$ where $p=60,48$ (stage 2) and $p=28,21$
(stage 1).

VII. Acknowledgements

The support of the U.S. Department of Energy (DE-FG05-ER45151) is gratefully acknowledged. The author gratefully acknowledges the many contributions of his former graduate students, especially J. Giergiel, and has benefited greatly from collaborations with K.R. Subbaswamy, G.D. Mahan, G. Dresselhaus and M.S. Dresselhaus.

References

1. S.A. Solin, Adv. Chem. Phys. 49, 455 (1982).
2. M.S. Dresselhaus and G. Dresselhaus, Adv. Phys. 30, 139 (1981).
3. M. Zanini, D. Grubisic and J.E. Fischer, Phys. Stat. Sol. 90, 151 (1978).
4. M.S. Dresselhaus and G. Dresselhaus in Topics in Applied Physics, Vol. 51, Light Scattering in Solids III, eds. M. Cardona and G. Güntherodt (Springer-Verlag, Berlin, Heidelber, 1982).
5. R. Al-Jishi and G. Dresselhaus, Phys. Rev. B 26, 4514 (1982).
6. M. Maeda, Y. Kuramoto and C. Horie, J. Phys. Soc. Japan 47, 337 (1979).
7. C. Horie, M. Maeda and Y. Kuramoto, Physica 99, 430 (1980).
8. S.Y. Leung, G. Dresselhaus and M.S. Dresselhaus, Solid St. Commun. 38, 175 (1981); Phys. Rev. B 24, 6083 (1981).
9. R. Al-Jishi and G. Dresselhaus, Phys. Rev. B 26, 4523 (1982).
10. R. Nemanich and S.A. Solin, Phys. Rev. B 20, 392, (1979).
11. B.S. Elman, M.S. Dresselhaus, G. Dresselhaus, E.W. Maby, H. Mazurek and G. Dresselhaus, Phys. Rev. B 24, 1027 (1981).,
12. R.J. Nemanich, G. Lucovsky and S.A. Solin, Solid St. Commun. 23, 117 (1977).
13. P.C. Eklund, M.H. Yang and G.L. Doll, Optical Properties of Donor GICs, in this volume.
14. P.C. Eklund, Synthesis of GICs, in this volume, p. 163.
15. C. Underhill, S.Y. Leung, G. Dresselhaus, and M.S. Dresselhaus, Solid St. Commun. 29, 769 (1979).
16. S.A. Solin, Mat. Sci. Engng. 31, 153 (1977).
17. D.M. Hwang and D. Guerard, Solid St. Commun. 40, 759 (1981).
18. W.A. Kamitakahara, J.L. Zarestky and P.C. Eklund, Syn. Met. 12, 301 (1985).
19. L. Pietronero and S. Strässler, Phys. Rev. Lett. 47, 593 (1981).
20. P.C. Eklund, C.H. Olk, F.J. Holler, J.G. Spolar and E.T. Arakawa, J. Mat. Res. 1, 361 (1986).
21. P.C. Eklund, Charge transfer and Electrochemistry, in this volume.
22. S.Y. Leung, C. Underhill, G. Dresselhaus and M.S. Dresselhaus, Solid State Commun., 33, 285 (1980).
23. G. Gualberto, C. Underhill, S.Y. Leung, G. Dresselhaus, Phys. Rev. B 21, 862 (1980).
24. C. Underhill, S.Y. Leung, G. Dresselhaus and M.S. Dresselhaus, Solid St. Commun. 29, 769 (1979).
25. J. Giergiel and P.C. Eklund, in Intercalated Graphite, eds M.S. Dresselhaus, G. Dresselhaus, J.E. Fischer and M.J. Moran, MRS Symposia Proc. Vol. 20, (North Holland, New York, Amsterdam, Oxford, 1982),
26. J. Giergiel, P.C. Eklund, R. Al-Jishi and G. Dresselhaus, Phys. Rev. B 26, 6881 (1982).
27. A. Magerl and H. Zabel, Phys. Rev. Lett. 46, 444 (1981).
28. M.V. Klein, in Light Scattering in Solids, ed. M. Cardona (Springer, New York, 1975).

29. Note the dramatic difference in the continuum scattering in LiC_6 in the following three Ref.'s: P.C. Eklund, G. Dresselhaus, M.S. Dresselhaus and J.E. Fischer, Phys. Rev. B 21, 4705 (1980), M. Zanini, L.Y. Ching and J.E. Fischer, Phys. Rev. B 18, 2020 (1978). G.L. Doll, P.C. Eklund and J.E. Fischer (submitted for publication).

30. P.C. Eklund and K.R. Subbaswamy, Phys. Rev. B 20, 5157 (1979).

31. M.S. Dresselhaus and G. Dresselhaus, in Physics and Chemistry of Materials with Layered Structures, ed. F. Levy (Dordrecht, Holland, 1978).

32. N. Caswell and S.A. Solin, Phys. Rev. B20, 2551 (1979).

33. P.C. Eklund, G. Dresselhaus, M.S. Dresselhaus, and J.E. Fischer, Phys. Rev. B16, 3330 (1977).

34. S.A. Solin, P. Chow and H. Zabel, Phys. Rev. Lett. 53, 1927 (1984).

35. G. Furdin, B. Carton, A. Herold and C. Zeller, Acad. Sci. Ser. B 280, 653 (1975).

36. D. Neumann, H. Zabel, J.J. Rush and N. Berk, Phys. Rev. Lett. 53, 56 (1984).

37. I. Ohana, D. Schmeltzer and Y. Yacoby, in Extended Abstracts of the 1984 MRS Symposium on Graphite Intercalation Compounds, ed. P.C Eklund, M.S. Dresselhaus and G. Dresselhaus, p. 121 (1984).

38. P.C. Eklund, G.D. Mahan, J.G. Spolar, E.T. Arakawa, J.M. Zhang and D.M. Hoffman, Solid St. Commun. 57, 567 (1986).

39. J. Blinowski, N. H. Hau, C. Rigaux, J.P. Vieren, R. Toullec, G. Furdin, A. Herold and J. Melin, J. Phys. (Paris) 41, 47 (1980).

40. K.K. -R Shung, Phys. Rev. B, in press.

41. J.M. Zhang and P.C. Eklund, Phys. Rev. B, (in press).

CHARGE TRANSFER AND ELECTROCHEMISTRY IN GRAPHITE INTERCALATION

COMPOUNDS

P.C. Eklund

University of Kentucky
Lexington, KY 40506

I.   Introduction

        Charge transfer in graphite intercalation compounds (GICs) refers to
the exchange of electrons between the graphitic carbon (C) and intercalate
(I) layers [1,2].   It is an important quantity which affects directly the
electrical, thermal and lattice dynamical properties.   If electrons are
"donated" to the carbon layers from the intercalate layers, the GIC is
referred to as "donor-type", on the other hand, if electrons are
transferred from the C layers and "accepted" by the I layers, the GIC is
termed "acceptor-type". The position of the Fermi level (E_F) in the carbon
π bands must, of course, be consistent with the degree of charge transfer.
From the point of view of the carbon sublattice, charge transfer can be
defined by the quantity f_C, which is the hole (electron) concentration
expressed as the number of free carriers per carbon atom.   Charge transfer
is also expressed in the literature from the point of view of the
intercalant. For example, in KC_{24}, $f_K \cong 1$, indicates one electron per K atom
is donated from the K atom.   For charge neutrality in this latter example,
we would have $24 f_C = f_K$, if charge is exchanged between the K(4s) and carbon
π states, <u>and no other states are involved</u>.   In stage 2 KC_{24} this is
thought to be the case, however in other GICs the situation may be more
complex. For example, (1) localized carbon states which can accept charge
may be formed as a result of intercalation or (2) several charged
intercalate species may coexist in the intercalate layer. Charge transfer
in GICs is studied using a variety of experimental techniques [1,2] which
may be sensitive to the mobile cariers in the carbon π bands (i.e.
sensitive to f_C), or sensitive to the excess charge localized on particular
intercalated molecules or atoms (i.e. sensitive to f_I, where f_I is the
excess charge residing on the intercalated species I). For reasons such as
those given above, caution must be exercised when comparing values obtained
for f_C and f_I. If these values are not consistent, it may indicate that all
the states participating in the charge transfer were not taken into
account.

        Charge transfer in acceptor and donor GICs is traditionally viewed
from different perspectives. In the case of acceptor-type GICs, charge
transfer is viewed in terms of electron transfer to form negatively charged
molecular anions.   It is presumed that the electrons in these anion states
are strongly localized. For example, in graphite-$SbCl_5$ (this notation
signifies the intercalation compound formed by the reaction of graphite and

SbCl$_5$) the anions SbCl$_4^-$ and SbCl$_6^-$ have been proposed by Boolchand et al. [3] on the basis of ^{121}Sb Mossbauer studies. This same work has reported evidence for neutral SbCl$_3$ and SbCl$_5$ in the I layers. Hwang et al. [4], using the x-ray fluorescence probe in a high resolution scanning transmission electron microscope (STEM), found evidence in graphite-SbCl$_5$ for islands (several hundred Å in diameter) of one stoichiometry (SbCl$_x$) embedded in a sea of another stoichiometry (SbCl$_y$). The charge transfer (which was presumed to correlate with the density of molecular anions) in the islands was found equal to that in the sea. To account for the fact that x≠y, Hwang et al. proposed that the island and sea contain different concentrations of the various antimony chloride molecular species listed above.

In the case of donor-type GICs, the charge transfer is viewed from a band perspective because the intercalate layers are typically comprised of metal atoms in densities comparable to ordinary metals. Charge transfer between carbon π- and intercalate-bands then depends on the band positions relative to E$_F$. Possible hybridization between π and intercalate bands further complicate the picture, but makes these GIC materials a challenging area of research which requires strong interaction between theoretical and experimental efforts.

In the study of charge transfer, GICs prepared electrochemically are very important because the electrochemical charge Q passed in the reaction can be related to the charge transfer. Electrochemical synthesis enjoys a second advantage over chemical synthesis in that the stoichiometry and/or stage of the compound is also controlled easily via Q by simply stopping the current flow in the electrochemical cell at the appropriate time. Early work in this phase of GIC synthesis was carried out by Rüdorff and Hofmann (solvated cation compounds) [5], Rüdorff and Schultze (graphite-H$_2$SO$_4$) [6] and Bottomley and Parry (graphite-H$_2$SO$_4$) [7]. During the electrochemical reaction, cations (A$^+$) or anions (X$^-$) are removed from the electrolyte and intercalated to form GICs. Neutral molecules may also co-intercalate. GIC stability in the electrolyte can be a problem, but a variety of interesting compounds have been synthesized in suitably chosen electrolytes. Electrochemical synthesis of GICs has been reviewed most recently by Bessenhard et al. [7].

In this paper we give an overview of the donor and acceptor GICs which have been prepared by the electrochemical technique. We also review some in situ studies on the electronic, structural and lattice dynamical properties of the prototypical GIC acceptor system-- graphite-H$_2$SO$_4$, which is prepared by the anodic oxidation of graphite in sulfuric acid.

II. Basic Electrochemistry of GICs

In Fig. 1 we show schematically an electrochemical cell suitable for liquid electrolyte reactions. The graphite sample is held in a Pt clip (sample electrode:SE) which provides both support and electrical contact. A constant current (I) power supply is used to drive the reaction, and is connected between the SE and a Pt counter electrode (CE). If quantitative cell potentials (V) are required, a third reference electrode (RE, e.g. standard calomel electrode) is included. The initial assumption, or electrochemical model, is that a single charged molecular (or atomic) species in solution will intercalate for every electron flowing in the external circuit. Therefore, the experimental quantity which determines the stoichiometry in this model is Q/M, where M is the initial mass of the graphite anode (or cathode, as the case may be) and Q is the total charge passed by the power supply during the reaction (Q=I∗time). Of course, the electrochemical model must be tested by subsequent experiments to verify if it is valid at all, or to determine at what Q/M it may break down.

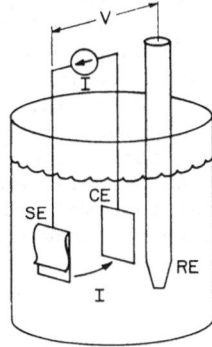

Fig. 1

Schematic representation of an electrochemical cell. The sample is held in the Pt clip (SE). A constant current (I) is passed between SE and a counter electrode (CE). Cell potentials $(V=V_{SE}-V_{RE})$ are obtained via the reference electrode (RE).

Electrochemical insertion of X^- or A^+ ions has been found to occur, with or without co-intercalated neutral molecules, depending on the solvents used in forming the electrolyte [8]. These neutral molecules are therefore often referred to as solvent molecules (solv). Graphite electrode reduction or oxidation leads to the following acceptor (C_n^+) or donor (C_n^-) compounds:

Reduction --------------→ $A^+(solv)_y C_n^-$ or $A^+ C_n^-$

Oxidation --------------→ $C_n^+(solv)_y X^-$ or $C_n^+ X^-$

where, for example, $A^+=K^+, Rb^+, N(Y)_4^+$ where $Y=$ Me, Bu (i.e. methyl, butyl) and $X^-=HSO_4^-, BF_4^-, PF_6^-$, and (solv) represents a neutral solvent molecule, usually present in the electrolyte which has co-intercalated with the anion or cation. Reactions resulting in these GICs are known to occur [8], and result in well-staged compounds. The electrolytes are formed by dissolving the salt (e.g., $N(Bu)_4 PF_6$) or alkali metal in a solvent which favors the stability of the resulting GIC. Solvents have been found to be stable against both oxidation and reduction: (1) oxidation: $SO_2C\ell F$, nitromethane (NM), acetonitrile (AN) and propylene carbonate (PC); and (2) reduction: dimethylsulfoxide (DMSO), ethers (e.g. 1,2-dimethoxyethane (DME)).

III. Synthesis, Characterization and Charge Transfer Effects in Graphite-H_2SO_4

Graphite-H_2SO_4 is prepared by the anodic oxidation of graphite in H_2SO_4. In the oxidation scheme discussed above, $X^-=(HSO_4)^-$ and solv=(H_2SO_4). This compound has been the subject of considerable study [9-11]. Graphite-H_2SO_4 has been shown to evolve through a series of well-staged compounds ($n=5\to4\to3\to2\to1$), where n is the stage index which refers to the number of carbon layers found between periodically stacked intercalate layers [1,2]. The oxidation process can be caused to occur slowly and reproducibly, by passing a low current I through the cell such as to produce a stage 1 compound in ~24 hours, or longer. A cell used to electrochemically produce graphite-H_2SO_4 and at the same time record the Raman spectra from the GIC c-face, is shown schematically in Fig. 2. The cell EMF is plotted vs Q/M in Fig. 3 for the anodic oxidation of graphite in H_2SO_4 and exhibits a staircase structure consistent with the evolution of stages. Plateaus occur during mixed stage regions. Across the top of the figure are the values of p in the chemical formula $C_p^+ HSO_4^- (H_2SO_4)_x$ which corresponds to values of Q/M on the ordinate. The regions of p (or Q/M) which have been identified with single stage material are shaded in the figure. The identification of the stage index in the deep bulk was made on the basis of in situ (00ℓ) x-ray studies [9,11]. The oxidation number p is given by $p=1/f_C$, where f_C is the charge transfer parameter. Of particular interest, are the wide single stage 2 and 1 regions in Fig. 3, referred to in the literature as "overcharging" intervals. In these intervals, for fixed stage index, we can vary continuously, and quite reproducibly, the charge transfer f_C by the internal conversion of neutral H_2SO_4 into HSO_4^- and the deintercalation of protons. This represents a minimal perturbation in the nature of the intercalate layer and allows a direct study of charge transfer effects in a prototypical acceptor GIC.

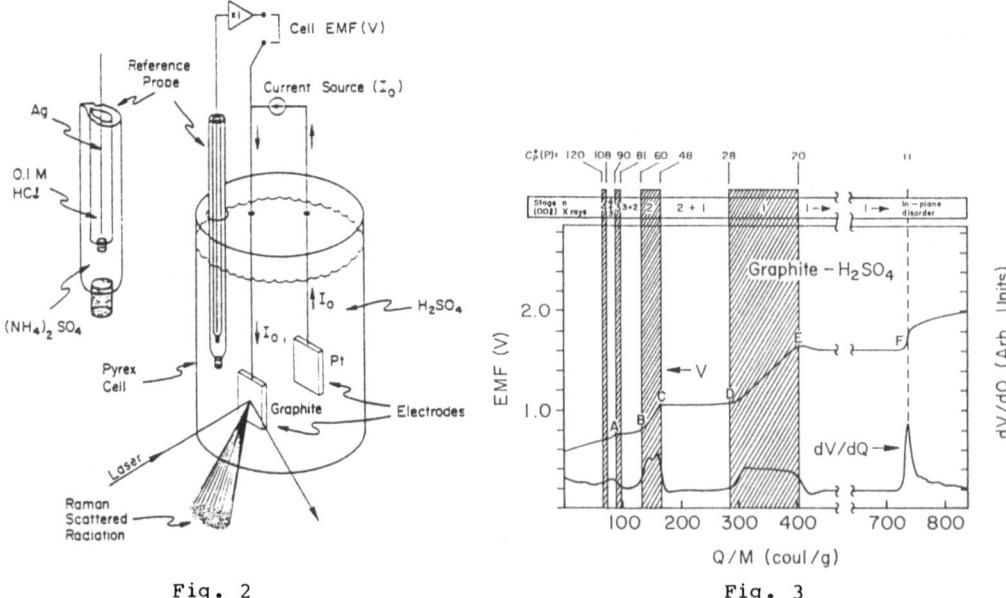

Fig. 2

Electrochemical Raman Cell. (From ref. 15.)

Fig. 3

Cell EMF (V) and dV/dQ vs Q/M for a Graphite-H_2SO_4 cell. (From ref. 15.)

Contraction of the In-plane C-C Bond

The contraction of the in-plane carbon-carbon (C-C) bond length d_{CC} of charge transfer is perhaps the most fundamental result. The contraction in d_{CC} was studied in situ by Kamitakahara et al. [10] using elastic neutron scattering. The samples were prepared using D_2SO_4 rather than H_2SO_4 because thermal neutrons are strongly absorbed by hydrogen. The positions of the Bragg (100) and (110) diffraction rings in graphite-H_2SO_4 based on polycrystalline highly-oriented-pyrolytic-graphite (HOPG) were monitored as a function of Q/M to determine d_{CC}. The results are shown in Fig. 4, where the change in d_{CC} is plotted as a function of time (or Q/M, since a constant current source was used). The upper scale is the charge transfer scale ($f_C=1/p$) obtained on the basis of the electrochemical model: one intercalated HSO_4^- molecule for every electron flowing in the external circuit. This model holds for p>21 [9], after which time the anode enters the regime known as "overoxidation". The shaded areas in Fig. 3 correspond to pure stage regions [9] , verified [10] by (00ℓ) neutron diffraction scans. In the mixed stage regions, the stage n and n-1 (hk0) peaks were not resolved and thus the data represent a composition-weighted-average value for d_{CC}. The solid curve in the figure is due to the theoretical

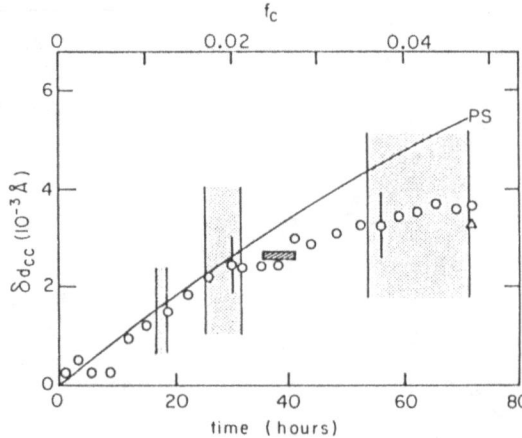

Fig. 4. Contraction in the carbon-carbon bond distance vs.
charge transfer per carbon atom f_c, or intercalation time for
D_2SO_4-graphite, compared with Pietronero-Strässler (PS) theory.
Open triangle and horizontal bar are measurements on stage-2
$SbCl_5$ and stage-1 H_2SO_4-graphite, respectively. [10]

calculations of Pietronero and Strässler [13]. Two separate data points on
the figure correspond to values obtained for graphite-$SbCl_5$ (stage 2) and
graphite-H_2SO_4 (stage 1). The fact that all the acceptor data lie on the
same curve gives credibility to the concept of a universal relationship
[10] between f_c and d_{cc} for acceptor GICs. Of course, several more GICs
need to be represented on the curve to test this concept. Theory and
experiment agree very well for f_c < .02, above this value both theory and
experiment depart from linearity, with the data exhibiting a stronger
nonlinear character than the theory.

Changes in the C-axis (00q) LA and LO Phonons

Using inelastic neutron scattering the c-axis phonon dispersion of
graphite-H_2SO_4 was studied by Eklund et al. [11] for the stage 1 and 2
compounds exhibiting the highest and lowest charge transfer. During the
"overcharging" intervals between these endpoint compounds, the c-axis
lattice parameter was observed [14] to undergo a continuous contraction
with increasing charge transfer, consistent with the increase in the
attractive coulomb force between the bounding carbon and intercalate
layers. The stage 1 and 2 c-axis (00q) LA and LO phonon data are shown in
Figs. 5 and 6, respectively. The high and low charge transfer data are
represented by the solid and open circles, respectively. The solid lines
are calculated using a one dimensional (1D) rigid ion model with
interactions including first and second nearest neighbors. The model also
treats the three layer sandwhich (0_2-S-0_2) as incompressible. The protons
have negligible mass and are ignored. Initial model calculations introduced
the appropriate 0_2-S force constant, consistent with Raman and IR mode
frequencies for H_2SO_4. The graphitic branches, such as shown in Figs. 5 and
6, were found to be insensitive to the addition of these strong 0_2-S
interlayer forces. The frequencies in the optic branches associated with
the internal c-axis motion of the (0_2-S-0_2) three layer sandwhich are very
high (ω>1400 cm^{-1}) and could not be observed using neutron scattering.
Considering the insensitivity of the graphitic branch frequencies to values
of the 0_2-S force constant, the calculations were then carried out in the
rigid intercalate layer approximation (i.e. 0_2-S force constant is
infinite). Calculated dispersion curves shown in the figures are the result
of least squares fits of the force constants (Table I) to the dispersion
data. Examination of the nearest neighbor C-C force constants in

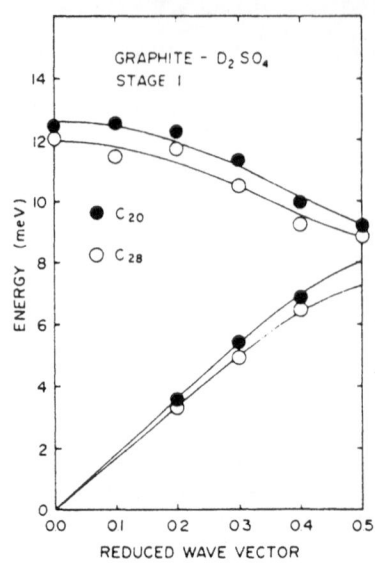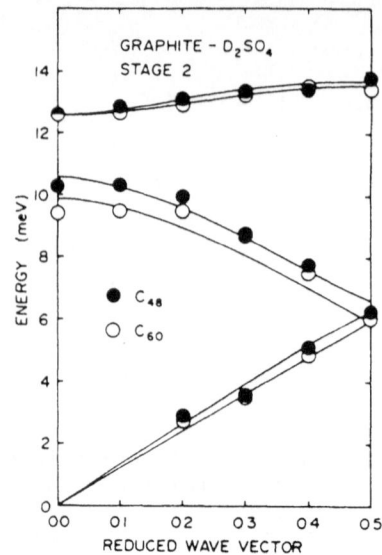

Fig. 5,6   LA and LO c-axis phonons in graphite-D_2SO_4.

Table I.   Interplanar Force Constants (10^3 dyn/cm) for $C_p^+HSO_4^-(H_2SO_4)_x$ per C-Atom.

Stage	P	Nearest Neighbor		Next Nearest Neighbor		
		C – C	C – SO_4	C – C	C – SO_4	SO_4 – SO_4
1	∿20		1.97	-0.010		-0.12
1	∿28		1.78	0.002		-0.18
2	∿48	2.65	2.01	0.064	-0.13	
2	∿60	2.81	1.82	-0.016	-0.18	
HOPG	∞	2.81		-0.041		

stage 2 indicates that this force constant differs by only a small amount from that found (Table I) for highly-oriented-pyrolytic-graphite (HOPG), where we have used the HOPG-data of Nicklow et al. [15]. The C-SO_4 force constants (Table 1) found for stage 1 and 2 are nearly equal, indicating that the charge transfer has a small effect on the harmonic expansion of the C-I interlayer potential. Finally, it was observed that the force constants did stiffen or soften in a manner consistent with both the increase in the charge per unit area and the sign of the charge on the respective layers.

Shift in the High-frequency Graphitic Intralayer Mode

Raman scattering experiemnts were recently carried out by Eklund et al. [11,16] to measure the charge transfer induced shifts in the high frequency graphitic intralayer phonons caused by the contraction in d_{CC}. The experiments were carried out for fixed stage index during the stage 1 and 2 overcharging intervals. In Fig. 7 we display the results of this in situ study of the bounding layer mode frequency vs. Q/M. The (+) and (x) indicate data collected above and below the acid level, respectively. The (+) and (x) are observed to superimpose during the overcharging periods (i.e. p=60→48 (stage 2) and p=28→21 (stage 1) when neutral H_2SO_4 is being electrochemically converted to HSO_4^- (proton diffusion), and differ during the periods when stage n-1 electrochemically grows in the bulk at the expense of stage n (sulfate diffusion). We note that the x-ray and neutron diffraction techniques probe the deep bulk, whereas the Raman experiments

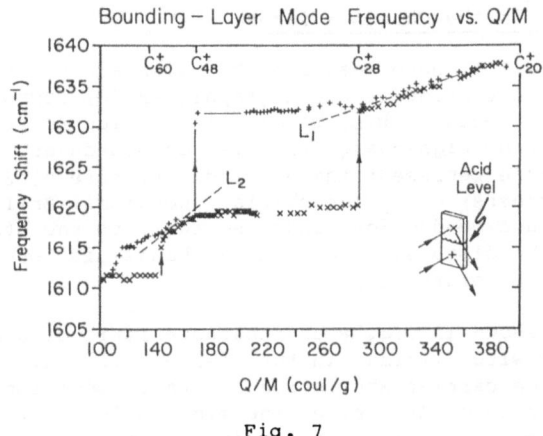

Fig. 7

Graphitic bounding layer mode frequency (E_{2g_2}) vs Q/M. From ref. [11].

probe the first ~1000 Å of the c-face. The superposition of the data in Fig. 7 during overcharging is testimony to the rapid diffusion of protons from below to above the acid level during these intervals.

As can be seen in the figure, the Raman frequency is observed to be linear in Q/M in the fixed stage overcharge intervals. From the slopes of the lines L_1 and L_2, Eklund et al. [11] calculate the derivative of the frequency with respect to charge transfer $f_c=1/p$: $\partial\omega/\partial f_c=460\pm30$ cm^{-1} (stage 1) and $\partial\omega/\partial f_c= 1050\pm120$ cm^{-1} (stage 2). These values can be used together with values obtained for $\partial d_{CC}/\partial f_C$ from Fig. 4 to obtain values for the change in bounding layer mode frequency with the contraction in d_{CC}. They arrive at values for $\partial\omega/\partial d_{CC}$ of $11,000\pm2000$ cm^{-1} Å^{-1} (stage 1) and $13,000\pm2000$ cm^{-1} Å^{-1} (stage 2) [11]. These values are quite similar and this suggests that the effect is dominated by the C-C intralayer forces and that interlayer interactions play a weak role, if any, in the process. Thus they conclude that a single layer theory for the effect of charge transfer on the intralayer mode frequency in stage 1 and 2 would be sufficient to model the results.

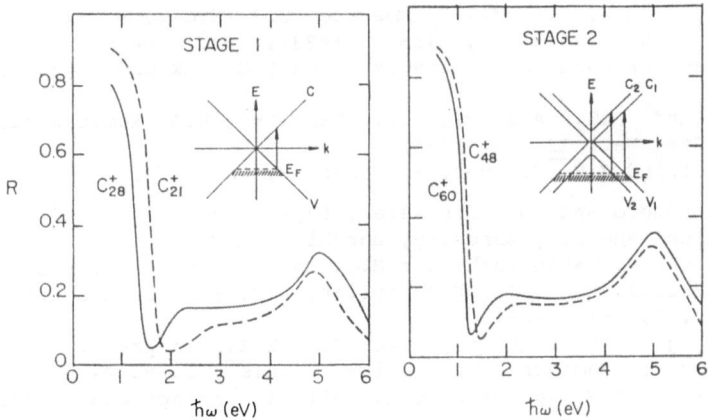

Fig. 8. Reflectance of stage 1 and 2 graphite-H$_2$SO$_4$. The energy band diagrams are from ref. [18].

Fermi Level Shifts in the Carbon π Bands

Zhang et al. [17] have measured the downward shift of E_F in the π valence band(s) of the stage 1 and 2 graphite-H_2SO_4 compounds via in situ optical (c-face) reflectance $R(\omega)$ studies in an electrochemical cell. The data for the low- and high-charge-transfer compounds are shown in Fig. 8. Insets to the figure represent the π band structure [18] near E_F at the K points (zone corners) of a (graphitic) hexagonal Brillouin zone. The number of valence and conduction bands is equal to the stage index of the GIC. An in-depth discussion of these bands is presented in another contribution to this volume [18].

The spectrum shown in Fig. 8 are characteristic of a metal, exhibiting sharp Drude edges with minima in the 1-2 eV region. These edges are associated with free carrier absorption and their position is a measure of f_C [18,19]. The rise of $R(\omega)$ after the minimum is due to a threshold for interband absorption. Transitions at the threshold are indicated by the vertical arrows between bands shown in the insets to the figure. Using a value of the principal band parameter [1,18,19] γ_0 = 2.93 eV, the hole density (f_C) obtained from analyses of the free carrier contribution to the reflectance spectra in Fig. 8 were found to be consistent with the electrochemical model for graphite-H_2SO_4 (p>21).

IV. Acknowledgements

This work was supported, in part, by the United States Dept. of Energy (#DE-FG05-ER45151). The author gratefully acknowledges collaborations with G.D. Mahan, E.T. Arakawa, W.A. Kamitakahara, J.L. Zarestky, J.M. Zhang and C.H. Olk on the in situ studies of graphite-H_2SO_4.

References

1. M.S. Dresselhaus and G. Dresselhaus, Adv. Phys. <u>30</u>, 139 (1981).
2. S.A. Solin, Adv. Chem. Phys. <u>49</u>, 455 (1982).
3. P. Boolchand, W.J. Bresser, D. McDaniel, K. Sisson, V. Yeh and P.C. Eklund, Solid State Commun. <u>40</u>, 1049 (1981).
4. D.M. Hwang, X.W. Qian and S.A. Solin, Phys. Rev. Lett. <u>53</u>, 1478 (1984).
5. W. Rüdorff and U. Hofmann, Z. Anorg. Allg. Chem. <u>238</u>, 1 (1938).
6. W. Rüdorff and E. Schulze, Angew. Chem. <u>66</u>, 305 (1954).
7. M.J. Bottomley, G.S. Parry, A.R. Ubbelohde and D.A. Young, J. Chem. Soc. 5674 (1973).
8. J.O. Bessenhard, H. Moewald and J.J. Nickl, Syn. Met. <u>3</u>, 187 (1981).
9. J.O. Bessenhard, E. Wudy, H. Moewald, J.J. Nickl, W. Biberacher and W. Foag, Synth. Met. <u>7</u>, 185 (1983), and refs. cited therein.
10. W.A. Kamitakahara, J.L. Zarestky and P.C. Eklund, Syn. Met. <u>12</u>, 301 (1985).
11. P.C. Eklund, E.T. Arakawa, J.L. Zarestky, W.A. Kamitakahara and G.D. Mahan, Syn. Met. <u>12</u>, 97 (1985).
12. A. Metrot and J.E. Fischer, Syn. Met. <u>3</u>, 201 (1981).
13. L. Pietronero and S. Strassler, Phys. Rev. Lett. <u>47</u>, 593 (1981).
14. P.C. Eklund and J.L. Zarestky, unpublished.
15. R. Nicklow, N. Wakabayashi and H.G. Smith, Phys. Rev. B <u>5</u>, 4951 (1972).
16. P.C. Eklund, C.H. Olk, F.J. Holler, J.G. Spolar and E.T. Arakawa, J. Mat. Res. <u>1</u>, 361 (1986).
17. J.M. Zhang and P.C. Eklund, Phys. Rev. B <u>34</u>, in press.
18. C. Rigaux, in another contribution to this volume, p. 235.
19. P.C. Eklund, M.H. Yang and G.L. Doll, in another contribution to this volume, p. 257.

RAMAN SCATTERING FROM CARBON AND INTERCALANT VIBRATIONS IN STAGE 1 C/AsF$_5$

I. Ohana and Y. Yacoby

Racah Institute of Physics
The Hebrew University of Jerusalem, Israel

Raman scattering has been extensively used to investigate the lattice dynamics of graphite and GICs.[1] Intercalant modes in GICs have however been observed with certainty[2] only in C/Br$_2$ and this is due to the fact that the measurement can be performed under resonance conditions. We present here Raman scattering measurements in stage 1 C/AsF$_5$.[3] The C–C vibration at 1638 cm^{-1} is resonantly enhanced in the blue–green region and has a pronounced distortion in the red region. Changing the excited surface from the c–face to a–face and using light polarized perpendicular to the graphite planes enables us to observe also the intercalant modes.

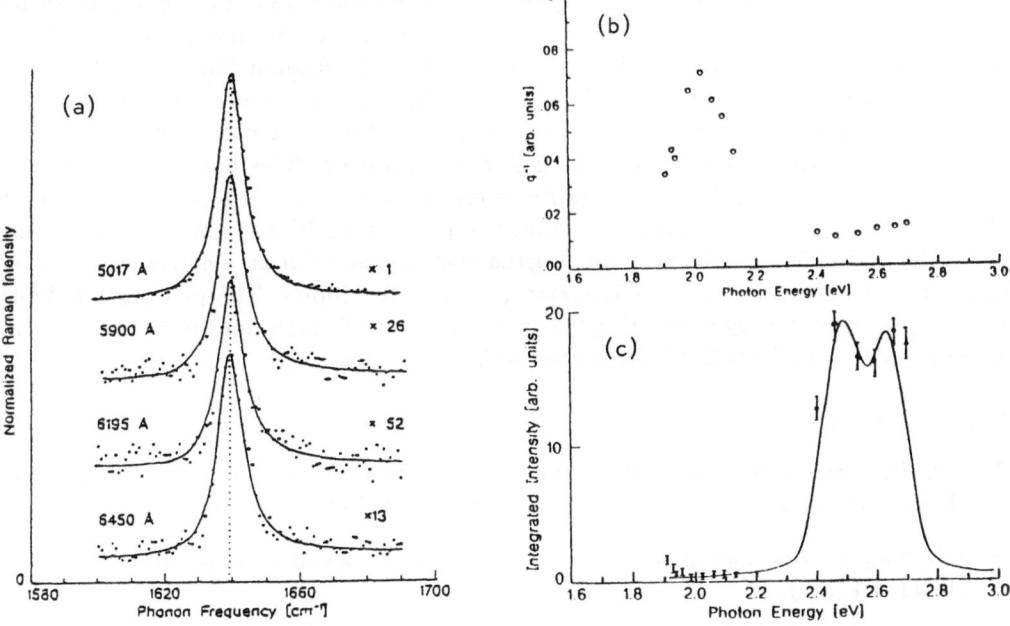

Figure 1: (a)Raman spectra of the E_{2g_2} mode in a stage 1 C/AsF$_5$ sample. The solid lines are fit to the experimental points and the exciting laser wavelengths are indicated. The dependence of the (b) Fano parameter q^{-1} and (c) Raman scattering efficiency on the exciting photon energy. The fit in (c) yields a threshold of 2.45 eV for direct interband transitions and 0.02 eV for the broadening parameter.[3]

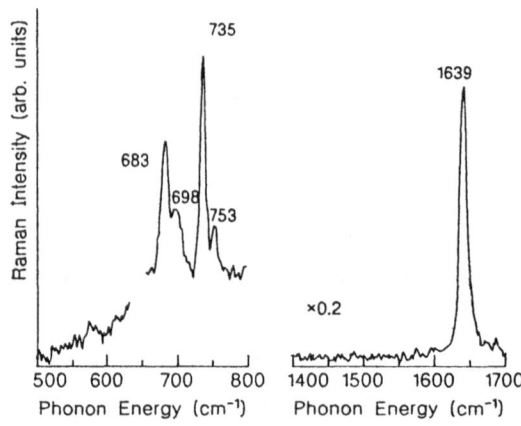

Figure 2: Raman spectra from stage 1 C/AsF$_5$. The spectra are taken on the a–face with $\vec{E} \parallel \vec{c}$ (Ref. 3).

The Raman line for the in–plane carbon mode with E_{2g_2} symmetry is shown in Fig. 1a. The spectra were taken with different exciting photon energies E_L and show enhancement in the blue–green region. A distortion from a Lorentzian lineshape is observed in the red region. The line shape is analyzed in terms of the Fano equation and yields the E_L dependence of the parameter q^{-1} with the result shown in Fig. 1b. The E_L dependence on the integrated intensity of the Raman line is presented in Fig. 1c. The Fano lineshape indicates that the one phonon excitation interacts with a continuum. A two–phonon excitation model does not agree with the observed E_L dependence of q^{-1}, suggesting that the continuum is due to electronic excitations.

The Raman measurement performed on the a–face with $\vec{E} \parallel \vec{c}$ is shown in Fig. 2. We observe 4 lines in the low frequency region. The strong 735 cm^{-1} mode is identified with the ν_1 breathing mode of the AsF$_5$ molecules surrounding the sample.[4] The other 3 modes are attributed to intercalate modes. The Raman line at 1639 cm^{-1} is the well known E_{2g_2} line of the C–C vibration. The observation of Raman lines for the intercalated molecules is possible because the penetration depth of the light in this configuration is much larger than in the $\vec{E} \perp \vec{c}$ configuration. The increased penetration depth is due to the fact that the graphite layers in stage 1 are well isolated from each other. Symmetry considerations show that the direct transitions between valence and conduction π–bands and the phonon assisted transitions, with an electronic excitation across these bands as an intermediate state, are also forbidden. The penetration depth ($\sim 10\mu$m) is estimated by comparing the intensity of the Raman lines from the external molecules with that from the intercalated AsF$_5$ molecules.

REFERENCES

1. M.S. Dresselhaus and G. Dresselhaus, in "Light Scattering in Solids, III", edited by M. Cardona and G. Güntherodt (Springer–Verlag, Berlin, 1982).

2. J.J. Song, D.D.L. Chung, P.C. Eklund and M.S. Dresselhaus, *Solid State Commun.* **20**, 1111 (1976).

3. I. Ohana, Y. Yacoby, and D. Schmeltzer, to be published; I. Ohana, Y. Yacoby, and D. Davidov, Proceedings of the Baden–Baden Conference, *Carbon'86*, p. 460.

4. L.C. Hoskins and R.C. Lord, *J. Chem. Phys.* **46**, 2402 (1967)

HOLE AND PHONON CONDUCTION IN LOW STAGES GRAPHITE ACCEPTOR INTERCALATION COMPOUNDS

Jean-Paul Issi

Unité de Physico-Chimie et de Physique des Matériaux
Place Croix du Sud 1, B-1348 Louvain-la-Neuve (Belgium)

Using recent experimental data on their temperature variation, the in-plane electrical and thermal conductivities of low stage graphite acceptor intercalation compounds are analyzed. A 2D model is used for the electrical resistivity to interpret the data and some features specific to 2D conduction are underlined. After introducing the various mechanisms responsible for thermal conduction, the effect of intercalation on the thermal conductivity of pristine graphites is discussed.

INTRODUCTION

In the study of electrical and thermal transport properties, one deals with a system of particles whose movement, initially isotropic, is directed in a preferred orientation under the action of external forces. The particles concerned are the charge carriers (electrons or holes), which may either carry the electrical or the heat current, and the quantized lattice vibrations, the phonons, which only carry heat energy. Collisions tend to bring the particle system back to equilibrium. The two external "forces" that will be considered here are those resulting from the electric or the thermal "field" (Fig. 1.1).

In a metallic or semimetallic system the electrons and holes are described by their Fermi surfaces and thus a model for this Fermi surface is an essential prerequisite for the study of the interactions of the electrons and/or hole system with the various perturbations of the lattice periodic potential. In graphite intercalation acceptor compounds (GAC), which are highly anisotropic systems, the Fermi surfaces are well approximated by circles (stage-1) or cylinders (higher stages) [1][2].

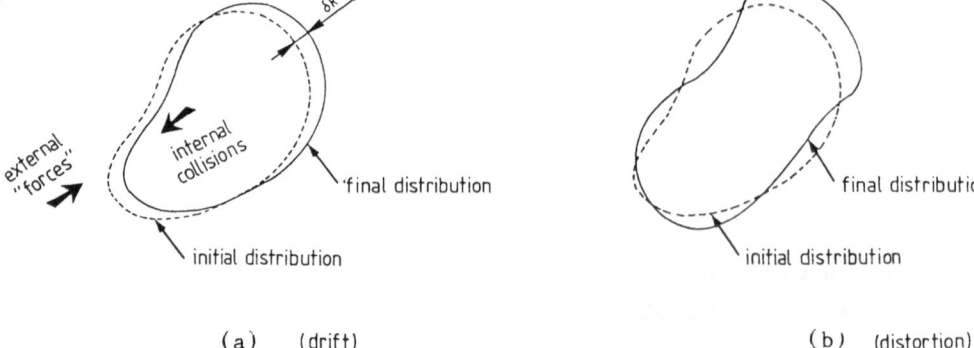

(a)   (drift)   (b)   (distortion)

Fig. 1.1 Schematic representation in momentum space of the distribution of a
system of particles which was initially at equilibrium and which is modified
under the action of external forces.
 In (a) we represent the case where the system drifts as a whole under the
action of an external force. This illustrates typically the effect of an
electric field on the distribution of charged particles (electrons or holes)
which is shifted bodily by an amount δk.
In (b) we represent the effect of a "force" which distorts the particle
distribution. This may be the case when a temperature gradient is applied to
charged particles (electrons or holes) (see Fig. 1.2). In crystalline solids the
symmetry of the system leads to particle distributions which are more
regular than the one we present schematically in this figure. In a 3D free
electron metal the particles are distributed in concentric spheres.

Under the action of an external electric field, the Fermi surface drifts as
a whole in k-space as if it were a rigid body. Collisions tend to bring the
Fermi surface into its initial position and the probability for the collisions
to take place, which determines the magnitude of the relaxation time τ, will
determine the quantitative amount of the shift δk in k-space at equilibrium
(Fig. 1.1). The shift of the Fermi surface, δk, corresponds to an electrical
current. The main interactions of the electronic particles are with the
phonons, the electrons or holes, the impurities, the static lattice defects and
the crystal boundaries.

Phonons are best described by their dispersion relations and are mainly
scattered by other phonons - through normal or umklapp processes -
impurities, lattice defects, crystal or grain boundaries and, in metallic
systems, by the charge carriers. The effect of a thermal gradient on the
Fermi surface is qualitatively different from that of the electric field since
it does not shift bodily the Fermi surface but only increases the thermal
smearing at the hot end with respect to that at the cold end (Fig. 1.2).

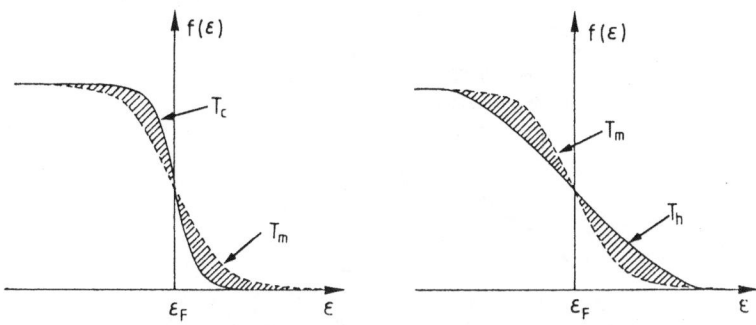

Fig. 1.2 Schematic representation of the effect of a temperature gradient on the top of the Fermi distribution, $f(\epsilon)$. T_m is the mean temperature of the sample while T_c and T_h are the temperatures of the cold and hot ends respectively. It is the redistribution in energies of the charge carriers around the Fermi level, ϵ_F, which is responsible for the electronic thermal conductivity of metallic systems.

The work on the transport properties of graphite intercalation compounds (GIC) was stimulated by the promise of realizing synthetic metals which would be candidates to compete with copper as practical electrical conductors. Besides, highly oriented pyrolytic graphite (HOPG) is for in-plane conduction the best known heat conductor with diamond around room temperature. On the other hand, electrical and thermal conductivity measurements are instrumental in gaining an insight into the electron and phonon interactions in solids as well as into their defect structures (Fig. 1.3). However, in order to use transport data as a tool to study the carrier interactions, it is essential that the measurements be performed over a wide temperature range and, for the electrical conductivity, that high resolution measuring systems be used. For the interpretation of the data, a realistic model for the electronic structure is also needed.

Nowadays, a few experimental data on the electrical resistivity are available for some HOPG-GACs from liquid helium temperature up to room temperature [3-5]. However, since the residual resistivity ratio – the ratio of the room temperature resistance to that at liquid helium – is rather small in these compounds (roughly 3 to 10), it is difficult to extract the ideal term in the low temperature range from the measured total resistivity in HOPG. In graphite fibers the defect structure may vary widely according to the type of fiber (vapor deposited, pitch-derived, pan-based, ...), the heat treatment temperature (HTT), and the quality of the precursor. In contrast to HOPG, because of the favorable length to cross section ratios, four-probe DC

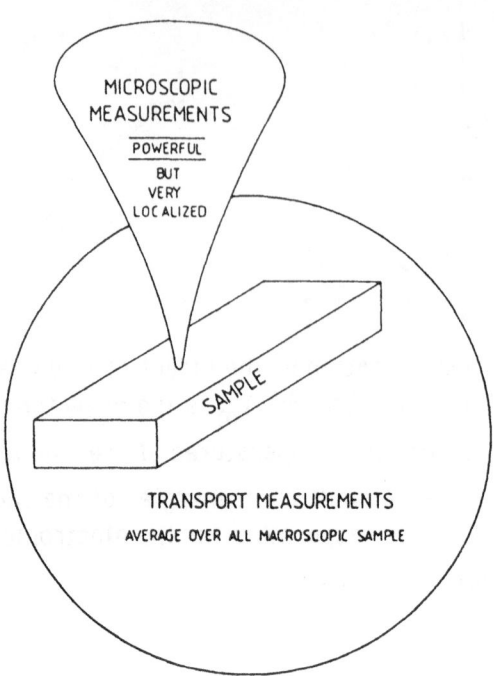

Fig. 1.3 Schematic representation illustrating the complementary nature of microscopic (e.g. high resolution microscopy) studies and transport measurements to characterize solid samples. This is particularly the case when lattice defects are investigated. Microscopic techniques are very powerful tools which allow these defects to be "seen" by the observer. However, only a tiny region of the sample may be scanned. Transport measurements, which probe the nature of the scattering of electrons and phonons may yield quantitative information about lattice defects provided that a correct model for the particle distribution is available. The information obtained from transport measurements provide a value averaged over the entire sample.

electrical measurements may be performed on graphite fibers. It is thus possible to perform high resolution electrical resistivity measurements on graphite fibers acceptor compounds and determine the ideal resistivity [6][7]. As a corollary, one may measure simultaneously the electrical and thermal conductivities on the same sample [8][9]. This allows one to extract directly the electronic component from the measured thermal conductivity in some temperature ranges using the Wiedemann-Franz relation (4.1).

One of the essential features of GACs is the very high anisotropy of the electrical conductivity [10] and, to a lesser extent, of the thermal conductivity [11][12]. In some GACs the ratio of the in-plane electrical conductivity to that along the c-axis may reach 10^6 at room temperature [10]. This is why it is justified to consider the hole system as a quasi two-dimensional (2D) gas. In addition to the high anisotropy, these compounds present essentially two marked differences when compared to other metals.

First, the 2D behavior is inherent to the distribution of the charge carriers, which are strongly localized in the graphitic planes and may be considered as free carriers only for motion along these planes. This is in contrast to metallic thin films where the quasi 2D behavior results from strong scattering of the charge carriers on the surfaces thus limiting the mean free path in the direction perpendicular to the surface. In the discussion of recently discovered 2D weak localization and interaction effects this will have an important bearing. It will also lead to differences in the energy dependence of the density of states.

Second, in GAC synthetic metals, the in-plane lattice properties are governed by the covalent nature of the graphitic planes while the electronic system originates from the charge transfer from the intercalate. This will introduce substantial differences when the situation is compared to that in ordinary metals, especially when treating the electron-phonon interactions [7]. In an ordinary metal it is the metallic character - i.e the metallic bond - which is responsible for the lattice properties. In other words, the large free carrier concentration of the Fermi gas, which governs the electronic transport properties, is also responsible for the cohesion of the material and will determine the intensity of the binding forces. The resulting cohesive energy, which though relatively high is still lower than for covalent solids, leads to Debye temperatures which rarely exceeds room temperature. This is to be compared with the much higher Debye temperatures for covalent materials (e.g. ≈ 2300 K for diamond, ≈ 640 K for silicon). In GACs the quasi 2D electron cloud gives only a minor contribution to the in-plane binding forces and the cohesive energy is almost entirely determined by the in-plane covalent forces. Conceptually, this is akin to the case of doped inorganic semiconductors where the lattice properties are governed by those of the host material, while, in the rigid band approximation, the doping elements govern the electronic properties with no significant influence on the lattice properties (phonon spectrum, band structure,..).This means that we should not expect to observe in GACs electron-phonon interactions similar to those observed in ordinary metals.

Using a 2D model for the electronic structure [1][2] and the Fermi energies determined from reflectivity measurements (§ 2), some recent electrical resistivity data, obtained for low stages GACs are analyzed (§ 3). Some features specific to 2D electrical transport will also be underlined. In § 4, after introducing the various mechanisms responsible for thermal conduction, the effect of intercalation on the thermal conductivity of the pristine host material is discussed.

THE ELECTRONIC STRUCTURE

The various electronic models which have been proposed for GACs are reviewed in this issue [1]. For the stages 1 and 2 acceptor compounds we shall use hereafter the Blinowski-Rigaux model [2], which has the advantage of being handled easily. Also this model, when applied to the experimental results concerning the electronic transport properties, gives consistent results.

For the carrier dispersion relations for a stage-1 GAC, a linear energy, ε_j, versus - wavevector, k, relationship is obtained

$$- \varepsilon_v(k) = \varepsilon_c(k) = \frac{3}{2} \gamma_0 \, b \, k \qquad (2.1)$$

where b and γ_0 are the in-plane nearest neighbor distance and overlap energy respectively and the indices c and v correspond to the conduction and valence bands respectively. For a stage 2-GAC, taking into account the nearest-neighbor out of plane overlap energy γ_1, the dispersion relations for bands 1 and 2 are given by [1][2]

$$- \varepsilon_{v1}(k) = \varepsilon_{c1}(k) = \frac{1}{2} [(\gamma_1^2 + 9 \gamma_0^2 b^2 k^2)^{\frac{1}{2}} - \gamma_1]$$

$$- \varepsilon_{v2}(k) = \varepsilon_{c2}(k) = \frac{1}{2} [(\gamma_1^2 + 9 \gamma_0^2 b^2 k^2)^{\frac{1}{2}} + \gamma_1] \qquad (2.2)$$

In this case a linear ε-k relation is obtained in the limit of large k and a parabolic ε-k relation for small k.

In a stage-1 compound the 2D hole density per square may be expressed in terms of the Fermi wave vector, k_F

$$N = k_F^2 / \pi \qquad (2.3)$$

and of the Fermi energy, ε_F,

$$N = 4 (9\pi \, \gamma_0^2 \, b^2)^{-1} . \varepsilon_F^2 \qquad (2.4)$$

and since the density of states is defined by

$$g(\varepsilon) = dN \, / \, d\varepsilon \qquad (2.5)$$

it gives at the Fermi energy

$$g(\varepsilon) = 8(9\pi \; \gamma_0^2 \, b^2)^{-1} \; \varepsilon_F \qquad\qquad (2.6)$$

The density of states varies linearly with energy, thus more rapidly than a 3D metallic system where it varies as $\varepsilon^{\frac{1}{2}}$. More generally, it may be shown that for higher stages the 2D hole density is given by

$$N = (1/\pi) \sum_j k_{Fj}^2 \qquad\qquad (2.7)$$

where k_{Fj} is the Fermi wave vector for the band j.

From these relations it may be seen that once the Fermi energy is known - e.g. as measured from optical reflectivity measurements - the electronic parameters are completely defined (Fig. 2.1). It is also the case for a 3D free electron system, but not for a real metal which has usually a complicated Fermi surface. This will facilitate to a great extent the interpretation of the electronic transport properties of GACs, which will in turn test the validity of the model. Note that for a stage-1 compound the Fermi energies are of the order of 1 eV [10].

THE ELECTRICAL RESISTIVITY

A general expression for the electrical conductivity of a 3D Fermi gas is given in standard solid state books [13][14]. It expresses the electrical conductivity in the form

$$\sigma_{3D} \propto \tau \int \underline{v}.\underline{dS} \qquad\qquad (3.1)$$

where $\underline{v}$ and τ are the electron velocity and relaxation time respectively and $\underline{dS}$ is an element of the area of the 3D Fermi surface. Thus the high electrical conductivity of 3D metallic systems is due to the high Fermi velocity of the charge carrier system.

Applying these ideas to a 2D circular Fermi surface leads naturally to the expression :

$$\sigma_{2D} = q^2 \, \tau \, (2\pi^2 \hbar)^{-1} \int_o^{2\pi k_F} \underline{v}.\underline{ds} \qquad\qquad (3.2)$$

where $\underline{ds}$ is an element of the 2D Fermi surface.

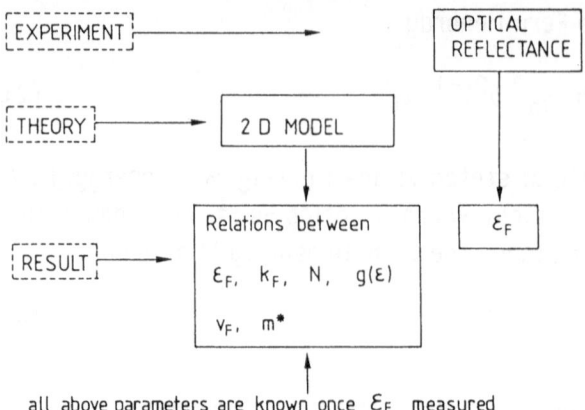

Fig. 2.1 Schematic diagram showing how in the framework of the Blinowski - Rigaux model [1][2] the experimental determination of one parameter, the Fermi energy ε_F, allows to obtain via the derived relations (see text) the main electronic parameters which describe the electronic distribution.

By integration (we shall drop hereafter the index 2D)

$$\sigma = (\pi \hbar)^{-1} . q^2 \, v_F \, \tau . k_F \tag{3.3}$$

where k_F is the Fermi wave-number, v_F the Fermi velocity and q the electronic charge. Expression 3.3 may be rewritten in the form :

$$\sigma = q^2 \, v_F \, \tau . (k_F^2/\pi) . (\hbar \, k_F)^{-1} \tag{3.4}$$

Since $\hbar \, k_F$ corresponds to $m v_F$ and combining relations (2.3) and (3.4) one finds for a given carrier group the universal expression of the electrical conductivity

$$\sigma = q^2 \, N \tau / \, m^* \tag{3.5}$$

where m* is the effective mass of the charge carriers. There is no great virtue in falling on such an expression (3.5), since one may demonstrate that whatever the model chosen the conductivity is always expressed in the same way. However, in order to understand the physics governing the conductivity mechanisms one should specify the real meaning of each parameter in expression (3.5). In the 2D model we are considering here, N is a carrier density per cm^2 (per cm^3 in 3D systems) and the conductivity is in units of Ω^{-1} ($\Omega^{-1}.cm^{-1}$ in 3D systems). Also, it is interesting to note that in contrast to a 3D free electron gas, where the Fermi velocity is energy and wave vector dependent and the mass m is the constant free electron mass (9.11 $10^{-28}g$), in our 2D system the Fermi velocity is energy and wave vector independent. Also the effective mass is a linear function of energy and wave vector. This may be considered as an extreme case of non parabolicity.

Since the carrier mean free path l is equal to $v_F \tau$, one may also write

$$\sigma = q^2 N l / \hbar k_F \qquad (3.6)$$

Macroscopically what is measured is the electrical resistance of a stacking of d/l_c parallel planes where d is the sample thickness and l_c the c-axis repeat distance. If R is the measured electrical resistance of a sample consisting of $1/l_c$ planes of width 1 cm and length 1 cm, then R is equivalent to a 3D resistivity,

$$R = l_c \rho \qquad (3.7)$$

where $\rho = \sigma^{-1}$ is the 2D electrical resistivity.

Up to now we have considered a single band and the results are directly applicable to a stage-1 compound where the 2D character of the graphite layers is also the most pronounced. However, for higher stages, more than one band should be considered each having its specific electronic parameters [1]. Note also that in GACs each band consists of many valleys distributed in k-space and the conductivity of each band is merely equal to the sum of the contributions of each valley.

Multiband conduction and scattering

In the case of multiband conduction (higher stages) the total electrical conductivity is given by the sum of the partial conductivities σ_j for each band j

$$\sigma = \sum_j \sigma_j \qquad (3.8)$$

The carriers from each band will experience various scattering events, which we shall denote by the subscript s. Hence applying Matthiessen's rule for each band

$$p_j = \sum_s p_{js} \qquad\qquad (3.9)$$

one may then write, combining (3.4) and (3.8) and (3.9)

$$\sigma = q^2/(\pi\hbar) . \sum_j [\, k_{Fj} \, (\sum_s l_{js}^{-1})^{-1} \,] \qquad\qquad (3.10)$$

which is the expression of the electrical conductivity for our system when the contribution of each band and the various scattering events of its population are taken into account.

Electrons and holes may be scattered by static defects leading to a relaxation time, and thus to a resistivity – the residual resistivity (s = r) – which is temperature independent. They may also be scattered by other electrons and by phonons and this leads to a temperature dependent resistivity – the ideal resistivity (s = i).

For the case of a stage-2 GAC since there are two hole bands (j = 1, 2) if we consider the case where the two scattering mechanisms r and i are both operative the total conductivity is given by

$$\sigma = (q^2/\pi\hbar) \, [k_{F1} \, (l_{1r}^{-1} + l_{1i}^{-1})^{-1} + k_{F2} \, (l_{2r}^{-1} - l_{2i}^{-1})^{-1}] \qquad (3.11)$$

This shows that the contribution to the conductivity from different carrier groups add, while it is the resistivity due to various scattering mechanisms that add and Matthiessen's rule (3.9) is applied for each band.

The residual resistivity

The residual resistivity is temperature independent since scattering is due to the static lattice defects. These include point defects, dislocations and surface defects such as sample boundaries for single crystals and grain boundaries for a polycrystalline material. If we consider a 2D material, a linear defect in the plane of the charge carrier propagation will be as effective a scatterer as a surface defect in 3D solids. Thus one would expect in GACs that, in addition to grain boundary scattering, the charge carriers will be scattered at low temperatures by the Daumas-Hérold domain boundaries. This will also be the case for phonons (§ 4). One case of practical importance is when large-scale defect scattering dominates, which

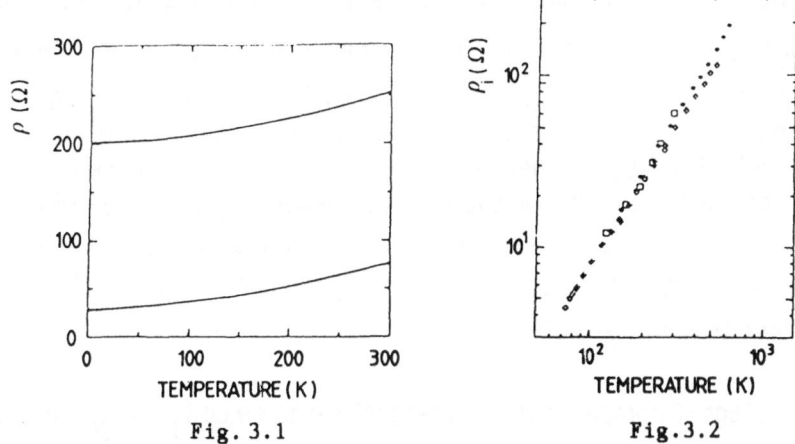

Fig. 3.1   Fig.3.2

Fig. 3.1 Typical results showing the temperature variation of the total 2D resistivity (residual + ideal) of a GAC with two different host materials. The intercalant in both cases is $CuCl_2$ and the compounds are of stage-2. In the lower curve the host material is a benzene-derived fiber heat treated at 3000°C and in the upper curve a pitch-derived fiber heat treated at about the same temperature (From reference 16)

Fig. 3.2 Typical result showing that the 2D ideal resistivity does not vary significantly for various intercalates and host structures. The results are relative to a stage-2 $CuCl_2$ intercalation compound with a benzene-derived fiber (◆) and a pitch-derived fiber (◇) as host materials, and a stage-1 $FeCl_3$ compound with a benzene-derived fiber (●) and HOPG (□) as host materials (From reference 7)

is believed to be the case for all GAC s in the liquid helium range [11]. In that case the mean free path is temperature and energy independent and one may use the residual resistivity as a tool to determine the size of these defects.

For a stage –1 compound when $\rho_{jr} \gg \rho_{ji}$ combining (2.1)(2.3) and (3.5) one may write

$$l_r = 3 \pi \hbar b \gamma_0 \, l_c \, (2 q^2 R_r \varepsilon_F)^{-1} \qquad\qquad (3.12)$$

Since in HOPG-GACs the equivalent 3D residual resistivity, R_r, is of the order of $10^{-6}\Omega.cm$, $l_c \approx 10^{-9}m$ and $\varepsilon_F \approx 1$ eV, the mean free path for defect scattering is of the order of a few $10^{-7}m$ (Table 3.1). This would give a higher limit for the size of large scale defects in stage-1 HOPG intercalation compounds[7]. For fibers, where the structure is generally more disordered than that of HOPG, R_r is higher and l_r is correspondingly smaller. For intercalated benzene-derived fibers (BDF), where the structure is the closest to HOPG, l_r is larger than for intercalated pitch-derived fibers (PDF) and is indeed close to l_r in HOPG (Table 3.1).

For a stage-2 compound in the residual range since $l_{ji} \gg l_{jr}$, relation (3.11) becomes

$$\sigma_r \cong (q^2/\pi\hbar)\,(k_{F1}\,l_{1r} + k_{F2}\,l_{2r}) \qquad (3.13)$$

and if we assume that $l_{1r} \approx l_{2r}$

$$\sigma_r = (q^2/\pi\hbar)\,(k_{F1} + k_{F2})\,l_{1r} \qquad (3.14)$$

and since $\sigma_r = l_c/R_r$

$$l_{1r} = (l_c/R_r)\,.\,(\pi\hbar/q^2)(k_{F1} + k_{F2})^{-1} \qquad (3.15)$$

Relation (3.15), as well as relation (3.12), are of practical importance for sample characterization since they show that the measurement of the low temperature resistance R_r and the knowledge of the Fermi energies or wave vectors allow one to estimate the size of large-scale defects. In Fig. 3.1 we present the temperature variation of the total 2D resistivity of a stage-2 $CuCl_2$ intercalation compound with a benzene-derived and pitch-derived fibers as host materials. The difference in residual resistivities for the two samples is clearly apparent.

The ideal resistivity

The main scattering mechanism for charge carriers which leads to a temperature dependent relaxation time is that by acoustical phonons. The electron-phonon interaction, which mainly determines the ideal or intrinsic resistivity of a solid, may occur within a single valley (intravalley scattering), or involve a transfer of momentum from one valley to another (intervalley scattering) or from one band to another (interband scattering). In

Table 3.1 Comparison of the hole mean free path for defect scattering, l_r, computed from R_r (cfr. relations 3.12 and 3.15) with the mean free path of phonons for boundary scattering, l_p, [8][11].

HOST GRAPHITE	INTERCALATE	STAGE	R_r $10^{-6}\Omega.cm$	l_r $10^{-8}cm$	l_p $10^{-8}cm$
HOPG	$FeCl_3$	2	0.7-0.88 (*)	5800-7300	5800
BDF	$CuCl_2$	2	3.4	1700	> 950
PDF (VSC-25)	$CuCl_2$	1 + 2	22	250	340

(*) From the values taken on four samples by E. Mc Rae and J-F. Marêché (private communication)

the case of GACs one should consider for scattering the in-plane and out-of plane graphitic phonons and the additional phonons introduced by the presence of the intercalate.

The quantitative description of the electron-phonon interaction is a delicate attempt even for a free electron system interacting with a phonon distribution within the framework of a Debye model. We refer to Ziman's [13] or Blatt's [14] books for a detailed account of the situation. A rough description of the state of the affairs in GAC's has been recently given [7]. We shall only state here the essential features which distinguishes the intravalley electron-phonon interaction in GACs from that of a 3D free electron system.

Generally, the effectiveness of a scattering event for a degenerate electron gas depends on the angle through which the charge carriers are scattered on the Fermi surface and on the number of phonons which are available for a given scattering angle. The main difference between a 3D free electron system and a 2D GAC is that below room temperature the charge carriers do not interact with the same class of phonons in the two systems. In the first case, the electrons are scattered by the Debye phonons around and above the Debye temperature and by phonons of energy $\approx k_B T$ below the Debye temperature. This leads to small relaxation times around room temperature and to large ones at low temperatures. For stage-1 and 2 GACs, the holes are scattered through large angles above room temperature by low energy subthermal phonons, which are not the dominant phonons at these temperatures. At low temperatures holes are scattered by phonons of energy $\approx k_B T$ through small angles. However, the recent findings concerning the low temperature behavior of these compounds [15][16] suggest that strong electron-electron interactions might be effective at low temperatures.

For the electron-phonon resistivity the Bloch-Gruneisen relation [13][14] predicts for 3D metallic systems a T^n law with n = 5 at very low temperature followed by a gradual decrease of n until it is equal to 1 around and above the Debye temperature. From ealier experimental data on the temperature variation of the electrical resisticity of GACs, it was pointed out that the temperature dependence of the ideal resistance could be expressed [10]

$$R(T) = BT + CT^2 \qquad (3.16)$$

where B and C are constants.

High resolution electrical resistivity measurements performed on low stage GACs confirmed the above relation in the temperature range 1.5 < T < 300K [6][7]. An almost linear variation of the ideal resistivity was observed in the low temperature range and a T^n behavior with n $\approx$ 2 around room temperature. These data are shown in Fig. 3.2 where the temperature variation of the 2D ideal resistivities of low stages GACs are presented for various intercalates and host materials. It may be seen that, contrary to what is observed in a 3D metal, we find a higher power law at higher temperature. It is hard to explain a T^1 variation if we consider only electron-phonon (acoustic) interactions in the low temperature small angle scattering region. One should probably invoke 2D electron-electron interactions to explain this anomalous dependence [Piraux et al., to be published].

If the nature of the temperature dependence of the resistivity still remains a matter of conjecture, there are however some interesting experimental facts which have recently been established. Indeed, one may note from Fig. 3.2 that the high temperature ideal resistivity of a stage-2 $CuCl_2$ intercalation compound with BDF or PDF as host materials is exactly the same for both systems while the residual resistivity is, as expected, quite different (Fig. 3.1). This applies also to $FeCl_3$ with BDF and HOPG (Fig. 3.2) and is also true with other intercalates ($SbCl_5$, $CoCl_2$). For almost all the systems which have been recently studied by Piraux and co-workers (to be published), the value of the room temperature ideal resistivity is (5 ± 1).

$10^{-6}\Omega$ cm. So one may reach an important conclusion. The electronic system (i.e. the Fermi energy) depends solely on the charge transfer from the intercalate while the scattering mechanism is mainly due to the phonon spectrum of the host graphite and its defect structure. The phonon spectrum is not too different from one graphite host to another provided it is relatively well ordered. Since, for a given stage, the Fermi energies are not too different for the various intercalates considered here, the electron-phonon interaction, and thus this part of the ideal resistivity, should not be too different from one sample to another at high temperature. On the other hand, the residual resistivity depends on the defect structure of the pristine host material and the additional defects introduced by intercalation. Thus it should vary from one host material to another and for the various intercalates.

Localization and interaction effects

The situation depicted in the preceeding sections, i.e. a resistivity consisting of additive residual and ideal contributions, was considered for a long time as a general feature of 3D metals and was applied to the case of GACs. In fact, from the experimental viewpoint a measurement performed with a $\approx 1\%$ accuracy will generally confirm the validity of Matthiessen's rule and reveal a constant resistivity in the lower temperature range. However, there is growing evidence now that in metallic systems this picture is incomplete and that interesting physics emerges when one looks more closely at the region where the resistivity is apparently flat. In fact, in quasi 2D metallic systems, if one is able to detect minute resistivity changes, a logarithmic increase of resistivity with decreasing temperature is observed [17][18]. This was also recently observed in GACs [15][16] where changes in ρ of the order of 0.1% were detected over a decade of temperature in the liquid helium range (Fig. 3.3). The logarithmic increase of resistivity with decreasing temperature was attributed to localization and/or interaction effects [15][16].

In the weak disorder limit, which is taken as $k_F l_r > 1$, the expression (3.5) of the conductivity, that we shall call herafter the Boltzmann conductivity, σ_B, should be corrected by a logarithmic quantum term $\delta\sigma$ such that the total conductivity reads

$$\sigma_t = \sigma_B + \delta\sigma \qquad (3.17)$$

The term $\delta\sigma$ accounts for localization and interaction effects which both predict a similar resistance -temperature anomaly.

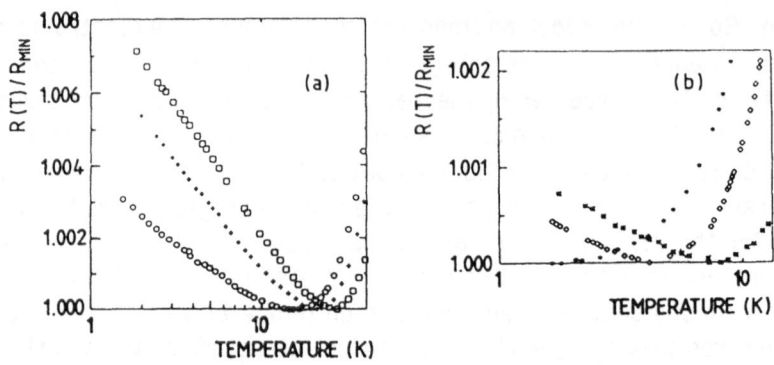

Fig. 3.3 Typical result relative to the low temperature variation of the electrical resistivity of GACs and showing localization and/or interaction effects for various intercalates and host materials with different defect structures. All host materials are pitch-derived fibers except one benzene-derived fiber (BDF)

a) VSC25-CuCl$_2$ (O), P100-4-CuCl$_2$ (◆) and VSC25-CoCl$_2$ (□)

b) BDF-SbCl$_5$ (●), PX5-CuCl$_2$ (◇) and PX5-SbCl$_5$ (■)

The data are normalized to the minimum value of the resistance (From reference 16)

It may be seen from Fig. 3.3 that this effect is apparent in most GAC s whatever the host material or intercalant. The temperature at which the minimum occurs in the ρ(T) curve is found to increase with the residual resistivity ρ_r thus suggesting that for the less disordered graphites decreasing the temperature will reveal the same behavior but with a minimum of resistivity below 1.5 K.

We think that GACs are the choice materials to study 2D localization and interaction effects. In metallic films, where the effect was first observed, the 2D character is associated with the anisotropy of the mean free path. GAC s and especially stage-1 compounds are natural 2D electronic systems since this character is inherent in their band structure. Also, the possibility of varying the defect structure of the host material over wide ranges and that of varying the stage and the intercalate allows large experimental possibilities to investigate the phenomena of localization and electron-electron interactions.

Also, one should consider what is the upper limit of temperature where weak disorder might exert an influence on the temperature dependent part of the resistivity. It should be interesting to check whether electron-electron interactions might explain the peculiar linear temperature variation of the resistivity.

THE THERMAL CONDUCTIVITY

The study of the thermal conductivity of GICs is particularly attractive and rewarding in many respects. It complements the information which may be obtained from electrical resistivity measurements either by using the same tool : the charge carriers, or by using a totally independent tool, the phonons.   Phonons are an alternative powerful means of probing the lattice defects and are the only way to investigate phonon-phonon interactions in these materials. On the other hand, a survey of the thermal conductivity of GICs associated with that of pristine graphite is a good introduction to the subject for the non initiated, since all the low temperature thermal transport mechanisms are operating in these materials and may be controlled.

Thermal conductivity mechanisms

The electronic thermal conductivity is directly related to the electrical conductivity σ through the Wiedemann-Franz law (WFL)

$$\kappa_E = L T \sigma \qquad (4.1)$$

where L is the Lorenz ratio which takes the value of the Lorenz number (L_0 = 2.44 $10^{-8}V^2K^{-2}$) for a degenerate free electron system which undergoes elastic collisions.

The lattice or phonon thermal conductivity κ_L is well approximated by the Debye formula

$$\kappa_L = \frac{1}{3} C_v v \, l \qquad (4.2)$$

where C_v is the lattice specific at constant volume, v is the phonon group velocity - the velocity of sound -, and l is the phonon mean free path, which is directly related to the phonon relaxation time τ

$$l = v \, \tau \qquad (4.3)$$

In dielectric solids heat is exclusively carried by phonons. For highly doped semiconductors, metallic alloys and group V semimetals the lattice thermal conductivity may be comparable to the electronic contribution in certain

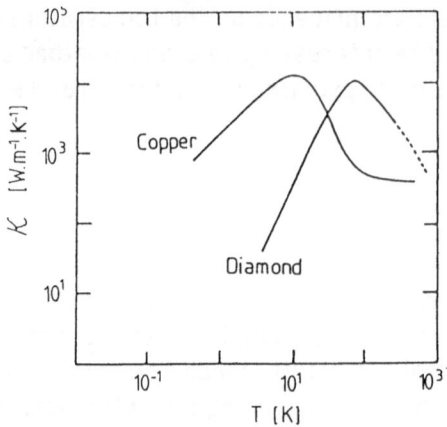

Fig. 4.1 Typical temperature dependence of the thermal conductivity of a metal (copper) and a dielectric (diamond). In the case of pure copper, the thermal conductivity is entirely electronic. It starts with a linear temperature variation at low temperature in the range where the electrical resistivity is constant, reaches a maximum and then decreases to level off at higher temperature, where the electrical resistivity increases linearly with temperature. Diamond exhibits a typical dielectric behavior. The low temperature thermal conductivity varies first as T^3, reaches a maximum and then decreases with temperature.

temperature ranges. In that case, the total thermal conductivity is expressed

$$\kappa = \kappa_E + \kappa_L \qquad\qquad (4.4)$$

Experimentally, it is possible to separate these two contributions by applying a large magnetic field. From the WFL (4.1) it may be seen that the application of a magnetic field to a sample with high magnetoresistance will deplete the electronic thermal conductivity so that $\kappa_E(H)$ may become negligible in comparison with κ_L. When $\kappa(H)$ saturates in the limit of high magnetic fields, $\kappa(H) \approx \kappa_L$ and κ_E and κ_L may be separated.

Before discussing the thermal conductivity of GICs we shall briefly introduce metallic thermal conduction   and lattice conduction,   which is well illustrated by the behavior of pristine graphite .

The electronic thermal conductivity

   A typical temperature variation of the electronic thermal conductivity is well illustrated by that of pure copper (Fig.4.1). It starts with a linear variation in the lowest temperature range, where the electrical resistivity is constant and is ascribed to the lattice static defects (see §3.2). At higher temperature, for a pure sample, κ_E reaches a sharp maximum then decreases and saturates to a constant value at higher temperatures (Fig.4.1). This saturation occurs in the range where the electrical resistivity is expected to vary linearly with temperature. The higher the purity and structural perfection of the sample, the sharper is the maximum of κ_E observed. In the limit of a very impure or a heavily damaged metal, the maximum may not be observed and the linear increase is directly followed by a temperature insensitive κ_E.

One must be cautious when transposing directly the electrical resistivity data to the κ_E behavior through the WFL. Solid state theory [13][14][19][20] states that relation 4.1 is only applicable when a charge carrier system is totally degenerate and when it may be assumed that the same relaxation time may be used for both electrical conductivity and electronic thermal conduction. This means that relation 4.1 is applicable in the lower temperature range where scattering is dominated by impurities and lattice defects, i.e. in the residual range where $\rho = \rho_r = $ constant. For 3D metals it is also applicable above the Debye temperature where large angle electron-phonon scattering dominates. This is when the size of the Fermi surface in k-space is comparable to that of the Debye sphere.

For GACs, since the size of the Fermi surface is smaller than that of the Debye sphere, it is an effective temperature that should be considered for electron-phonon interaction instead of the Debye temperature [7]. This means that around room temperature one should expect quasi-elastic scattering to occur in GACs. Also, since most GACs have very low residual resistivity ratios, the residual resistivity dominates up to relatively high temperatures. Thus the validity of the WFL in the residual range is expected to be maintained up to temperatures higher than those for pure 3D metals.

The thermal conductivity of various graphites

The thermal conductivity of HOPG was extensively studied in the sixties and the situation has been recently reviewed by Kelly [21]. In pure HOPG the in-plane thermal conductivity is exclusively due to the phonons above the liquid helium range [22][23]. The reason for that is the relatively low electrical conductivity of this semimetal as compared to metals and the very high lattice thermal conductivity as compared to dielectric materials, if we exclude diamond. Indeed, one may think in a naive way that the strong covalent bonding and the small mass of the carbon atoms facilitates the

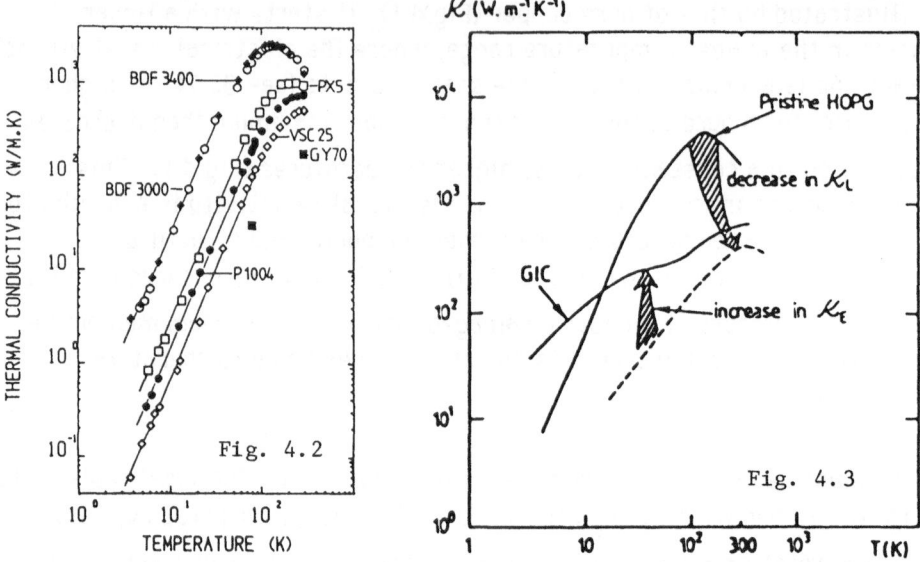

Fig. 4.2 Temperature variation of the thermal conductivity of various pristine graphite fibers showing the influence of large scale defects on the lattice thermal conductivity. The temperature variation of the thermal conductivity of pristine HOPG is exactly the same as that of a benzene-derived fiber (BDF) heat treated at the same temperature (here 3000 and 3400 °C). The PX5, P1004 and VSC25 are pitch-derived fibers and the PX5 has the highest structural perfection while the VSC25 has the lowest. GY70 is a pan-based fiber (From reference 25)

Fig. 4.3 Schematic representation of the effect of intercalation on the thermal conductivity of pristine HOPG. Intercalation has two main effects. First, the lattice defects introduced during the intercalation process decrease the phonon mean free path leading to a decrease in the lattice thermal conductivity, κ_L, in the intercalated sample relative to pristine HOPG (lower dashed curve). Second, because of charge transfer, the charge carrier density increases and this increase is accompanied by a decrease in carrier mobility. However, this decrease is less important than the enhancement of the charge carrier density and the overall effect is an increase in the electronic thermal conductivity, κ_E, as it is the case for the electrical conductivity. The final result is schematically illustrated by the curve GIC.

transmission of their vibrational motion resulting in a very high in-plane lattice thermal conductivity. On the other hand, the weakness of the interplanar forces leads to a very low c-axis lattice thermal conductivity which is, at room temperature, two orders of magnitude smaller than in-plane. Also, the low electrical conductivity in the c-axis direction is responsible for the negligible contribution of the charge carriers to the thermal conductivity in this direction.

The Debye formula (4.2) is a useful relation to discuss qualitatively the observed lattice thermal conductivity results. For the case of pristine graphite, whether HOPG or fibers, it gives a first insight into the mechanisms governing the temperature dependence of the observed thermal conductivity. As is the case for diamond - a typical covalent 3D dielectric -, the Debye formula shows that in the lowest temperature range the temperature variation is mainly that of the lattice specific heat. The mean free path is constant and limited by boundary scattering. For a diamond single crystal, boundary scattering is caused by phonon reflections at the crystal walls, while for graphite, phonons are scattered at the crystallite boundaries. Since the velocity of sound is almost temperature independent, the temperature variation of κ_L reflects directly that of C_v (Fig.4.1).

In the higher temperature range, where the lattice specific heat is temperature independent, κ_L decreases when the temperature increases. This reflects the variation of the phonon mean free path due to resistive phonon-phonon umklapp scattering [13][19][20][24]. Between these two regions lies the dielectric maximum whose magnitude and corresponding temperature depends on the structural perfection and size of the sample.

The effect of structural perfection on the in-plane lattice thermal conductivity of HOPG has been investigated in detail [21]. Recently, these studies were extended to carbon fibers of various origins and of different structural perfection in the temperature range 2 < T < 300 K [25]. It was found that, for a given precursor and up to a certain temperature, increasing the heat treatment temperature yields a higher thermal conductivity with a sharper maximum shifted to lower temperature ([Fig. 4.2]. This behavior is what one expects from the thermal conductivity of dielectric solids when the sizes of the large scale defects vary in a given material [19].

In well-ordered pristine graphite the temperature dependence of the thermal conductivity below the maximum follows a T^n law with n $\approx$ 2.3. This temperature variation is stronger than that of the specific heat from 13 K to 54 K where n = 2 [21][26]. For 3D dielectrics as diamond n = 3 (Fig.4.1). In the boundary scattering regime the lattice thermal conductivity of well-ordered pristine graphites, whether HOPG or fibers, may thus be expressed [27]

$$\kappa_L = A L_a T^n \qquad\qquad (4.5)$$

where A is a constant, which has roughly the same value for all samples, n ≈ 2.3 and L_a is directly related to the in-plane crystallite size. So a measurement of κ_L at low temperature allows a direct estimation of the crystallite size or more generally of the size of large-scale defects. This correlation may be used to estimate the effect of intercalation on the in-plane coherence length in the graphitic planes.

Now, if we need quantitative information about the effect of the various scattering mechanisms on κ_L one should introduce some refinements to the Debye approach. One has to take into account all the resistive mechanisms as point defects, large scale defects and phonon-phonon umklapp scattering processes (U-processes). Also normal processes (N-processes) should be considered. Though they do not contribute directly to the thermal resistance, N- processes have an indirect influence since they lead to the creation of higher-frequency phonons, which are more apt to undergo resistive processes such as umklapp scattering or point defect scattering.

Callaway [28] developed a method which allows us to take into account these various mechanisms in a quantitative way. He expresses the total lattice thermal conductivity κ_L as a sum of two terms :

$$\kappa_L = \kappa_1 + \kappa_2 \qquad\qquad (4.6)$$

where κ_1 considers all collisions including N- processes as resistive mechanisms. κ_2 is then a correction term which makes allowance for the fact that N-processes do not contribute directly to the thermal resistance. The Callaway treatment which is well described by Berman [19] was recently applied to GICs [11]. Though we refer to [11] and [19] for a detailed discussion, we shall point out here what are the main phonon scattering mechanisms which are of interest to graphites and GICs.

Phonon scattering mechanisms

When discussing charge carrier scattering in a degenerate electron gas, the carriers of interest at any temperature are those whose energies are close to the Fermi energy. For phonon conductivity at a given temperature one should take into account all the spectrum with the different phonon frequencies. Also, up to the Debye temperature, the phonon spectrum varies with temperature.

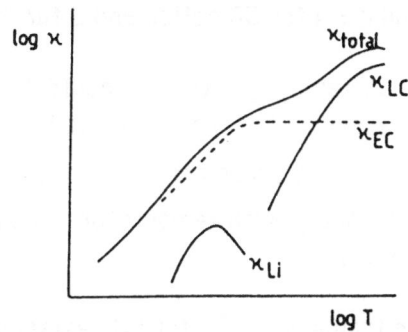

Fig. 4.4 Log-Log schematic representation of the various contributions to the thermal conductivity of a GIC below 300 K. The maximum total thermal conductivity is of the order of a few hundreds W m^{-1} K^{-1}. At low temperatures the thermal conductivity of the intercalated graphite increases linearly with temperature, then an additional contribution κ_{Li} shows up.

Around room temperature the thermal conductivity of the intercalated graphite is due to the lattice, κ_{LC}, and electronic, κ_{EC}, contribution of the graphitic planes while at the lowest temperatures it is entirely electronic for very good samples (From reference 32).

The two main phonon scattering events are those with static lattice defects and those with other phonons. For lattice defects one should distinguish between :

- large scale defects whose sizes are much larger than the phonon wavelength, such as crystal boundaries or for GICs the Daumas-Hérold domains [29]. These lead to relaxation times independent of the phonon frequencies

$$\tau_B = L \, / \, v \qquad\qquad (4.7)$$

where the boundary scattering length L is, within a numerical factor which is usually small, equal to the distance between such defects.

- point defects whose sizes are small compared to the phonon wavelength. The relaxation times for this type of scattering vary strongly with the phonon frequencies ω

$$\tau_D = D \, \omega^r \qquad\qquad (4.8)$$

where D is a constant and r = 4 for 3D solids and 3 for 2D solids.

For phonon-phonon scattering events one should distinguish between normal and umklapp processes which are described by a relaxation time τ_N and τ_U respectively. Both relaxation times are frequency and temperature dependent and may vary strongly with temperature in some temperature ranges [cfr. e.g reference 19].

The total relaxation frequency τ_R^{-1} for all resistive processes is then

$$\tau_R^{-1} = \tau_U^{-1} + \tau_B^{-1} + \tau_D^{-1} \qquad (4.9)$$

and if we want to take into account resistive as well as N-processes

$$\tau_C^{-1} = \tau_N^{-1} + \tau_R^{-1} \qquad (4.10)$$

where τ_C^{-1} is the relaxation time which enters into the calculation of κ_1 in relation (4.6).

The effect of intercalation on the in-plane thermal conductivity of graphites

One would expect a decrease of the in-plane lattice thermal conductivity in GICs with respect to the pristine material owing to the defects introduced by intercalation which reduce the phonon mean path. Also, the main effect of intercalation is to increase the electrical conductivity as a result of the charge transfer : an increase in the carrier density accompanied by a decrease, but in a smaller proportion, of the electronic mobility. An increase in electrical conductivity will directly lead to an increase in the electronic thermal conductivity (relation 4.1).

The result of these two competing effects is that the thermal conductivity of all intercalation compounds is decreased at high temperature and increased at low temperature with respect to the pristine material (Fig. 4.3). In Fig. 4.4 a typical behavior is illustrated, while in Fig.4.5 we present the various results obtained on the temperature variation of the thermal conductivity of HOPG-GICs and in Fig. 4.6 that of intercalated fibers.

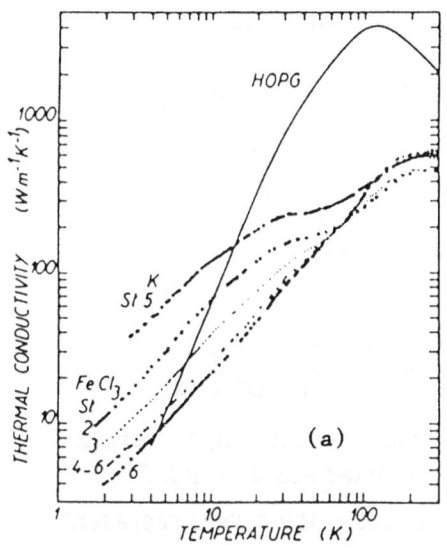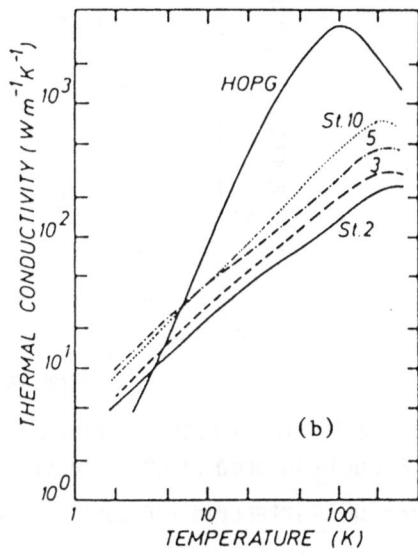

Fig. 4.5 Temperature variation of the in-plane thermal conductivity of various graphite intercalation compounds compared to that of pristine HOPG. a) for pure stages 2, 3 and 6, and a mixed stage (4-6) FeCl$_3$ GAC [11]. Data for a stage-5 potassium donor intercalation compound are also presented. b) for pure stages of SbCl$_5$ compounds [31].

The fact that in the lowest temperature range the thermal conductivity of HOPG-GICs varies linearly with temperature and that when applying the WFL a reasonable value for the electrical resistivity was found [11], suggested that the electronic thermal conduction dominates at low temperature. However, the best method to separate κ_E and κ_L in a quantitative manner is to apply a high magnetic field. This separation, which was done for some intercalation compounds, confirmed that κ_E is dominant for very good samples [30][31].

Also κ_L could be determined and the effect of intercalation on lattice defects estimated [11]. In table 3.1 the values of the phonon mean free paths in the boundary scattering regime are presented for three GACs and compared to the hole mean free paths in the residual range [7]. It may be seen that the agreement is fair and that the two independent tools, the holes and the phonons, used to estimate the dimensions of the large scale defects give similar results. This confirms the information obtained from microscopic studies (cfr. M.S. Dresselhaus and J.S. Speck, this issue).

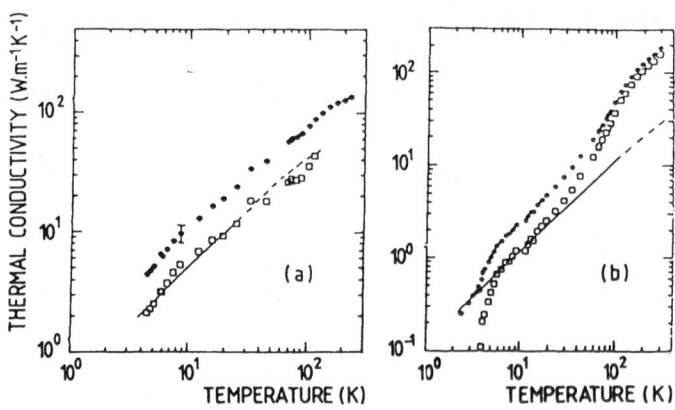

Fig. 4.6 The temperature variation of the thermal conductivity of a stage-2 BDF-CuCl$_2$ intercalation compound (a) and of a mixed stage (1 and 2) PDF-CuCl$_2$ intercalation compound (b). In both cases the curves represent the electronic thermal conductivities κ_E computed from the data of the electrical resistivities measured on the same sample. The dark circles represent the values of the measured total thermal conductivity κ and the open squares represent the difference between κ and κ_E. All fibers were heat treated at 3000 °C (From references 8 and 9).

Dimensionality and phonon conduction

Because of the exceptionally high anisotropy of the electrical conductivity in GACs ($\approx 10^6$ at room temperature), and though the interlayer interactions may still exert an influence on the electronic properties, it was found justified for most practical situations to use a 2D model for the analysis of electrical resistivity results. This is also applicable to the electronic thermal conductivity. However, the situation is quite different with regard to the lattice thermal conductivity. For lattice waves which belong to the crystal as a whole, the graphite layers are too thin to be regarded as individual systems. The lattice vibrations of the layers must be coupled in some way in order to insure mechanical cohesion in the c-axis direction. Also, the anisotropy of the lattice thermal conductivity of pristine HOPG reaches "only" $\approx 10^2$ around room temperature and decreases down to a factor of ≈ 3 in the liquid helium range. In GIC's the situation is almost the same, though the anisotropy of the total thermal conductivity remains around 10^2 in the liquid helium range. However this high anisotropy at low temperature is due to the electronic contribution κ_E and not to κ_L [12].

If in order to get a rough qualitative picture of phonon conduction around room temperature we still persist in describing the high temperature phonon properties in a naive way, then one may invoke that an anisotropy of 10^2 justifies a quasi 2D model . In that case, as it was proposed for the electrical conductivity of GAC s [10] we may consider that the lattice conductance per sandwich is equal to the sum of the conductances of the constituent graphite and intercalate layers[11][32]. This facilitates to a great extent the qualitative interpretation of the high temperature data [11].

One may understand the difference in anisotropy of electrical and phonon conductivities from simple arguments. A survey of the room temperature thermal conductivity values of known crystalline solids shows that between the best and the worst heat conductors there are merely about three orders of magnitude. For the electrical resistivity there are 24 orders of magnitude. Thus a material with no free electrons will be a perfect electrical insulator. In GACs, the intercalate layers may play the role of an electrical insulator and one may understand that electrical conduction may then be very low in the c-axis direction as revealed by the very high anisotropy in that case. The fact that the situation is not the same for heat conduction is due to phonon conduction which exists regardless of the presence of electrons. Phonon conduction is related directly to the atomic binding energies and the anisotropy of the interatomic forces is limited to a few orders of magnitude in a crystalline material. This leads to smaller anisotropies of the lattice properties.

REFERENCES

1. C. Rigaux, this issue p. 235.
2. J. Blinowski, H.H. Nguyen, C. Rigaux, J;P. Vieren, R. Le Toullec, G. Furdin, A. Hérold and J. Melin, J. Phys. 41 (1980) 47
3. L.A. Pendrys, T.C. Wu, C. Zeller, H. Fuzellier and F.L. Vogel, Ext. Abstr. of the 14th Biennial Conf. Carbon, (1979), edited by P.A. Thrower, p. 306
4. E. Mc Rae, Thèse d'Etat, Nancy (1982)
5. J.F. Marêché, E. Mc Rae, N. Nadi and R. Vangelisti, Synthetic Metals, 8, 163(1983)
6. L. Piraux, J-P. Issi, L. Salamanca-Riba and M.S. Dresselhaus, Synthetic metals (in press)
7. J-P. Issi and L. Piraux, Annales de Physique, Colloque n°2, supplément au n°2, vol.11 (1986)
8. L. Piraux, B. Nysten, J-P. Issi, J-F. Marêché and E. Mc Rae, Sol. St. Comm. 55, 517 (1985)
9. L. Piraux, B. Nysten, J-P. Issi, L. Salamanca-Riba and M.S. Dresselhaus,

Solid Stat. Comm. <u>58</u>, 265 (1986)

10. M.S. Dresselhaus and G. Dresselhaus, Advances in Physics <u>30</u>, 139 (1981)

11. J-P. Issi, J. Heremans and M.S. Dresselhaus, Phys. Rev. <u>B27</u>, 1333 (1983)

12. J-P. Issi, Mat. Res. Soc. Symp. Proc. vol. 20, p.147, Elsevier Science Publishing Co., Inc. (1983)

13. J-M. Ziman, <u>Electrons and Phonons</u>, Clarendon Press Oxford (1960)

14. F.J. Blatt, <u>Physics of electronic Conduction in Solids</u> (Mc Graw Hill, New York 1968)

15. L. Piraux, J-P. Issi, J-P. Michenaud, E. Mc Rae and J-F. Marêché, Sol. St. Comm. <u>56</u>, 567 (1985)

16. L. Piraux, V. Bayot, J-P. Michenaud, J-P. Issi, J-F. Marêché and E. Mc Rae, Sol. St. Comm. (in press)

17. C. Van Haesendonck, L. Van den Dries, Y. Bruynseraede and G. Deutscher, Phys. Rev. <u>B25</u>, 5090 (1982)

18. G. Bergmann, Phys. Rep. <u>107</u>, 1 (1984)

19. R. Berman, <u>Thermal Conduction in Solids</u> (Clarendon Press, Oxford, 1976)

20. P.G. Klemens, <u>Encyclopedia of Physics</u>, S. Flügge ed. (Springer-Verlag, Berlin, 1956) volume XIV, p. 198

21. B.T. Kelly, <u>Physics of Graphite</u> (Applied Science Publishers, London, 1981)

22. M.G. Holland, C.A. Klein and W.D. Straub, J. Phys. Chem. Solids <u>27</u>, 903 (1966)

23. C.K. Chau and S.Y. Liu, J. Low Temperature Physics <u>15</u>, 447 (1974)

24. P.G. Klemens, <u>Thermal Conductivity</u>, R.P. Tye ed. (Academic Press, New York 1969) vol. 1, P. 69

25. B. Nysten, L. Piraux and J-P. Issi, Proc. of the 17th International Thermal Conductivity Conference, Cookeville (1985), in press

26. W. de Sorbo and W.W. Tyler, J. Chem. Phys. <u>21</u>, 1660 (1953)

27. B.T. Kelly, Carbon <u>5</u>, 247 (1967), Ibid. <u>6</u>, 71 (1968), Ibid. <u>6</u>, 485 (1968)

28. J. Callaway, Phys. Rev. <u>113</u>, 1046 (1959)

29. N. Daumas and Hérold A., C.R. hebd. séance Acad. Sci. Paris <u>C268</u>, 373 (1969)

30. J. Heremans, M. Shayegan, M.S. Dresselhaus and J-P. Issi, Phys. Rev. <u>B26</u>, 3338 (1982)

31. M. Elzinga, D.T. Morelli and C. Uher, Phys. Rev. <u>B26</u>, 3312 (1982)

32. J-P. Issi, J. Heremans and M.S. Dresselhaus, <u>Physics of Intercalation Compounds</u>, L. Pietronero and E. Tosatti eds. (Springer-Verlag, Berlin 1981) p. 310

WEAK LOCALIZATION AND COULOMB INTERACTION IN GRAPHITE ACCEPTOR INTERCALATION COMPOUNDS

Luc Piraux

Unité de Physico-Chimie et de Physique des Matériaux
Place Croix du Sud 1, B-1348 Louvain-la-neuve (Belgium)

The transport properties of two-dimensional (2D) weakly disordered electronic systems have been the subject of great interest in the last few years. Theoretical and experimental investigations of the low-temperature electrical resistivity of a variety of 2D systems have led to the observation of new and anomalous non metallic effects. The prominent feature of all these systems is a small logarithmic increase in resistance with decreasing temperature at very low temperature. For reviews on the subject see references 1, 2 and 3. This non-metallic behavior may be explained theoretically, taking into account the presence of disorder, by invoking two different models: the first is weak-electron localization due to constructive quantum interference between elastically back-scattered electron waves (the usual Boltzmann transport theory neglects interference between the scattered electron waves); the second involves electron-electron many-body interactions which are enhanced in the presence of scattering on lattice static defects. Low magnetic field magnetoresistance measurements are the most widely used method to discriminate with some success between these two approaches. Indeed, the application of a low transverse magnetic field tends to suppress the weak-localization effect –and as a result produces negative magnetoresistance–, while the interaction model predicts no magnetoresistance.

Recently, Piraux et al (4,5) reported experimental evidence for localization and interaction effects in low stage acceptor graphite intercalation compound (GAC). A logarithmic increase of the resistivity with decreasing temperature has been observed on a set of GACs based on two different host graphite materials and intercalants as shown elsewhere in this volume by Prof. J-P Issi. The large number of different types of graphite host materials that may be intercalated provides a means to control the amount of disorder. Thus GACs constitute an attractive class of 2D disordered materials and are ideally suited for the study of weak localization and interaction phenomena. In figure 1 we have plotted the theoretical values of the temperature T_{min} of the resistivity minimum versus the values of the resistivity ρ_0 at this minimum considering either the localization or the interaction theoretically predicted. Figure 2 shows the magnetoresistance curves obtained at 4.2K in a magnetic

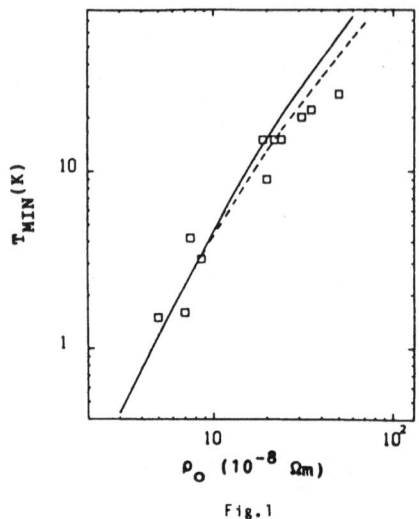

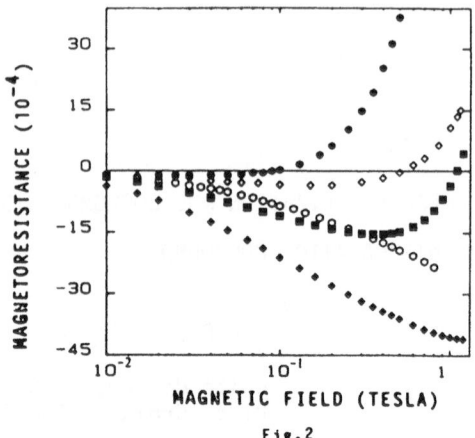

Fig.1

Fig.2

Fig.1   Relation between the temperature T_{min} of the resistivity minimum and the value of the resistivity ρ_o at this minimum. The square symbols refer to experimental observations while the full and the dashed lines are calculated considering localization and interaction effects respectively.

Fig.2   Transverse magnetoresistance as a function of magnetic field for a set of 5 samples at 4.2K: BDF-SbCl$_5$ (●), PX5-CuCl$_2$ (◇), PX5-SbCl$_5$ (■), VSC25-CuCl$_2$ (○) and P100-4-CuCl$_2$ (◆)

field perpendicular to the fiber axis for five samples of the same origin as those reported elsewhere in this volume by J-P Issi. The observed negative magnetoresistance indicates the presence of localization effects. However, the relative contribution to the logarithmic temperature dependence of the resistance from the localization and interaction effects may only be estimated by measuring the temperature dependence of the magnetoresistance (2). Preliminary results, that will be presented elsewhere (6), show that elctron-electron interactions are also present and may even dominate in GACs. When experimental data are compared to theoretical expressions, we see that both resistance and magnetoresistance measurements are consistent with the coexistence of both a 2D weak-localization and electron-electron interaction model.

REFERENCES

1. Localization, Interaction and Transport Phenomena, B. Kramer, G. Bergmann and Y. Bruynseraede (Eds), Series in Solid-State Sciences (1985)
2. G. Bergmann, Phys. Rep. 107, 1 (1984)
3. P.A. Lee and T.V. Ramakrishnan, Rev. Mod. Phys. 57, 287 (1985)
4. L. Piraux, J-P Issi, J-P Michenaud, E. McRae and J-F Marêché, Solid State Comm. 56, 567 (1985)
5. L. Piraux, V. Bayot, J-P Michenaud, J-P Issi, J-F Marêché and E. McRae, Solid State Comm. 59, 711 (1986)
6. L. Piraux, V. Bayot, J-P Michenaud and J-P Issi, to be published

$\vec{c}$ AXIS CONDUCTIVITY IN GRAPHITE INTERCALATION COMPOUNDS

Edward McRae, Jean-François Marêché and Albert Hérold

Laboratoire de Chimie du Solide Minéral, UA 158, Service de Chimie Minérale Appliquée, B.P.239 54506 Vandoeuvre-les-Nancy Cedex (France)

While various groups over the past years have furnished numerous data on the basal plane conductivity, σ_a ($\equiv 1/\rho_a$) of a wide variety of GIC's over the 4.2-300 K range, $\vec{c}$ axis studies are much more restricted[1], the only quite complete investigation being that on the SbCl5 family (stages, n=1-4,6,10)[2]. Over this temperature (T) range, our work has recently extended the results to encompass new acceptor and donor binary and ternary compounds[3,4,5].

Such studies have received motivation from several directions, not the least of which is the most complete sample characterization possible. A more fundamental reason is to determine the dimensionality of these layered materials. Indeed at room temperature, the conductivity anisotropy A ($\equiv \rho_c/\rho_a$) spans about 5 orders of magnitude from the binary alkali metal compounds, MC8 (A $\sim$ 20) to low stage AsF5 GIC[6] (A $\sim$ 10^6). Dimensionality often plays a fundamental role in transport phenomena and in evaluating to what extent a given 2D or 3D model may be applicable. The interpretation of experimental data from other experimental probes such as conduction electron spin resonance and thermal conductivity is also dimension-dependent and conductivity data are a necessary complement. In their own right, these conductivity studies should help in determining the mechanisms involved in this process, a point which is not yet clear in graphite itself.

Figure 1 shows typical curves of ρ_c versus T for 3 samples of high quality HOPG[7]. The variation is characterized by a $\rho_c = C \exp(BT)$ relationship from 300 K down to a sample dependent temperature of $\sim$ 50 K. Below, ρ_c is T-independent in most cases, though as seen in this figure for the sample possessing the largest in-plane domain size, ρ_c may further decrease. The 300 and 4.2 K values of A are of the order of 2.10^3 and 4.10^4 respectively.

Figure 2 illustrates the dramatic changes that can be provoked by intercalation, in this specific case for the Rb-Bi ternary donor GIC's[3,4]. Other than the existence of a maximum in $\rho_c(T)$ for the stage 1 and 2 materials (see below), the qualitative features are representative of almost all CIC's. At 300 K, ρ_c is primarily intercalant-dependent, the stage dependence becoming increasingly marked as the temperature drops. Over a wide interval, the richest compounds have "metallic" character in the sense that $d\rho/dT > 0$, the slope becoming negative as the stage index

Table 1. $\vec{c}$ axis resistivity and anisotropy (A) at 295 K and 4.2 K for a selection of donor and acceptor GICs

Intercalant	n	295 K		4.2 K		ref.
		$\rho_c(\Omega$ cm$)$	A	$\rho_c(\Omega$ cm$)$	A	
HOPG	–	5.5–$9.7\ 10^{-2}$	1.3–$2.2\ 10^3$	0.2–0.3	2.4–$4.8\ 10^4$	7
Li	1	5–$6.7\ 10^{-5}$	10–15	–	–	10
K	1	2–$3\ 10^{-4}$	21–33	–	–	10,11
Rb–Bi	1α	0.2–0.3	2–$2.2\ 10^4$	0.03–0.05	5.5–$6.5\ 10^4$	3,4
	7β	0.3	$3.8\ 10^4$	1.3	$1.7\ 10^6$	"
$FeCl_3$	2	1.2	2–$2.4\ 10^5$	0.45	5–$7\ 10^5$	5
	4	0.8	$1.5\ 10^5$	1.4	$1.3\ 10^6$	"
$GaCl_3$	1	4.3	$9.5\ 10^5$	0.4	$4.4\ 10^5$	"
	4	4.3	$1.1\ 10^6$	2.4	$3.1\ 10^6$	"
$SbCl_5$	1	~ 0.9	–	~ 0.3	–	2
	10	~ 0.3	–	~ 0.8	–	"
AsF_5	1	–	$> 10^6$	–	–	6

further rises. This thermal variation in ρ_c implies that A increases as T decreases for all but saturated compounds. Some representative values are given in Table 1.

Two theories of $\vec{c}$ axis conductivity have recently been put forward to explain the T- and n-dependence[8,9]. Sugihara[8] has proposed a combination of phonon-assisted (PA) and impurity-assisted (IA) hopping processes ; the former is suggested as predominating in the poorer compounds, the latter in the richer GIC's since these presumably contain a greater number of defects. Shimamura's theory[9] similarly involves two complementary mechanisms: a process due to the existence of interlayer conducting paths (CP) as well as a PA process, the analysis of which was done using a model somewhat simpler than Sugihara's. Analysis of the equations leads to the following expected T-dependences of ρ_c :

Mechanism		Low T	High T
Sugihara	– PA	T^{-2}	T-independent
	– IA	T-independent	increases with T
Shimamura	– PA	negative temperature coefficient	
	– CP	ρ_c varies as ρ_a (+ temp. coeff.)	

It is quantitatively difficult at the present time to evaluate the validity of one model over the other and it may well be that all three processes are involved. In the equations developed by Sugihara, for example, there are over a dozen parameters, the exact values of which are unknown for a given material. Even his own analysis applied to the available $SbCl_5$, K and $FeCl_3$ GIC data remained qualitative as regards the T-dependence of ρ_c. In his semiquantitative room temperature treatment of a "typical" 2nd stage GIC, many of the parameter values utilized were those of graphite.

Both theories, however include a negative temperature coefficient PA process, consistent with the experimental findings in the poorer (high stage) materials. The richest GIC always manifest a positive temperature coefficient as predicted for the CP or IA mechanisms. The maximum in $\rho_c(T)$ illustrated in Fig. 2 suggests that in certain compounds there is a crossover temperature below which the CP or IA process predominates and above which the PA mechanism becomes dominant. The T-dependence of the maximum would not be inconsistent with Sugihara's suggestion since the IA process would give way to the PA process at a higher temperature for stage 1 than stage 2. This is indeed found, not only for the Rb-Bi ternaries, but also for the Cs-Bi GIC.

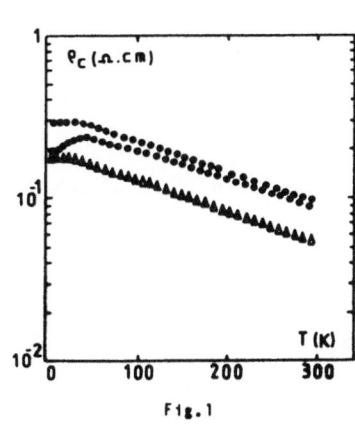

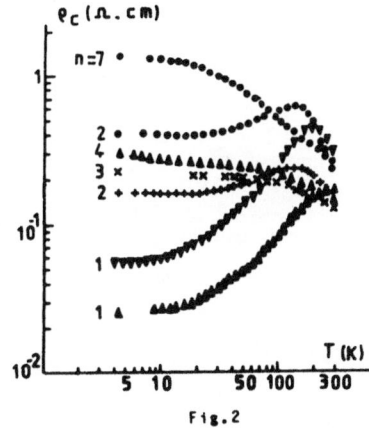

Fig. 1. ρ_c versus T for 3 pristine HOPG samples (ref. 7)

Fig. 2. ρ_c versus T for ternary Rb-Bi GIC. All are α except 7th stage (β) (ref. 3,4).

Work is in progress to more explicitly analyze the stage and temperature dependence of ρ_c and to extend the results to a wider range of GICs.

REFERENCES

1. J.P. Issi, B. Poulaert, J. Heremans and M.S. Dresselhaus, Sol. State Commun. 44 : 449 (1982)
2. D.T. Morelli and C. Uher, Phys. Rev. B 27 : 2477 (1983)
3. E. McRae, J-F. Marêché, A. Bendriss-Rerhrhaye, P. Lagrange and M. Lelaurain, Ann. Phys. 11 : colloq. 2, 13 (1986)
4. J-F. Marêché, E. McRae, A. Bendriss-Rerhrhaye and P. Lagrange, J.Phys. Chem. Solids,in press
5. N.E. Nadi, E. McRae, J-F. Marêché, M. Lelaurain and A. Hérold, Carbon, in press
6. G.M.T. Foley, C. Zeller, E.R. Falardeau and F.L. Vogel, Sol. State Commun. 24 : 371 (1977)
7. D. Marchand, C. Frétigny, M. Lagues, A.P. Legrand, E. McRae, J-F. Marêché and M. Lelaurain, Carbon 22 : 497 (1984)
8. K. Sugihara, Phys. Rev. B 29 : 5872 (1984)
9. S. Shimamura, Synth. Metals 12 : 365 (1985)
10. J.E. Fischer, in "Molecular Metals", W.E. Hatfield, ed., Plenum, N.Y. (1979)
11. A.R. Ubbelohde,Proc. Roy. Soc. Lond. A 327 : 289 (1972)

EVIDENCE FOR A TWO-DIMENSIONAL METAL-INSULATOR TRANSITION IN POTASSIUM-AMMONIA GRAPHITE

Y.Y. Huang, X.W. Qian*, and S.A. Solin

Department of Physics and Astronomy
Michigan State University
East Lansing, MI   48824-1116

J. Heremans and G.G. Tibbetts

Physics Department
General Motors Corporation
Warren, MI   48090-9055

We have studied the ammonia pressure dependence and composition dependence of the a-axis electrical resistivity of the potassium-ammonia ternary graphite intercalation compounds $K(NH_3)_xC_{24}$, $0 < x < 4.33$. Our results show evidence of the 2D analog of the well-studied bulk 3D metal-insulator transition [1] in bulk K-NH$_3$ solutions.

In order to probe the electrical properties of the intercalate layer in $K(NH_3)_xC_{24}$, we measured the in-plane a-axis resistivity as a function of ammonia pressure [2]. For convenience, we used a benzene-derived onion skin-like graphite fiber [3] rather than HOPG. Thus, the fiber axis is approximately coaxial with the cylindrical graphite planes. From the (001) x-ray diffraction patterns of a single fiber (see Fig. 1) we determined the basal spacing of the pristine fiber (d_a = 3.35Å), of stage-2 KC_{24} potassium binary GIC (d_b = 8.74Å) and of the stage-1 ternary GIC $K(NH_3)_{4.33}C_{24}$ (d_c = 6.64Å). These results indicate that the graphite fiber has the same Gibbs phase diagram [4] as HOPG.

A four-probe measurement technique was developed for monitoring the pressure dependence of the a-axis relative resistance (R/R_o) of $K(NH_3)_xC_{24}$, using a pressure up-quenching technique. [Here $R(R_o)$ is the a-axis resistance of the ammoniated (KC_{24}) compound.] The result is plotted in Fig. 2. Note that the relative resistance ratio remains a constant below one atmosphere and then starts to increase in the stage-2 to stage-1 phase transition region at ≈0.5 atm. [5]. The increasing resistivity in a stage-2 to stage-1 phase transition is a common feature of binary GIC's [5]. But when NH$_3$ is added to KC_{24}, some delocalized electrons in the carbon layer are back-transferred to the intercalate layer [6] so that the conductivity of the carbon layers decreases. However, R/R_o starts to decrease dramatically at about 3-4 atm. at which pressure x is greater than four [6]. This phenomenon may be a consequence of a 2D metal-insulator transition. When x = 4, there are enough NH$_3$ molecules to completely solvate potassium and form a 4-fold coordinated K-NH$_3$ clusters [7]. Higher NH$_3$ concentration leads to a sufficient amount of electron back-transfer

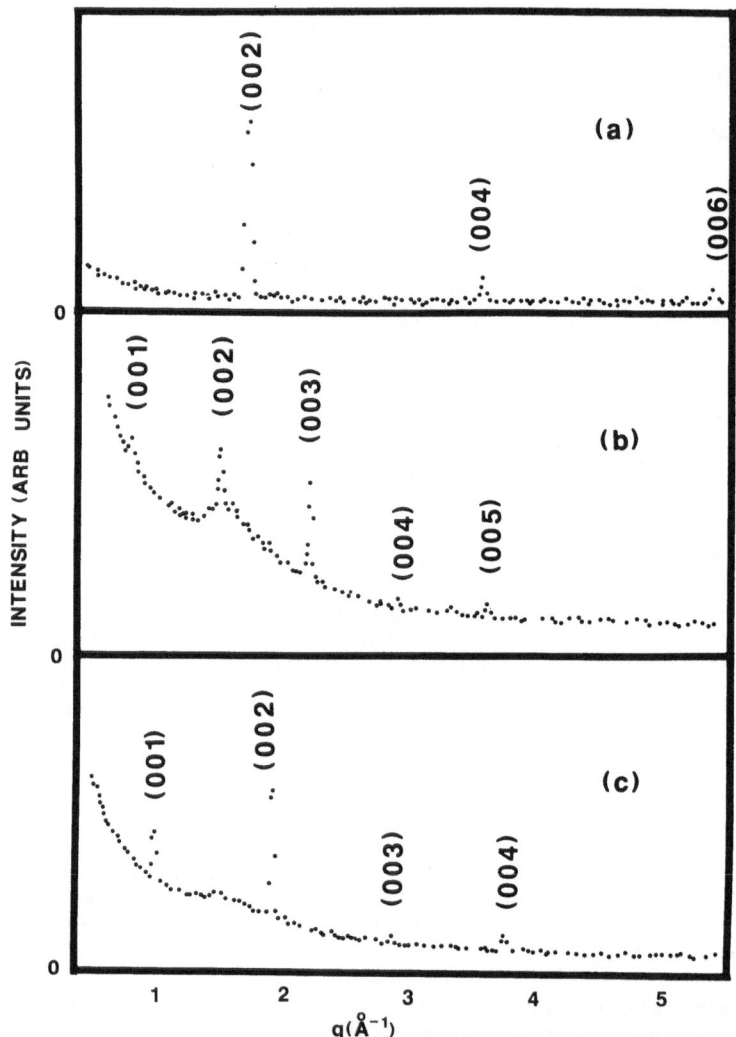

Fig. 1. The (00ℓ) x-ray diffraction patterns of (a) a single pristine cylindrical onion skin-like graphite fiber, (b) the same fiber as in (a) after intercalation with potassium to form the binary GIC KC_{24}, and (c) the same fiber as in (b) after ammoniation to form the ternary GIC $K(NH_3)_{4.3}C_{24}$. The diffraction patterns were recorded at room temperature using MoKα radiation. The background continuum is diffuse scattering from the glass envelope and from air in the beam path.

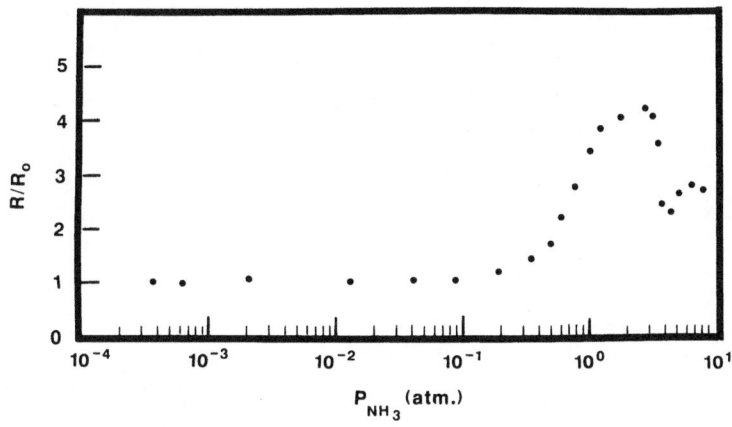

Fig. 2. The room temperature relative resistance a-axis ratio R/R_o of potassium-ammonia-graphite as a function of ammonia pressure. Here R_o and R correspond respectively to the resistance of KC_{24} and $K(NH_3)_xC_{24}$.

from the graphite layers to the intercalate layers to induce a hopping-type metal-insulator transition [8]. When there is sufficient solvation to cause in-plane overlap of the electron wave function, the intercalate layer starts to conduct in parallel with the carbon layers and the a-axis resistance drops as is observed in Fig. 2. The study of the composition dependence of the dielectric constant of $K(NH_3)_xC_{24}$ also provides evidence of a metal-insulator transition at $x \approx 4$.

ACKNOWLEDGEMENTS

We thank Y.B. Fan for useful discussions. This work was supported by the National Aeronautics and Space Administration under grant NAG-3-595 and in part by the U.S. National Science Foundation under grant 85-17223.

*Exxon Fellow

REFERENCES

1. J.C. Thompson, Electrons in Liquid Ammonia (Clarendon, Oxford, 1976).
2. Y.Y. Huang, S.A. Solin, J. Heremans, and G.G. Tibbetts, to be published.
3. J. Tsukamoto, K. Matsumura, T. Takahashi, and K. Sakoda, Synthetic Metals 13, 255-264 (1986).
4. Y.B. Fan and S.A. Solin, Proceedings of the Materials Research Society Meeting, Boston, (1985).
5. M.S. Dresselhaus and G. Dresselhaus, Adv. Phys. 30, 139 (1981).
6. B.R. York and S.A. Solin, Phys. Rev. B31, 8206 (1985).
7. X.W. Qian, D.R. Stump, B.R. York, and S.A. Solin, Phys. Rev. Lett. 54, 1271 (1985).
8. See reference 1 and references therein.
9. J.M. Zhang, P.C. Eklund, Y.B. Fan, and S.A. Solin, to be published.

STRUCTURE AND IN–PLANE CONDUCTIVITY STUDIES OF GRAPHITE FLUORINE INTERCALATION COMPOUNDS

I. Palchan* and M. Talianker[†]

*Racah Institute of Physics
The Hebrew University of Jerusalem, Israel
†Department of Materials Science
Ben–Gurion University, Beer Sheva, Israel

Although covalent graphite–fluorine compounds have been known for a long time, conductive graphite–fluorine intercalation compounds were discovered only recently.[1] These new air–stable GICs exhibit high in–plane conductivity which, per unit weight, is twice that of copper. The repeat distances for stages $n \geq 1$ can be described by the formula:

$$I_c = 6.05 + (n-1)3.35 \quad \text{Å}.$$

For stage I compounds two versions of I_c were found; one with $I_c = 11 \pm 0.5$Å, i.e., almost double the expected value.

Figure 1 presents the x–ray diffraction pattern for a mixed stage II plus stage I $C_{3.6}F$ GIC with two versions of the stage I c–axis repeat distances. TEM patterns of a stage I $C_{2.3}F$ specimen were obtained at room temperature using a JEOL–200 KeV TEM and are shown in Fig. 2. Two features are observed in Fig. 2: a) the existence of a superlattice with a commensurate $(2 \times 2)R0°$ in–plane unit cell; b) a change of

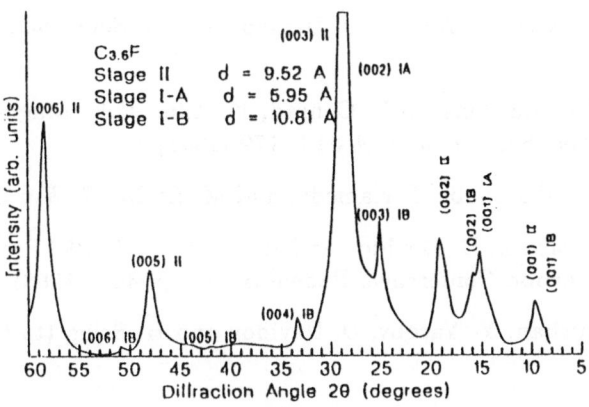

Figure 1: X–ray (00ℓ) diffraction patterns of $C_{3.6}F$.

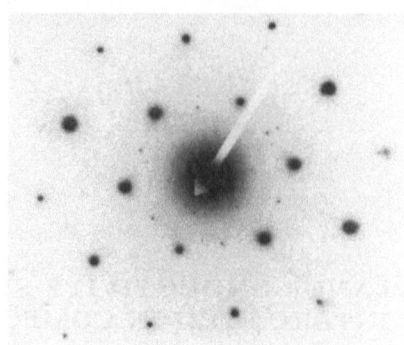

Figure 2: TEM diffraction patterns of stage 1 $C_{2.3}F$.

the symmetry of the graphite matrix from hexagonal to monoclinic with the following parameters: $a = 2.29\text{Å}$, $b = 11.2\text{Å}$, $c = 2.45\text{Å}$, $\beta = 119°$ (here b is $\perp$ to the graphite planes). We found that each stage exists over a wide range of fluorine concentrations, but the maximum filling of a pure stage n compound by F is in accordance with the formula $C_{2n}F$.

We know also from previously published Raman studies[2] that stage II C_nF GICs may also exhibit doubling of the in–plane unit cell. So it may be suggested that a saturated intercalated layer has the same structure for different stages. Also it was found by T. Mallouk, et al.[3] that even in stage I C_2F compounds graphite–like C–C bonds exist together with C–F and C–F_2 bonds.[3] However, C–F covalent bonds have been observed even in dilute stage III $C_{10}F$ compounds.[4]

The low in–plane conductivity of a stage I compound[5] may possibly be understood on the basis of band structure changes due to the doubling of the unit cell and changes in lattice symmetry. However, our recent preliminary reflectivity studies[6] fail to show a drastic difference between the free charge density in stage II and stage I compounds, suggesting that the defect scattering may be preferable as a mechanism for the decrease in conductivity.

The authors wish to thank Profs. D. Davidov and H. Selig for many useful discussions and D. Vaknin, I. Ohana, V. Shargorodsky and M. Palchan for help with the work.

REFERENCES

1. I. Palchan, D. Davidov and H. Selig, *J. Chem. Soc. Chem. Comm.* 657 (1983).

2. I. Ohana, I. Palchan, Y. Yacoby, D. Davidov and H. Selig, *Solid State Comm.* **56**, 505 (1985).

3. T. Mallouk, B.L. Hawkins, M.P. Conrad, K. Zilm, C.E. Magiel, and N. Bartlett, *Phil. Trans. Roy. Soc. London, A* **314**, 179 (1985).

4. T. Nakajima, N. Watanabe, I. Kameda, and M. Endo, *Carbon* **24**, (1986).

5. I. Palchan, D. Vaknin, D. Davidov and H. Selig, *Carbon'86* Proceedings of the 4th International Carbon Conference, Baden–Baden, p. 462 (1986).

6. I. Ohana, I. Palchan, Y. Yacoby, D. Davidov and H. Selig, (to be published).

SUPERCONDUCTIVITY IN GRAPHITE INTERCALATION COMPOUNDS

G. Dresselhaus and A. Chaiken

Massachusetts Institute of Technology
Cambridge, MA 02139, USA

INTRODUCTION

Anisotropic superconductivity has been observed in a variety of layered materials including deliberately structured materials,[1] transition metal dichalcogenides intercalated with large organic molecules,[2] and graphite intercalation compounds (GICs). Because of the long range of the superconducting coherence distance ξ, the observation of 2D superconductivity focuses on samples with superlattice repeat distances I_c large compared with the superconducting coherence distance to achieve the condition $I_c > \xi$. Molecular beam epitaxy and magnetron sputtering offer the greatest promise for quantitative studies of 2D superconductivity and the 2D–3D crossover because of the flexibility of these synthesis techniques for the generation and control of large superlattice repeat distances in deliberately structured superconductors.[1] Historically, early work on this subject was carried out with transition metal dichalcogenides intercalated with large organic molecules, where repeat distances of $I_c \approx 60$ Å were achieved for example by intercalation of n–octadecylamine into TaS_2.[3] Though intercalate repeat distances greater than the superconducting coherence length have not yet been achieved in GICs, the superconducting GICs have nevertheless provided an interesting system for the study of anisotropic superconductivity phenomena. In this paper, we briefly review superconductivity in GICs and provide some comparison with anisotropic superconducting behavior in deliberately structured materials and intercalated transition metal dichalcogenides.

Superconductivity in graphite intercalation compounds was first observed by Hannay et al.[4] who identified superconducting transitions in the stage 1 heavy alkali metal compounds with the intercalants K, Rb and Cs. Subsequently the superconducting properties of C_8K were characterized in greater detail by the magnetic susceptibility measurements of Koike et al.[5] More recently, a number of ternary donor compounds have been found to show superconducting phases.[6,7,8,9,10,11] A list of superconducting GICs and their transition temperatures T_c is given in Table 1. There is no known acceptor GIC which has shown evidence for superconductivity.

Particular interest in superconductivity in GICs relates to the dimensionality of the superconductivity phenomena and to the occurrence of superconductivity in a compound such as C_8K ($T_c \sim 140$ mK) where neither constituent is itself superconducting. Shown in Fig. 1 is the transition from the normal to superconducting state in first stage C_8K as measured by the ac susceptibility method. The dimensionality effects relate to the stage dependence of the critical temperature T_c and the anisotropy of the critical field $H_c(T)$. Whereas the stage 1 compounds exhibit 3D behavior, 2D behavior is expected (but not yet clearly demonstrated) for higher stage compounds in which the c–axis superconducting coherence length $\xi_\parallel$ would be less than the c–axis repeat distance ($\xi_\parallel < I_c$).

The superconducting transition temperature in GICs tends to be higher for intercalants that are themselves superconducting, e.g., KHg and RbHg. It is expected that T_c should decrease with increasing stage, consistent with the expected decrease in the density of electronic states. Notable exceptions are the stage 1 compounds C_4KHg and C_4RbHg which have lower reported transition temperatures than the second stage compounds C_8KHg and C_8RbHg (see Table 1).[6,7] The MBi (M = K, Rb, Cs) intercalation compounds (see Table 1) may also show similar behavior.

In most, if not all GICs, a variety of superconducting transition temperatures T_c have been reported; for example, T_c values ranging from 0.72 to 1.65 K have been reported for stage 1 KHg–GICs.[8,11,12] In some samples, a broad transition between 0.9 and 1.42 K was observed, which became narrow and increased to 1.49 K at a hydrostatic pressure of 0.8 kbar.[14] Samples containing mostly well ordered regions with the $(2 \times 2)R0°$ in–plane structure showed transition temperatures of ~ 1.5 K.[11] Possible reasons for the wide range of T_c values observed in this compound (see Table 1) may be due to deviations in the stoichiometry, especially the Hg content, or to the formation of different in–plane structures under different sample preparation conditions. By adding small concentrations of hydrogen, it has been found (as is shown in Fig. 2) that the transition temperature becomes stabilized at the maximum T_c (~ 1.5 K), independent

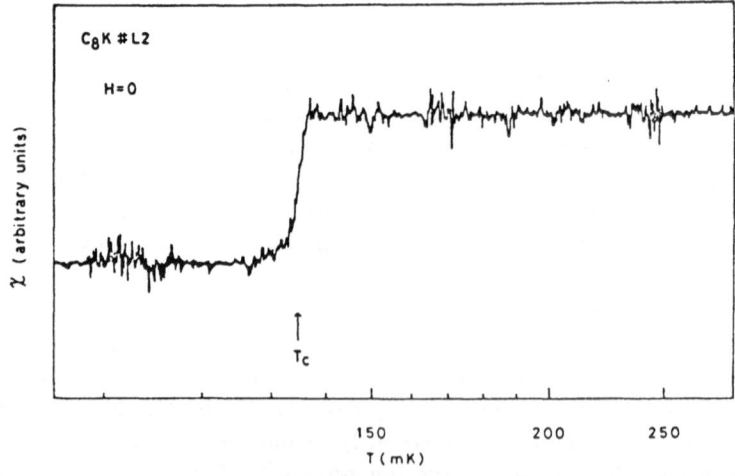

Figure 1: The a.c. magnetic susceptibility of a C_8K sample as a function of temperature at nearly zero d.c. field. The abrupt change in the magnetic susceptibility near 134 mK is identified with the superconducting transition.[5]

Table 1: Graphite intercalation compounds which have been reported to show super-conductivity.

Stage	Compound	T_c (K)[a]	Comment
1	C_8K	0.128–0.198	Ref. 5
1	C_8Rb	0.026	Ref. 9
1	$C_8KH_{0.19}$	0.2	Ref. 10
1	$C_4KTl_{1.5}$	2.56	Ref. 8
2	$C_8KTl_{1.5}$	1.53	Ref. 8
1	C_4KHg_x	0.72–1.65	$x \simeq 0.7$, Ref. 8
1	C_4RbHg_x	0.99	$x \sim 0.7$, Ref. 8
1	$C_4KHg_xH_\epsilon$	1.53	$x \simeq 0.7$, $\epsilon < 0.1$, Ref. 12
2	C_8KHg_x	1.90	$x \simeq 0.7$, Ref. 8
3	$C_{13}KHg_x$	1.8	$x \simeq 0.62$, Ref. 11
2	C_8RbHg_x	1.41	$x \sim 0.7$, Ref. 8
2	$C_8RbBi_{0.6}$	1.25–1.5	α phase, Ref. 13
1	$C_4CsBi_{0.55}$	4.05	α phase, Ref. 13
1	$C_4CsBi_{0.55}$	2.3	β phase, Ref. 13
2	$C_8CsBi_{0.55}$	2.7	β phase, Ref. 13

[a]The transition temperature often varies from one sample to another. Presumably the sample dependence is associated with differences in stoichiometry and in–plane order of the compound.

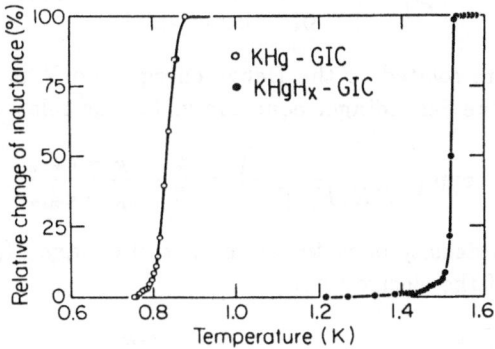

Figure 2: Effect of hydrogen doping on the superconducting transition of a stage 1 KHg–GIC sample.[12]

of the transition temperature of the compound prior to the addition of hydrogen.[12] Another striking example of the importance of sample preparation on the value of T_c is provided by C_8K for which early measurements[4] yielded a value of ~ 0.55 K, while extensive subsequent work showed T_c to be a factor of three smaller.[8]

THEORETICAL CONSIDERATIONS

Initial attempts to understand superconductivity in GICs were based on the BCS theory.[15] The BCS theory is based on a collection of Cooper pairs[16,17] of electrons with weak attractive interaction between electrons, H', resulting from phonon exchange between pairs of electrons. If the interaction is small, then there is a minimum energy $\epsilon_{min} = (\hbar^2 q_{min}^2)/2m$ below which the effect of H' can approximately be ignored, as then the first–order perturbation energy correction $[|\langle 1|H'|2\rangle|^2/(\epsilon_1 - \epsilon_2)]^2$, will be negligible. The reason is that ϵ_2 must be greater than or equal to the Fermi energy, E_F, for the $T = 0$ electron gas. By the same sort of reasoning, there should also be a maximum energy, $\epsilon_{max} = (\hbar^2 q_{max}^2)/2m$ above which H' is unimportant. However, between ϵ_{min} and ϵ_{max} the perturbation can be appreciable. Since we know that in real superconductors the electron–electron interaction is phonon–mediated, we assume that the minimum energy for which the perturbation H' is non–negligible is about $(E_F - \hbar\omega_D)$, and the maximum energy is about $(E_F + \hbar\omega_D)$, where $\hbar\omega_D$ is the Debye energy which represents the approximate peak in the phonon density of states.

The wavefunctions of the electron pairs are taken as plane waves

$$\phi(\vec{k}_1, \vec{k}_2; \vec{r}_1, \vec{r}_2) = \frac{1}{\sqrt{\Omega}} e^{i(\vec{k}_1 \cdot \vec{r}_1 + \vec{k}_2 \cdot \vec{r}_2)} = \frac{1}{\sqrt{\Omega}} e^{i(\vec{K} \cdot \vec{R} + \vec{k} \cdot \vec{r})} \tag{1}$$

where Ω is the normalization volume and where center of mass coordinates have been introduced. The spins of the two electrons in question must be antiparallel, so that their overall wavefunction will be antisymmetric under interchange of their coordinates, as is required for fermions.

Since the perturbation H' couples electrons only in a small shell around the Fermi surface, we make the approximation that

$$\langle \vec{k} \mid H' \mid \vec{k}' \rangle \approx \left\{ \begin{array}{cc} -|V|, & q_{min} \leq k \; ; \; k' \leq q_{max} \\ 0, & \text{otherwise} \end{array} \right\}, \tag{2}$$

where q_{min} and q_{max} are related to the Debye energy and V is the average attractive potential. Solution of the Schrödinger equation yields the relation

$$\exp\left(\frac{1}{N(K, E_F)|V|}\right) = \frac{E - \epsilon_K - \epsilon_{max}}{E - \epsilon_K - \epsilon_{min}} \tag{3}$$

where $N(K, E_F)$ is the density of states at the Fermi energy. Upon simplification, Eq. (3) yields the energy of the electron pair

$$E = \epsilon_K + \epsilon_{min} - \frac{2\hbar\omega_D}{\exp\left(\frac{1}{N(K,E_F)|V|}\right) + 1}. \tag{4}$$

There are several interesting points to be made about Eq. (4). The first is that the energy of the electron pair is lower than that of the individual electrons, and the energy of the pair is lowest if ϵ_K, the center–of–mass energy, is zero. This implies that the electrons have both opposite spin and momenta; such a state is known as a Cooper pair. The energy of such a pair is lower than that of the individual electrons by an amount Δ defined by

$$\Delta = \frac{2\hbar\omega_D}{\exp\left(\frac{1}{N(K,E_F)|V|}\right) + 1} \approx 2\hbar\omega_D \exp\left(\frac{-1}{N(K,E_F)|V|}\right). \tag{5}$$

If $\Delta \approx 2k_B T_c$ where k_B is the Boltzmann constant, then we recover the famous BCS equation for T_c. Because Δ goes to zero when the wavevector for the center of mass motion, K, approaches $(q_{max} - q_{min})$, we expect that excitations of the pair will tend to split the pair and destroy the superconducting state rather than give the pair more translational energy.

According to the BCS theory of superconductivity, the transition temperature T_c and the Debye temperature Θ_D are related by

$$T_c = 0.85\Theta_D \exp\left(\frac{-1}{N(0)V}\right) \tag{6}$$

where $N(0)$ is the density of states at the Fermi energy and V is an interaction energy. The explicit inclusion of many body effects results in the McMillan formula given by

$$T_c = \frac{\Theta_D}{1.45}\exp\left(\frac{-1.04(1 + \lambda)}{\lambda - \mu^*(1 + 0.62\lambda)}\right) \tag{7}$$

where λ is the electron–phonon coupling constant and μ^* is the effective Coulomb repulsion parameter. To date, almost all measurements on superconducting GICs have been interpreted in terms of the BCS theory using the above formulae. As we review the properties of superconducting GICs, particular emphasis well be given to those properties which cannot be explained by the standard BCS theory, but require refinements of this theory.

We have already described some unusual properties of GICs, relevant to their superconductivity which are not readily understood on the basis of standard BCS theory. These include: (1) the occurrence of superconductivity in stage 1 C_8K where neither constituent is superconducting in its pristine crystalline state, (2) the absence of superconductivity in stage 1 C_6Li, another first stage alkali metal GIC, (3) the absence of superconductivity in stage 2 and higher stage alkali metal GICs though stage 1 is superconducting, (4) the absence of superconductivity in acceptor GICs, (5) the observation of a higher T_c in stage 2 C_8KHg_x and C_8RbHg_x relative to their stage 1 counterparts (see Table 1), though the density of states at the Fermi level is believed to be higher in stage 1 than in stage 2, and (6) the large value of λ (the electron–phonon coupling) obtained by fitting Eq. (7) to measured T_c values and taking $\mu^* = 1$, as suggested by McMillan. We discuss below an extension of the standard BCS theory which accounts for these and other unusual phenomena in GICs.

CRITICAL FIELD BEHAVIOR

Two types of superconductivity are distinguished by their behavior in an external magnetic field as is shown in Fig. 3. Both types of superconductivity are found in GICs.

The type I superconductor shows a complete Meissner effect up to a critical field H_c at which the transition to a normal metal occurs. The type II superconductor shows a complete Meissner effect up to a lower critical field H_{c1} above which penetration of the magnetic field occurs in the form of current vortices. The material remains superconducting to the upper critical field designated H_{c2} at which field the transition to the normal state occurs. Referring to Fig. 3, we see that $\chi = \partial M / \partial H$ is constant up to H_c (in Fig. 3a) or to H_{c1} (in Fig. 3b), where a discontinuity in the slope of M occurs; and hence a singularity in χ is expected at H_{c1}, as magnetic flux begins to penetrate the superconductor.

The anisotropic electronic structure of GICs gives rise to an anisotropy in the superconducting behavior upon application of a magnetic field, and the anisotropic properties are conveniently described in terms of the angle θ between an applied magnetic field H and the c–axis. Interesting behavior is generally found for the angular dependence $H_{c2}(\theta)$ and for the temperature dependence of $H_{c2}(T)$, the upper critical field. The behavior for C_8K is especially unusual. Through measurement of the magnetic susceptibility χ as a function of H for various magnetic field orientations, Koike et $al.$[5] found that for $0° \leq \theta \leq 65°$, C_8K behaves as a type I superconductor. This is shown in Fig. (4a) for $\theta = 0°$, where full magnetic flux exclusion is found for most of the magnetic field range, with H_c denoting the thermodynamic critical field and H_{sc} a supercooled critical field. In contrast, for $65° \leq \theta \leq 90°$, C_8K is found to behave as a type II superconductor, showing an incomplete Meissner effect, as illustrated in Fig. (4b) for $\theta = 90°$. The upper critical field H_{c2} (where $H_{c2\perp} \approx 10$ G for $\vec{H} \parallel$ c–axis, i.e., $\theta = 0°$) is indicated on the figure.

For all superconducting GICs that have been investigated, the critical magnetic fields are highly anisotropic (i.e., strongly dependent on the angle θ). This is illustrated in Fig. 5 for the stage 2 compound C_8KHg for two values of temperature $T < T_c$.[8] Anisotropy in H_{c2} can also be seen in Fig. 4 for C_8K.[5]

According to a generalization of the simple Ginzburg–Landau (GL) theory to anisotropic media by Lawrence and Doniach,[18] a simple relation for the angular dependence

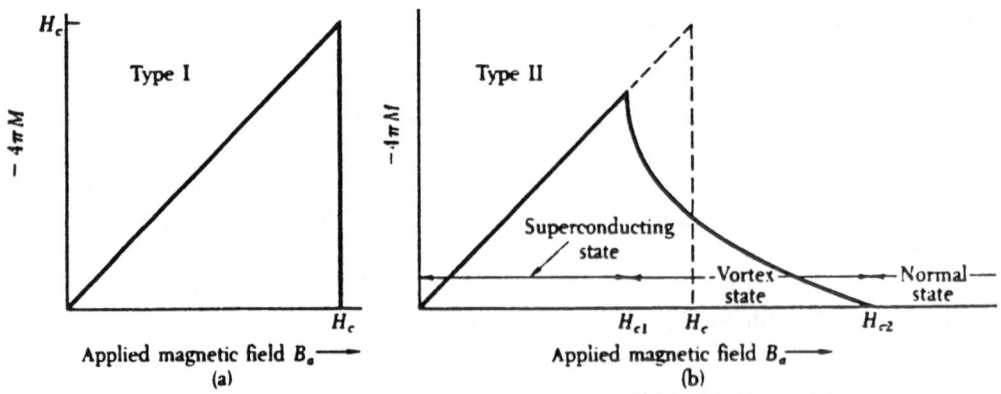

Figure 3: Schematic dependence of the magnetization on applied field for (a) type I superconductors with a complete Meissner effect and (b) type II superconductors where the field begins to penetrate the sample at H_{c1} and the sample finally becomes normal at H_{c2}.[16]

of H_{c2} is obtained

$$H_{c2}(\theta) = \frac{\Phi_0}{2\pi\xi_\perp^2}\left(\cos^2\theta + \epsilon^2\sin^2\theta\right)^{-1/2}, \qquad (8)$$

in which $\Phi_0 = hc/2e = 2.07 \times 10^{-7}$ gauss–cm^2 is the elementary flux quantum. Here the parameter ϵ characterizes the anisotropic behavior and is given by

$$\epsilon^2 = m_\perp/m_\parallel = \left(H_{c2\perp}/H_{c2\parallel}\right)^2 = \left(\xi_\parallel/\xi_\perp\right)^2 \qquad (9)$$

where ξ denotes the coherence length and m denotes the effective mass in the $\parallel$ and $\perp$ directions. We denote by $H_{c2\perp}$ the upper critical field for $\theta = 0°$ whereby the diamagnetic screening currents lie in the layer planes. Similarly we denote by $H_{c2\parallel}$ the upper critical field for $\theta = 90°$ whereby the diamagnetic screening currents must also traverse

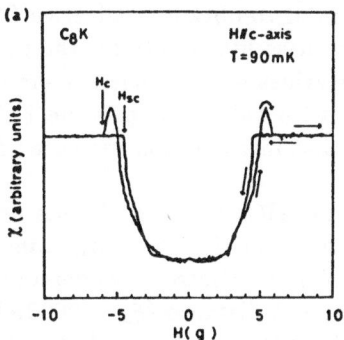 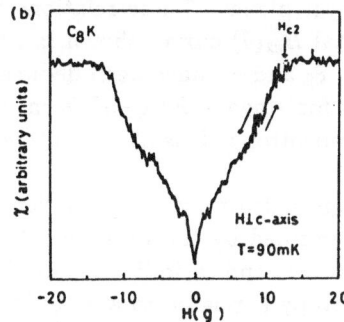

Figure 4: AC magnetic susceptibility at 90 mK as a function of DC magnetic field H for a first stage C_8K sample: (a) for $\vec{H} \parallel \vec{c}$–axis showing type I superconducting behavior, (b) for $\vec{H} \perp \vec{c}$–axis showing type II superconducting behavior.[5] Whereas Fig. (4a) shows hysteresis, Fig. (4b) does not.

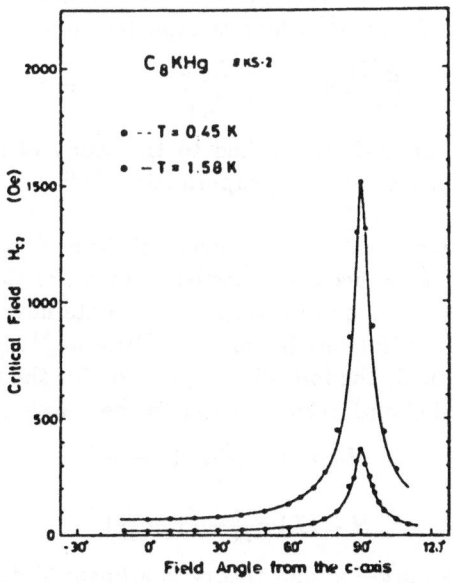

Figure 5: Angular dependence of the critical field for the stage 2 compound C_8KHg.[8]

Table 2: Superconducting parameters of stage 1 KHg–GICs and pristine KHg.

Parameter	Stage 1 KHg–GIC	Pristine KHg
I_c (Å)	10.22	–
T_c (K)	1.53	0.98
$H_{c2\perp}$ (Oe)	83	244
$H_{c2\parallel}$ (Oe)	850	244
$\xi_\perp(0)$ (Å)	1960	1135
$\xi_\parallel(0)$ (Å)	192	1135
$\xi_\perp/\xi_\parallel$	10.2	1

different layer planes. These relations have been used to obtain qualitative fits to the experimental $H_{c2}(\theta)$ curves shown in Fig. 5. From these fits, values for the parameters $H_{c2\parallel}$, $H_{c2\perp}$, $\xi_\parallel$ and $\xi_\perp$ have been deduced. Specific values of these parameters are given in Table 2 for stage 1 KHg–GICs in comparison with values for pristine KHg which shows no anisotropy. It is also of interest to note that intercalation increases T_c.

The large anisotropy of the critical field H_{c2} for GICs (see Fig. 5) has been qualitatively described by the Lawrence–Doniach extension of the Ginzburg–Landau (GL) theory, Eq. (8), and is indicative of three–dimensional anisotropic superconductivity. This anisotropy is related to a large in–plane coherence distance ($\xi_\perp \sim 2000$Å) relative to the c–axis coherence distance ($\xi_\parallel \sim 100$Å). Since the in–plane coherence distance is relatively insensitive to stage index while the c–axis coherence distance decreases rapidly with increasing stage, the preparation of higher stage superconducting GICs could result in materials exhibiting 2–dimensional superconductivity phenomena (see Fig. 6).

Although no GICs exhibiting 2D superconductivity have yet been synthesized, we include here the Tinkham formula[17] which is expected to apply in this limit given by

$$\left|\frac{H_{c2}(\theta)}{H_{c2\perp}}\cos\theta\right| + \left[\frac{H_{c2}(\theta)}{H_{c2\parallel}}\sin\theta\right]^2 = 1. \tag{10}$$

Equation (10) has been successfully applied to the study of 2D superconductivity in deliberately structured superconducting superlattices.[19,20]

Also of interest is the temperature dependence of the critical fields $H_{c2}(T)$. Measurements of $H_{c2}(T)$ on GICs have been interpreted in terms of the anisotropic Ginzburg–Landau theory which provides a qualitative fit to the data near T_c. Assuming an elliptical fluxoid to describe the anisotropy in coherence lengths,[21] the anisotropic Ginzburg–Landau theory (Eq. (8)) yields the following expression for the temperature dependence for the critical fields parallel and perpendicular to the c–axis:

$$H_{c2\parallel}(T) = \frac{\Phi_0}{2\pi\xi_\parallel\xi_\perp}\left(1 - \frac{T}{T_c}\right)$$

$$H_{c2\perp}(T) = \frac{\Phi_0}{2\pi\xi_\perp^2}\left(1 - \frac{T}{T_c}\right). \tag{11}$$

Characteristic of the Ginzburg–Landau theory is a linear T dependence of H_{c2} near T_c for both $H_{c2\perp}$ and $H_{c2\parallel}$. The anisotropy in Eq. (11) comes in through the prefactor coefficients. Critical field measurements on stage 2 KHg–GICs for both $H \parallel$ c–axis and $H \perp$ c–axis yield a linear T dependence near T_c in accordance with Eq. (11) as shown in Fig. 7.[14] Fits to these data yield values for the parameters $\xi_\parallel$ and $\xi_\perp$, consistent with values obtained from fits to the angular dependence of $H_{c2}(\theta)$ discussed above.

A more complete theory for $H_{c2}(T)$ has been proposed by Klemm, Luther and Beasley for anisotropic superconductors[22] and has been applied to explain the different functional forms observed for $H_{c2\perp}$ and $H_{c2\parallel}$ in 2D superconducting deliberately structured superlattices.[19] This theory has not yet been applied to GICs. As higher stage GICs are synthesized and accurate H_{c2} data are obtained near T_c, it will be interesting to apply this more complete theory to GICs.

PRESSURE EFFECTS

The effect of hydrostatic pressure on the transition temperature of GICs is also of special interest. For GICs generally, the transition temperature T_c decreases by application of pressure (see Fig. 8) and shows pressure coefficients for T_c of the same order of magnitude as found in typical type I BCS superconductors such as Hg or Sn.[14] Prediction of the effect of pressure on T_c is usually written in terms of the BCS expression for T_c given by Eq. (6), so that

$$T_c \sim \Theta_D e^{-1/g}, \tag{12}$$

where

$$g = N(0)V = \frac{N(0)\langle I^2 \rangle}{M\Theta_D^2} \tag{13}$$

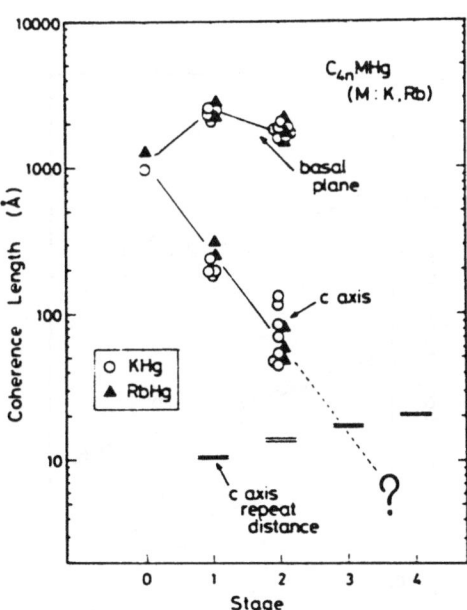

Figure 6: The stage dependence of the superconducting coherence distance $\xi_\parallel$ and $\xi_\perp$ in the alkali metal–mercury–GICs. Stage zero denotes the pristine intercalant materials KHg and RbHg. The c–axis stacking period is also shown for comparison. The extrapolation suggests that 2–D behavior might occur above about stage 3.[8]

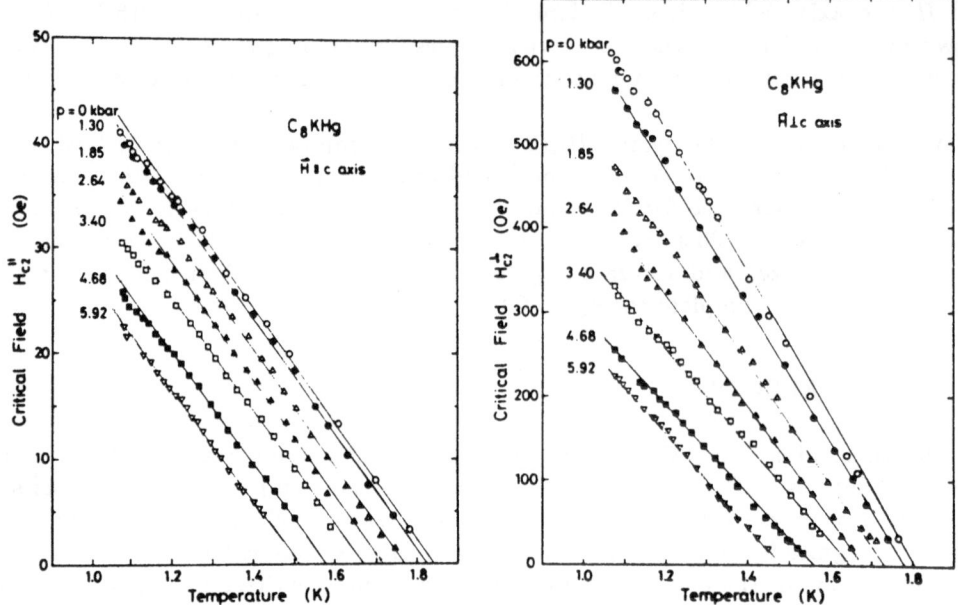

Figure 7: The H_{c2} vs. T relations for stage 2 C_8KHg for the magnetic field parallel to the c–axis (left figure, $H_{c2\perp}$), and perpendicular to the c–axis (right figure, $H_{c2\parallel}$).[14] Results are given for ambient pressure ($p = 0$) and for different values of the applied pressure.

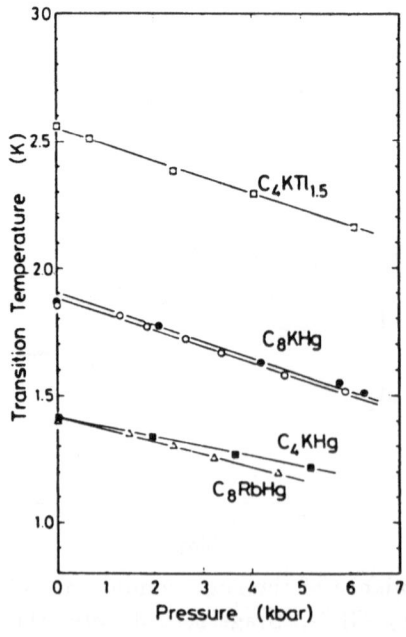

Figure 8: Pressure dependence of T_c for several ternary GICs.[14]

and $N(0)$ is the density of states at E_F, $\langle I^2 \rangle$ is the matrix element for the electron-phonon interaction averaged over the Fermi surface, and M is the ion mass. Although the Debye temperature Θ_D occurs in both the prefactor and the exponential in Eq. (12), the influence of the exponential factor g on T_c is usually more important. Since the phonon frequency increases with pressure, T_c is expected to decrease with increasing pressure, in agreement with the results shown in Fig. 8.

The temperature dependence of $H_{c2\parallel}$ and $H_{c2\perp}$ has also been explored for a range of externally applied pressures (see Fig. 7) and a linear temperature dependence of H_{c2} is observed near T_c for both field orientations for stage 2 KHg-GICs,[8] in accordance with Eq. (11). Thus the functional form of Eq. (11) is retained upon application of pressure, though the parameters T_c, $\xi_\parallel(0)$, and $\xi_\perp(0)$ may all be pressure dependent.

In some classes of materials, pressure causes a drastic change in the superconducting properties associated with a structural phase changes. The ternary compounds however show lesser pressure dependent structural phase changes than the binary compounds. The pressure dependence of superconductivity in the stage 1 heavy alkali metal GIC C_8K has very recently been investigated in connection with the structural phase transitions occurring in these compounds as a function of pressure.[23,7]

MODEL FOR SUPERCONDUCTIVITY IN GICS

The standard BCS model with only occupied π bands cannot explain the observed transition temperatures because the π-band density of states is too small. On the other hand, a model with only heavy mass intercalate-related bands cannot explain the large observed anisotropy in the superconducting properties of GICs. Al-Jishi[24] therefore proposed a model in which both π and heavy mass intercalate bands are occupied and which accounts for most of the current observations in GICs. Thus the occurrence of superconductivity in graphite intercalation compounds depends on the presence of both graphitic π-electrons and electrons associated with partial occupation of the intercalate band (incomplete charge transfer from the intercalant to the graphite host).[24]

The Al-Jishi model starts with the BCS theory and forms a Bogoliubov-like Hamiltonian from pairs of electrons consisting of a heavy intercalant electron and a light graphitic π electron. The heavy electron is connected with a band having little anisotropy. The standard BCS equation for the transition temperature (Eq. (6)) is generalized by Al-Jishi to give

$$T_c \approx \Theta_D \exp\left(\frac{-1}{|V|\sqrt{N_s(0)N_\pi(0)}}\right), \qquad (14)$$

where $N_s(0)$ and $N_\pi(0)$ are respectively, the densities of states at the Fermi level of the heavy mass and the graphitic π bands. Hence, the large anisotropy in the superconductivity of GICs is associated with the highly anisotropic graphitic π-band Fermi surface. Another theory by Shimizu and Kamimura[25] for superconductivity in C_8K is based on the tight binding band structure of Ohno et al.[26] which connects superconductivity to simultaneous partial occupation of alkali s bands and graphitic π bands. The same basic conclusion however is reached,[25] namely, that it is necessary to have at least two types of occupied bands or strong hybridization between them to produce superconductivity in GICs.

These models predict both the transition temperature and anisotropy of H_{c2} for GICs and are generally consistent with the observed data. They explain the absence of superconductivity in acceptors which lack occupied heavy mass intercalant levels. The models show that the value of the superconducting transition temperature is more sensitive to changes in the π–band electronic density than to changes in the heavy mass band. On this basis, Al–Jishi is able to explain[24] the lower T_c of stage 1 Rb– and Cs–intercalated graphite in comparison to C_8K. The weak dependence of T_c on the intercalate concentration in stage 1 K–GICs[27] is explained by assuming that the fractional charge transfer increases with a decrease in the intercalate concentration. The models further explain the absence of superconductivity in C_6Li and the higher stage alkali–metal GICs where there are no occupied intercalate bands. The large increase in transition temperature under pressure for C_8K is explained by an increase in the fractional charge transfer from the intercalate to the graphite layer under pressure.

One key to a detailed understanding of superconductivity has been through in-depth investigations of the tunneling characteristics which yield direct information on the density of states. Such experiments have not yet been performed on GICs. With the development of improved methods for fabricating tunneling devices and of the scanning tunneling microscope,[1] it may soon be possible to carry out tunneling experiments on superconducting GICs.

SUPERCONDUCTIVITY IN INTERCALATED DICHALCOGENIDES

The earliest reports of superconductivity in layered transition metal dichalcogenides intercalated with organic molecules[28] focused on the increase in T_c which occurred through intercalation. This increase is now believed to be associated with the suppression of charge density wave formation rather than some exotic mechanism associated with two–dimensional d–band superconductivity.[2] Nevertheless, the effect of lower dimensionality was observed in this work, especially in systems such as TaS_2 intercalated with n–octadecylamine where a separation of nearly 60 Å was achieved between sequential Ta layers. This work was important in stimulating theoretical calculations of 2D superconductivity.[22] In comparison to GICs, the transition metal dichalcogenides intercalated with large organic molecules have thus far been more suitable for studying lower dimensional superconducting phenomena.

Perhaps the best example of 2D–3D crossover in the superconducting behavior of an intercalation compound (whether graphite or dichalcogenide) has been observed in $TaS_2(aniline)_{3/4}$ with $T_c = 2.9$ K by Prober et al.[29] The TaS_2 and $TaS_{1.6}Se_{0.4}$ materials in this study were intercalated with organic molecules collidine, pyridine and aniline, achieving c–axis periodicities ranging from 9.7 Å for the collidine to 12.0 Å for the pyridine to 18.1 Å for the aniline. All of these compounds showed very high critical fields for $\theta = \pi/2$ (e.g., 10^5 Oe at low temperatures) and very high anisotropy ratios $H_{c2\parallel}/H_{c2\perp} \sim 35$. The critical fields $H_{c2\perp}$ were relatively low. They found that the intercalated compounds (especially the $TaS_2(aniline)_{3/4}$) showed some two–dimensional behavior at low temperatures (see Fig. 9), while the behavior near T_c was three dimensional. The 2D–3D crossover effect in the $TaS_2(aniline)_{3/4}$ intercalation compound was documented by extensive comparison of the experimental data for the temperature dependence of $H_{c2\|}$ and the angular dependence of $H_{c2}(\theta)$ near $\pi/2$ with the microscopic theory of Klemm, Luther and Beasley.[22] Evidence for the cross over to two–dimensional behavior at low temperature is seen most clearly in the critical field *vs.* temperature data for $TaS_2(aniline)_{3/4}$ shown in Fig. 9. Here the departure of the $H_{c2\|}(T)$ curve for $\theta = 90°$ from those for $H_{c2\perp}(T)$ for $\theta = 0°$ is interpreted in terms of a Josephson tunneling mechanism for coupling two–dimensional sheets.[29] Note that the data for $H_{c2\|}(T)$ are divided by large scale factors because of the large magnitude of this critical field. Near T_c, the coherence length $\xi_\|$ becomes large so that quasi–3D behavior is observed, independent of the magnetic field direction.

SUPERCONDUCTING SUPERLATTICES

Of the various topics in low–dimensional physics that have been studied with metallic superlattices, most progress has been made with the study of low–dimensional superconductivity, in part because the superconducting coherence distance is large compared with lattice dimensions, thereby imposing less stringent requirements on the sharpness of the interfaces than would for example be the case for synthesizing *magnetic superlattices*. Though a new research field, the metallic multilayers that have been synthesized thus far already provide better examples of anisotropic superconductivity, 2D superconductivity and 3D–2D cross–over phenomena than almost any of the previous types of samples that have been studied, including layered dichalcogenides and intercalation compounds based on them, and superconducting GICs. Nevertheless, GICs are of particular interest for studying superconducting superlattices because of their atomically abrupt interfaces and the long range 3D periodicity that can be achieved.

We shall see below that because of inter–diffusion phenomena, a deliberately struc-

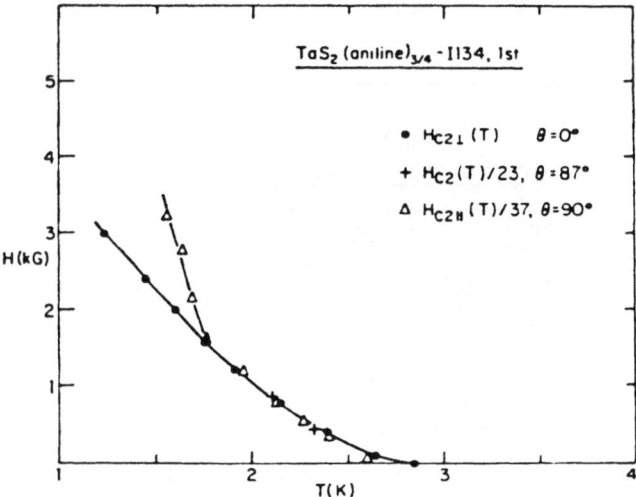

Figure 9: Upper critical fields H_{c2} near T_c for various angles θ with respect to the c–axis for $TaS_2(aniline)_{3/4}$. The upward rise of $H_{c2\|}$ at low temperatures has been identified with a Josephson tunneling mechanism of two–dimensional sheets.[29] To emphasize the functional form the $\theta = 87°$ and $90°$ data are scaled as indicated.

tured superlattice, such as shown in Fig. 10 for alternating layers of Nb and Ge, can only be fabricated over a limited range of layer thicknesses. Over the range of thicknesses where a crystalline superconducting phase can be synthesized ($d_{Nb} \geq 30$ Å for the Nb/Ge superlattice shown in Fig. 10), we shall see below that a wealth of information on anisotropic 3D superconductivity, 2D–3D crossover phenomena and 2D superconductivity phenomena can be obtained.[19,20]

There are however ranges of d_{Nb} and d_{Ge} that are inaccessible by magneton sputter deposition because of the limits on the abruptness of the Nb/Ge interfaces that can be synthesized (see Fig. 10(b) and (c)). It is however precisely this limit which is accessible by GICs where monolayer thicknesses of each constituent are possible. Thus superconducting GICs and superconducting/metal superlattices can play a complementary role in the exploration of low–dimensional superconductivity phenomena.

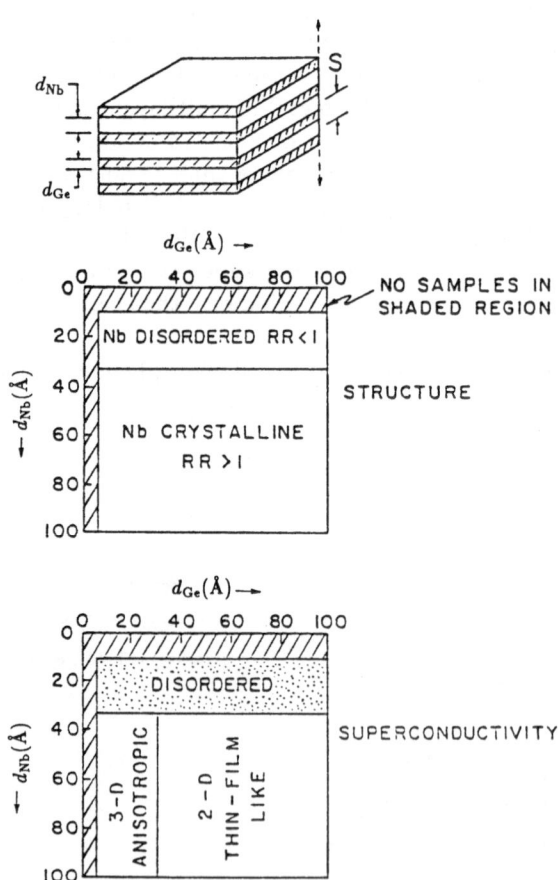

Figure 10: Structure and dimensionality of the Nb/Ge multilayer composites.[30] (a) Schematic diagram showing the thicknesses of as–deposited Nb and Ge layers. S denotes the effective thickness of the superconducting layer. (b) Regimes of ordering (characterized by the resistance ratio, RR) for various ranges of layer thicknesses. (c) Regimes of superconductivity for various ranges of layer thicknesses.

There are four basic types of deliberately structured superconducting superlattices that can be prepared: (1) two different metals, both of which are superconducting (the S/S' system), (2) two different metals, one of which is superconducting, while the other is normal down to 0 K (the S/N system), (3) two different metals, one of which is superconducting and the other magnetic (the S/M system), and (4) two different types of materials, one of which is superconducting, the other insulating or semiconducting (the S/I system).

Most of the superconducting superlattices that have been fabricated to date have been prepared by sputter deposition under modest vacuum conditions ($\sim 10^{-6}$ to 10^{-7} Torr), which is not such a serious limitation for general studies of superconductivity, since superconductivity is relatively insensitive to nonmagnetic impurities. As experience in this field has been gained, the quality and control of the sample preparation has been constantly improving to yield sharper interfaces, better control of the d_1 and d_2 thicknesses of the superlattices and better homogeneity in composition and crystalline orientation both perpendicular and parallel to the layers. Since many of the type II superconductors of interest require high deposition temperatures, the use of MBE techniques may not be straightforward for achieving sharp interfaces.

The S/I systems are particularly attractive for studying anisotropy in the superconducting coherence length ξ, two–dimensional superconductivity and the cross–over between 3D and 2D behavior. A number of lengths which are all temperature–dependent are of importance for these studies, including the superconducting coherence lengths $\xi_\perp$ in the layer planes and $\xi_\parallel$ along the superlattice growth direction, the magnetic flux penetration depth λ. Other scale parameters are the superlattice constants d_S and d_I, denoting the layer thicknesses for the superconductor and insulator, respectively. The superconducting properties shown in Fig. 11 are expected to depend strongly on the relative magnitudes of these lengths.

For this superlattice the layer thicknesses are found from x–ray diffraction measurements.[1] Since the germanium layers are amorphous and the Nb layers contain small crystallites with the (110) planes aligned parallel to the layer surface, the diffraction patterns are primarily sensitive to the Nb layers and the superlattice periodicity.[19] The x–ray analysis for a variety of sample thicknesses indicates that the interface between the Nb and Ge is ~ 30 Å, due to inter–diffusion of the Nb and Ge species during the sample preparation (see Fig. 10). The in–plane conductivity measurements indicate that the resistivities of the Nb and interface material are 9.8 $\mu\Omega$–cm and 114 $\mu\Omega$–cm, respectively.[1,19]

The observation of anisotropic superconducting behavior is best illustrated through measurements of the temperature dependence of the upper critical field $H_{c2}(T)$, as shown in Fig. 11 for the Nb/Ge superlattice, which is of the S/I type. If $d_I \ll \xi_\parallel$ where d_I is the thickness of the insulator (or semiconductor), then the superconducting wavefunctions can extend along the length of the superlattice, so that bulk anisotropic 3D superconducting behavior is expected. This is illustrated for the sample for which $d_{\mathrm{Nb}}/d_{\mathrm{Ge}} = 45\text{Å}/7\text{Å}$ where $H_{c2\parallel}$ depends linearly on the reduced temperature $t = T/T_c$.

In the opposite limit where $d_I \gg \xi_\parallel(0)$, the individual layers retain their bulk properties from the point of view of the superconducting order parameter and other microscopic superconducting properties. For the magnetic field perpendicular to the layer planes, the temperature dependence $H_{c2\perp}(T)$ is essentially the same for $d_I \gg \xi_\parallel(0)$ as for the $d_I \ll \xi_\parallel(0)$ case, almost independent of layer thickness (see Fig. 11), as expected since the screening currents lie in the layer planes. However, the behavior for $H_{c2\parallel}(T)$ is highly dependent of d_I as shown in Fig. 11. Very close to T_c, a linear dependence of $H_{c2\parallel}(T)$ on $(T_c - T)$ is found even for $d_I \gg \xi_\parallel(0)$ because $\xi_\parallel(T) \to \infty$ as $T \to T_c$. However, for reduced temperature $t = (T/T_c) \leq 0.98$, then $d_I \gg \xi_\parallel(T)$, and two–dimensional behavior is found whereby

$$H_{c2\parallel}(T) \sim (1 - t)^{\frac{1}{2}}. \tag{15}$$

The intermediate region where $d_I \approx \xi_\parallel(T)$ is called the cross–over region between 3D and 2D behavior. In Fig. 11 for the Nb/Ge superlattice, 3D behavior is demonstrated for the 45Å/7Å superlattice periodicity, 2D behavior for the 45Å/50Å periodicity, and 2D/3D cross–over behavior for the 65Å/35Å periodicity. Referring to Fig. 11, we note that the same sample for which $d_{\mathrm{Nb}}/d_{\mathrm{Ge}} = 65\text{Å}/35\text{Å}$ shows 2D behavior at low T, cross–over behavior at intermediate T and 3D behavior close to T_c.

Of particular interest for the interpretation of the temperature dependence of $H_{c2\parallel}(T)$ for a highly anisotropic superconductor is the formula derived by Klemm–Luther–Beasley[22] and given by

$$H_{c2\parallel}(T) = \frac{\Phi_0}{2\pi d^2 (m_\perp/m_\parallel)^{\frac{1}{2}} [1 - d^2/2\xi_\parallel{}^2(T)]^{\frac{1}{2}}} \tag{16}$$

in which d is the layer spacing (I_c for GICs) and $m_\perp$ and $m_\parallel$ are, respectively, the effective mass components perpendicular and parallel to the layers. Referring to Eq. (11)

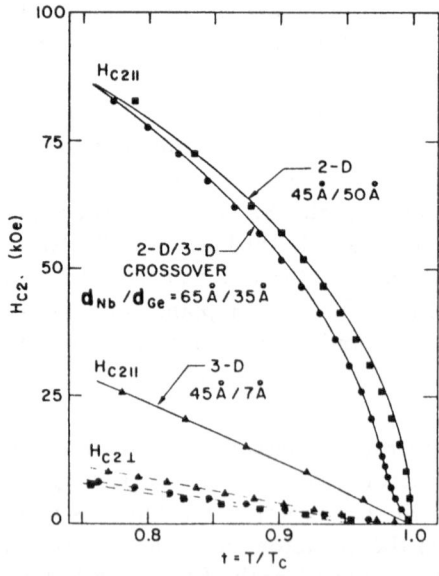

Figure 11: Upper critical fields near T_c for Nb/Ge multilayers. Systematically increasing the Ge–layer thickness between 2D Nb layers results in a progression from anisotropic 3D superconducting behavior [where $H_{c2\parallel}(T) \sim (1 - t)$], to "crossover" behavior and finally to decoupled, 2D superconducting behavior [$H_{c2\parallel}(T) \sim (1-t)^{1/2}$]. Solid lines are fits from the Klemm–Luther–Beasley theory.[19]

we note that $(\xi_\parallel/\xi_\perp)^2 = m_\perp/m_\parallel$. Equation (16) defines a cross–over temperature T^* where $H_{c2\parallel}(T)$ diverges. Thus T^* is determined from the relation $\xi_\parallel(T^*) = d/\sqrt{2}$. This singularity strongly influences the temperature dependence of the upper critical field for H perpendicular to the c–axis. For $T^* < T < T_c$ the interlayer coherence length $\xi_\parallel(T)$ extends over many layers and the bulk–like anisotropic GL theory applies. As T falls below T^* and $\xi_\parallel(T)$ remains above $d/\sqrt{2}$, then Eq. (16) applies (see Fig. 12). The physical picture for this limit is that the normal cores of the current vortices can effectively fit between the superconducting layers in the Ge layers where the vortices have no pair–breaking effect. These various regimes of superconductivity are summarized in Fig. 12.[20] According to the Klemm–Luther–Beasley theory, below T^*, 2D Josephson coupled superconductivity occurs. The temperature T^* corresponds to the 2D–3D cross–over, and anisotropic 3D behavior is expected above T^*. According to Fig. 12, the theory predicts 3D and 2D superconducting fluctuations above T_c. The experiments in Fig. 11 generally support the $H_{c2\parallel}(T)$ behavior predicted by Eq. (16) although the agreement with the angular dependence of $H_{c2\parallel}$ is not as good. Since $\xi_\parallel(0) \ll \xi_\perp(0)$, higher critical fields are obtained for $H_{c2\parallel}(T)$ than for $H_{c2\perp}(T)$ as seen in Fig. 11.

The Nb/Cu deliberately structured superlattices (S/N) represent another quasi–two–dimensional superconducting system where interlayer Josephson coupling is possible. The immiscibility of the two metal constituents of the superlattice, Nb and Cu, inhibits interface diffusion so that relatively small interface thicknesses of ~ 10 Å are achieved.[20] For the S/N system the superconducting coupling takes place via the proximity effect by which the superconducting wave function extends into the normal layer. For the S/N case, much longer characteristic Josephson coupling lengths are observed. This result is clearly evident in Fig. 13 where $H_{c2\parallel}(T)$ is plotted $vs.$ reduced temperature T/T_c for various superlattice periodicities, $d_{Nb} = d_{Cu} = d$. For $d = 171.5$ Å there is an upturn in the critical field at low temperatures, qualitatively like that expected from 3D–2D dimensional crossover. For the Nb/Cu multilayers the T_c values are observed to decrease as the layer thickness d is decreased, perhaps due to more effective

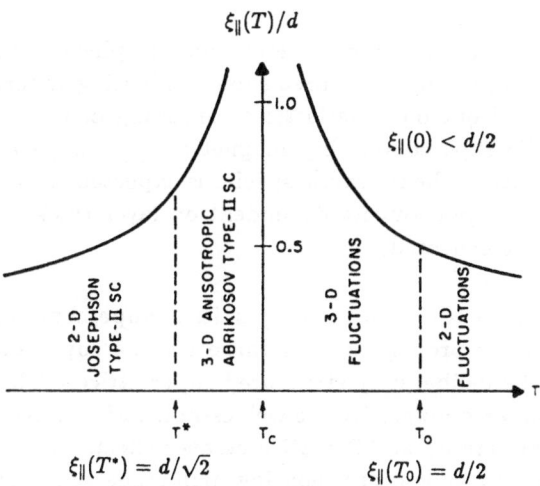

Figure 12: Theoretical superconducting behavior of a quasi–2D system.[20,30] Plot of the coherence length $\xi_\parallel(T)$ scaled by the layer repeat distance d. The dimensionality of each region is indicated.

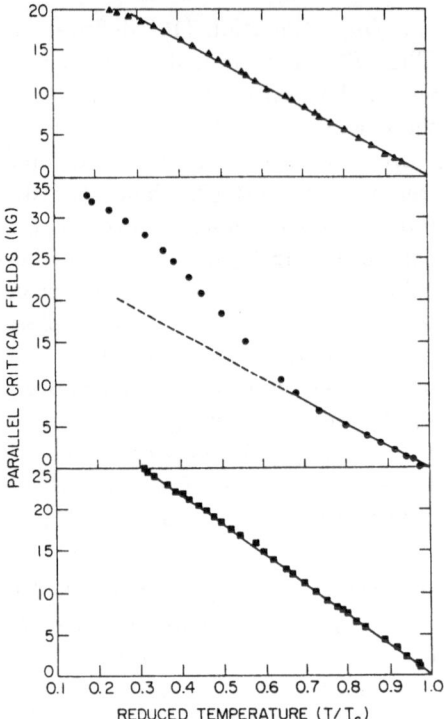

Figure 13: Plot of $H_{c2\parallel}$ of Nb/Cu multilayers as a function of reduced temperature for various layer periodicities.[31] ($\blacktriangle$, $d = 42.3$ Å), ($\bullet$, $d = 171.5$ Å), and ($\blacksquare$, $d = 867.5$ Å).

pair–breaking by the normal metallic layer. For layer thicknesses where the Nb layers are decoupled, the authors[31] find that $H_{c2\parallel}/H_{c2\perp} \to 1$, presumably due to quenching of surface superconductivity on the individual Nb layers by the pair–breaking effects of the Cu. In contrast, no quenching occurs in the S/I superlattices because the insulating layer does not cause pair breaking.

This summary of low–dimensional superconducting phenomena covers but a few of the possible types of deliberately structured superconducting superlattices.[20] Some work has already been carried out on superlattices consisting of two superconductors S/S'. Also of interest are the superconducting/magnetic (S/M) superlattices. For the S/M superlattice, the presence of the magnetic species is expected to act as a pair breaker for superconductivity. Thus, phenomena dependent on layer thickness of the superlattice constituents can also be expected.

In this chapter we have reviewed the properties of superconducting donor GICs. We have also compared their properties with those for the superconducting intercalated dichalcogenides and the deliberately structured superlattices. The theoretical concepts of two–dimensional superconductivity are well established.[20] However the study of two–dimensional superconductivity and 3D–2D crossover effects in GICs requires resolution of the difficult synthesis problem of preparing high stage superconducting GICs. Nevertheless, the observations of superconductivity in GICs reported thus far give insights into the Fermi surface and band structure of the donor GICs, such as indicating occupation of a Γ–point energy band with s–type character near the Fermi level in C_8K.[24]

ACKNOWLEDGMENTS

The authors wish to thank Professors M.S. Dresselhaus and P.C. Eklund for many important discussions. This work was supported by AFOSR contract #F49620–83–C–0011.

REFERENCES

1. M.S. Dresselhaus, (this volume) p. 1.

2. W.Y. Liang, (this volume) p. 31.

3. F.R. Gamble and T.H. Geballe, "Treatise on Solid State Chemistry, Inclusion Compounds", (Plenum, New York, 1976), Chapter 3.

4. N.B. Hannay, T.H. Geballe, B.T. Matthias, K. Andres, P. Schmidt and D. MacNair, *Phys. Rev. Lett.* **14**, 255 (1965).

5. Y. Koike, H. Suematsu, K. Higuchi and S. Tanuma, *Solid State Commun.* **27**, 623 (1978).

6. Y. Iye and S. Tanuma, *Synthetic Metals* **5**, 257 (1983).

7. R. Clarke and C. Uher, *Adv. Phys.* **33**, 469 (1984).

8. Y. Iye, *Intercalated Graphite*, edited by M.S. Dresselhaus, G. Dresselhaus, J.E. Fischer and M.J. Moran, (Elsevier Science, North Holland, Amsterdam, 1983), **20**, p. 185.

9. M. Kobayashi, T. Enoki, H. Inokuchi, M. Sano, A. Sumiyama, Y. Oda and H. Nagano, *Synthetic Metals* **12**, 341 (1985).

10. S. Kaneiwa, M. Kobayashi and I. Tsujikawa, *J. Phys. Soc. Japan* **51**, 2375 (1982).

11. G. Timp, B.S. Elman, M.S. Dresselhaus, and P. Tedrow, *Proceedings of the Materials Research Society* **21**, 201 (1983).

12. G. Roth, A. Chaiken, T. Enoki, N.C.Yeh, G. Dresselhaus and P. Tedrow, *Phys. Rev.* **B32**, 533 (1985).

13. P. Lagrange, A. Bendriss–Rerhrhaye, J.F. Marêché and E. McRae, *Synthetic Metals* **12**, 201 (1985).

14. L.E. Delong and P.C. Eklund, *Synthetic Metals* **5**, 291 (1983).

15. J. Bardeen, L. N. Cooper, and J. R. Schrieffer, *Phys. Rev.* **108**, 1175 (1957).

16. C. Kittel, *Introduction to Solid State Physics*, Sixth Edition (J. Wiley & Sons, New York) p. 631 (1985).

17. M. Tinkham, *Introduction to Superconductivity*, Robert E. Krieger Publishing Company, (1980).

18. W.E. Lawrence and S. Doniach, in "Proceedings of the Twelvth International Conference on Low Temperature Physics", edited by E. Kanda (Academic Press of Japan, Kyoto, 1971), p. 361.

19. S.T. Ruggiero, T.W. Barbee, Jr. and M.R. Beasley, *Phys. Rev.* **B26**, 4894 (1982).

20. S.T. Ruggiero and M.R. Beasley, in "Synthetic Modulated Structures", edited by L.L. Chang and B.C. Giessen, (Academic Press, Orlando, Florida, 1985), p. 365.

21. R.C. Morris, R.V. Coleman, and R. Bhandar, *Phys. Rev.* **B5**, 895 (1972).

22. R.A. Klemm, A. Luther and M.R. Beasley, *Phys. Rev.* **B12**, 877 (1975).

23. L.E. Delong, V. Yeh, V. Tondiglia, P.C. Eklund, S.E. Lampert, and M.B. Maple, *Phys. Rev.* **B26**, 6315 (1982).

24. R. Al-Jishi, *Phys. Rev.* **B28**, 112 (1983).

25. A. Shimizu and H. Kamimura, *Synthetic Metals* **5**, 301 (1983).

26. T. Ohno, K. Nakao and H. Kamimura, *J. Phys. Soc. Japan* **47**, 1125 (1979).

27. M. Kobayashi and I. Tsujikawa, *Physica* **105B**, 439 (1981); *J. Phys. Soc. Jpn.* **50**, 3245 (1981).

28. F.R. Gamble, F.J. DiSalvo, R.A. Klemm, and T.H. Geballe, *Science* **168**, 568 (1970).

29. D.E. Prober, R.E. Schwall and M.R. Beasley, *Phys. Rev.* **B21**, 2717 (1980).

30. S.T. Ruggiero, Ph.D. Thesis, Stanford University, (unpublished), 1981.

31. I. Banerjee, Q.S.Yang, C.M. Falco and I.K. Schuller, *Phys. Rev.* **B28**, 5037 (1983).

MAGNETISM IN GRAPHITE INTERCALATION COMPOUNDS

G. Dresselhaus and M.S. Dresselhaus

Massachusetts Institute of Technology
Cambridge, MA 02139, USA

INTRODUCTION

Graphite intercalation compounds (GICs)[1] provide a unique system of compounds which are of interest for fundamental studies of low dimensional magnetism. Both donor (e.g., stage 1 C_6Eu) and acceptor (e.g., all stages of certain transition metal chlorides) GICs have been found to exhibit low temperature magnetic phases. The possibility of preparing single isolated magnetic layers either by intercalation or by atomic deposition processes has stimulated a great deal of experimental activity in the observation of highly anisotropic and 2–dimensional (2D) magnetic phenomena. We start our discussion with a brief review of our current state of knowledge about quasi–2D magnetism resulting from studies of magnetic GICs. Complementary information is provided by samples prepared by atomic deposition processes[2] or by intercalated transition metal dichalcogenides.[3]

The main interest in fundamental studies of magnetism in GICs is the ability to form quasi–2D magnetic compounds in which the magnetic species are arranged in layers separated by a controlled number of non–magnetic graphite layers to form a superlattice perpendicular to the layer planes with atomically abrupt interfaces between magnetic and non–magnetic layers. This superlattice formation is made possible by the staging phenomenon. For all known magnetic GICs, the magnetic species is confined to a single intercalate layer, though in acceptor compounds the magnetic layer is generally sandwiched between two non–magnetic halogen layers, each of which lies between the magnetic intercalate layer and a graphite bounding layer (the graphite layer adjacent to an intercalate layer).[1] Thus, by varying the separation between sequential magnetic layers through the intercalation mechanism, the interplanar magnetic coupling can be reduced in a controlled manner to very small values, thereby providing a convenient system for studying the cross–over between 3D and 2D magnetic systems. In addition, by proper choice of intercalant, the spin dimensionality can be varied, as discussed below.

In contrast, for transition metal dichalcogenide host materials, the intercalation of donor cations is relatively easy to achieve, while the intercalation of acceptor anions is difficult.[3] A second striking difference is that a large number of elemental magnetic intercalants can be introduced, in contrast to the case of GICs where most of the magnetic transition metals are introduced as magnetic salts, forming acceptor compounds. A third difference relates to the staging phenomenon characteristic of GICs but not generally found in intercalated magnetic transition metal dichalcogenides. A fourth difference is that the transition metal dichalcogenide host material can itself be magnetic, though the interlayer coupling in this case is quite large and the magnetic interactions are 3–dimensional. Some common host materials for magnetic intercalation compounds include TaS_2, NbS_2, and $NbSe_2$, and some of the common intercalants include V, Cr, Mn, Fe, Co, and Ni. Easy magnetic axes are usually either $\parallel$ or $\perp$ to the layer planes. Magnetic ordering occurs roughly in the liquid nitrogen temperature range while for GICs the ordering temperatures are an order of magnitude lower. Interplanar coupling is thought to be due to an RKKY interaction[4] through the conduction electrons or by superexchange involving appropriate empty chalcogen orbitals.[3]

For deliberately structured materials, a variety of superlattices with magnetic constituents have been synthesized. The great advantage of these materials is the wide choice of magnetic species and structures that can be synthesized. A major disadvantage of the deliberately structured superlattices is the difficulty in preparing isolated single magnetic layers with abrupt interfaces. Thus, deliberately structured magnetic superlattices and magnetic GICs are complementary systems for the study of low dimensional magnetic interactions. We review here briefly some aspects of recent developments in deliberately structured magnetic superlattices, with particular emphasis on their relation to magnetic GICs.

Consider the magnetic Hamiltonian

$$
\begin{aligned}
H = \ & -J \sum_{\langle i,j \rangle} \vec{S}_i \cdot \vec{S}_j + J_A \sum_{\langle i,j \rangle} S_{iz}\, S_{jz} \\
& + J' \sum_{\langle i,k \rangle} \vec{S}_i \cdot \vec{S}_k - J_A' \sum_{\langle i,k \rangle} S_{iz}\, S_{kz}
\end{aligned}
\tag{1}
$$

in which J and J_A are the *in–plane* exchange and anisotropy energies and J' and J_A' are the *inter–plane* exchange and anisotropy energies. To realize quasi–2D magnetic behavior, sequential magnetic layers must be sufficiently separated by non–magnetic graphite layers through the staging phenomenon so that $J' \ll J$. The exchange coupling between three dimensional spins $\vec{S}$ on a single magnetic plane (2D spatial system) has several possibilities. When the anisotropy term J_A is small, then the exchange couplings are essentially isotropic and a Heisenberg spin system results; the $FeCl_3$–GICs are believed to be an example of such a system. If on the other hand, J_A is negative and of comparable magnitude to the in–plane exchange coupling J, then only the x and y spin components contribute to Eq. (1) and the system is described by an XY model whereby the spins lie in the magnetic plane; examples of a 2D–XY system are the $CoCl_2$–GIC and $NiCl_2$–GIC systems. Finally, if J_A is large and positive, then the z components of the spin dominate in Eq. (1) and the system is described by an Ising model whereby the spins lie perpendicular to the magnetic plane; the $FeCl_2$–GIC system is believed to be an example of a 2D–Ising system. As the temperature and magnetic field are varied, unusual magnetic phase transitions are observed for the various types of spin alignments that can arise in these low dimensional systems. Most of the magnetic phase transitions in GICs have been observed at low temperatures, but over a wide range of magnetic fields. Of particular interest are firstly the high stage compounds which exhibit 2D behavior, and secondly the stage 1 magnetic GICs, where cross–over behavior between 3D and 2D spatial dimensionality dominates. Figure 1 shows the temperature dependence of the AC magnetic susceptibility of $CoCl_2$–GICs, and highlights the difference in behavior between the stage 1 and higher stage compounds.[5]

The number of magnetic GICs that have been synthesized to date is small, and most of these have relatively low magnetic transition temperatures. Both donor (Eu) and acceptor ($FeCl_3$, $FeCl_2$, $CoCl_2$, $NiCl_2$, $MnCl_2$, $MoCl_5$, etc.) magnetic GICs have been synthesized. The magnetic transition temperatures in the intercalation compounds tend to be significantly lower than in the pristine parent materials because the number of magnetic nearest neighbors is reduced by the planar geometry; in the intercalation compound, the c–axis nearest neighbors are at much larger distances. Even though the magnetic behavior in the incommensurate acceptor GICs can be complex, the in–plane magnetic interactions in the intercalation compounds are closely related to those of the parent magnetic material prior to intercalation, especially in the transition metal chloride compounds where the crystal structure of the intercalate layers remain essentially identical to the corresponding layers in the pristine material.

Of the various types of magnetic GICs, the ferromagnetic 2D–XY systems have been most widely studied, because of the special role these systems play in theoretical studies of critical phenomena. Also of consequence is our ability to prepare well–staged GIC systems exhibiting 2D–XY behavior. The absence of long range order in ideal 2D systems results in short range correlations in the form of vortices, and the binding of these vortices is described by the Kosterlitz–Thouless (KT) phase transition.[6] Of particular interest is the extent to which magnetic GICs approximate an ideal 2D–XY system.

TWO–DIMENSIONAL XY MAGNETIC PHENOMENA IN GICS

Of the various magnetic GICs, the $CoCl_2$–GICs exhibit the largest c–axis anisotropy ($J_A/J = 0.56$), so that these compounds are expected to provide the best approximation to a 2D–XY model for GIC systems. Furthermore, it is possible to synthesize

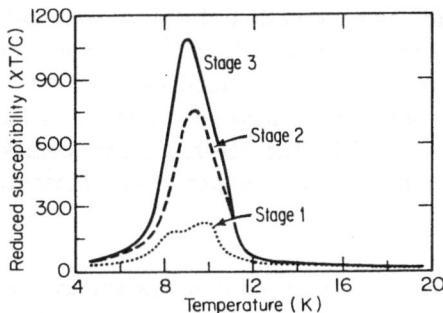

Figure 1: Reduced susceptibility ($\chi T/C$) of stage 1, 2, and 3 $CoCl_2$–GICs versus temperature. C is the Curie constant determined by high temperature susceptibility measurements ($C = 3.36$ K emu/mole).[5]

CoCl$_2$–GICs for a variety of stages, and thus to investigate the stage dependence of the observed magnetic phenomena. Bi–intercalation compounds have also been synthesized[7] with a layer sequence along the c–axis of [G–CoCl$_2$–G–GaCl$_3$–G] for example, giving separations of over 18 Å between magnetic layers, where G denotes a graphite layer. Moreover, the structural in–plane coherence lengths in the intercalate islands are large enough to contain a statistically meaningful number of spins ($\approx 10^3$ spins/island), and yet small enough to permit Monte Carlo calculations for specific spin arrangements. It is for these reasons that the CoCl$_2$–GICs have been studied extensively.[5,8,9,10]

The experimental work on magnetic GICs exhibiting 2D–XY behavior dates back to the early 1970s when Karimov and others first discovered anomalies in the magnetic susceptibility of a variety of transition metal chloride–GICs.[8] In this pioneering work, attention was also given to the identification of temperature ranges where spontaneous magnetization could be found and where anomalies in the specific heat could be correlated with anomalies in the susceptibility. For all the acceptor 2D–XY systems that have been studied, two critical temperatures were identified (T_{cl} and T_{cu}), thereby delineating three temperature regimes, each exhibiting distinct magnetic behaviors: (1) T $< T_{cl}$, (2) $T_{cl} < T < T_{cu}$ and (3) $T > T_{cu}$. At high temperature ($T \gg T_{cu}$), Curie–Weiss behavior was found.[5,8,9]

Though the presence of these two magnetic phase transition temperatures can be observed in many properties,[9] one of the most sensitive and most widely studied is the measurement of the ac differential magnetic susceptibility χ. The susceptibility χ (for the magnetic probing field in the layer planes and perpendicular to the c–axis) is shown in Fig. 1 for stages 1, 2, and 3 of the CoCl$_2$–GIC system.[5] From studies of the dependence of the susceptibility on temperature and magnetic field, the transition at T_{cu} has been identified with the Kosterlitz–Thouless (KT) transition. The temperature T_{cl} has been identified with a 2D–3D crossover driven by the in–plane symmetry breaking fields,[11] with interplanar correlation effects becoming important at low temperature $T < T_{cl}$.

For this interpretation, the spins in the high stage CoCl$_2$–GICs lie in the basal plane in a bound vortex configuration in the temperature range $T_{cl} < T < T_{cu}$, with strong in–plane spin–spin correlation but negligible interplanar correlation, consistent with neutron scattering studies which show no dependence of the scattering intensity on wave vector along the c–direction in this temperature range.[9] The absence of a remanent magnetization and the observation of a linear dependence of magnetization on magnetic field identifies the phase above T_{cu} with paramagnetism, though the spins are thought to exhibit short range order described by free vortices. The transition at T_{cu} is thus identified with the binding of these spin vortices. Clear evidence that the high stage CoCl$_2$–GICs are described by the 2D–XY model comes from the fit of the high temperature series expansion for χ to the experimental measurements taken on a stage 3 CoCl$_2$–GIC sample.[5,10]

Weak perturbations, such as the 6–fold in–plane symmetry–breaking crystal field H_6 and the interplanar magnetic interaction J' become important at lower temperatures[11] and give rise to a 2D–3D cross–over transition which is identified with T_{cl}. Below T_{cl} the spontaneous magnetization due to coupling of the spins to H_6 gives rise to ferromagnetically ordered in–plane domains. These spins are coupled antiferromagnetically by an interplanar exchange giving rise to an antiferromagnetic stacking arrangement of ferro-magnetic planar domains, as implied by neutron scattering experiments.[9] Recent studies of the temperature dependence of elastic neutron diffraction scattering intensities[12] (see Fig. 2) are strongly suggestive of the KT bound vortex state. The (00ℓ) scans show both a half order peak ($\ell = 1/2$) associated with a magnetic doubling of the c–axis unit cell (below T_{cu}) and a broad background (between T_{cu} and T_{cl}).

Since the magnitudes of these magnetic perturbations are very small, weak mag-netic fields have a major effect on the magnetic properties of the CoCl$_2$–GICs. For example, magnetic fields of a few Oe are sufficient to suppress the KT transition and to establish long range spin order, consistent with experimental observations.[9,13,14] At lower temperatures, small externally applied magnetic fields also compete with the small 6–fold in–plane anisotropy field in establishing the magnetic direction of the magnetic domains. Different field and temperature dependences are found at T_{cl} and T_{cu} for the various magnetic properties such as the magnetic susceptibility, magnetization and heat capacity, indicative of the different origins of the phase transitions at T_{cl} and T_{cu}.

Measurements of the magnetic field dependence of χ, plotted for example for a stage 1 CoCl$_2$–GIC sample for various temperatures in Fig. 3 provide direct information about the magnetic phase diagram of the CoCl$_2$–GIC system[13] for the magnetically ordered phases below T_{cl} (see Fig. 4a). As shown in Fig. 3, these curves clearly exhibit two peaks, denoted by H_{c1} and H_{c2}. For $H = 0$, an angle of $\theta = \pi$ is maintained between the mag-netization vectors of domains on adjacent planes (see Fig. 4b). Application of a field perpendicular to the c–axis causes θ to decrease. At the field H_{c1}, an interplanar spin-flop phase is achieved. For H_{c1} along an easy in–plane direction, the spins on adjacent

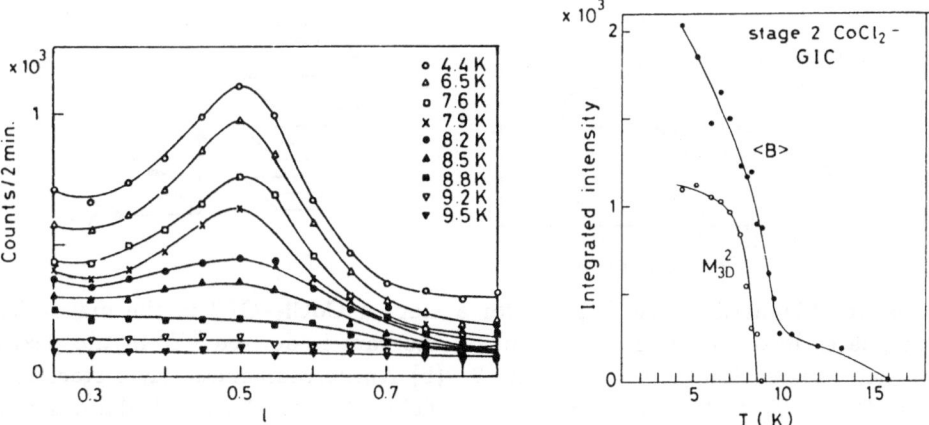

Figure 2: Magnetic neutron scattering from a stage 2 CoCl$_2$–GIC sample. (a) The peak occurring at $(0,0,\frac{1}{2})$ is consistent with an antiferromagnetic spin arrangement on adjacent planes. (b) Plot of the temperature dependence of the $(0,0,\frac{1}{2})$ peak intensity (M^2_{3D}), and of the background $\langle B \rangle$ from spectra such as in (a). The magnetic scattering background $\langle B \rangle$ may be identified with scattering from bound spin vortices.[12]

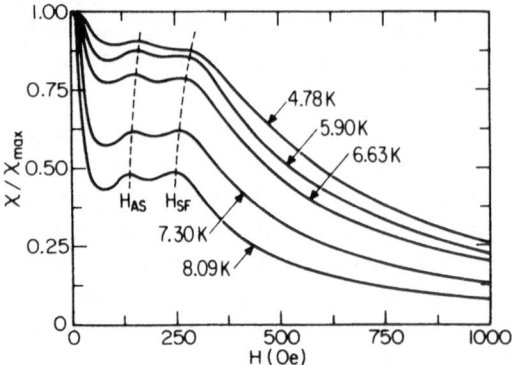

Figure 3: Magnetic field ($\vec{H} \perp \vec{c}$–axis) dependence of the magnetic susceptibility for various values of temperature. The two χ peaks in the figure are identified with magnetic phase transitions shown in Fig. 4a.[13]

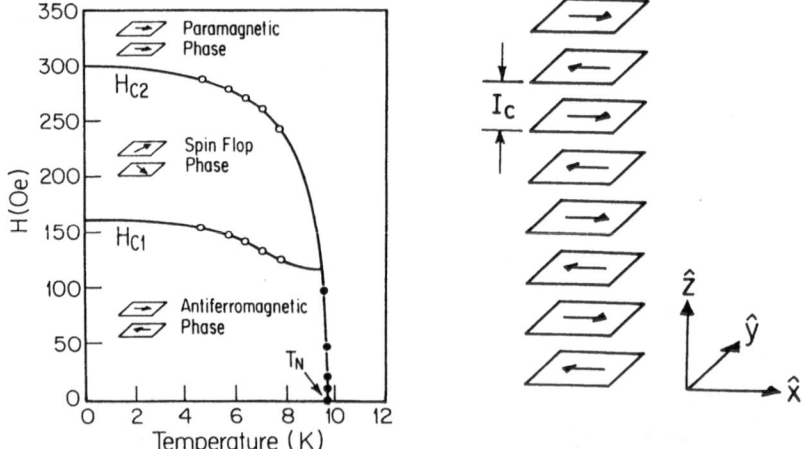

Figure 4: (a) Magnetic phase diagram for a stage 1 $CoCl_2$–GIC in the magnetically ordered phases below $T_{cl} = T_N$. This phase diagram is derived[13] from the peaks in the magnetic susceptibility shown in Fig. 3. (b) Antiferromagnetic layer stacking for $CoCl_2$–GICs for $T < T_{cl}$ and zero magnetic field. Monte Carlo calculations[14] as a function of field yield the spin arrangements shown in (a).

planes become canted and lie along easy in–plane axes defined by the 6–fold symmetry breaking field, with the spins on adjacent planes making an angle $\theta = 2\pi/3$ with respect to each other. The angle θ becomes zero at H_{c2} where interplanar ferromagnetic domain alignment along the applied field is achieved. By measuring the temperature dependence of H_{c1} and H_{c2}, the magnetic phase diagram shown in Fig. 4a is obtained. Estimates for H_6 in $CoCl_2$–GICs based on field dependent susceptibility, magnetization and neutron scattering experiments yield $H_6 \sim 10$ Oe while estimates for J' for a stage 1 compound yield ≈ 160 Oe or $(J'/J) \sim 4 \times 10^{-3}$. The antiferromagnetic interplanar ordering of the magnetic domains below T_{cl} is a 3D magnetic effect. A schematic diagram of the spin alignment at zero field is shown in Fig. 4b. Model calculations in the harmonic approximation and Monte Carlo simulations of the spin arrangements both establish the magnetic spin alignments shown in Figs. 3 and 4. Studies of these magnetic field induced phase transitions provide values for H_6 and J' which can then be used to study 2D–3D crossover effects above T_{cl}. Unfortunately, the magnetic field–induced phase transitions for stages $n \geq 2$ have not been clearly resolved in the magnetic susceptibility studies, so that a magnetic phase diagram such as shown in Fig. 4a is not available for stages $n \geq 2$.

Other magnetic acceptor GICs have been studied and all show magnetic behavior which is significantly different from the pristine materials. The $NiCl_2$–GICs are very similar[15] to the $CoCl_2$–GICs described above. The $MnCl_2$–GICs are however very different because the nearest neighbor in–plane interaction is antiferromagnetic and the triangular array of magnetic ions gives rise to spin frustration.[16] The $FeCl_3$– and $FeCl_2$–GICs are more difficult to prepare in a pure valence state, thus making the measurements more sample dependent. The $MoCl_5$–GICs are interesting because they can be prepared for a variety of stages, but their crystal structure (monoclinic) is more complicated than for the $CoCl_2$–GICs which have the theoretically interesting triangular in–plane magnetic lattice. The low dimensional magnetism of these compounds has not yet been modeled in detail. Synthesis of high quality samples is however essential for the success of all of these studies, and remains a major challenge to progress in this field.

A macroscopic model of 2D superparamagnetism has been suggested by Rancourt[10,17] to explain the low temperature behavior of some of the magnetic GICs. This model makes use of the islandic morphology of the intercalate layer and the domain structure of the intercalant.[18] The $CoCl_2$–GICs have recently been synthesized with higher filling factors, which may result in larger island sizes. A detailed comparison of such data with the size dependences in the Rancourt model and the finite size KT model may be used to decide which model (or neither, or both) fits the experimental data.

MAGNETIC FIELD–INDUCED PHASE TRANSITIONS IN THE DONOR C_6Eu

The magnetic properties of donor GICs are expected to be distinctly different from those of the acceptor compounds because of differences in the basic magnetic coupling mechanism occurring in the pristine magnetic intercalants (which are metals) and in the intercalation compounds (where metallic behavior is associated only with the graphite π–bands). In the magnetic donor compounds, the mechanism for the magnetic exchange is by the RKKY mechanism through the conduction band electrons.[4]

The only magnetic donor compound that has been studied in depth for its magnetic

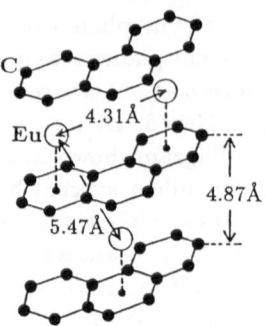

Figure 5: Crystal structure of stage 1 C_6Eu. The Eu atoms lie in a planar triangular lattice and the Eu layers show $\alpha\beta$ stacking.[19]

properties is the first stage compound with Eu, which has the stoichiometry C_6Eu.[19] This compound is of particular interest as an XY spin system insofar as the Eu spins align within the magnetic planes. The unusual features of the observed magnetic behavior of C_6Eu are the large anisotropy in magnetic properties observed for $\vec{H} \perp \hat{c}$ and $\vec{H} \parallel \hat{c}$, the difference in magnetic properties of Eu resulting from intercalation, the large number of magnetic phases that are observed as a function of field and the observation of an unusual metamagnetic (ferrimagnetic) phase. To account for this ferrimagnetic phase, explicit calculations for the XY model and Monte Carlo simulations both show that in addition to the usual two–spin $\vec{S}_i \cdot \vec{S}_j$ interaction terms, 4th order terms in S are required in the Hamiltonian to account for the experimental magnetization results.[19,20]

The intercalate layer of C_6Eu is commensurate with the adjacent graphite layers and has an in–plane unit cell with the $(\sqrt{3} \times \sqrt{3})R30°$ structure. Each Eu atom in the triangular planar lattice has 6 nearest–neighbor Eu atoms at a distance of 4.31Å within the layer plane and 6 next nearest neighbor Eu atoms lie at a distance of 5.47Å in adjacent planes (see Fig. 5). The next nearest–neighbor Eu–Eu in–plane distance is 7.46Å, indicating that higher stage compounds are needed to demonstrate quasi–2D behavior. To date, no higher stage donor compounds have been synthesized.

The configuration of the Eu^{2+} ion is $^8S_{7/2}$ which arises from a $(4f)^7$ electronic configuration, giving a total spin of 7/2 with no orbital angular momentum. Because the nearest–neighbor exchange is antiferromagnetic, the spins form an antiferromagnetic frustrated triangular spin lattice, but lie in the intercalate layer.[19] This spin system can easily be modified by introducing weak interactions, so that a variety of magnetic phases can be realized by variations of the temperature and magnetic field.

As the sample is cooled in zero magnetic field, the susceptibility χ shows only weak structure at the magnetic ordering temperature of 40 K, characteristic of antiferromagnetic systems, while the specific heat shows a strong anomaly at 40.0 ± 1.0 K, which is confirmed by direct magnetization measurements.[19] The most remarkable behavior of C_6Eu is found in the magnetization[19] and magnetoresistance[20] measurements. The magnetoresistance curves taken up to 40 Tesla at a variety of temperatures between 4.2 and 34 K clearly show 5 distinct magnetic phases, delineated by the 4 magnetic phase transition fields H_{c0}, H_{c1}, H_{c12} and H_{c2} (see Fig. 6). From the temperature dependence of these critical field values the magnetic phase diagram in Fig. 7 is constructed. The

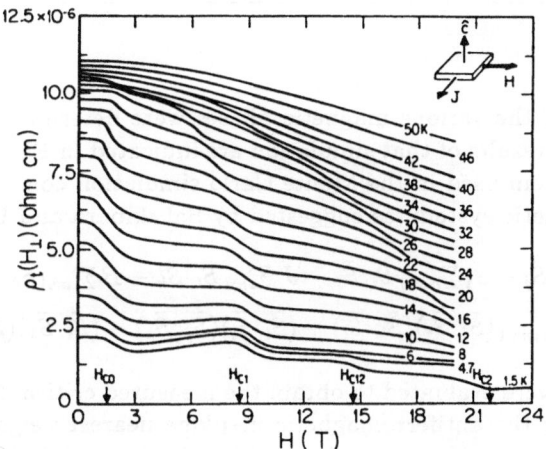

Figure 6: Transverse magnetoresistance for C_6Eu for various temperatures for the magnetic field orientation ($\vec{H} \perp \vec{c}$–axis and $\vec{H} \perp \vec{j}$).[20] The anomalies in the magnetoresistance are identified with magnetic phase transitions (see text).

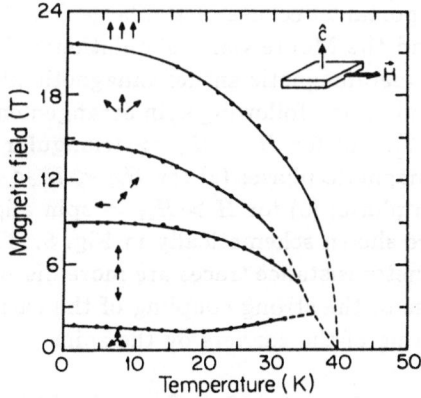

Figure 7: C_6Eu phase diagram with the spin configuration shown for the various magnetic phases.[20]

Table 1: Parameters for magnetic interactions in C_6Eu.

	$J_0(K)$	$J_1(K)$	$J'(K)$	$BS^2(K)$	$KS^2(K)$
Sakakibara & Date's Model[19]	−0.5	0.4	0.1	0.02	0.05
Monte Carlo simulation[20]	−0.656	0.525	0.131	0.0262	0.0656

spin arrangements of the various magnetic phases were determined by a Monte Carlo simulation[20] and the results of that simulation are indicated in Fig. 7. The Hamiltonian for the magnetic system used in the Monte Carlo simulation considered the Eu ions to form a 3D–XY magnetic system, as suggested by Sakakibara and Date[19]

$$\mathcal{H}_0 = -J_0 \sum_{nn} \vec{S}_i \cdot \vec{S}_j - J_1 \sum_{nnn} \vec{S}_i \cdot \vec{S}_j - J' \sum_{nn} \vec{S}_i \cdot \vec{S}_k - B \sum_{nn} (\vec{S}_i \cdot \vec{S}_j)^2 - g\mu_B \sum_i \vec{S}_i \cdot \vec{H}$$
$$+ K \sum_{4-spin\ ring} \{(\vec{S}_i \cdot \vec{S}_j)(\vec{S}_k \cdot \vec{S}_l) + (\vec{S}_i \cdot \vec{S}_l)(\vec{S}_j \cdot \vec{S}_k) - (\vec{S}_i \cdot \vec{S}_k)(\vec{S}_j \cdot \vec{S}_l)\}$$

(2)

and the parameters were evaluated to obtain the measured critical fields. The first term in Eq. (2) represents the antiferromagnetic in–plane nearest neighbor exchange interaction ($J_0 < 0$). The second term represents the ferromagnetic in–plane next nearest neighbor exchange interaction ($J_1 > 0$). The third term is the ferromagnetic inter-plane nearest neighbor exchange interaction ($J' > 0$). The fourth term is the in–plane nearest neighbor biquadratic exchange interaction. The fifth term is the Zeeman energy due to the external magnetic field. The last term is the in–plane four–spin ring cyclic exchange interaction. Values of the parameters used to fit the magnetic fields for the magnetic phase changes observed in the magnetization measurements and magnetoresistance measurements (see Fig. 6) are given in Table 1.

Although the parameters B and K are very small compared with J_0, the four spin terms are nevertheless important because of the large S value ($S = \frac{7}{2}$). In fact, both the model calculation[19] and the Monte Carlo simulation[20] show that the 4–spin terms are essential to realize the metamagnetic antiferromagnetic phase between H_{c0} and H_{c1}. According to this Hamiltonian, the following spin arrangements were identified for each magnetic phase (see Fig. 7): (a) for $H < H_{c0} \Rightarrow$ triangular spin or Δ–phase; (b) for $H_{c0} < H < H_{c1} \Rightarrow$ ferri–magnetic phase; (c) for $H_{c1} < H < H_{c12} \Rightarrow$ canted phase; (d) for $H_{c12} < H < H_{c2} \Rightarrow$ fan phase; (e) for $H > H_{c2} \Rightarrow$ spin aligned paramagnetic phase. These magnetic phases are shown schematically in Fig. 8. The magnetic phase transitions observed in the magnetoresistance traces are more distinct than those observed in the magnetization, because of the strong coupling of the carriers to the magnetic ions, leading to a strong scattering of the carriers by the spins.

Other rare earth GICs (such as with Sm, Tm and Yb) have been synthesized[21] but have not yet been studied with regard to their magnetic properties. In view of the very interesting magnetic properties of C_6Eu, the other rare earth GICs are also expected to exhibit interesting magnetic properties. Of particular interest would be the synthesis of higher stage Eu–GICs to study the effect of the reduction of J' on the magnetic behavior.

The most serious factor impeding progress in the study of 2D magnetism in magnetic

GICs is the synthesis of well–staged magnetic GICs of the appropriate intercalants and stages. In addition, further theoretical developments will be needed to treat a variety of perturbations causing 3D–2D crossover effects in magnetic GICs, including multiple symmetry–breaking fields and finite size effects. Although these theoretical problems are very difficult, significant progress has been made for 2D–XY systems with ferromagnetic nearest neighbor interactions.[22] But many unanswered questions remain concerning the phase diagrams of magnetic GICs.

DELIBERATELY STRUCTURED MAGNETIC SUPERLATTICES

Complementary information is provided by superlattices prepared by atomic deposition processes and by intercalation, each offering advantages and disadvantages. The greatest advantage of the atomic deposition process relates to the flexibility in the choice of materials and periodicities.[2] The greatest control of the layer thickness and in–plane homogeneity is provided by molecular beam epitaxy (MBE), though metal organic chemical vapor deposition (MOCVD) offers almost as much control and greater sample throughput. With MBE, it is possible, in principle, to prepare a superlattice from layers of magnetic species A and non–magnetic species B, or even from layers of two different magnetic species. Layers of almost any magnetic element or compound can be deposited, thereby allowing, in principle, the study of 2D ferromagnetic, antiferromagnetic, ferrimagnetic and spin glass behavior. Likewise, the non–magnetic species can be chosen to be insulating, semiconducting or metallic. With MBE, the layer thicknesses d_A and d_B can be controlled over wide ranges (5 Å $\leq d \leq$ 200 Å), and samples with several hundred superlattice unit cells $d = d_A + d_B$ can be fabricated.

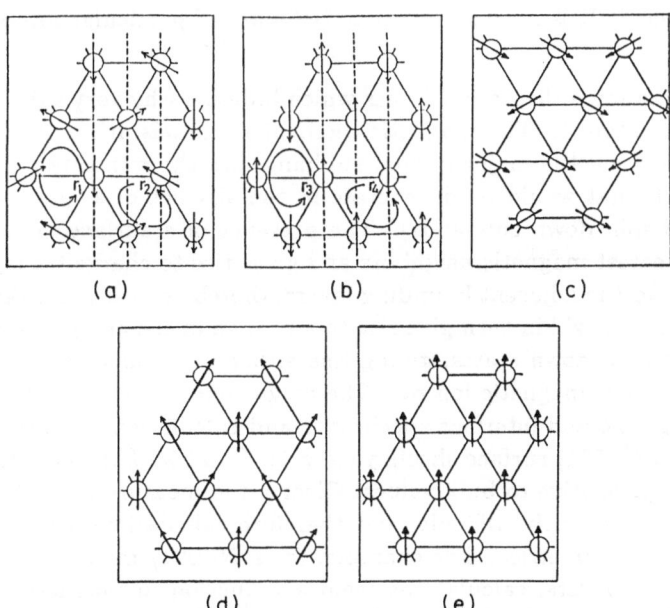

Figure 8: Triangular spin arrangement of C_6Eu for the various magnetic phases: (a) triangular spin frustrated phase, (b) ferrimagnetic (three sublattice) phase, (c) canted phase, and (d) fan phase, and (e) spin aligned paramagnetic phase.[20]

417

The special advantages of intercalation compounds for studying 2D magnetism relate to their unique structural features, allowing preparation of ordered *isolated* magnetic layers. Synthesis techniques for the atomic deposition process for metallic superlattices are not sufficiently developed to form a crystalline in–plane structure with long range order. In contrast, GICs have been prepared with in–plane crystalline coherence distances in excess of 1 μm.[23] A high degree of crystalline in–plane ordering can however be achieved in diluted magnetic semiconductor (DMS) superlattices,[24] as for example those based on $Cd_{1-x}Mn_xTe$. In this case, strained layer superlattices[25] are formed, so that the lattice constants are weakly dependent on the relative layer thicknesses d_A/d_B. This dependence on d_A/d_B implies that the lattice constants must be measured for each sample. At the same time, this dependence offers the flexibility of varying the lattice constants over small ranges. Because of the substitutional nature of the dilute magnetic cations, the magnetic interactions in DMS superlattices are mainly random, and spin–glass ground states are observed.[24]

Deliberately structured magnetic superlattices offer the possibility of modifying the symmetry of the magnetic constituents, introducing new symmetries, lowering the coordination number of magnetic species near the interfaces, and introducing the possibility of interface states. Because of these modifications, new and exotic magnetic phenomena are expected and new device applications are likely to follow. A major goal of research on magnetic superlattices will be the preparation of systems exhibiting 2–dimensional magnetism in a controlled way. Although there is much excitement in this field, only a few magnetic superlattice systems have been prepared and studied to date. Because magnetic interactions are mainly short range, very thin magnetic layers (e.g., magnetic monolayers) are of particular interest, thus imposing severe constraints on the synthesis of the superlattices. For this reason, sample preparation of suitable coherent crystalline superlattices with sharp interfaces remain a major challenge for materials scientists working in this field. It is in this limit that GICs are of particular importance.

Let us consider some theoretical issues which indicate why very thin magnetic layers are of particular interest. The magnetic moment of metals arises from the difference in occupation between the majority spin subband and the minority spin subband of a Kramers doublet, that would be degenerate in the absence of a magnetic field (called the spin up and spin down subbands). The presence of a surface or interface reduces the number of nearest magnetic neighbors and gives rise to charge transfer between the magnetic layer and the adjacent bounding layers, thereby causing the Fermi level E_F to shift. The shift in E_F will in turn give rise to a change in the energy difference between the spin up and spin down states, giving rise either to an increase or decrease in the magnetic moment per magnetic ion m_s. The magnitude and sign of this effect depends on both the magnetic and interface species.[26] Band calculations for free–standing BCC Fe(001) and FCC Ni(001) surface sheets yield a 30% and 20% increase in magnetic moment, respectively, relative to bulk values. Clearly the measurement of a free–standing magnetic layer is extremely difficult, and therefore calculations have also been made for more favorable geometries. For example, for a Ni(001) monolayer sandwiched between two Cu(001) layers, calculations yield a reduction in magnetic moment m_s by about 50% due to charge transfer from the Cu to the Ni. Since these additional charges preferentially occupy the minority d–band, a reduction in magnetic moment follows. However, the same calculations[26] find that two Ni layers sandwiched between Cu layers show an enhanced moment. These surface magnetism calculations lead to two main conclusions: (1) that large magnetic effects can occur at magnetic interfaces and (2) that the high sensitivity of the magnitude and sign of these changes in magnetic moment to the number of magnetic layers and the nature of the interface, makes it very difficult to carry out quantitative experiments on these systems. A multitude of experimental problems hamper progress to measure such small effects including topological problems (inter–diffusion, surface roughness, island formation), chemical contamination, and crystallographic difficulties (lattice matching and the lattice distortion due to lattice mismatch, surface reconstruction, grain growth, etc.).

Of all the metallic magnetic superlattices, the Cu/Ni superlattice has been studied most extensively.[27] For this system both Cu and Ni crystallize in the FCC structure with almost the same lattice constant so that good lattice matching is achieved. Furthermore, a great deal is known about the magnetic properties of Cu_xNi_{1-x} alloys. Because of the high degree of miscibility of Cu and Ni, the preparation of Cu/Ni superlattices is complicated by interdiffusion and clustering effects. Measurements of the static magnetization M and anisotropy field H_K (see Fig. 9) show that variation of the Cu thickness d_{Cu} for a fixed Ni thickness $d_{Ni} = 10$ Å does not affect either M or H_K.[28] However decreasing d_{Ni}, decreases M and increases H_K with functional forms of approximately $M \approx M_B - A/t$ where M_B is the bulk magnetization and $H_K \approx B/t$ for $t \leq 30$ Å (see Fig. 9). For $t \geq 30$ Å, bulk behavior is observed.[28] An explanation of these observations suggests an RKKY exchange coupling between Ni layers via conduction electrons.[28]

These interesting experimental results are closely connected with recent calculations by Freeman and coworkers[26] showing strongly enhanced ferromagnetic moments for 3d BCC transition metal overlayers, sandwiches and superlattices with Au and Ag. Because of the reduced coordination number and lower symmetry of the monolayer magnetic film,

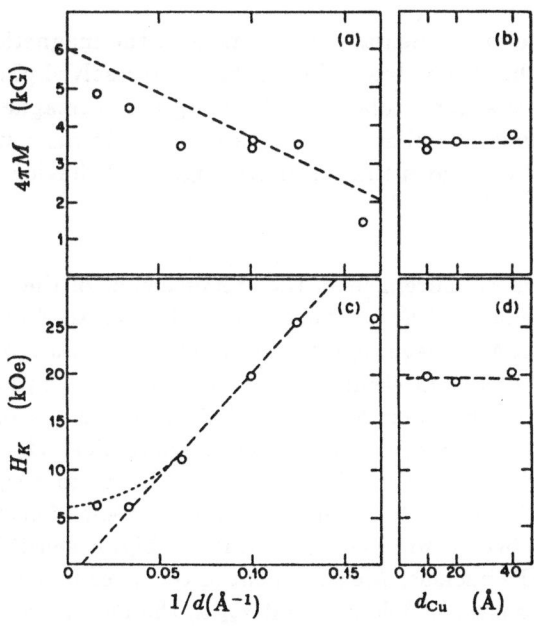

Figure 9: Static magnetization M and anisotropy field H_K for a series of Cu/Ni multilayers: (a) $d_{Ni} = d_{Cu} = d$, (b) $d_{Ni} = 10$ Å, (c) $d_{Ni} = d_{Cu} = d$, (d) $d_{Ni} = 10$ Å.[28]

enhanced magnetic moments are found. For example, a free–standing Cr(001) film is found to be *ferromagnetic* with a magnetic moment $m_s = 4.12\mu_B$ in contrast to bulk Cr which is *antiferromagnetic* with $m_s = 0.59\mu_B$ on each of the two magnetic sublattices. When placed in contact with an Au substrate (interface) or between two Ag layers, the value of m_s is only slightly decreased to $3.70\mu_B$ for 1 Cr/Au(001) overlayer and to $3.10\mu_B$ for a Au/Cr/Au(001) sandwich. The reason that the decrease in m_s is small is because the Au d–band lies far below the Fermi level (by ~ 4.5eV). When two Cr(001) layers are in contact with an Au(001) surface then antiferromagnetic behavior occurs, still with large moments on each of the layers ($2.90 \mu_B$ on the surface layer (s) and $-2.30 \mu_B$ on the second layer $(s-1)$), but now the moments are of opposite sign, establishing the antiferromagnetic behavior observed in the bulk. The results for Au/3 Cr/Au(001) yield $m_s = 1.75 \mu_B$ for the s layers and $m_s = -1.07\mu_B$ for the middle Cr layer, while the Au/5Cr/Au(001) structure yields $m_s = 1.55 \mu_B$, $-0.77 \mu_B$ and $0.67 \mu_B$ for the interface, subinterface and central Cr layers, respectively, with the central layer value close to that for bulk Cr metal. Also of interest is the small magnetization of the interface Au layer due to spin polarization of the d–electrons; a magnetic moment of $m_s = 0.14 \mu_B$ was found for the surface gold atoms. The theoretical results for Fe free monolayers, overlayers, interfaces all show ferromagnetic behavior. Enhancement of m_s over the bulk value for BCC Fe metal ($m_s = 2.15 \mu_B$) was found in all cases that were calculated.[26] One remarkable finding of these calculations is the case of Vanadium which is normally non–magnetic. An isolated monolayer of V(001) is also found to be non–magnetic. However, these calculations show that an overlayer of V/Au(001) is magnetic with $m_s = 1.75 \mu_B$. Ferromagnetic behavior of the interface V is retained for the 2V/Ag(001) overlayer structure. However the vanadium sandwich Ag/V/Ag is non–magnetic, so that a superlattice based on V is also expected to be non–magnetic.[26] In all cases the formation of a magnetic state leads to a lower energy than the paramagnetic state. For the noble metal/transition metal interface, the charge transfer is small ($< 0.1e$), reflecting the small hybridization of the d–bands of the two species.

The most sensitive experimental measurement of the magnetic moment of Cr in an overlayer of Cr on Ag has been obtained with the angle resolved photoemission (ARPES) technique,[29] showing that a monolayer of Cr on Ag is ferromagnetic, not antiferromagnetic. A large experimental enhancement was reported in m_s, consistent with theory.[26] These experiments also confirm that antiferromagnetic behavior is achieved by two Cr overlayers on Ag.

The most sensitive measurement of the enhancement of the magnetic moment of a ferromagnetic surface layer of Fe has been obtained on clean Fe(110) surfaces by means of conversion–electron Mössbauer spectroscopy (CEMS) on a monolayer of ^{57}Fe strategically placed under ultra high vacuum conditions within 21 layers of Fe(110) deposited on a W(110) substrate. By substituting the ^{57}Fe layer for any one of the first 11 ^{56}Fe layers, the layer dependence of the hyperfine magnetic field B_{hf} of the Mössbauer experiment was found,[30] yielding good agreement with the corresponding layer by layer calculation of B_{hf}.[31] With this technique, the enhancement of m_s for the surface layer of ferromagnetic Fe has been confirmed, exploiting the highly sensitive local probe, the hyperfine field of the ^{57}Fe nucleus. Measurement of the layer by layer quadrupole splitting ϵ yielded values of the electric field gradient V_{zz} at the Fe surface,[30] again in good agreement with V_{zz} calculated from the band model.[31] CEMS experiments were also carried out on similar samples with 21 layers of Fe(110) on a W(110) substrate but now covered by a single monolayer of Ag; these experiments showed the change in B_{hf} of the surface layer to decrease by a factor of 5 with an Ag monolayer present, also consistent with the calculations. The temperature dependence of B_{hf} was found to follow the spin wave relation $B_{hf}(T) = B_{hf}(0)(1 - bT^{3/2})$ for both surface and bulk layers, where b is layer dependent. In fact b at the surface was found to be exactly twice the value in the bulk, consistent with the interpretation that the surface probes the square amplitudes of all thermally excited spin waves in the antinodes, while the bulk probes their mean values.[30]

Recently, superlattice layers of ferromagnetic (Fe)/antiferromagnetic (Cr) films have been fabricated by electron beam evaporation on a sapphire substrate. The evaporation was done onto substrates at relatively low temperature (280°C) and under UHV conditions (10^{-9} torr). The magnetization and hysteresis behaviors of these superlattices have also been examined.[32] In this study, all superlattice samples contained 5 layers of Fe, but the thickness of the Cr was varied from 2 to 30. From the x–ray analysis, it was concluded that the film growth was along the (110) direction and that the interfaces remained relatively sharp (to within 2 atomic layers) and that little c–axis inter–diffusion occurred. The magnetic results showed that the saturation magnetization for only 5 atomic layers of Fe remained close to that for pristine Fe, despite the presence of intervening Cr layers. Of particular interest is the inverse Cr layer thickness dependence and inverse temperature dependence of the coercivity field (H_C) and of the anisotropy field H_K. Though both H_C and H_K decrease with decreasing Cr film thickness, H_C decreases more rapidly than H_K (see Fig. 10). In contrast, H_C for a pure Fe film is temperature independent. Thus the preparation of magnetic monolayers gives rise to new magnetic effects. An important research objective of studies of deliberately structured magnetic superlattices will be the discovery of extreme magnetic properties (such as

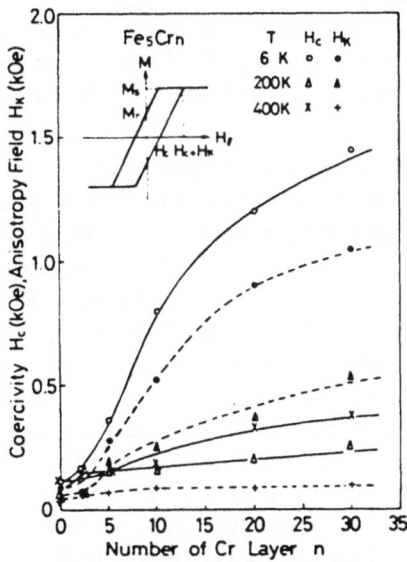

Figure 10: Coercivity (H_c) and anisotropy field (H_K) vs. number of Cr layers in a Fe/Cr superlattice with 5 Fe layers and n Cr layers (see legend). A hysteresis curve is shown in the inset where $H_\parallel$ is the magnetic field in the layer planes.[32]

magnetization, coercivity, magnetic anisotropy, magnetostriction, etc.) connected with the layering and with magnetic size effects in the superlattice.

Using an MBE system, Kwo et al.[34] were able to prepare superlattices with rare earth ferromagnetic/non–magnetic (e.g., Gd/Y) with a high degree of in–plane crystallinity, good periodicity along the z–axis of the superlattice and interfaces of 2 atomic layers thickness. The sharpness of the interface was deduced from spatial fluctuations (~ 5 Å) in the x–ray linewidth analysis. The results show that the magnitude of the magnetic moment, temperature dependence of the magnetization and the Curie temperature T_c all depend on d_{Gd} and d_Y as the number of atomic layers was varied between 5 and 20. The magnetization curves follow the behavior expected for an ideal ferromagnetic thin film with weak anisotropy, namely a square loop for H in the plane of the film, and a linear dependence of the magnetization on H for the perpendicular field orientation for fields less than the saturation magnetization $4\pi M_0$. Some reduction in the Curie temperature T_c was found for thin films (~ 5 layers thick). The high quality of the materials and structures that have been prepared by the MBE technique holds great promise for interesting new magnetic phenomena and device physics in the next few years.

Another interesting magnetic limit that has received attention is a system exhibiting a spiral spin configuration where the period of the spin configuration is less than or greater than d_M the thickness of the magnetic species in the superlattice. The explicit system that was studied was the Dy/Y system. In this work,[33] the thickness of the non–magnetic species Y was varied from 5 Å to 40 Å, and the sample preparation was by MBE. It was found that in the magnetic superlattice, the spiral spin configuration is retained, but since the spiral period is limited by the layer thickness of the superlattice, a net magnetization for the superlattice results. The magnetization is also found to be sensitive to whether the sample is cooled in a high magnetic field or not. The layers tend to freeze into a non–magnetic state at low temperature in the absence of an external field. Study of the effect of the superlattice periodicity on the periodicity of spiral spin configurations is clearly an important research area for the future.

The synthesis of deliberately structured magnetic superlattices offers great promise for the synthesis of materials with novel and controllable anisotropic magnetic properties. Another goal may be the synthesis of magnetic multilayer materials, with enhanced properties relative to the pristine constituents. The ability to control both thickness and composition of the multilayers should provide a powerful tool for the study of magnetic coupling mechanisms (direct exchange, dipole coupling, RKKY through conduction electrons, or indirect exchange) between two sequential magnetic films separated by a thickness d of a variety of materials: a non–magnetic insulator or semiconductor, a magnetic insulator or semiconductor, a nonmagnetic metal, a magnetic metal with the same or different type of spin alignment, a superconductor. From an applications point of view, we can expect a variety of non–reciprocal devices emerging from this technology. One factor which is especially favorable to device technology is the good lattice match between the ($1\bar{1}0$) zinc blende GaAs face ($a/2 = 2.827$ Å) and the ($1\bar{1}0$) BCC Fe face ($a = 2.866$ Å). This lattice match should facilitate the use of magnetic elements to control semiconductor components in monolithic integrated circuits.[35] In the overall picture of magnetic superlattices, magnetic GICs represent the limiting case of a magnetic monolayer separated by a controlled number of nonmagnetic layers. Implementing the promise of magnetic superlattices for producing new classes of magnetic materials with unique anisotropic properties will require major advances over the next few years in materials synthesis. The analysis of magnetic properties in the limit of very small magnetic layer thickness will clearly stimulate further theoretical work.

ACKNOWLEDGMENTS

The authors wish to thank Drs. S.T. Chen, J. Nicholls, K. Sugihara, and K.Y. Szeto for many important discussions. This work was supported by AFOSR contract #F49620–83–C–0011.

REFERENCES

1. M.S. Dresselhaus and G. Dresselhaus, *Advances in Physics* **30**, 139 (1981).

2. M.S. Dresselhaus, (this volume) p. 1.

3. W.Y. Liang, (this volume) p. 31.

4. H.J. Zeiger and G.W. Pratt, "Magnetic Interactions in Solids", Clarendon Press, Oxford (1973).

5. K.Y. Szeto, S.T. Chen, and G. Dresselhaus, *Phys. Rev.* **B32**, 4628 (1985).

6. J.M. Kosterlitz and D.J. Thouless, in *Progress in Low Temperature Physics*, edited by D.F. Brewer (North–Holland, Amsterdam, 1978) **7B**, 395 (1978).

7. I. Rosenman, F. Batallan, Ch. Simon, and L. Hachim, *J. de Physique* (to be published).

8. Yu.S. Karimov, *Zh. Eksp. Teor. Fiz.* **68**, 1539 (1975), [*JETP Lett.* **41**, 772 (1976)].

9. M. Suzuki, H. Ikeda, and Y. Endoh, *Synthetic Metals* **8**, 43 (1983).

10. J.T. Nicholls and G. Dresselhaus, (this volume) p. 425.

11. J.V. José, L.P. Kadanoff, S. Kirkpatrick, and D.R. Nelson, *Phys. Rev.* **B16**, 1217 (1977).

12. M. Suzuki, D.G. Wiesler, P.C. Chow, and H. Zabel, *J. Mag. Mag. Materials* **54–57**, 1275 (1986).

13. K.Y. Szeto, S.T. Chen, and G. Dresselhaus, *Phys. Rev.* **B33**, 3453 (1986).

14. S.T. Chen, Ph.D. Thesis, Massachusetts Institute of Technology (unpublished) (1985).

15. M. Suzuki and H. Ikeda, *J. Phys. C* **14**, L923 (1981).

16. K. Koga and M. Suzuki, *J. Phys. Soc. Jpn.* **53**, 786 (1984).

17. D.G. Rancourt, B. Hun, and S. Flandrois, *Annales de Physique* **11**, 107 (1986).

18. S. Flandrois, A.W. Hewat, C. Hauw, and R.H. Bragg, *Synthetic Metals* **7**, 305 (1983).

19. T. Sakakibara, K. Sugiyama, M. Date, and H. Suematsu, *Synthetic Metals* **8**, 165 (1983).

20. S.T. Chen, G. Dresselhaus, M.S. Dresselhaus, H. Suematsu, H. Minemoto, K. Ohmatsu, and Y. Yosida, *Phys. Rev.* **B34**, 423 (1986).

21. M. El Makrini, D. Guérard, P. Lagrange, and A. Hérold, *Physica* **99B**, 481 (1980).

22. K.Y. Szeto and G. Dresselhaus, *Phys. Rev.* **B32**, 3186 (1985); *Phys. Rev.* **B32**, 3142 (1985); *J. Phys.* *C* **19**, 2063 (1986).

23. A. Erbil, A.R. Kortan, R.J. Birgeneau, and M.S. Dresselhaus, *Phys. Rev.* **B28**, 6329 (1983); S.G.J Mochrie, A.R. Kortan, R.J. Birgeneau, and P.M. Horn, *Phys. Rev. Lett.* **53**, 985 (1984).

24. J.K. Furdyna, *J. Appl. Phys.* **53**, 7637 (1982).

25. G.C. Osbourn, *J. Appl. Phys.* **53**, 1586 (1982); G.C. Osbourn, R.M. Biefeld, and P.L. Gourley, *Appl. Phys. Letts.* **41**, 172 (1982).

26. A.J. Freeman, H. Krakauer, S. Ohnishi, D.S. Wang, M. Weinert, and E. Wimmer, *J. Mat. Mag. Mat.* **38**, 269 (1983); C.L. Fu, A.J. Freeman, and T. Oguchi, *Phys. Rev. Lett.* **54**, 2700 (1985).

27. C.M. Falco and I.K. Schuller, in "Synthetic Modulated Structures", edited by L.L. Chang and B.C. Giessen, (Academic Press, Orlando, Florida, 1985), p. 339.

28. E.M. Gyorgy, J.F. Dillon, Jr., D.B. McWhan, L.W. Rupp, Jr., L.R. Testardi, and P.J. Flanders, *Phys. Rev. Lett.* **45**, 57 (1980).

29. L.E. Klebanoff, S.W. Robey, G. Liu, and D.A. Shirley, *Phys. Rev.* **B31**, 6379 (1985).

30. J. Korecki and U. Gradmann, *Phys. Rev. Lett.* **55**, 2491 (1985).

31. A.J. Freeman, private communication.

32. C. Sellers, Y. Shiroishi, N.K. Jaggi, J.B. Ketterson, and J.E. Hilliard, *J. Mat. Mag. Mat.* **54-57**, 787 (1986).

33. S. Sinha, J. Cunningham, R. Du, M.B. Salamon, and C.P. Flynn, *J. Mat. Mag. Mat.* **54-57**, 773 (1986).

34. J. Kwo, E.M. Gyorgy, F.J. DiSalvo, M. Hong, Y. Yafet, and D.B. McWhan, *J. Mat. Mag. Mat.* **54-57**, 771 (1986); J. Kwo, E.M. Gyorgy, D.B. McWhan, M. Hong, F.J. DiSalvo, C. Vettier, and J.E. Bower, *Phys. Rev. Lett.* **55**, 1402 (1985).

35. G.A. Prinz and J.J. Krebs, *Appl. Phys. Lett.* **39**, 397 (1981).

MAGNETIC PHASE TRANSITIONS IN ACCEPTOR GICs

J.T. Nicholls and G. Dresselhaus

Massachusetts Institute of Technology
Cambridge, MA 02139, USA

INTRODUCTION

Graphite intercalation compounds (GICs) prepared from transition metal chlorides have the advantage of inserting many graphite layers between sequential magnetic layers, without significantly affecting the structure of these layers. The effects of the graphite layers are to greatly increase the interlayer distance, and reduce the interplanar exchange interaction J'. With a significant reduction of J', especially in the high stage compounds, we may have the right conditions for observing 2D magnetism.[1]

It has been shown theoretically that a Kosterlitz–Thouless (KT) type magnetic phase transition can occur in an ideal two dimensional XY magnetic system.[2] Kosterlitz predicts that for an infinite 2D XY system the functional form for the ac differential magnetic susceptibility is $\chi \sim \exp(bt^{-0.5})$, where $t = (T - T_{KT})/T_{KT}$ and T_{KT} is the vortex unbinding temperature.[2] With susceptibility measurements it has been possible to identify two critical temperatures T_{cl} (lower) and T_{cu} (upper). In the low temperature region $T < T_{cl}$, the temperature dependence of the magnetic susceptibility is three-dimensional for all stages measured. It is suggested that for the CoCl$_2$–GICs, planes of Co^{2+} spins become ferromagnetically aligned along the easy in–plane axes and that these ferromagnetic sheets become stacked antiferromagnetically as T is lowered below T_{cl}.[1] Further evidence from neutron scattering experiments[3] supports this 3D → 2D mechanism for the CoCl$_2$–GICs; the integrated intensities of the 3D antiferromagnetic Bragg peaks disappear as the temperature is raised above T_{cl}. This suggests that in the temperature range $T_{cl} < T < T_{cu}$ there is no interplanar correlation and there is a possibility for the existence of a 2D XY phase. This 2D XY phase will however be modified by finite size effects due to the islandic structure of the CoCl$_2$ intercalant and the sixfold symmetry–breaking crystal field h$_6$. From studies of the dependence of the susceptibility on temperature and magnetic field of high stage ($n > 2$) CoCl$_2$–GICs, the transition at T_{cu} is identified with T_{KT}. The merits of both the KT model and the macroscopic supermoment model recently introduced by Rancourt[4] are here discussed with particular reference to CoCl$_2$–GICs.

EVIDENCE FOR KOSTERLITZ–THOULESS TRANSITION

Fits to the Kosterlitz's functional form have been tried in the region $T \to T_{KT}^+$, but it may never be possible to fit experimental data to this form because this equation is calculated for an ideal system with no perturbations or finite–size effects. It is also derived in the limit where there is a small concentration of vortex pairs, which is only true just above T_{KT}. A better fit may be obtained using high temperature series expansions (HTSE) which provide approximate expressions for thermodynamic quantities in the critical region, *i.e.* for temperatures $T > T_c$, where T_c is the critical temperature. The expansions have been carried out for two dimensional Hamiltonians for several lattices (*e.g.*, square, triangular etc.). If J is the nearest neighbor intraplanar ferromagnetic exchange constant, then the expansions are calculated in powers of $J/k_B T$, and the susceptibility for a 2D XY system of spins on a planar triangular lattice becomes:[5]

$$\chi = \chi_{\text{Curie}}\left(1 + 6(J/k_B T) + 30(J/k_B T)^2 + ... + 1738252.392(J/k_B T)^{10} + ...\right) \qquad (1)$$

The coefficient for each power of $J/k_B T$ grows as the power increases. The difficulty in obtaining higher order terms in the series is a result of the large number of diagrams that must be counted to calculate their coefficients. In the high temperature limit, we have $T \gg J/k_B$, so that $\chi \to \chi_{\text{Curie}} = C/T$.

For the high stage $CoCl_2$–GICs, we can assume that the in–plane magnetic properties of the $CoCl_2$ are unaffected by intercalation, so that J is known from measurements on pristine $CoCl_2$.[6] By carrying out careful weight and absolute susceptibility measurements to calculate the Curie constant C, a comparison to the HTSE can be made. Such a comparison has been made for $CoCl_2$–GICs with HTSE for 2D XY, 2D Ising and 2D Heisenberg models.[7] With no adjustable parameters, the data for the stage 3 $CoCl_2$–GIC system is found to fit the high temperature series expansion of the 2D XY model,[7] as shown in Fig. 1. Unfortunately, there are no absolute susceptibility measurements available for any of the other high stage acceptor GICs. Such measurements would make it possible to determine the type of spin–ordering and magnetic Hamiltonian pertinent to each compound. The assignment of a universality class may not be quite so simple in all cases. For example, from the studies of pristine $FeCl_2$ and $FeCl_3$, we might expect 2D Ising and 2D Heisenberg systems in the high stage GICs respectively. It is however very difficult to make samples of single valency for these intercalants.

RANCOURT MODEL

Recently Rancourt[4] has proposed a macroscopic model with which he was able to reproduce qualitatively correct susceptibility curves for low stage $CoCl_2$–GICs. In this model it is argued that at temperatures below $T = J/k_B \approx 28K$ (where J is the nearest neighbor exchange constant), the spins within each island are ferromagnetically aligned. The island diameter is typically 100–150 Å, giving a supermoment of about 1500 spins. The supermoment spins are assumed to lie down in the xy plane, and will interact with other supermoments within the plane with a ferromagnetic coupling energy E_F, and with those in adjacent planes with an antiferromagnetic coupling energy E_A. To reproduce the experimental susceptibility curves for $CoCl_2$–GICs, Rancourt introduces an additional interaction parameter energy E_0 which describes the variation in the intercalate concentration along the c–axis. A fourth parameter in this model is the energy

426

E_H due to the interaction of the supermoment and the applied in–plane field H. Using a mean field theory based on each of the sublattices, two consistent equations are solved numerically at each temperature T to obtain the magnetization. Then it is possible to calculate susceptibility curves as shown in Fig. 2, which exhibit the double–peaked structure familiar in the stage 1 $CoCl_2$ experimental data. This low stage compound is believed to exhibit strong interplanar magnetic correlations that eliminate the possibility of a 2D phase transition. We note that without the parameter E_0 a single rather than double peaked curve is obtained. From a comparison with the experimental ratio of the susceptibility peaks $\chi_1^{max}/\chi_2^{max}$ in a field of 10 Oe, where χ_1^{max} and χ_2^{max} are the peak heights at T_{c1} and T_{c2} respectively, it should be possible to set the scale between the parameter E_H and the applied field H in the simulations. Some estimate for the island size is also obtained ($\sim$ 150 Å for stage 1 $CoCl_2$–GICs).[7]

DISCUSSION

The two proposed models approach the $CoCl_2$–GIC system from very different extremes. In the KT mechanism, we examine the spin interaction within an island. We assume that despite the small size of the statistical system, it will be large enough to show a vortex unbinding transition at T_{KT}. The island size will be an upper bound to ξ_a, the in–plane spin–spin correlation length. More convincing evidence for this model would come from an experiment that could detect these vortices directly. One such experiment proposed by Szeto involves muon decay effects.[8] The Rancourt approach uses a mean field model, and has been successful in describing the overall features of the susceptibility anomalies. It uses the island as the basic unit of the statistical system, fixing the size of the supermoment and imposing a lower limit on ξ_a.

By neutron scattering studies, one can probe ξ_a and perhaps see which model, if either or both, is appropriate for a particular stage. Investigations on a stage 2 $CoCl_2$–

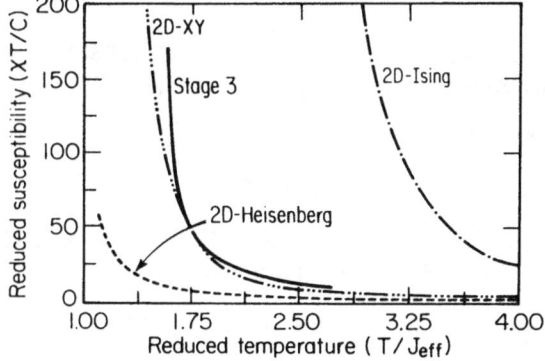

Figure 1: The experimental data for the susceptibility of stage 3 $CoCl_2$–GIC (solid curve) are compared to the high temperature series expansion (HTSE) for different models.[7]

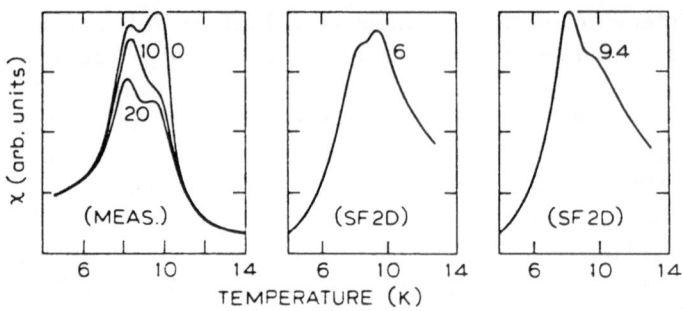

Figure 2: The experimental susceptibility $\chi(T)$ of a stage 1 $CoCl_2$–GIC compared with the simulations produced by the Rancourt model[4] for various values of applied field.

GIC[3] show that in the narrow temperature region between T_{cl} and T_{cu} there are only 2D magnetic correlations present. Further high resolution lineshape experiments are needed to establish the temperature variation of ξ_a and the critical exponent η near T_{cu}. As we proceed to higher stage compounds the likelihood of observing a KT transition increases. Therefore it would be interesting to carry out detailed neutron scattering experiments on a stage 3 $CoCl_2$–GIC or on one of the new bi–intercalation compounds with a large separation between magnetic layers.

ACKNOWLEDGMENTS

The authors wish to thank Prof. M.S. Dresselhaus for many useful discussions. This work was supported by AFOSR contract #F49620–83–C–0011.

REFERENCES

1. G. Dresselhaus and M.S. Dresselhaus, (this volume) p. 407.

2. J.M. Kosterlitz and D.J. Thouless, *J. Phys.* C6, 1181 (1973); J.M. Kosterlitz, *J. Phys.* C7, 1046 (1974).

3. M. Suzuki, D.G. Wiesler, P.C. Chow and H. Zabel, *J. Mag. Mag. Materials* 54-57, 1275 (1986).

4. D.G. Rancourt, *J. Mag. Mag. Materials* 51, 133 (1985); D.G. Rancourt, B. Hun and S. Flandrois, *Annales de Physique* 11, 107 (1986).

5. M. A. Moore, *Phys. Rev. Lett.* 23, 861 (1969).

6. M.J. Hutchings, *J. Phys.* C6, 3143 (1973).

7. K.Y. Szeto, Ph.D. Thesis, M.I.T. (1985); K.Y. Szeto and G. Dresselhaus, *Phys. Rev.* B32, 3186 (1985); *Phys. Rev.* B32, 3142 (1985); *J. Phys.* C 19, 2063 (1986).

8. G. Dresselhaus and M. Laguës, (this volume) p. 271.

ANOMALOUS TRANSPORT PROPERTIES IN ACCEPTOR–TYPE MAGNETIC GICs

N–C. Yeh, K. Sugihara, J. T. Nicholls, and G. Dresselhaus

Massachusetts Institute of Technology
Cambridge, MA 02139, USA

INTRODUCTION

Magnetic phase transitions have been studied in both donor and acceptor–type magnetic GICs using AC magnetic susceptibility χ and magnetization M measurements.[1] However, C_6Eu is the only GIC for which a magnetic phase diagram has been successfully determined using magnetoresistance (transport) measurements.[2] The strong coupling between graphite layers and Eu layers in C_6Eu results in strong scattering effects on the graphite conduction π–electrons by the magnetic excitations of the Eu ions. Therefore dramatic changes in the magnetoresistance of C_6Eu are observed as the system undergoes magnetic phase transitions. In contrast, acceptor–type magnetic GICs have a much weaker magnetic coupling between intercalate and graphite layers because of the relatively larger spacing between the graphite and the magnetic species.[3,4,5,6] Therefore the change in magnetoresistance due to changes in magnetic ordering is expected to be smaller than for the case of C_6Eu. In this work we show that transport–property measurements (magnetoresistance and temperature–dependent zero–field resistivity) can yield measurable effects associated with magnetic phase transitions for the acceptor–type magnetic stage–1 and stage–2 $CoCl_2$–GICs. Thus studies on the transport properties of acceptor magnetic GICs provide a complementary way to investigate the magnetic phase diagrams.[1] In addition, such measurements provide information on the π–d electron coupling constants ($J_{\pi-d}$) and the interplane exchange coupling constants (J').[7]

EXPERIMENTAL RESULTS

The zero–field temperature–dependent resistivity of stage–1 and stage–2 $CoCl_2$–GICs are shown in Fig. 1(a) and Fig. 1(b). For both stages, two scattering mechanisms contribute to the resistivity $\rho(T)$: scattering by phonons and lattice defects (as in all GICs), and in addition spin scattering effects associated with excitation of the magnetic states, which are important for $T < T_c$, where T_c is the magnetic ordering temperature. Well above T_c (9.7 K for stage–1 and 9.2 K for stage–2) where $\rho(T)$ exhibits the usual functional form for GICs: $\rho(T) = A + BT + CT^2$ except that the value of A is anomalously high due to spin disorder scattering above T_c. Similar phenomena were also observed in C_6Eu.[2] As shown in Fig. 1(a) for the stage–1 compound, an anomalous *increase* in $\rho(T)$ is observed as T is lowered below T_c. (Here T_c corresponds to the higher temperature peak for the measured low T susceptibility.[1]) This is in contrast to normal magnetic metallic systems which generally exhibit a *decrease* in resistivity below T_c as the spin disorder scattering is quenched by the onset of magnetic order.

In this context, the behavior of the stage–2 $CoCl_2$–GIC is as expected (see Fig. 1(b)), where a sudden decrease in $\rho(T)$ is found below $T_{cl} \sim 9.2$ K. It is noteworthy that the phase transition at $T_{cu} \sim 10.6$ K which has been associated with the unbinding of magnetic vortices does not have a significant effect on $\rho(T)$. The dramatic difference in the behavior near T_c between stages 1 and 2 is of particular interest.

Magnetoresistance measurements on a stage–1 sample at several temperatures are shown in Fig. 2. Below $T_c \simeq 9.7$ K, as the magnetic field is increased the resistivity $\rho(T, H)$ remains essentially constant at the zero field value $\rho(T, 0)$ until a field of ~ 300 Oe $\simeq H_{c2}$ is reached where H_{c2} is the field at which the transition to the spin aligned paramagnetic phase is seen in the $\chi(H)$ data.[1,3] At higher fields, $\rho(T, H)$ decreases and then saturates at a lower value depending on T. Monte Carlo calculations have identified the peak in $\chi(H)$ at H_{c2} with the transition from a spin flop phase to a spin aligned paramagnetic phase.[4] Note that there is no significant change in the resistivity at H_{c1} where a peak in $\chi(H)$ is also observed and identified by Monte Carlo calculations with a transition from an antiferromagnetic to a spin flop phase.[1,3,4]

The observation of a decrease in $\rho(T < T_c, H)$ as the spin aligned paramagnetic phase is established with increasing field is consistent with the increased magnetic order and with a resulting decrease in spin disorder scattering. The magnitude of the anomalous negative magnetoresistance is very large at low temperatures $T < T_c$ (e.g., $\Delta\rho(H = 1200\ Oe)/\rho(H = 0) \sim -11.8\%$ at 4.2 K). Above 9.7 K, a much smaller change of resistivity is observed: $\Delta\rho(H = 1200\ Oe)/\rho(H = 0) \sim -1.5\%$ at 10 K. This negative magnetoresistance at temperatures higher than T_c can be attributed to a spin–disorder scattering effect and is related to the π–d electron coupling constant $J_{\pi-d}$. The magnetic phase diagram extracted from the temperature and field dependent resistivity for stage–1 $CoCl_2$–GICs is shown in the inset of Fig. 2.

In contrast, the magnetoresistance data of stage–2 $CoCl_2$–GICs do not exhibit a clear plateau at low magnetic fields. This is consistent with a much weaker interplane coupling J' in stage–2 samples.[3,4,5,6] Specifically the change of resistance for $T < T_{cl}$ (e.g., $\Delta\rho(H = 1200\ Oe)/\rho(H = 0) \sim -1.4\%$ at 4.2 K) for stage 2 compounds is one order of magnitude smaller than that for stage–1. Our measurements for stage–2 at $T < T_{cl}$ are therefore not sufficiently accurate to yield a magnetic phase diagram. For $T > T_{cl}$, a negative magnetoresistance is still observed (e.g., $\Delta\rho(H = 1200\ Oe)/\rho(H = 0) \sim -1.1\%$ at 10.0 K), which can be attributed to spin–disorder scattering. The similar magnitude of the negative magnetoresistance for both stage–1 and stage–2 $CoCl_2$–GICs at $T > T_c$ is consistent with a similar π–d electron coupling strength between the bounding graphite layers and the intercalate layers, as is expected from the similar spatial separation of the π–electrons from the magnetic species in both stages.[7]

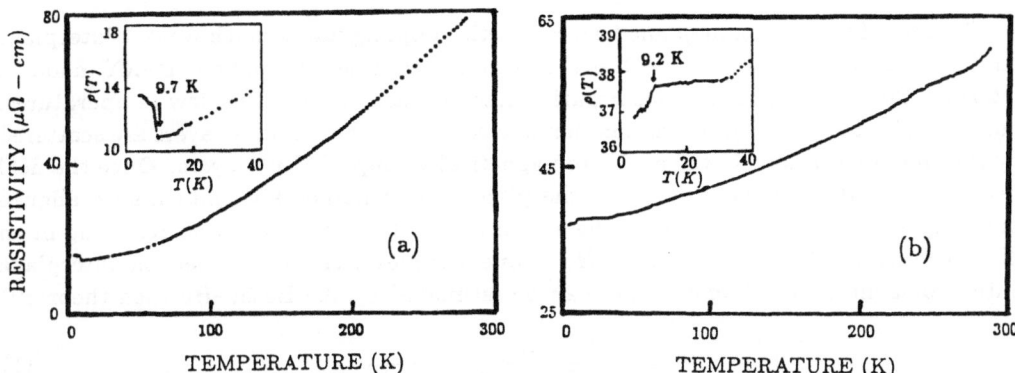

Figure 1: Zero–field temperature dependent resistivity (a) of a stage–1 $CoCl_2$–GIC, where an anomaly occurs at $T_c = 9.7$ K; and (b) of a stage–2 $CoCl_2$–GIC, where an anomaly occurs at $T_c = 9.2$ K. The insets show the behavior near T_c on an expanded scale.

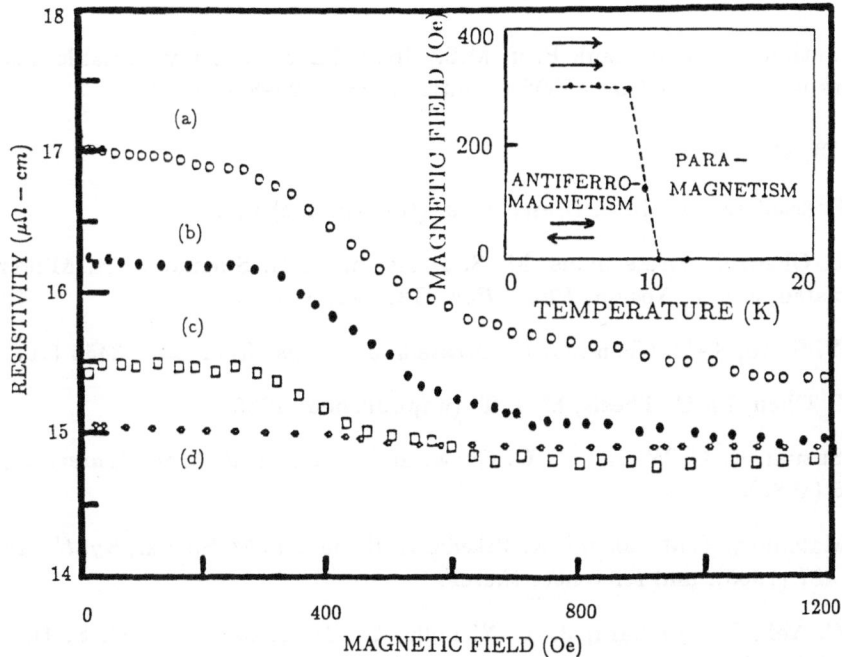

Figure 2: Transverse magnetoresistivity ($\vec{H} \perp$ c–axis and $\vec{H} \perp \vec{j}$) of a stage–1 $CoCl_2$–GIC sample at various temperatures: ($\circ$, $T = 4.2$ K), ($\bullet$, $T = 6.0$ K), and ($\square$, $T = 8.0$ K), and ($\diamond$, $T = 10.0$ K). The inset gives the magnetic phase diagram indicated by the magnetoresistance data.

DISCUSSION

Because of the strong in–plane ferromagnetic coupling and a much weaker interplane antiferromagnetic coupling of $CoCl_2$–GICs, one can assume a simplified 1D–XY model to interpret the observed anomalous negative magnetoresistance at very low temperatures. For $H < H_{c2}$ and $T < T_{cl}$, the conduction π–electrons of the graphite layers are scattered by the spin fluctuations in the antiferromagnetically coupled Co^{2+} layers. Once the field reaches the critical value $H_c = H_{c2}$, the planar supermoments go into a spin aligned paramagnetic state, thus reducing the scattering of the π–electrons and resulting in an abrupt decrease of $\rho(T, M)$ for $H > H_c$. If one assumes a 1D–XY model, the interplane antiferromagnetic coupling constant can be estimated by the Bethe–Hulthen theory:[7,8]

$$J' = \frac{g_{eff}\mu_B H_c}{2S(S + (2\ell n2 - 1))} \tag{1}$$

where g_{eff} ($= 5.4$) is the effective g factor and $S = 1/2$. The value of $(J')_{stage-1}$ is estimated to be ~ 0.056 K. This J' value is consistent with that obtained from a classical Monte Carlo calculation ($J' \sim 0.1$ K).[4] Our recent calculations[7] using RKKY theory relate J' to $J_{\pi-d}$ yielding an estimate of $J_{\pi-d} \sim 10^{-2}$ eV for stage-1 $CoCl_2$–GICs, which is one order of magnitude smaller than the $J_{\pi-d}$ for C_6Eu. This is expected from the large spatial separation of the π orbitals from the magnetic species in $CoCl_2$–GICs.[7]

ACKNOWLEDGMENTS

The authors wish to thank Prof. M.S. Dresselhaus for many valuable discussions. This work was supported by AFOSR contract #F49620–83–C–0011.

REFERENCES

1. G. Dresselhaus and M.S. Dresselhaus, (this volume) p. 407.

2. S.–T. Chen, G. Dresselhaus, M. S. Dresselhaus, H. Suematsu, H. Minemoto, K. Ohmatsu, and Y. Yosida, *Phys. Rev.* **B34**, 423 (1986).

3. K.–Y. Szeto, S.–T. Chen and G. Dresselhaus, *Phys. Rev.* **B33**, 3453 (1986).

4. S.–T. Chen, Ph.D. Thesis, M. I. T. (unpublished) 1985.

5. M. Suzuki, D. G. Wiesler, P. C. Chow, and H. Zabel, *J. Mag. Materials*, **54–57**, 1275 (1986).

6. M. Matsuura, Y. Murakami, K. Takeda, H. Ikeda, and M. Suzuki, *Synthetic Metals*, **12**, 427 (1985), and references therein.

7. N.–C. Yeh, K. Sugihara, J. T. Nicholls, G. Dresselhaus and M. S. Dresselhaus, unpublished.

8. L. Hulthen, *Arkiv. Mat. Astron. Fysik*, **26A**, No. 11 (1938).

RECENT DEVELOPMENTS IN MAGNETIC RESONANCE STUDIES OF

GRAPHITE INTERCALATION COMPOUNDS

Dan Davidov* and Henry Selig

Racah Institute of Physics* and Institute of Chemistry
The Hebrew University of Jerusalem
Jerusalem 91904, Israel

Magnetic resonance has been extensively used to study various aspects of solid state physics and chemistry over the last four decades. It is just natural, therefore, that these powerful techniques should have been applied to the investigation of graphite intercalation compounds (GIC). These are layered materials which are formed by insertion of atoms or molecules between the graphite planes.[1] The intercalation process is usually accompanied by charge transfer between the intercalant species and the graphite layers. Similar to doped semiconductors they are classified as "donor" GIC or "acceptor" GIC depending on whether the inserted species donate or accept an electron. The intercalation process usually occurs without disrupting the integrity of the carbon sheets, but the interplanar bonds are weak and easily broken. This is believed to be the origin of some of the unique properties of GIC including the dramatic anisotropic conductivity,[1] the "staging" phenomena,[2,3] the 2-dimensional order-disorder and commensurate-incommensurate phase transitions,[4] etc.

We provide here a short review on recent developments in the field of ESR and NMR of GIC with emphasis on some selected subjects more related to the former technique. Although Conduction Electron Spin Resonance (CESR) of graphite single crystals and GIC with alkali metals was discovered as early as the sixties by Wagoner[5] and Muller and Kleiner,[6] very little work was done till the late seventies. The field has received renewed impetus with the discovery that upon intercalation with arsenic penta-fluoride, AsF_5, the in-plane electrical conductivity dramatically increases, approaching that of copper.[7]

This paper is organized as follows: Section I reviews recent theories and experiments on the CESR of GIC with emphasis on the lineshape (A/B ratio). We demonstrate how the variation of the A/B ratio in in-situ CESR experiments can yield information on the process of diffusion and intercalation. A comparison with two dimensional diffusion coefficients as measured by NMR is given. We also discuss various mechanisms for the CESR linewidth and phase transitions as probed by CESR. Section II provides some examples of how ESR and NMR of the intercalant species can provide information on the charge transfer and on local symmetries. We demonstrate the existence of anisotropic Korringa relaxation in the ESR of paramagnetic intercalant species and in ^{13}C NMR. This provides information on the local density of states which is complementary to that obtained via the NMR-ESR Schumacher-Slichter technique. Some concluding remarks are given in Section III.

CONDUCTION ELECTRON SPIN RESONANCE (CESR)

g-Factor and Anisotropy

The CESR of graphite single crystals was discovered by Wagoner.[5] The spectra exhibit a single "metallic" resonance with an anisotropic g-value of $g_{\parallel}$ = 2.048 and $g_{\perp}$ = 2.002 for magnetic fields parallel and perpendicular to the c-axis, respectively. The anisotropic g-value as well as its temperature dependence was explained using the theory of Elliot.[8] According to this theory a large g-shift, δg, is expected in semiconductors for which the Fermi level has a nearby band degeneracy; δg, in this case, strongly depends on the location of the Fermi energy, E_F, and on the temperature. Indeed, the band structure calculations of Slonczewski and Weiss (SW)[7] do show the existence of such a degenerate band edge near E_F along the HKH direction in the Brillouin zone.[7] A complete theory for the anisotropic g-value was given by McClure and Yafet[10] using the band model of SW.[7]

Unlike pure graphite, for which $\delta g \sim 0.05$, GIC based on Highly Oriented Pyrolytic Graphite (HOPG) show only little anisotropy in the g-value. This was recognized, first, by Muller and Kleiner[6] for the donor type graphite-alkali metal compounds. This observation could be explained in the framework of the theory of McClure and Yafet[10] by lifting the band-degeneracy and shifting of the Fermi level to accomodate the charge transfer. Since the pioneering work of Muller and Kleiner,[6] CESR was observed for a large class of donor and acceptor type GIC with well defined stages and even on residue compounds.[11-19] In all cases the spectra exhibit a single Dysonian lineshape around $g_{\parallel}$= 2.002 (free electron value) with a very small anisotropy in the field for resonance. A typical ESR lineshape is shown in Fig. 1. No clear systematics of this anisotropy across the various systems exist.

Some confirmation for the g-shift calculations of McClure and Yafet[10] can be found in a very recent in-situ ESR study of the intercalation of HNO_3 into HOPG.[20] Fig. 2 exhibits the ESR spectra for different exposure

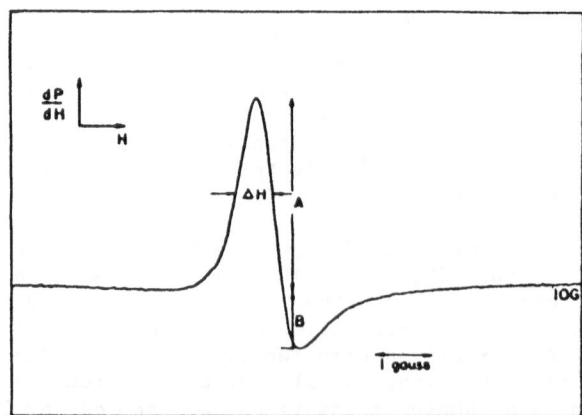

Fig. 1. ESR spectra of C_8AsF_6 (stage I) at room temperature. The parameters A,B and H are defined in the Figure (after Khanna et al. refs. 12,1)

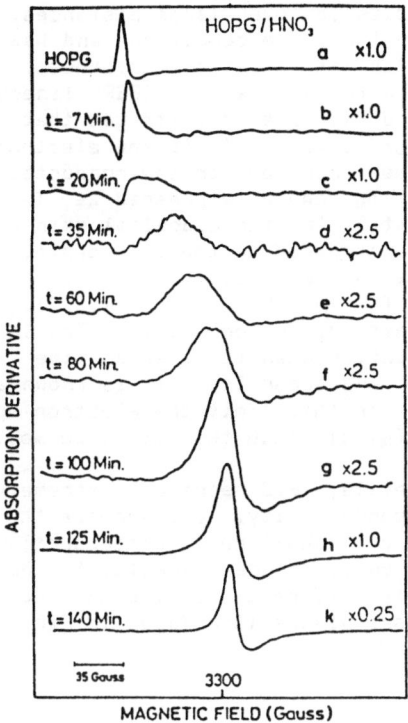

Fig. 2. ESR spectra of HOPG/HNO₃ as a function of HNO₃ exposure time (after Palchan et al, ref. 20)

times of HOPG to "fuming" HNO_3 (partial pressure ~300 torr). The magnetic field is parallel to the c-axis. Before exposure, g = 2.05 appropriate to HOPG. Immediately after exposure, there is a "phase reversal" of the ESR lineshape (but see below) accompanied by a decrease of the ESR intensity. After 20 minutes, the resonance associated with HOPG actually disappears and a new broad line appears roughly at the same position. This line continuously shifts toward g = 2. After 140 minutes we see a single metallic resonance line at g = 2.002, appropriate to HNO_3/GIC. In the framework of the rigid band model, the continuous shift in Fig. 2 can be explained as associated with the shift of the Fermi energy to accomodate the holes produced by the charge transfer. We note, however, that in the dilute doping regime the guest species strongly distort the graphite layers, particularly in their vicinity.[1,21] In the presence of such strains the spins (holes) might be partially localized and the simple rigid band model might be incorrect.[21] Evidence for charge and spin localization at the very beginning of the intercalation is seen in the temperature dependence of the resonance spectra produced by HOPG intercalated with fluorine.[21]

CESR Lineshape - Theoretical Considerations

As in most "thick" metals[22] the CESR lineshape in GIC is asymmetric. The line asymmetry (A/B) is defined as the ratio of the maximum peak height to the minimum peak height, both measured with respect to the baseline of the resonance derivative (Fig. 1). As demonstrated first by Dyson,[23] this "metallic" lineshape is due to non-uniform microwave penetration (skin effects) and to the diffusion motion of the electrons which carry the magnetization over large distances. Dyson has solved the Maxwell equations under the above conditions and has calculated the CESR lineshape for a flat slab with arbitrary thickness. Feher and Kip[22] have used the theory of Dyson to analyze the CESR lineshape in metals. For a bulk sample Feher and Kip have demonstrated that the A/B ratio strongly depends on the ratio T_D/T_2, where T_2 is the electron spin relaxation time and T_D is the time required for the conduction electrons to diffuse across the skin depth. T_D can be expressed as $T_D = \delta^2/2D$ where D is the diffusion constant. In the classical skin effect regime and for a sample thickness large compared to the skin depth, Feher and Kip have shown that the A/B ratio dramatically varies from A/B = 19 for $T_D \ll T_2$ to A/B = 3 for $T_D \gg T_2$ (see Fig. 7 in ref. 22). The limit $T_2 \gg T_D$ is known as the "fast diffusion" limit. The latter limit ($T_D \gg T_2$) corresponds to the situation when it takes a time longer than T_2 to cross the skin-depth and accordingly it is known as the "diffusionless" limit in the theory;[22] in this limit the electron spins are "localized" on the time scale of T_2, although they might be mobile.

Unlike simple metals, GIC represent extremely anisotropic systems; i.e., the basal plane conductivity, σ_a, exceeds the c-axis conductivity, σ_c, by many orders of magnitude.[1] Consequently, the skin depth, δ_c, for microwave penetration via planes parallel to the c-axis (b-c faces in Fig. 3) is completely different than the skin depth, δ_a, for microwave penetration via the basal planes (a-b faces in Fig. 3) according to the relation

$$\delta_i = \frac{C}{\sqrt{2\pi\omega_o\sigma_i}} \qquad (i = a, c) \qquad (1)$$

Here ω_o is the microwave frequency and C is the velocity of light. Using (1), we estimate $\delta_a = 3\mu$ and $\delta_c = 0.16$ mm for pristine HOPG as an example. Due to the large anisotropic skin depths and conductivities, as well as possible anisotropic carrier diffusion, the application of the simple Dyson-Feher-Kip theories to the case of GIC is not obvious. This was pointed out by Blinowski et al[24] and Saint-Jean et al.[25] Following Kaplan[26] they have written the following, Bloch type, equation of motion for the magnetization as

$$\frac{d\vec{M}}{dt} = \frac{g\mu_B}{h} [(\vec{M}+\vec{M}_o) \times (\vec{H}_o+\vec{H})] - \frac{M_\parallel}{T_1} - \frac{M_\perp}{T_2} + D_a \left(\frac{d^2\vec{M}}{dy^2}\right) + D_c\left(\frac{d^2\vec{M}}{dz^2}\right) \qquad (2)$$

where $M_\parallel$ and $M_\perp$ are the longitudital and the transverse components of M; the oscillating magnetic field and magnetization are H exp{-iwt} and M exp{-iwt}, D_a and D_c are diffusion coefficients along the planes and c-axis, respectively, and M_o is the equilibrium magnetization. The coordinates are defined by Fig. 3. Neglecting quadratic and higher orders in the expansion of M_x and using the standard boundary conditions that the component of the gradient in (2) vanishes at the surface of the sample, Blinowsky[24] and Saint-Jean[25] have obtained a solution for the transverse part of magnetization $M_x(w)$. The power absorption, I, was calculated and the A/B ratio was estimated from dI/dH. The results are given in Fig. 4 as a function of the dimensionless parameters $q = c/\delta_a$ and $p = a/\delta_c$ (c and a are defined in Fig. 3) for the following limiting cases:

1) Very slow diffusion limit: $T_D^{(c)} \gg T_2$ and $T_D^{(a)} \gg T_2$. Here $T_D^{(c)}$ and $T_D^{(a)}$ are the diffusion times along the c-axis and basal plane, respectively ($T_D^{(c)} = \delta_a^2/2D_c$; $T_D^{(a)} = \delta_c^2/2D_a$).

2) Very rapid diffusion along the c-axis and the basal plane $T_2 \gg T_D^{(c)}$; $T_2 \gg T_D^{(a)}$.

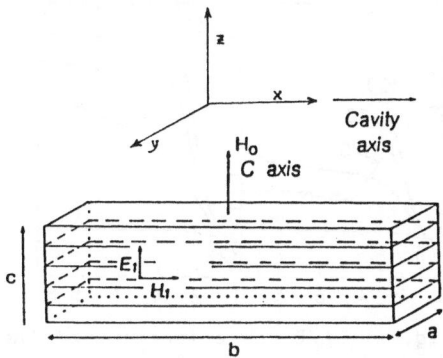

Fig. 3. The orientation of the HOPG slab with respect to the external magnetic field Ho and the cavity axis. H_1 and E_1 are the magnetic and electric components of the rf field. The coordinate system is also defined.

Saint-Jean et al[25] have demonstrated that for typical sample dimensions of a = 3 mm, c = 0.5 mm, q/p > 100 for most acceptor GIC and, therefore, the Dyson formula with $\sigma_c(\delta_c)$ and D_a is adequate for the analysis of the CESR.

It should be stressed, however, that if both the skin depth δ_a, δ_c and the diffusion constants D_a and D_c are determined by the a-axis and c-axis conductivities, we expect the diffusion time to be direction independent; namely,

$$T_D = T_D^{(a)} = T_D^{(c)} = \frac{N(E_F)e^2c^2}{2\pi\omega} \, \rho_a\rho_c \tag{3}$$

Here $N(E_F)$ is the density of state of the carriers at the Fermi level, $\rho_a(\rho_a = 1/\sigma_a)$ and $\rho_c(\rho_c = 1/\sigma_c)$ are the resistivities along the a-axis and c-axis, respectively.

Equation (3) provides some justification for the use of Dyson-Feher-Kip type of analysis for an anisotropic medium as T_D is independent of the direction of microwave penetration.

CESR Line-Shape Comparison With Experiment

Both the theories of Dyson-Feher-Kip,[22,23] equation (3) and Blinowsky et al,[24] clearly demonstrate that the asymmetry parameter is uniquely determined by the sample dimensions, ρ_a, ρ_c, $N(E_F)$ and T_2. Many attempts were made[11,12,13,16,17,27,28] to extract the conductivities and density of states from CESR studies using (3). The usual procedure is as follows: the measured A/B ratio is used, together with Fig. 7 of ref. 22, to extract the ratio T_D/T_2. The relaxation time T_2 is estimated from the CESR linewidth. Using (3), $N(E_F)$, ρ_c or ρ_a can be calculated provided that two out of the three parameters are known. For example, using this procedure of analysis, Stein, et al have argued that ρ_c for stage I AlCl$_3$/HOPG is activated.[18] Saint-Jean, et al[25] have argued that

437

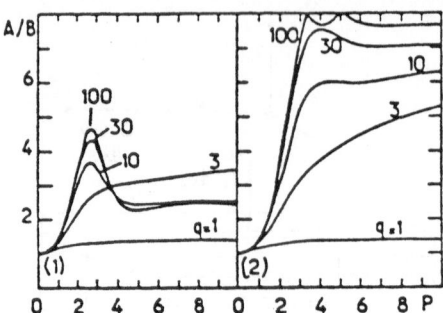

Fig. 4. A/B as a function of the dimensionless parameter p ($p=a/\delta_c$) for various q ($q=c/\delta_a$) for: (1) slow diffusion limit ($T_2 \ll T_D$), (2) fast diffusion limit ($T_2 \gg T_D$) (Saint-Jean et al[25]).

in some of these studies[18,19] the basic assumption was that the microwave penetrates via the basal-plane, which is inconsistent with the theoretical arguments. Consequently they have suggested reanalysis of some of the data.

There are several experiments including those of Saint-Jean et al[25] and Vaknin et al[26] that should be mentioned. Saint-Jean et al have studied the CESR of graphite HOPG/AsF$_5$ stage I as a function of temperature and for several different samples having different widths. They found that the A/B ratio depends on the sample width along the basal planes and not on the sample thickness (along the c-axis), suggesting that the CESR resonance is mainly associated with microwave penetration via planes parallel to the c-axis. Furthermore, their experiment suggests the applicability of the in-plane slow diffusion limit. However the experimental A/B ratios are somewhat larger than the theoretical prediction for the slow diffusion limit. Saint-Jean, et al have demonstrated, that the A/B ratio dramatically decreases upon increasing the microwave power and the frequency of the modulation, indicating saturation and passage effects. This suggests that is necessary to work at very low modulation frequencies and at low microwave power.

The conclusion that the CESR in acceptor fluorides is mainly associated with microwave penetration via planes parallel to the c-axis was reached also by Vaknin et al.[26] These authors have used the fact that both σ_a and σ_c vary dramatically as a function of fluorine concentration in HOPG/fluorine GIC.[26] They have "designed" a sample for which δ_c is of the order of the sample size along the basal plane, but δ_a is significantly smaller than the thickness along the c-direction. The observation of A/B ~ 1.5 confirms the expectations. Finally, Vaknin et al have critically checked the theory of Feher-Kip and the validity of (3) by measuring the A/B ratio, ρ_a and ρ_c versus temperature on the same samples. The diffusion time, T_D, was extracted from Fig. 7 of Feher and Kip[22] using the experimental A/B ratio; T_2 was estimated from the linewidth at room temperature. No correlation between the temperature dependence of T_D and the product $\rho_a \rho_c$, as expected according (3), could be seen. There are several possibilities for this "discrepancy" including:
(a) Surface effects dominate the lineshape (but see below).
(b) The simple analysis of Feher and Kip is invalid and the modification introduced by Saint Jean et al is necessary.

(c) The CESR linewidth does not present a good measure of T_2. The observation of saturation effects[25] supports this possibility.

Diffusion and the Kinetics of the Intercalation Process

There is much interest recently in the dynamics of pattern formation during the process of intercalation.[29-32] Kirczenow[27,30] was able to obtain real space simulations of the intercalant distribution in various galleries by using three dimensional Monte-Carlo computer calculations. His model is based on the concept of "Elementary Islands" (EI) interacting via repulsive interlayer forces and local elastic effects. Particularly, Kirczenow has demonstrated that the growth, of say stage II structure, can occur as a single domain with relatively sharp intercalation fronts which propagate through the graphite host or, alternatively, it can grow as an array of separate domains. The growth mode depends on kinetic constraints as well as on the nature of the guest species.

In this section we shall demonstrate that CESR lineshape studies by <u>in-situ</u> measurements during the process of intercalation yield valuable information on the diffusion of the intercalation fronts. The method is based on the fact that the ESR lineshape of magnetic species in a substrate is strongly affected by the presence of a nearby surface metallic layer. A comprehensive theoretical review of the ESR lineshape of such layer-substrate system was given by Zevin and Zuss.[33] These authors have checked their theory critically by performing extensive experimental studies in which germanium layers with controlled conductivities and thickness were grown on ruby substrates.[33]

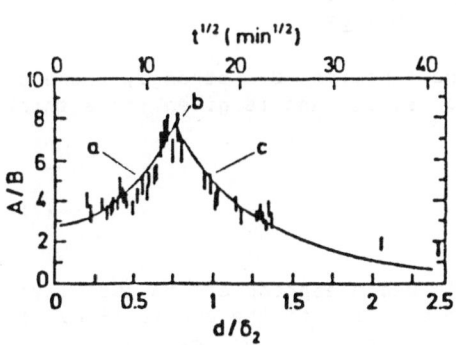

Fig. 5a
Experimental A/B ratio versus d/δ_2 and the exposure time, $\sqrt{t}$. Solid lines give theoretical fits with $d^2 = 2Dt$ and $D = 4.4 \times 10^{-7}$ cm²/min.

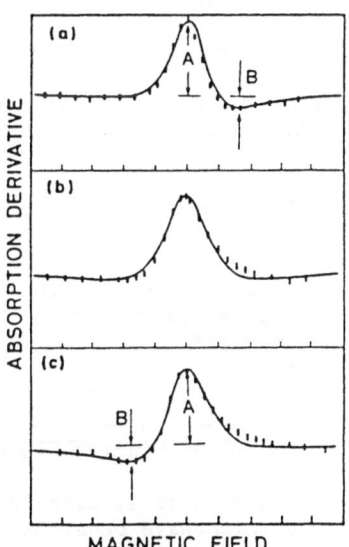

Fig. 5b
CESR lineshape of HOPG upon exposure to HNO_3 (pressure of 80 torr). Points (a), (b) and (c) corresponds to A/B ratios and exposure times given in Fig. 5a (after Palchan et al)[20]

The _in-situ_ ESR study was performed by exposing HOPG to very low vapor pressures of fluorine, HNO_3 or AsF_5.[34,35] We emphasize here results on HNO_3 intercalation. Specially cut samples guarantee that the microwave penetration is via planes parallel to the c-axis (Fig. 3). The specially prepared HOPG samples exhibit A/B ratios between 3-4 before exposure. This A/B ratio is "normal" in the sense that the maximum peak height of the resonance derivative occurs at smaller magnetic field. Upon exposure to HNO_3 vapor pressure the A/B increases initially reaching a maximum value of $A/B \cong 8$ (Fig. 5). Upon further exposure to the inter- calant atmosphere, the A/B is "reversed" (i.e., minimum peak height of the signal derivative occurs at lower field) and its magnitude decreases. Fig. 5b describes the CESR lineshape for various exposure times to an HNO_3 atmosphere (partial pressure $\sim$ 80 torr). Fig. 5a yields the experi- mental A/B ratio as a function of exposure time. Similar results were obtained for the intercalation with fluorine.[34,35]

In analyzing the ESR lineshapes it is assumed that the intercalant species occupy part of the galleries and form a "macroscopic" layer of GIC (ordered or disordered) with thickness **d**. The resonance associated with the bulk (substrate) HOPG is strongly affected by the presence of this layer due to attenuation and reflection of the microwave field caused by this layer. For the configuration described in Fig. 3, the power absorption is given by[33-35]

$$I = \frac{c}{16\pi} \langle H_1^2 \rangle_{z=0} Z^{mag} \tag{4}$$

where $\langle H_1^2 \rangle_{z=0}$ is the square of the oscillating magnetic field at the sample-vacuum interface. The magnetic part of the surface impedance, Z^{mag}, was calculated to be

$$Z^{mag} \approx [Re\emptyset + Im\emptyset]\chi' + [Re\emptyset - Im\emptyset]\chi'' \tag{5}$$

where Re and Im refer to the real and imaginary parts, and χ' and χ'' are the normalized lineshapes given by

$$\chi' = \frac{1}{\pi} \frac{Ho - H}{(Ho-H)^2 + \Delta^2}; \quad \chi'' = \frac{1}{\pi} \frac{\Delta}{(Ho-H)^2 + \Delta^2} \tag{6}$$

where Δ is the resonance linewidth. The function $\emptyset$ strongly depends on the layer thickness d and the ratio σ_1/σ_2 and is given for a thick sample by[34,35]

$$\emptyset = \frac{1}{[\cos K_1 d - i\sqrt{\sigma_1/\sigma_2} \sin K_1 d]^2} \tag{7}$$

where $K_1 = (1+i)/\delta_1$ and δ_1 is the skin depth appropriate to the layer GIC. For microwave penetration via planes parallel to the c-axis (Fig. 3), the skin depth is determined by the c-axis conductivities, σ_1 and σ_2 for the layer GIC and that of the HOPG substrate, respectively.

The values of $(Re\emptyset + Im\emptyset)$ and $(Re\emptyset - Im\emptyset)$ as a function of the dimensionless parameter d/δ_2 (δ_2 is the skin depth for the HOPG substrate) for $\sigma_1/\sigma_2 = 0.2$ are plotted in Fig 6. As seen $(Re\emptyset - Im\emptyset)$ is negative for a certain range of d/δ_2. This leads to "negative" absorption in (5) and is the origin for the phase reversal phenomena. We note that (5) was derived for localized spins. However, as demonstrated[20,34] the

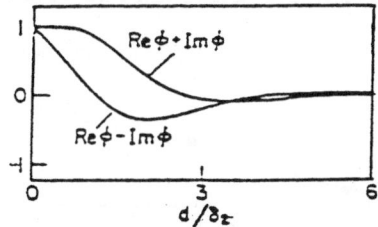

Fig. 6. (Re∅+Im∅) and (Re∅-Im∅) versus d/δ_2 (δ_2 is the skin depth of HOPG for microwave penetration via planes parallel to the c-axis; σ_1/σ_2 = 0.2. (After Zevin, ref. 33).

relatively large skin depth δ_2 = 0.16 mm guarantees a slow diffusion time, T_D, such that the diffusionless limit is satisfied.

The best fit of the ESR lineshape to the theory, (4) and (5), yields the values of d and σ_1/σ_2. The fitting procedure is less sensitive to σ_1/σ_2 than to d. The most striking feature is that d depends on the exposure time as d^2 = 2Dt (Fig. 7) where D is the diffusion constant for the intercalant. We found D = 4 x 10^{-7} cm²/min., for HNO_3 diffusion under the experimental conditions described in ref. 20. For fluorine diffusion D = 1.57 x 10^{-7} cm²/min. was found (see Fig. 7).

Finally, diffusion coefficients during the process of intercalation were measured by weight uptake techniques and very recently by one dimensional NMR imaging.[36] Particularly, NMR images of the ^{19}F concentration profile as a function of time yield information about AsF_5 diffusion during the intercalation.[36] A comparison of diffusion coefficients in the different experiments might be meaningless as the diffusion strongly depends on the intercalant, its partial pressure, purity and quality (and shape) of the HOPG substrate.[34,35]
It should be noted that Fick's law of diffusion in a semi-infinite one-dimensional system with the concentration of diffusing species held fixed at the boundary predicts that the filling coefficient grows as $\sqrt{t}$, t being the exposure time. This is in agreement with our observations. The simulations of Kirczenow deviate somewhat from the $\sqrt{t}$ law.

Comparison with Diffusion Coefficients from NMR: 2 Dimensional Diffusion

The diffusion coefficient found by our in-situ ESR study during the intercalation process depends not only on the nature of the intercalant species and the intercalant phase (solid, liquid, etc.) but also on surface barriers and the elastic strain field which acts during the opening of the graphite galleries. As such, it might be completely different from the diffusion coefficient of the intercalant in the equilibrium state. Diffusion coefficients in the quasi-equilibrium state have been measured using inelastic neutron scattering or NMR techniques.[37]

Recently a theory for the spin lattice relaxation time, T_1, in the presence of diffusion was developed.[38,39] It was shown that T_{1D} depends strongly on the resonance frequency ω_0, the correlation time for diffusion τ_D and the dimensionality, d. Particularly for the high temperature limit;

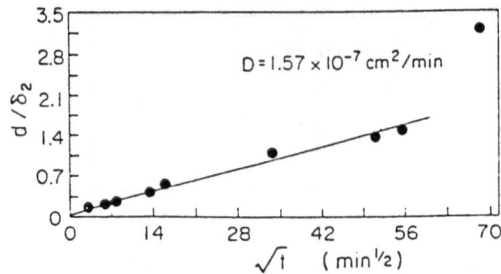

Fig. 7. d/δ_2 for fluorine diffusion into HOPG versus $\sqrt{t}$, t is the exposure time. d/δ_2 was estimated from the CESR lineshape [34]. The linearity suggests the relation $d^2 = 2Dt$. The slope yields the value of D.

i.e., for $1/\tau_D \gg \omega_0$, the theory predicts for the spin-lattice relaxation time, T_{1D}, the following dependence on dimensionality

$$\frac{1}{T_{1D}} \cong \left\{ \begin{array}{ll} (\tau_D/\omega_0)^{1/2} & d = 1 \\ \tau_D \ln (1/\omega_0 \tau_D) & d = 2 \\ \tau_D & d = 3 \end{array} \right. \qquad (8)$$

The correlation time for diffusion, τ_D, is related to the diffusion coefficient, D, as $D = d^2/2\tau_D$. The results in (8) can be understood noting that $1/T_{1D}$ is proportional to the power spectrum which is just the Fourier transform of the temporal correlation function, $\emptyset(t)$. For low dimensions and long times, $\emptyset(t)$ can be approximated by $\emptyset(t) \sim |t|^{-d}$ (d = 1 or 2).[40,41]

Kume and coworkers[37] have studied the NMR spin lattice relaxation T_1 of ^{19}F in stage II HOPG/AsF_5. The observed relaxation rate could be shown to originate from the diffusion relaxation term $1/T_{1D}$ and the rotational relaxation term $1/T_{1R}$. By subtracting $1/T_{1R}$ from the measured values, Kume and collaborators[37] could find $1/T_{1D}$ to follow the relation (for high temperatures): $1/T_{1D} = (1/D)\ln[D/\omega_0]$ (see Fig. 8). The thus obtained diffusion coefficient, D, is activated according to the relation $D = D_0 \exp[-E_D/K_B T]$ with $D_0 = 7 \times 10^{-5}$ cm²/sec and $E_D = 770K$. A similar logarithmic behavior was observed for $NH_3 - TaS_2$ and $NH_3 - TiS_2$.[42]

The experiment of Kume and his co-workers[37] provides a most beautiful demonstration for two dimensional diffusion in GIC. One-dimensional diffusion which obeys the law $1/T_{1D} \sim \omega_0^{-1/2}$ was observed only recently in the ESR and NMR of polyacetylene (see for example the paper of Robinson, et al[43]). Generally speaking, one would expect also 2-dimensional character of the CESR lineshape in GIC. However, fits with the experiment indicate Dysonian lineshapes.

Mechanisms for CESR Line Broadening in GIC

There is enough evidence now that the CESR linewidth in GIC increases significantly with the increase of the atomic number of the intercalant species. This was observed by Muller and Kleiner[6] for alkali metal GIC and recently for the XF_n fluoride GIC[27] (Fig. 9). It implies that the carriers are not confined to the graphite layers and that conduction

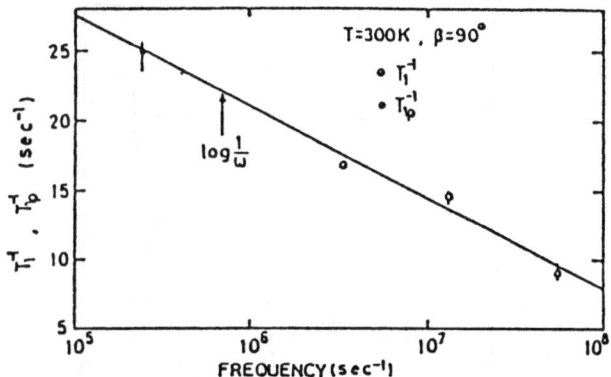

Fig. 8. Frequency dependence of the spin
lattice relaxation rate, $1/T_{1D}$, and that
in the rotating frame, $1/T_1$, β is the
angle between the c-axis and the magnetic
field (after ref. 37).

electron relaxation due to spin-orbit scattering by the intercalant
species is an important mechanism. It may explain also why no CESR was
oberved in any of the GIC with heavy intercalant species belonging to
the 5d series such as IrF_6/HOPG, OsF_6/HOPG, and PtF_6/HOPG.[44]

Relaxation of conduction electrons via spin-orbit coupling on nonmagnetic impurities was suggested for semiconductors[8] and metals.[45] It
was demonstrated, also, that the excess relaxation due to these impurities
is correlated with the excess resistivity.[46] This suggests that in simple
metals the Elliot relation[8] is valid; namely,

$$\Delta H \cong \frac{(\Delta g)^2}{\tau} \sim \rho_a \qquad (9)$$

where Δg is the g-shift of the conduction electrons and τ is the
momentum scattering time; τ is related to the basal plane resistivity,
ρ_a.

However, recent systematic studies of the ESR linewidth and the
residual resistivity in a large class of GIC/fluorides[44] reveal no
clear relation between ΔH and the residual resistivity, as expected
according to the Elliot relation.[8] Although no theory of CESR relaxation exists, we believe that the origin of the CESR relaxation mechanism
(Fig. 9) is due to spin-orbit interaction on intercalant species in the
vicinity of Daumas-Herold type of domain-walls.

Another interesting feature is the dependence of the linewidth on the
stage of GIC. In most GIC the linewidth versus stage exhibits a minimum
for stage II compounds.[25] Fig. 10 gives, as an example, the linewidth,
ΔH, and the conductivity, σ_a, at room temperature for HOPG/AsF_5. The
results show a minimum in ΔH but a maximum in σ_a for stage II suggesting
the validity of the Elliot relation.[7] However, a possible alternative
explanation for this behavior is due to different g-values for the CESR
on bounding and non-bounding graphite layers (evidence for different
Knight shifts associated with bounding and nonbounding layers has been
observed already[47]). This would give rise to line broadening associated
with a distribution of g-values for high stage GIC (stages III, IV) but
not for stages I, and II, for which only bounding graphite layers exist.

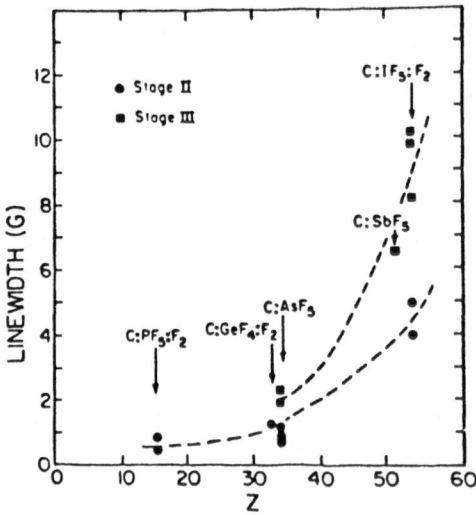

Fig. 9. CESR linewidth versus the atomic number, Z, for various fluorides (after Davidov, ref. 27)

Phase Transitions

The idea that the carriers are not confined to the graphite layer and can relax via spin-orbit scattering on the intercalant species (Fig. 9), may suggest that phase transitions within the intercalation layers (such as 2D order-disorder, commensurate-incommensurate, etc.) can be probed using the CESR technique. There have been several reports on phase transitions in CESR experiments.[12,18,19] Khanna et al[12] have noticed "anomalous" behavior in the CESR of HOPG/AsF$_5$ versus temperature and have attributed this anomaly to an "order-disorder" phase transition. However, no anomalous behavior was observed in the CESR study of the same GIC system by Saint-Jean et al.[25] Similarly, no signature for a phase transition in the CESR of HOPG/AsF$_5$ (stages I, II) was seen by the Jerusalem group.[40] It should be stressed, however, that recent resistivity studies of HOPG/AsF$_5$ versus temperature by Vaknin[44] reveal no indication for a phase transition in the basal-plane resisitivity but a clear phase transition anomaly in the c-axis resistivity. Actually, this seems to be a general feature of GIC. In a large class of XF$_n$/GIC (e.g., IrF$_6$/HOPG, MoF$_6$/HOPG, etc.) one clearly sees signatures for phase transition anomalies (including hysteresis behavior in the vicinity of the phase transition) in the c-axis resistivity, but no anomalies in the in-plane resistivity. The "failure" to observe phase transitions in the in-plane resistivity of HOPG/AsF$_5$ can partially explain why no evidence for such a phase transition is seen in the CESR study.

Nevertheless, Rettori and his collaborators[19] were able to see a clear signature for a phase transition in HOPG/AlCl$_3$ (stage 7). The CESR linewidth as a function of temperature for this system is given in Fig. 11 As seen the linewidth behavior exhibits irreversible behavior (namely, cooling and heating yield different linewidths) in the temperature range 140K ≤ T ≤ 190K but reversible behavior above T = 190K. The low temperature linewidth depends on the thermal treatment and "thermal history". Stein, et al[19] have explained these remarkable kinetics by assuming that below the transition temperature, the system can be "frozen" in different "states".

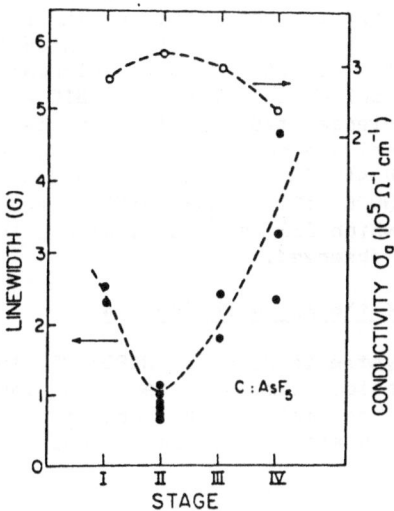

Fig. 10. CESR linewidth and the in plane conductivity, σ_a, for various stages of HOPG/AsF$_5$ (Milo, ref. 48)

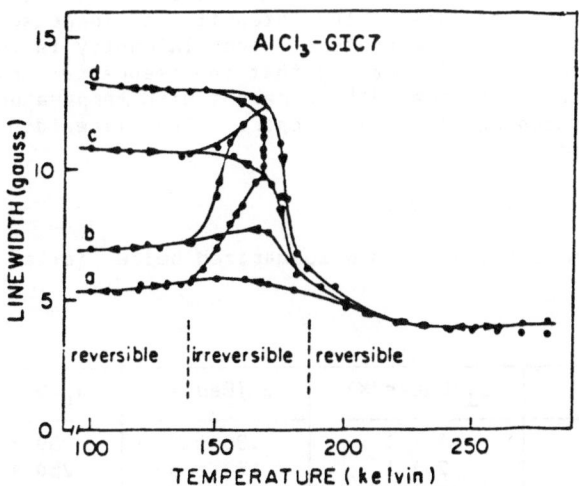

Fig. 11. The temperature dependence of the CESR linewidth of HOPG/AlCl$_3$, stage 7. Curves a, b, c, d correspond to different thermal treatment (after Stein et al. ref. 19).

RESONANCE OF LOCALIZED MOMENTS ON THE INTERCALANT SPECIES - COMPARISON WITH NMR AND ^{13}C NMR

While CESR gives information on the dynamic behavior of the carriers, ESR and NMR of the intercalated species can provide information about the nature of these species and, thus, on the chemical reaction taking place upon intercalation. This is directly related to the charge transfer and the conductivity variation upon intercalation. In the following we shall restrict ourselves to a family of GIC with 4d and 5d metal hexafluorides. The intercalated species in several cases are magnetic (due to charge transfer) and can serve as a magnetic probe for the ESR experiment. To the best of our knowledge (beside ESR studies of the magnetic GIC HOPG/FeCl₃, HOPG/CoCl₂, HOPG/NiCl₂, and HOPG/Eu [47]) these are the only cases for which ESR associated with a localized moment on the intercalant species was observed.

ESR Studies: Identification of the Magnetic Species

Probably the best understood system to date is HOPG/OsF₆.[50] ESR studies of various stages and even residue compounds reveal a single anisotropic line below 50K with asymmetry parameter A/B between 3 and 5. The anisotropy of the resonance position can be expressed in terms of its effective g value as

$$g^2_{eff} = g_{||}^2 \cos^2\theta + g_\perp^2 \sin^2\theta \qquad (10)$$

where θ is the angle between the magnetic field H and the c-axis; $g_{||}$ and $g_\perp$ are the resonance g-values for $H \parallel c$ and $H \perp c$, respectively. Experimentally, we find $g_{||} = 1.65 \pm 0.05$ and $g_\perp = 3.35 \pm 0.05$. Note that $g_\perp/g_{||} = 2$, independent of the stage of the GIC. Associated with the anisotropic g-value there is also an anisotropic linewidth. ESR signals as a function of temperature for $H \parallel c$ and $H \perp c$ are shown in Fig. 12. As seen there is a significant decrease of the intensity and increase in the broadening as a function of temperature. The intensity is inversely proportional to the temperature, indicating that the resonance originates with localized moments. The linewidth increases with temperature (Fig. 13), but the thermal broadening is anisotropic. The linewidth versus temperature can be fitted to

$$H_\alpha = a_\alpha + b_\alpha T \qquad (\alpha = ||, \perp) \qquad (11)$$

The results for the various stages are summarized below (after Vaknin et al[51])

Stage	$b_{||}$(Gauss/K)	$b_\perp$(Gauss/K)	$a_{||}$(Gauss)	$a_\perp$(Gauss)
I	8 ± 1	4 ± 1	180 ± 50	250 ± 50
II	8 ± 1	2 ± 1	150 ± 50	250 ± 50
III	8 ± 2	2.5 ± 1.5	140 ± 50	300 ± 70
IV	7.5 ± 1	2 ± 1.5	140 ± 50	300 ± 70

The anisotropic g-value and the resonance position can be completely explained in terms of a quartet ground state of Os^{+5}, 5d³ configuration, in a slightly distorted octahedral crystal field.[50] Such a quartet ground state can be described by the following Spin-Hamiltonian with an effective spin S = 3/2 (2S + 1 = 4)[50]

$$\mathcal{H} = g\mu_B[H_z S_z + 1/2(H_+S_- + H_-S_+)] + D[S_z^2 - 1/3\,S(S+1)] \qquad (12)$$

or in matrix form[51]

$$\mathcal{H} = \begin{vmatrix} D-(3/2)g\mu_B H_\| & (\sqrt{3}/2)g\mu_B H_\perp & 0 & 0 \\ (\sqrt{3}/2)g\mu_B H_\perp & -D-(1/2)g\mu_B H_\| & g\mu_B H_\perp & 0 \\ 0 & g\mu_B H_\perp & -D+(1/2)g\mu_B H_\| & (\sqrt{3}/2)g\mu_B H_\perp \\ 0 & 0 & (\sqrt{3}/2)g\mu_B H_\perp & D+(3/2)g\mu_B H_\| \end{vmatrix} \qquad (13)$$

Diagonalization of the 4 x 4 matrix (13) yields the energy levels of the quartet. Particularly, for $H\|c$ one obtains

$$\begin{aligned} E_\|^{(\pm 1/2)} &= -D \pm (1/2)g\mu_B H \\ E_\|^{(\pm 3/2)} &= +D \pm (3/2)g\mu_B H \end{aligned} \qquad (14)$$

For $H\perp c$ one finds, in the first approximation

$$\begin{aligned} E_\perp^{(\pm 1/2)} &= -D \pm g\mu_B H - (3/8)^2 \mu_B^2 H^2/D \\ E_\perp^{(\pm 3/2)} &= +D + (3/8)g^2\mu_B^2 H^2/D \end{aligned} \qquad (15)$$

Thus, for H = 0, the quartet splits into two doublets (±1/2 and ±3/2) with energy splitting 2D. The g-values can be easily calculated using (14) and (15) to be $g_\| = g$ and $g_\perp = 2g$. Thus $g_\perp = 2g_\|$ in agreement with the experimental results.

Let us examine (14) and (15) more closely. For $H\|c$ at low temperatures (T << 2D), only the ground state ±1/2 levels, are populated. This can be described by an effective magnetic moment, $\mu_{eff}^{(L.T)}$, given as $\mu_{eff}^{(L.T)} = g\sqrt{S(S+1)}\mu_B = 1.4\ \mu_B$ using g = 1.65 and S = 1/2. However, for T >> 2D, both the ±1/2 and the ±3/2 levels are populated and the (high temperature) effective moment $\mu_{eff}^{H.T}$, is $\mu_{eff}^{(H.T)} = g\sqrt{S(S+1)}\mu_B = 3.2\ \mu_B$ using g = 1.65 and S = 3/2. The ratio $(\mu_{eff}^{(H.T)}/\mu_{eff}^{L.T})^2$ equals ~5 and this should be the ratio of the slopes of $1/\chi_\|$ (the inverse susceptibility for $H\|c$) at high and low temperatures. Indeed, susceptibility studies[50] exhibit different slopes for $1/\chi_\|$ at high and low temperatures with the correct ratio (Fig. 14).

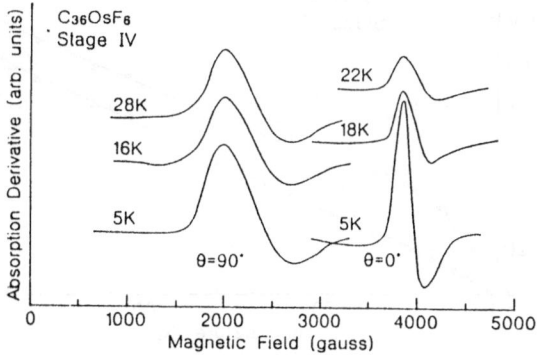

Fig. 12. ESR signals as a function of temperature for $H\|c$ ($\theta = 0$) and $H\perp c$ ($\theta = 90°$) at X band frequency. The sample is $C_{36}(OsF_6)$ stage IV (after Vaknin et al, ref. 51).

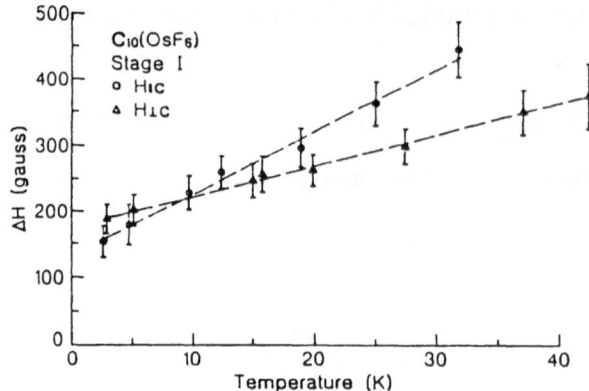

Fig. 13. ESR linewidth, ΔH, versus temperature for $C_{10}(OsF_6)$ stage I. The slopes yield the anisotropic thermal broadenings to be $b_{\parallel} = 8$ G/K and $b_{\perp} = 4$ G/K (after Vaknin et al, ref. 51).

For $H \perp c$ the excited $\pm 3/2$ levels are almost non-magnetic and consequently no change in the slope of $1/\chi_{\perp}$ versus temperature is expected, in agreement with experiment (Fig. 14).

The combined ESR and susceptibility studies of HOPG/OsF$_6$ suggest that the magnetic moment per intercalant species is 3.2 μ_B corresponding to an OsF$_6^-$, $5d^3$ configuration; i.e. there is charge transfer of X = 1 electron per OsF$_6$ molecule <u>independent</u> of OsF$_6$ concentration and stage. The results provide also evidence for a well defined crystal field and a well

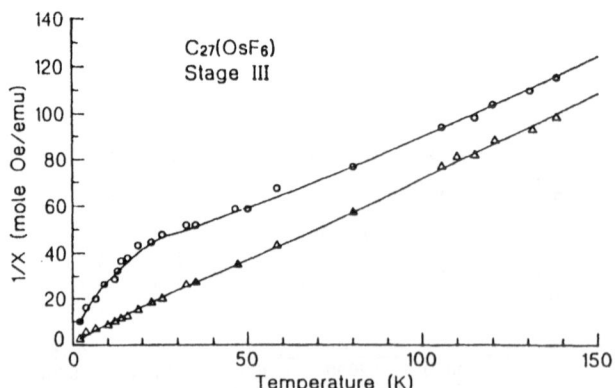

Fig. 14. The inverse susceptibility $1/\chi_{\parallel}$ (circles) and $1/\chi_{\perp}$ (triangles) as a as a function of temperature. Note the change in the slope of $1/\chi_{\parallel}$ at high and low temperatures (after Vaknin et al, ref. 51).

defined orientation of the OsF_6^- molecules within the graphite layers. We note that no CESR signal could be detected for HOPG/OsF₆ probably due to a spin-orbit relaxation mechanism and the "heavy" intercalant species. This is expected according to the general behavior of Fig. 9.

Another system to be investigated was that of HOPG/MoF₆.[52] The HOPG/MoF₆ system differs significantly from OsF₆ in its magnetic properties. In contrast to the latter where in the ESR only one signal arising from a localized species is observed, the MoF₆ system displays two signals for stages I and II and an additional weak signal for stage III (Fig. 15) labelled d, c_1 and c_2, respectively. The latter (c_1 and c_2) are attributed to carriers on bounding and non-bounding graphite layers.

The two main signals are believed to arise from two different species, because they show different temperature dependences. Signal d appears at $g = 1.60 \pm 0.05$. It shows slight anisotropy, but its intensity varies roughly as $1/T$ as expected for a paramagnetic species. It is assigned to an intercalated $5d^1$ species corresponding to an unresolved hyperfine interaction with the ^{95}Mo and ^{97}Mo nuclei.

Signal c_1, at approximately $g \approx 1.89$, is angularly independent. However, it shows a slight temperature dependence in g-value changing from $g = 1.89$ at 4K to $g = 1.99$ at 300K. This signal is believed to be associated with conduction carriers. Normally this would not show intensity variations with temperature. Indeed it is still visible at room temperature in contrast to signal d. However, the temperature dependence of c_1 can be explained by exchange with the localized species (d). The estimated magnetic moment from ESR ions (assuming $S = 1/2$ and $g_d = 1.6$) is significantly larger than the effective moment observed in the susceptibility study.[52] This suggests that only 20% of the intercalated MF₆ species are in a reduced form. Another interesting feature of the MoF₆ system is the observation of a weak additional signal for stage III at $g \approx 2$. This signal is attributable to a graphite interior layer not adjacent

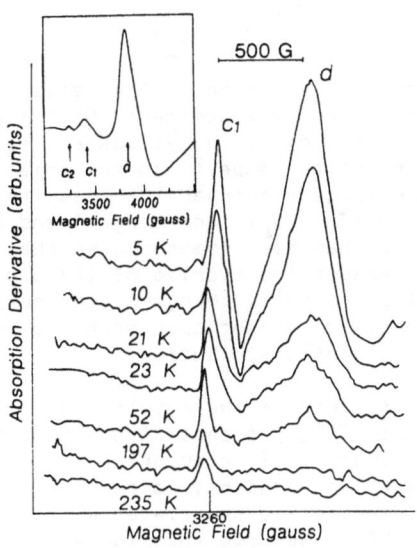

Fig. 15. ESR spectra of HOPG/MoF₆ (stage I). Insert: ESR spectrum of stage III. X band frequency (after Vaknin et al, ref. 52).

to an intercalate layer. The ability to resolve the signals c_1 and c_2 is probably due to the spatially short range exchange field which shifts the position of c_1 but not c_2, as well as the relatively slow interlayer hopping.

Anisotropic Relaxation and N(E_F)

The results of Vaknin et al[51] clearly indicate that the linear thermal broadening strongly depends on the direction of the magnetic field with respect to the c-axis. Vaknin, et al[51] have suggested anisotropic exchange interaction as the origin of this anisotropic Korringa relaxation of the form

$$H_{ex} = J_{\parallel} S_z \cdot s_z + J_{\perp}(S_x s_x + S_y s_y) \tag{16}$$

where S is the localized spin on the osmium and s is the spin of the π-like electron on the bounding graphite layer, $J_{\parallel}$ and $J_{\perp}$ are components of the exchange interaction. The total relaxation rate was calculated using the Bloch-Redfield[53] kinetic equations. It was demonstrated that in the presence of anisotropic exchange,[16] the transverse and the longitudinal fluctuations contributions to the total relaxation rate are different and depend on the quantization axis. Generally speaking, the relaxation rate for $H \parallel c$ and $H \perp c$ could be written as[51]

$$(1/T_2)_{\alpha} = A_{\alpha} N^2(E_F) K_B T \qquad (\alpha = \parallel, \perp) \tag{17}$$

where N(E_F) is the density of states at the Fermi level, the coefficient $A\alpha$ ($\alpha = \parallel, \perp$) depends on the exchange parameters as follows

$$A_{\parallel} = \pi/2h \ (J_{\parallel}^2 + J_{\perp}^2) \qquad ; \qquad (H \parallel c) \tag{18A}$$

$$A_{\perp} = \pi/4h \ (J_{\parallel}^2 + 3J_{\perp}^2) \qquad ; \qquad (H \perp c) \tag{18B}$$

Using (17,18) the ratio $b_{\parallel}/b_{\perp}$ can be expressed as follows

$$\frac{b_{\parallel}}{b_{\perp}} = \left[\frac{g_{\perp}}{g_{\parallel}}\right] \frac{(1/T_2)_{\parallel}}{(1/T_2)_{\perp}} = \frac{g_{\perp}}{g_{\parallel}} \frac{2J_{\parallel}^2 + 2J_{\perp}^2}{J_{\parallel}^2 + 3J_{\perp}^2} \tag{19}$$

Relation (19) allows the estimate of the anisotropy $J_{\parallel}/J_{\perp}$ provided that $b_{\parallel}/b_{\perp}$ are known from the experiment. Unfortunately the relatively large error in the determination of $b_{\perp}$ yields a large error in $J_{\parallel}/J_{\perp}$ and no conclusive decision with respect to the anisotropy can be reached. Nevertheless, within the large error-bar the results of Vaknin suggest almost isotropic exchange ($J_{\parallel} = J_{\perp}$) for stage I and some anisotropy ($J_{\parallel} > J_{\perp}$) for higher stages. It has been demonstrated that the anisotropy of the exchange occurs because the overlap of the orbital charge densities depends on the orientation of these charges with respect to the interatomic axis.[53,54] Thus, the anisotropic exchange is directly related to the nature of the π-d charge distribution.

A most interesting feature of the results [51] is the near independence of $b_{\parallel}$ of stage and OsF_6 concentration (note the value of $b_{\parallel}$ was determined with high accuracy). This suggests that the density of states, N(E_F), is nearly stage independent, according to (17). Vaknin has suggested[51] that this behavior is associated with the following dispersion relation: $E \sim k^2$. We note that for a two dimensional spherical Fermi surface such a dispersion relation yields a constant N(E_F) independent of the location of E_F.

Anisotropic Korringa relaxation has been observed also by Kume and his coworkers using high resolution ^{13}C NMR of GIC prepared from HOPG. They were able to observe completely resolved ^{13}C spectra consisting of absorption lines with several ppm width. Examples for the ^{13}C spectra of Kume for various stages are given in Fig. 16. As seen, the spectra of stage II exhibit two resolved lines which were attributed to the two carbon sites on bounding graphite layers, the spectra of stage III exhibits 4 lines; i.e., two doublets associated with bounding layer and nonbounding layer, respectively, etc. The relaxation times, T_1, for the ^{13}C nuclei at all the different sites were measured and found to be shorter than that of pristine graphite and linearly dependent on temperature. Furthermore, T_1 was found to be extremely anisotropic such that T_1 for the magnetic field parallel to the c-axis $(T_1)_{||}$ was 5-6 times longer than $(T_1)_{\perp}$ for the magnetic field perpendicular to the c-axis. Kume et al have attributed the position of the peaks to a Knight shift and the relaxation T_1 to a Korringa-type process due to coupling between the spin of the metallic π electrons and that of the ^{13}C nuclei. T_1 could be expressed by a relation similar to (17). However, unlike the case of ESR the coupling constant A_α originates from two different mechanisms; namely (1) the isotropic core polarization interaction and (2) the non-isotropic dipolar interaction.

The different relaxation rates, $1/T_1$, associated with the different doublets in Fig. 16, are due to different density of states, $'N(E_F)$, on bounding and non-bounding graphite layers. Kume, et al were able to explain the anisotropy of T_1 and extracted $N(E_F)$ for the different graphite layers. Fig. 17 yields the density of states $N(E_F)$ for bounding and interior graphite layers for K-GIC and various stages. The average value of $N(E_F)$ is compared with specific heat data and the agreement is good (Fig. 17).

The elegant method of Schumacher and Slichter should also be mentioned. This method uses a NMR CW spectrometer to measure simultaneously and on the same sample the NMR absorption intensity, I_N and the ESR

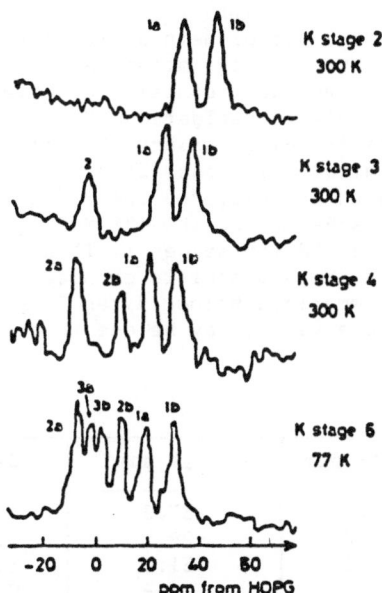

Fig. 16. ^{13}C NMR spectra of K/HOPG (after Kume et al, ref. 37).

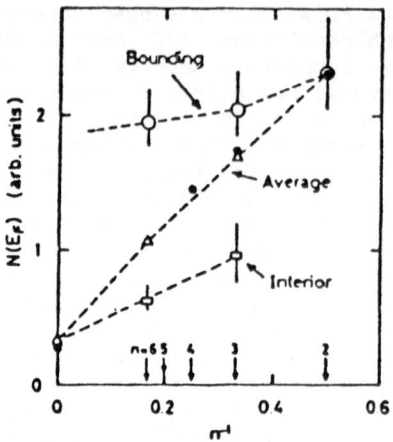

Fig. 17. The density of states $N(E_F)$ for bounding and non-bounding layers (after ref. 37)

absorption intensity, I_c. A proper scaling yields[15,55] $N(E_F) = N\mu_N I(I+1) I_c/6K_B T\mu_e I_N$ where N is the number of nuclei, μ_N and μ_e are the magnetic moments of the nuclei and the electrons, respectively, and I is the nuclear angular momentum. The method has been extensively used[15,16,55] to study $N(E_F)$ in various acceptor and donor GIC. However, the Schumacher-Slichter method, like the specific heat method, gives information on the "average" density of state while relaxation methods yield the "local" $N(E_F)$ on the various graphite layers.

The Identification of Intercalated Species by Nuclear Magnetic Resonance Spectroscopy

NMR is one of the more useful methods for determining the degree of charge transfer per intercalated molecule. This technique is of course applicable only if the intercalant contains a nucleus suitable for NMR. Some of the most thoroughly investigated GIC's are the acceptor compounds with inorganic fluorides.[56] Fingerprints characterizing a given species are: (1) its NMR chemical shift and (2) hyperfine and quadrupole splittings. Here we show, as an example, the identification of the intercalant species in graphite-BF_3-F_2 using the ^{19}F NMR spectrum. The NMR spectra for H ∥ c are shown in Fig. 18 for stages I, II and III. We identify BF_3, BF_4^- and F^- species and the appropriate chemical shifts, δ (measured with respect to C_6F_6) are summarized below (after Brusilovsky, et al[57]). Discrepancies with literature values may be due to different environments.

$\delta(F^-)$ ppm*	$\delta(BF_4^-)$ ppm*	$\delta(BF_3)$ ppm*	Stage
----	-22.5	-37.3	I
-42.0	-20.8	-29.2	II
-45.0	-21.0	-31.0	III
----	-20.0	----	$C_{64}^+ BF_4^-·2.85\ CH_3NO_2$
-40.0	-13.0	-29.6	Literature values

* relative to C_6F_6, external standard [Note: $\delta_{lit.}(B_2F_7^-)$ is -18 ppm].

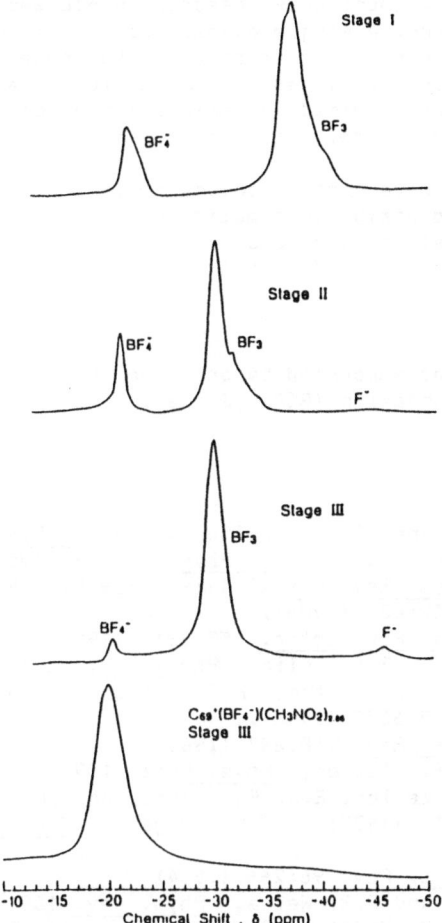

Fig. 18. NMR spectra of HOPG/BF$_3$/F$_2$ for H∥c at room temperature (after (Brusilovsky et al, ref. 57).

The plausibilities of the assignments are strengthened by the fact that the intensity ratio BF$_4^-$/BF$_3$ increases with time indicating release of the neutral BF$_3$ from the lattice. Furthermore, authentic BF$_4^-$ obtained from a reaction of HOPG with NO$_2$BF$_4$ in nitromethane gives essentially the same signal as that assigned to BF$_4^-$ in HOPG/BF$_3$/F$_2$.

Because part of the BF$_3$ is evolved spontaneously, calculation of the BF$_4^-$/BF$_3$ ratio from the relative peak intensities would be unreliable. Reflectivity measurements on stage I give a charge per carbon of ∼0.03. If this is balanced by intercalated BF$_4^-$, a combination with weight uptake yields an approximate composition C$_{32}^+$BF$_4^-$(BF$_3$)$_4$ for stage I.[57]

Concluding Remarks

The last decades witnessed significant progress in the development and applications of more sophisticated resonance techniques such as Electron Spin Echo (ESE), Electron Nuclear Double Resonance (ENDOR) and NMR Imaging. However, almost no studies of GIC using these more developed techniques exist today. Most CESR research on GIC was done using CW conventional speccetrometers. A single asymmetric line is usually observed and limited information can be extracted. The observation of saturation effects[29] opens new possibilities for ESE studies at low temperatures. Such investigations might yield valuable information about the electronic states, wave functions and relaxation processes.

Nevertheless, we have clearly demonstrated that both ESR and NMR techniques can provide useful information on the dynamics of the carriers, the intercalation kinetics, the nature of the intercalant species, charge transfer, density of states and anisotropic relaxation.

Acknowledgements

This research was supported by grant No. 84-00192 from the US-Israel Binational Science Foundation (BSF), Jerusalem, Israel.

REFERENCES

1. M.S. Dresselhaus and G. Dresselhaus, Adv. Phys. 30:139 (1981); R. Clarke and C. Uher, Adv. Phys. 33:469 (1984).
2. S.A. Safran, Phys. Rev. Letts. 44:937 (1980); G. Kirczenow: Phys. Rev. Letts. 52:437 (1984).
3. G.Kirczenow, Phys. Rev. Letts. B55:2810 (1985).
4. R. Clarke, N. Wada, S.A. Solin, Phys. Rev. Lett. 44:1616 (1980); A. Erbil, A.R. Kortan, K.J. Birgeneau and M.S. Dresselhaus, Phys. Rev. B28:6329 (1983).
5. G. Wagoner, Phys. Rev. 118:647 (1960).
6. K.A. Muller and R. Kleiner, Phys. Lett. 1:98 (1962).
7. G.M.T. Foley, C. Zeller, E.R. Falardeau and F.L. Vogel, Solid State Comm. 24:371 (1977); F.L. Vogel, Jour. Mat. Sci. 12:982 (1977).
8. R.J. Elliot: Phys. Rev. 96:266 (1954)
9. J.C. Slonczewski and P.R. Weiss, Phys. Rev. 109:272 (1958). .
10. J.W. McClure and Y. Yafet, in "Proceedings of the Fifth Conference on Carbon," 1961 Pergamon, New York (1962) Vol. 1, page 22.
11. P. Lauginie, H. Estrade, J. Conard, D. Guerard, P. Lagrange, M. El Makrini, Physica, 99B:514 (1980).
12. S.K. Khanna, E.R. Falardeau, A.J. Heeger and J.E. Fischer, Sol. State Comm. 25:1059 (1978).
13. P. Pfluger, K.A. Muller, W. Berlinger, V. Geiser, H.J. Guntherodt, Synth. Met. 8:15 (1983).
14. H. Estrade-Szwarckopf, Helvetica Physica Acta, 58:139 (1985).
15. B.R. Weinberger, J. Kaufer, A.J. Heeger, E.R. Falardeau and J.E. Fischer, Sol. State Comm. 27:163 (1978).
16. S. Ikehata, J.W. Milliken, A.J. Heeger and J.E. Fischer, Phys. Rev. B25:1726 (1982).
17. D. Davidov, O. Milo, I. Palchan and H. Selig, Synth. Met. 8:83 (1983).
18. R.M. Stein, L. Walmsley and C. Rettori, Phys. Rev. B32:4134 (1985).
19. R.M. Stein, L. Walmsley, G.M. Gualberto and C. Rettori, Phys. Rev. B32:4774 (1985).
20. I. Palchan, D. Davidov, V. Zevin, G. Polatsek and H. Selig, Synt. Met. 12:413 (1986).
21. I. Palchan, F. Mustachi, D. Davidov and H. Selig, Synth. Met. 10:101 (1984/1985).
22. G. Feher and A.F. Kip, Phys. Rev. 98:337 (1955).
23. F.J. Dyson, Phys. Rev. 98:349 (1955).

24. J. Blinowski, P. Kacman, C. Rigaux and M. Saint-Jean, Synthetic Metals, 12:419, in Proceedings of the International Symposium on GIC, Tsukuba, Japan (1985).
25. M. Saint-Jean, C. Rigaux, B. Clerjand, J. Blinowski, P. Kacman and O. Furdin, to be published (1986).
26. J.I. Kaplan, Phys. Rev. 115:575 (1959).
27. D. Davidov, O. Milo, I. Palchan and H. Selig, Synth. Met. 8:83 (1983).
28. D. Vaknin, I. Palchan, D. Davidov, H. Selig and D. Moses, submitted to Synth. Met. (1986).
29. G. Kirczenow, Phys. Rev. Lett. 55:28100 (1985); Synth. Met. 12:143 (1985).
30. G. Kirzenow, Submitted to Phys. Rev.
31. S.E. Ulloa and G. Kirczenow, Phys. Rev. B33:1360 (1986).
32. S. Miyazima, Synth. Met. 12:155 (1985).
33. V. Zevin and J.T. Suss, Submitted to Phys. Rev.
34. I. Palchan, D. Davidov, V. Zevin, G. Polatsek and H. Selig, Phys. Rev. B32:5554 (1985).
35. I. Palchan, D. Davidov, V. Zevin, G. Polatsek and H. Selig, Synth. Metals 12:413 (1985).
36. G.C. Chingas, J. Milliken, H.A. Resing and T. Tsang, Synth. Met. 12:131 (1985).
37. D. Kang, K. Nomura and K. Kume, Synth. Metals, 12:395 (1985).
38. P. M. Richards, in "Physics of Superionic Conductors," M.B. Salamon, ed., Springer-Verlag, Berlin, Heidelberg, New York (1979).
39. C.A. Sholl, J. Phys. C14:447 (1981).
40. P.M. Richards and M.B. Salamon, Phys. Rev. B9:32 (1974).
41. L.P. Kadanoff and P.C. Martin, Ann. Phys. (N.Y.) 24:419 (1963).
42. B.G. Silbernagel and F.R. Gamble, Phys. Rev. Lett. 32:1436 (1979); R.L. Kleinberg and B.G. Silbernagel, Sol. State Commun. 33:867 (1980).
43. B.H. Robinson, et al, Mol. Cryst. Liq. Cryst. 117:421 (1985).
44. D. Vaknin, Ph.D. Thesis, Jerusalem, Israel. (Unpublished)
45. J.R. Asik, M.A. Ball and C.P. Slichter, Phys. Rev. Lett. 16:740 (1966); ibid, 181:645 (1969); E.K. Cornell and C.P. Slichter, Phys. Rev. 180:358 (1969); D. Davidov, C. Rettori, R. Orbach, A Dixon and E.P. Chock, Phys. Rev. 11:3546 (1975); C. Rettori, D. Davidov, R. Orbach, E.P. Chock and B. Ricks, Phys. Rev. 11:1 (1973); Y. Yafet, Jour. Appl. Phys. 42:1564 (1971); F. Beuneu and P. Monod, Phys. Rev. B13:3424 (1976); N.A.W. Holzwarth and M.J.G. Lee, Phys. Rev. B13:2231 (1976).
46. F. Beuneu and P. Monod, Phys. Rev. B13:3424 (1976); P. Monod and F. Beuneu, Phys. Rev. B19:911 (1979).
47. K. Kume, K. Nomura, Y. Hiroyama, Y. Maniwa, H. Suematsu and S. Tanuma, Synth. Met. 12:307 (1985).
48. Orna Milo, M.Sc. Thesis, Jerusalem, Israel 1983. (Unpublished)
49. M. Suzuki, K. Koga and Y. Jinizaki, in Graphite Intercalation Compounds - Progress of Research in Japan, S. Tanuma and H. Kamimura, eds., (1985) page 18 and reference therein.
50. D. Vaknin, D. Davidov, H. Selig, V. Zevin, I. Felner and Y. Yeshurun, Phys. Rev. B31:3212 (1985).
51. D. Vaknin, D. Davidov, V. Zevin and H. Selig, Submitted to Phys. Rev.
52. D. Vaknin, D. Davidov, H. Selig and Y. Yeshurun, J. Chem. Phys. 83:3859 (1985).
53. F. Bloch and P. Wangsness, Phys. Rev. 89:728 (1953); F. Bloch, ibid. 102:104 (1956); A. Redfield, ibid. 98:787 (1955).
54. P.M. Levy, Phys. Rev. 177:509 (1969) and reference therein.
55. S. Ikehata, T. Morimoto, H. Suematsu and S. Tanuma, Synth. Met. 12:313 (1985).

56. See for example G.R. Miller, H.A. Resing, M.J. Moran, L. Banks, F.L. Vogel, A. Pron and D. Billaud, <u>Synth. Met.</u> 8:77 (1983); G.R. Miller, H.A. Resing, F.L. Vogel, A. Pron, T.C. Wu and D. Billaud, <u>Jour. Phys. Chem.</u> 84:3333 (1984).

57. D. Brusilovsky, D. Vaknin, I. Ohana, H. Selig and D. Davidov, to be published (1986).

NMR CHARACTERIZATION OF HEAVY ALKALI METAL-ORGANIC MOLECULE

GRAPHITE COMPOUNDS PREPARED FROM HIGHLY-ORIENTED PYROLYTIC GRAPHITE

M.F. Quinton,[*] F. Beguin,[**] and A.P. Legrand[*]

[*]Laboratoire de Physique Quantique, UA 421, E.S.P.C.I. 10 rue Vauquelin, 75231 Paris Cedex 05, France
[**]C.N.R.S.-C.R.O.S.I. 1b rue de la Férollerie, 45045 Orleans Cedex, France

Many different organic molecules can be intercalated into binary heavy alkali metal-GIC's [1]-[6]. It is expected that the electronic properties of the initial binary GIC will be modified due to the complexation of the alkali ions by the organic molecules [7][8]. Powdered samples of some ternary GIC's of this kind have already been studied by ^{1}H NMR [9]-[12] and ^{13}C NMR [13]. In order to have more precise and reliable information, we have carried out ^{1}H and ^{13}C NMR studies with well-characterized samples prepared from HOPG.

Table 1. Characteristics of the samples. [*][THF] represents the numbers of tetrahydrofuran molecules per 24 graphitic C-atoms.

Sample code	Alkali atom	Initial binary GIC		Final ternary GIC		
		stage	I_c (Å)	stage	I_c (Å)	[THF]
K1T2	K	2	8.71∓0.01	1	8.83∓0.01	2.45∓0.05
R1T2	Rb	2	9.02∓0.01	1	8.86∓0.01	2.40∓0.03
C1T1	Cs	2	9.35∓0.01	1	7.07∓0.01	1.58∓0.02

Characteristics of the three samples studied are presented in table 1. The materials were prepared by exposing binary alkali-GIC to tetrahydrofuran (THF) vapor. The K-THF and Rb-THF samples are very similar, with very similar values of c-axis repeat distance (I_c), and amount of THF ([THF]). The Cs-THF sample is quite different. The ^{1}H NMR spectra at room temperature confirm this fact. When the sample orientation in the magnetic field is varied, the linewidth obeys a $|3 \cos^2 \theta - 1|$ law for the three samples, where θ is the angle between the c-axis and the static magnetic field B_o. This linewidth is characteristic of "flat" THF molecules with pseudo-rotation, which is a characteristic torsional motion [14], for the Cs-THF sample. For the K-THF and Rb-THF samples, the smaller linewidth is attributed to molecules "standing-up" between the graphitic planes, and rotating about the sample c-axis.

The ^{13}C spectra at room temperature are similar for the three samples (fig.1). The ^{13}C THF lines are visible only at $\Phi \simeq 54°$, corresponding to $3\cos^2\theta - 1 = 0$ (magic angle). This fact confirms the mobility of THF, with a motion where the sample c-axis is a symmetry axis. The shift of the ^{13}C graphitic lines with θ obeys the same law for the three sample (fig.2). By comparison with the initial binary alkali-GIC [15]-[16], we conclude that

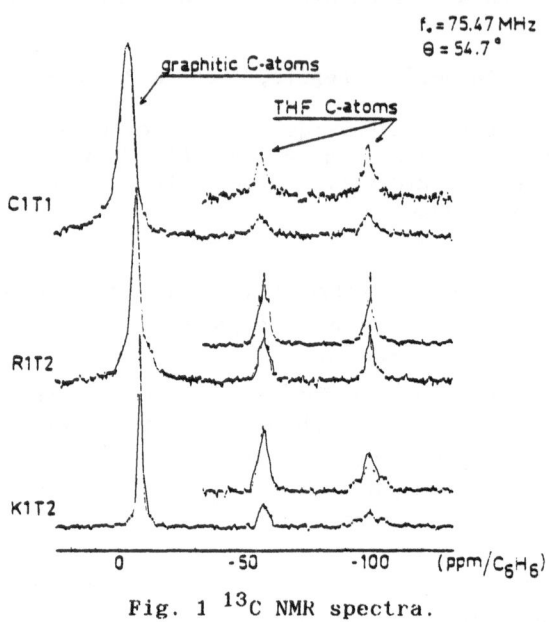

Fig. 1 ^{13}C NMR spectra.

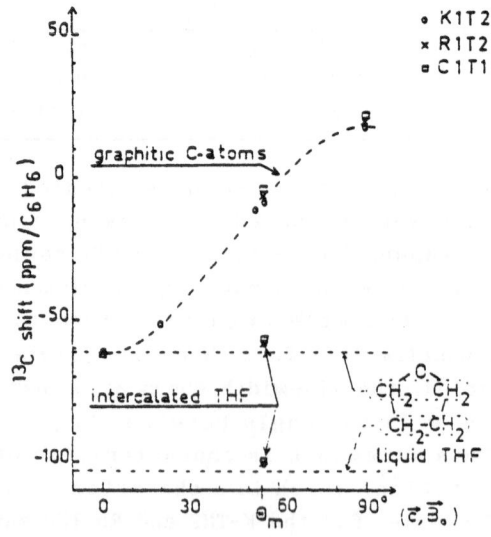

Fig. 2 Sample orientation dependence of the ^{13}C NMR spectra position.

the THF molecules form an efficient "screen" between the alkali ions and the graphitic planes. Furthermore, the sign of the anisotropy of the ^{13}C graphitic lineshift is the same as in the second stage binaries, and opposite to that of the first stage binaries. We conclude that, in these first stage ternary GIC's, the graphitic planes have the same electronic surroundings as in the second stage binaries.

By lowering the temperature of the K-THF at $\theta = \theta_m$, the ^1H and ^{13}C spectra show a "phase transition" at about 250 K, related to the lower mobility of THF inside the intercalated layer. The measurement of the graphitic ^{13}C spin lattice relaxation time exhibits a Korringa law behavior between 118K and 293K. By comparison with the result obtained with KC_{24} (17), we can conclude, to a first approximation, that the density of states at the Fermi level is the same in both compounds.

References

[1] F. BEGUIN, R. SETTON, A. HAMWI, P. TOUZAIN, Mater. Sc. Eng. 40: 167 (1979)

[2] L. BONNETAIN, P. TOUZAIN, A. HAMWI, Mater. Sc. Eng. 31: 45 (1977)

[3] F. BEGUIN, R. SETTON, L. FACCHINI, A.P. LEGRAND, G. MERLE, C. MAI, Synth. Met. 2: 161 (1980)

[4] Yu. V. ISAEV, Yu. N. NOVIKOV, M.E. VOL'PIN, Synth. Met. 5: 23 (1982)

[5] J. JEGOUDEZ, C. MAZIERES, R. SETTON, Synth. Met. 7: 85 (1983)

[6] Yu. ISAEV, Yu. N. NOVIKOV, M.E. VOL'PIN, I. RASHKOV, I. PANAYOTOV, Synth. Met. 6: 9 (1983)

[7] J. AMIELL, P. DELHAES, F. BEGUIN, R. SETTON, Mater. Sc. Eng. 31: 243 (1977)

[8] A.P. LEGRAND, L. FACCHINI, D. BONNIN, J. BOUAT, M.F. QUINTON, F. BEGUIN, Synth. Met. 12: 175 (1985)

[9] M.F. QUINTON, L. FACCHINI, F. BEGUIN, A.P. LEGRAND, Rev. Chim. Min. 19: 407 (1982)

[10] A.L. BLUMENFELD, E.I. FEDIN, Yu. V. ISAEV, Yu. N. NOVIKOV, Synth. Met. 6: 15 (1983)

[11] A.L. BLUMENFELD, Yu. V. ISAEV, Yu. N. NOVIKOV, Synth. Met. 10: 193 (1985)

[12] M.F. QUINTON, F. BEGUIN, A.P. LEGRAND, Synth. Met. 14: 179 (1986)

[13] L. FACCHINI, Thesis, Paris (1983)

[14] W.J. LAFFERTY, A.W. ROBINSON, R.V. SAINT LOUIS, J.W. RUSSEL, H.L. STRAUSS, J. Chem. Phys. 42: 2915 (1965)

[15] J. CONARD, H. ESTRADE, P. LAUGINIE, H. FUZELIER, G. FURDIN, R. VASSE, Physica 99 B: 521 (1980)

[16] K. KUME, Y. MANIWA, H. SUEMATSU, Y. IYE, S. TANUMA, Synt. Met. 8: 69 (1983)

[17] Y. MANIWA, K. KUME, H. SUEMATSU, S. TANUMA, J. Phys. Soc. of Japan 54, 2: 666 (1985)

INTERCALATED GRAPHITE FIBERS

M.S. Dresselhaus

Massachusetts Institute of Technology
Cambridge, MA 02139, USA

INTRODUCTION

The novel geometry of graphite fibers facilitates the study of various intercalation phenomena and is of particular importance for practical applications of GICs. Also in their pristine form, graphite fibers offer unique advantages over other forms of graphite for the study of novel phenomena and for a variety of practical applications. The synthesis, structure and properties of intercalated graphite fibers have much in common with intercalated bulk graphite host materials, such as kish graphite and highly oriented pyrolytic graphite (HOPG),[1] which are extensively discussed in this volume. We therefore briefly review here the structure and properties of intercalated graphite fibers and the differences between GICs prepared from fiber and bulk graphite host materials.

Of particular significance to the properties of intercalated graphite fibers is the structure and structural order of the pristine fibers, which in turn is strongly dependent on the precursor material and on the heat treatment process of the pristine fibers.[2,3] For all graphite fibers, the a–axis is along the fiber axis, yielding a stiff, high strength anisotropic material. Graphite fibers based on polyacrylonitrile (PAN), mesophase pitch and vapor grown precursor materials have all been successfully intercalated to produce both donor and acceptor type GICs. For the PAN fibers (diameters 7–10μm), the preferred c–axis orientation favors the radial direction perpendicular to the fiber axis, while for most of the pitch fibers (diameters 10–20μm) the graphitic layer planes themselves have a preferred orientation in radial directions. For the vapor grown fibers, the layer planes have an annular configuration as the growth rings inside a tree (see Fig. 1).

Both PAN and pitch based fibers are prepared by an initial spinning process, forming a continuous filament of the precursor, followed by a preoxidation step to increase the thermal stability of the fibers and finally by a carbonization process during which H_2O and N are evolved.[3] As a result, carbon ribbons are formed with layer planes generally aligned along the fiber axis. Heat treatment to temperatures greater than 2000°C results in further growth of the carbon ribbons and partial graphitization. During the

PAN   Pitch
                                              Vapor grown

Figure 1: Schematic diagram of the fiber morphology, showing the arrangement of carbon hexagon planes in "dog–bone" PAN fiber, "PAC–man" mesophase pitch fiber, and vapor grown fiber.

heat treatment, defects in the carbon hexagonal layer structures (missing atoms, bond disorder, impurities, etc.) are reduced and the density of voids is decreased, leading to a higher mass density of the fibers. At the same time, the size of the graphite ribbons increases in both length and width, while the average number of layers stacked together in the ribbon increases. At sufficiently high heat treatment temperatures $T_{HT}(\geq 2200°C)$, the layers begin to lock in an ordered sequence, and three–dimensional ordering of the layers is initiated.

Intercalation is facilitated by a high degree of structural order of the graphite host material and is retarded by defects, structural imperfections and disorder. Staging is only achieved in hosts with a high degree of structural perfection. Thus basic scientific studies have focused on highly ordered graphite host materials. In contrast, the requirements of specific commercial applications of intercalated graphite fibers may involve rather different issues, such as the amount of intercalate uptake, quality control of the intercalation process, long term stability of the intercalated fibers under a variety of environmental conditions, and the availability of continuous intercalated fiber lengths. Since the focus of the present discussion is on scientific studies, we give particular attention to fiber host materials of high structural perfection. For commercially available fibers, the highest structural perfection is achieved with PAN and mesophase pitch fibers heat treated to $\sim 2800°C$; the vapor grown fibers are not yet widely available commercially.

For the same heat treatment temperatures (e.g., 2800°C), a higher degree of crystallinity (larger in–plane and c–axis coherence lengths L_a and L_c) is achieved for the mesophase pitch–based fibers than for the PAN fibers. Both of these host fiber materials have been successfully intercalated by many intercalant species. However for both of these continuous precursor fiber materials, the long–range crystalline order has in most cases been insufficient to synthesize well–staged graphite intercalation compounds.

In contrast, vapor grown fibers, heat treated to $\sim 3000°C$, exhibit crystalline coherence lengths of $L_a \sim 1000$ Å from which donor and acceptor compounds with well established staging can be synthesized.[2] The vapor grown fibers are prepared by passing a gaseous mixture of a hydrocarbon (e.g., benzene or methane) and hydrogen over a heated substrate ($\sim 1100°C$) containing small (~ 100Å) catalyst particles of Fe or Ni. A large variety of growth geometries, catalysts, hydrocarbons and growth conditions have been used successfully.[4,5] In the vapor growth process, hollow carbon filaments grow as the carbon precipitates from the catalyst particles. These filaments are subsequently thickened by deposition of carbon from the decomposition of the hydrocarbons in the gas stream.[4] By this method, fibers can be prepared with diameters ranging from 0.1 μm to 100 μm.[5] Submicron filaments have also been prepared by ion bombardment of carbon surfaces.[6] Structurally these vapor grown graphite fibers exhibit a tree ring arrangement of the graphite layers, but with some faceting (see Fig. 1). The faceting flattens the graphite layers, enhancing the interplanar site correlation of the ideal graphite crystal structure. The role of heat treatment in enhancing the crystallinity of the vapor grown fibers can be seen in the plot of the resistivity as a function of heat treatment temperature T_{HT} (see Fig. 2).[7] This figure clearly shows that for $T_{HT} \sim 3000°$C, nearly single crystal behavior is achieved.

For PAN and pitch–based fibers, the average misorientation of the layer planes with respect to the fiber axis decreases with increasing T_{HT}, but is also strongly influenced by the tension under which the fibers are held during the processing steps.[3] With increasing T_{HT}, the c–axis interplanar distance decreases toward that of single crystal graphite ($c_0/2 = 3.35\text{Å}$). Thus, measurement of the positions and linewidths of the (00ℓ) diffraction peaks sensitively characterizes the structural perfection of the carbon fibers; investigation of the (10ℓ) peaks provides an even more powerful characterization tool because of the sensitivity of the (10ℓ) peaks to interplanar site correlations. Also highly sensitive to the structural perfection are the elastic properties, so that the Young's modulus increases with increasing structural perfection as the carbon layers stiffen and are able to slide past one another more easily. Comparison of the Young's modulus E of the fibers is made to the very high value characteristic of single crystal graphite ($E = 1020$ GPa). Commercial Celion (Celanese) PAN fibers ($E \sim 200$ GPa) are useful for structural applications because of their higher strain to failure and higher stress to failure characteristics.

Raman spectroscopy provides a highly sensitive probe for determination of the in–

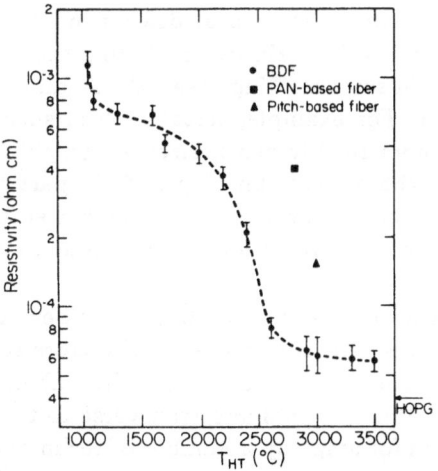

Figure 2: Heat–treatment temperature T_{HT} dependence of the room–temperature resistivity for vapor grown (benzene–derived, BDF) fibers. Also indicated on the figure are resistiviy values for PAN– and pitch–based fibers and for HOPG.[7]

plane crystallite ordering in pristine graphite fibers.[2,8,9] It is the E_{2g_2} intralayer mode that is primarily used for the characterization of graphite fibers. When disorder is introduced into the graphite lattice, the Raman–active E_{2g_2} line at 1582 cm^{-1} is broadened and a new disorder–induced line is observed near $\sim$ 1360 cm^{-1}, where there is a high phonon density of states.[8,10] A second disorder–induced line near $\sim$ 1620 cm^{-1} can also be observed under appropriate conditions, again corresponding to a high density of states.[8] Because of the small wave–vectors for visible light, the Raman–allowed lines all correspond to lattice modes at the center of the Brillouin zone ($q = 0$). The introduction of disorder into the lattice allows contributions to the Raman spectra from phonons throughout the Brillouin zone. Thus, both disorder–induced lines at $\sim$ 1360 cm^{-1} and $\sim$ 1620 cm^{-1} arise from the high density of phonon states near these phonon frequencies.[2,8,10] Thus, to characterize the extent of the structural disorder in pristine graphite fibers, the intensity of the disorder–induced line relative to that for the Raman–allowed line I_{1360}/I_{1582} is measured. Other sensitive characterization parameters are the linewidths and peak frequencies of both the disorder–induced and Raman–allowed lines. Valuable information about the defect density in carbon fibers is provided by studying both the first and second order Raman spectra.[10]

In this review, the physical properties of intercalated graphite fibers are discussed, with particular emphasis given to the properties of the most highly ordered fibers which are most easily, homogeneously, and reproducibility intercalated. Attention is also given to new physics that can be pursued with intercalated fibrous materials.

FIBER INTERCALATION, STAGING AND CHARACTERIZATION

Graphite fibers are normally intercalated by techniques similar to those used for the corresponding HOPG–based GICs,[11] though the specific intercalation conditions will be different in detail. In the case of vapor grown fibers, intercalation is initiated at the free ends of the fibers and then proceeds along the fiber length. Heat treatment of the fibers to high temperatures ($\approx$ 3000°C) prior to intercalation increases the structural ordering of the graphite planes, leading to enhanced intercalation and more uniform staging. In most cases, the higher density of structural defects in the fibers tends to retard the intercalation process relative to bulk HOPG, including the threshold pressure for vapor phase intercalation.[12] Yet in some striking cases, the small dimensions of the fibers lead to enhanced intercalation. For example, certain GICs, such as stage 1 FeCl$_3$–GIC or stage 1 NiCl$_2$–GIC are more readily synthesized in graphite fiber hosts than in bulk graphite, at least within the optical skin depth.[13] Of particular significance for both scientific studies and practical applications is the higher stability of intercalated vapor grown fibers with regard to deintercalation under ambient conditions.[14,15]

For comparison of the properties of GICs based on fibers and on bulk graphite hosts, knowledge of the stage of the GIC is important. Therefore considerable effort has been directed to the characterization of intercalated graphite fibers with regard to both stage index and staging fidelity (longitudinal and transverse to the c–axis). The intercalated fibers can be characterized for stage by a number of techniques. For example, the staging of a bunch of intercalated graphite fibers can be conveniently measured using the Debye–Scherrer x–ray diffraction technique. This technique has been applied to characterize the staging in intercalated PAN, mesophase pitch and vapor grown graphite fibers. An example illustrating the degree to which staging can be achieved in a vapor grown graphite fiber is shown in Fig. 3, where a Debye–Scherrer pattern for stage 2 $FeCl_3$–GIC fiber sample is shown.[7] The densitometer trace for this diffraction pattern shows some small admixture of stage 3 regions. Single fibers have been characterized for staging using high resolution x–ray techniques.

To identify the staging in a single fiber and to determine the staging variation along the length of the fiber, characterization can be carried out using either lattice imaging techniques with a high resolution Transmission Electron Microscope (TEM)[13,16] or Raman spectroscopy[17] with a Raman microprobe.[18] The Field Emission Scanning Transmission Electron Microscope (FESTEM) can be used to determine the compositional variation along the fiber length, based on the x–ray fluorescence technique.[19]

Lattice fringes for vapor grown graphite fibers heat treated to temperatures $T_{HT} \geq$ 2800°C show defect–free parallel lattice planes for distances extending at least 1000 Å (see Fig. 4). The effect of intercalation is illustrated in Fig. 5 where layer sequences with $n = 1$ and $n = 3$ are readily identified in a dominantly stage $n = 2$ $CuCl_2$–intercalated vapor grown fiber ($T_{HT} = 2900°C$).[2,13,20] This identification is made possible by the different atomic numbers and hence different in–plane electron densities in the intercalate layers relative to the graphite layers, thereby showing contrast between electrons scattered by the intercalate and by the graphite layers. These contrasts vary from one intercalant to another, thereby giving rise to characteristic visual patterns for each intercalant. It is significant that the extent of the long, defect–free regions of graphite intercalation compounds is strongly dependent on the perfection of the host crystal, the characteristics of the intercalate species and the conditions of intercalation. Commensurate GICs using intercalants such as KHg and Br_2 also tend to exhibit the largest defect–free regions in the corresponding GIC.[21]

In this context, fibers provide a convenient host material for studying c–axis structural phenomena, because the lattice imaging method can be applied generally to thin fibers (diameter $\leq 1\mu m$), without further thinning. In contrast, c–axis lattice images are more difficult to observe in bulk graphite hosts since their observation requires the

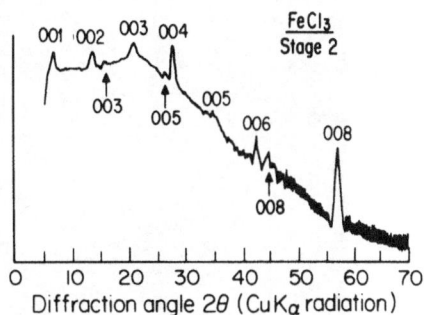

Figure 3: Debye–Scherrer x–ray diffraction pattern taken with Cu Kα radiation and the corresponding microdensitometer trace for a stage–2 $FeCl_3$–intercalated vapor grown fiber. The x–ray pattern shows characteristic lines associated with a well–staged intercalation compound. The (00ℓ) reflections are indexed, with labels above the trace for the stage 2 reflections, and below the trace for the small admixture of stage 3 regions.[7]

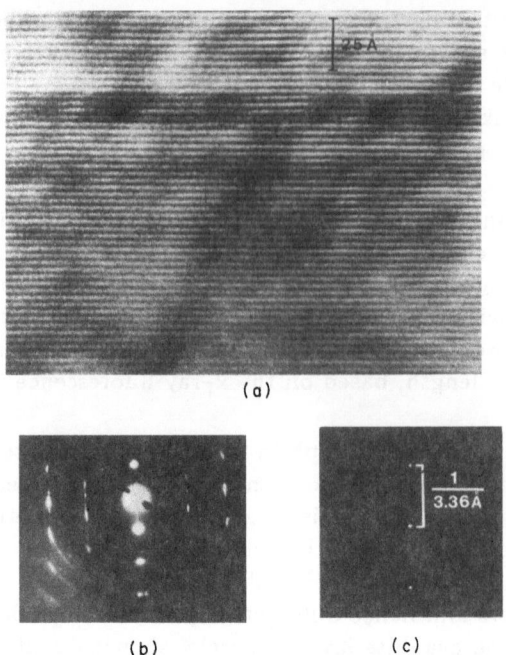

(a)

(b)   (c)

Figure 4: Lattice fringe image of a pristine vapor grown fiber heat treated to 2900°C:[13] (a) lattice fringe pattern; (b) selected area diffraction pattern corresponding to (a); and (c) optical diffractogram corresponding to (a).

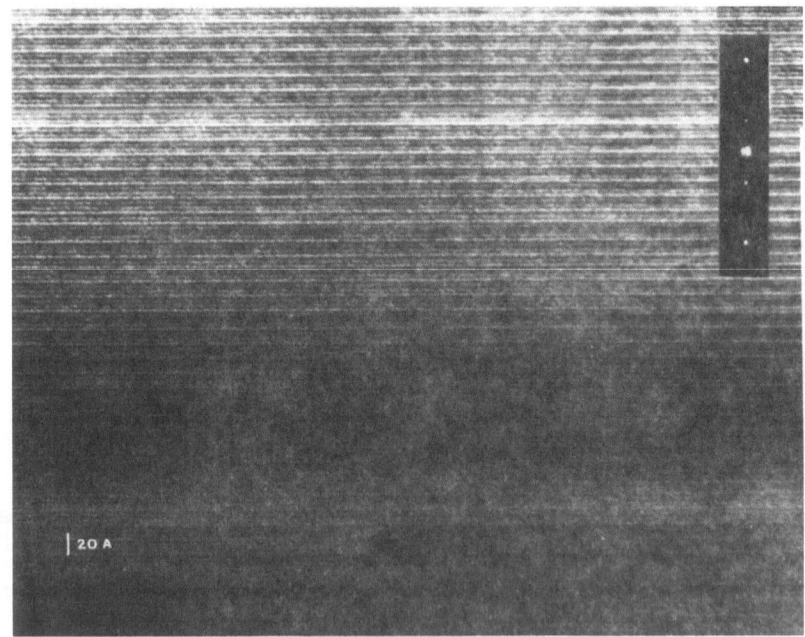

Figure 5: Lattice fringe image of a stage 2 $CuCl_2$–intercalated benzene–derived fiber ($T_{HT} = 2900°C$) and the optical diffractogram (upper right) taken using the photographic film of the lattice fringes as a diffraction grating. Note the presence of stage 1 and stage 3 sequences occurring in a nominally stage 2 fiber. Also note the high degree of parallelism of the graphite and intercalate planes over several hundred Å.[2,20]

sample edges to be turned up to achieve the proper geometry for c–axis lattice imaging. Thus graphite fibers provide a convenient host material for studying lattice damage introduced by various means, as for example, by ion implantation.[22]

The use of Raman spectroscopy for the stage characterization of intercalated graphite fibers is essentially identical to that for intercalated graphite generally.[8] Since the stage dependence of Raman–active frequencies for both the graphite interior layers $\omega(E_{2g_2}^{\circ})$ and graphite bounding layers $\omega(\hat{E}_{2g_2})$ is the same for HOPG and graphite fiber–based GICs, the measured Raman frequencies can be used to yield the stage of a given donor or acceptor fiber.[8] The relative intensities of the Raman lines associated with the graphite bounding layers and the graphite interior layers can also be used to provide information on the stage of the graphite fibers.[8,10] As the intercalate uptake increases, and the number of graphite bounding layers relative to the number of graphite interior layers increases, so does the intensity of the bounding layer mode $\hat{E}_{2g_2}$ relative to that for the interior layers $E_{2g_2}^{\circ}$ mode. We note that for stage 1 and 2 compounds, all the graphite layers are adjacent to intercalate layers, so that the intensity for the $E_{2g_2}^{\circ}$ mode due to the graphite interior layers vanishes.[8]

Narrow Raman lines occur only for well–staged samples. Thus, use of the small Raman mode frequency shifts to provide stage identification for graphite fibers requires the fibers to be well–staged. In the case of intercalated PAN and mesophase pitch fibers, the Raman spectra for the $\hat{E}_{2g_2}$ and $E_{2g_2}^{\circ}$ lines are often very broad, and only qualitative information on the staging can be obtained from the peak frequencies or the relative intensities of the $\hat{E}_{2g_2}$ and $E_{2g_2}^{\circ}$ Raman lines. For vapor grown graphite fibers heat treated to $T_{HT} > 2900°C$, well–staged fibers can be prepared, showing Raman linewidths comparable to those for HOPG intercalated with the same intercalant and to the same stage (see Fig. 6),[7] thereby allowing definitive stage characterization from the Raman frequencies for the graphite bounding layer mode.[17] This feature is utilized in the application of the Raman microprobe for the stage characterization of single graphite fibers,[18] as described below.

A Raman microprobe (beam size of $\sim 2\mu m$) can be used to determine the stage index of single fibers and the staging fidelity along the length of a fiber. Using the Raman microprobe, stage determination is achieved by measurement of both the mode frequencies and the mode intensities for the graphite bounding layers relative to those for the graphite interior layers. The stage determination of *single* intercalated graphite fibers by a non–destructive technique is important because it allows quantitative and reproducible measurements to be made of the intercalation–induced modifications to the properties of graphite fibers. To illustrate typical variations in the staging of a fiber, we show in Fig. 7 Raman microprobe spectra taken along the length of a nominally stage 3 FeCl$_3$–intercalated vapor grown graphite fiber ($T_{HT} = 2900°C$). The spectra show variations from one region to another, indicative of some admixture of stage 4 regions.[18] The variation in the spectra along an intercalated fiber (as demonstrated by the microprobe scans) tends to be significantly larger than for similar scans taken along an HOPG–based sample with the same intercalant and nominal stage.[18] The same conclusions are reached by lattice fringing studies which show individual staging defects (e.g., one sequence of stage 3 followed by several stage 2 sequences as shown in Fig. 5). In contrast, the Raman microprobe provides staging information averaged over a small

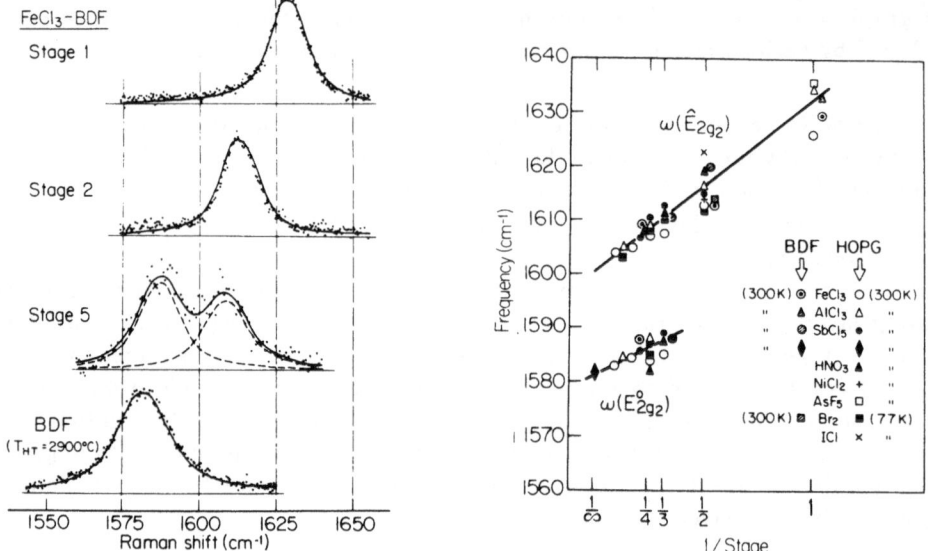

Figure 6: (a) First–order Raman spectra for several stages of $FeCl_3$–intercalated vapor grown fibers ($T_{HT} = 2900°C$). The solid lines are fits to the experimental points and determine the central frequency and linewidth of each Raman line. The spectrum for the E_{2g_2} mode of pristine graphite fibers (BDF) is also shown for comparison.[2] (b) The stage of the compound can be determined from the frequencies of the bounding and interior layer E_{2g_2} modes for acceptor GICs.[10]

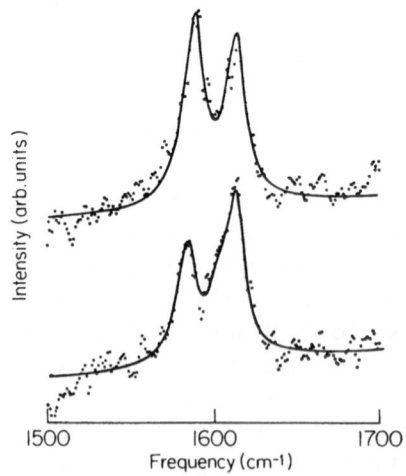

Figure 7: Raman microprobe spectra of a nominal stage 3 $FeCl_3$–intercalated vapor grown graphite fiber, showing adjacent regions with different mixtures of stages 3 and 4. The two spectra were taken from regions separated by 7.5 μm along the length of the fiber.[18]

(2 μm) distance, which nevertheless contains many unit cells. While the Raman micro-probe provides a convenient tool for rapid, non–destructive characterization of single fibers, the lattice fringing technique provides complementary microscopic information on individual defects. The lattice fringing method however cannot be used for rapid, non–destructive characterization of fibers which are to be subsequently used for other experiments.[20]

The field emission scanning transmission electron microscope (FESTEM) has also been used to monitor the homogeneity of the intercalant concentration along the fiber length. Specific studies have been made for a partially desorbed Br_2 intercalated pitch-based fiber by mapping the intensity of the Br $K\alpha_1$ x-ray line while scanning the fiber length.[19] The FESTEM technique has also been applied to study the lateral homogeneity of the bromine distribution across the fiber cross section, showing that the bromine concentration is approximately uniform along the cross section for bromine–rich regions and follows a Gaussian distribution over the bromine–poor regions. This microscopic information was then used to model the macroscopic residual resistivity of the fiber.[19]

TRANSPORT PROPERTIES

The large magnitude of the ratio of the fiber length to cross sectional area (L/A) is highly favorable for reliable and accurate transport measurements. Transport measurements on graphite and its intercalation compounds have in the past presented a host of difficulties due to the large anisotropy of the conductivity.[23] Hence graphite fibers in their pristine and intercalated forms have recently been exploited for high resolution electrical conductivity, thermal conductivity, thermopower and magnetoresistance measurements.[23] Especially interesting have been the measurements on single fibers.

In the past few years, attention has been focused on the intercalation–enhanced electrical conductivity of graphite fibers for the fabrication of practical conductors.[24,25] The fiber geometry (especially for the vapor grown fibers) offers advantages relative to HOPG for increasing the compositional stability of the GIC under ambient conditions.[13]

The conductivities achieved by the intercalation of graphite fibers are generally lower than for the same intercalant and stage in an HOPG host because of the greater structural disorder in graphite fibers, consistent with the lower conductivities of the fiber hosts prior to intercalation. However, in some cases the intercalation–enhanced conductivity of the fibers is comparable to the corresponding increase in HOPG.[14] The most extensive effort to produce a high conductivity graphite fiber was carried out with AsF_5 as the intercalant,[14] because AsF_5 intercalation yields the highest conductivity in HOPG–based GICs.[23] Several important conclusions were reached in the conductivity studies on AsF_5 intercalated into vapor grown fibers.[14] In these experiments the conductivity was monitored as the intercalation proceeded and the intercalated fibers were subsequently characterized by SEM and TEM for structural order and by electron microprobe analysis with regard to the distribution of AsF_5. These measurements confirmed that intercalation is enhanced by increased structural order, which was achieved by choice of the most graphitizable precursor fiber and heat treatment of the fiber to the maximum T_{HT} (3300°C). The experiments conclusively confirm that for these fibers intercalation starts from the fiber ends and proceeds along the fiber length. A maximum room temperature conductivity (9×10^5 S/cm) for the AsF_5 intercalated fibers was obtained (see Fig. 8), significantly above that for copper (5.9×10^5 S/cm). These high values for σ_a were obtained while the fiber was surrounded by AsF_5 gas in the intercalation ampoule. Steps in the conductivity *vs.* time curve may be identified with staging transitions. It was further found that the structural organization of the vapor grown fibers inhibited AsF_5 desorption, both in vacuum and under ambient laboratory conditions. The physical basis for this effect is the great reduction of surface area where intercalate edge planes are exposed and could lead to intercalate desorption. Similar results for the inhibition of intercalant desorption were also obtained for Br_2–intercalation of vapor grown fibers.[15] Long term stability (hundreds of days) was achieved for residue compounds of AsF_5 intercalated fibers having conductivities in excess of 10^5 S/cm, high enough for useful application as a lightweight conductor. For the highly ordered fibers, the correlation between the conductivity before and after intercalation was reproducible, as were the absolute values of the conductivity for specified conditions of intercalation.[14] Significant intercalant uptake was also observed for the best ordered mesophase pitch fibers (Union Carbide P100), though the intercalation–induced increase in conductivity was lower and the sample to sample variation was much larger than for the corresponding $CuCl_2$ intercalation into HOPG.[26] Also shown in Fig. 8 is information on the kinetics.[21] Before heat treatment of the fibers there is negligible enhanced conductivity of the fibers. When the ends of heat treated fibers are sealed, again the increase in σ is very small, consistent with a small intercalant uptake. This indicates that intercalation starts at the fiber ends, consistent with the visual appearance of the color front during the intercalation process.

Studies of the temperature dependence of the conductivity are of particular interest with regard to identification of the conduction mechanism. Low stage intercalated graphite fibers typically show a metallic temperature dependence of σ (see Fig. 9), as is found for bulk host graphite materials.[23] Values of the residual resistance ratio ($RRR = R_{300K}/R_{4.2K}$) provide a figure of merit of the defect concentration of the fibers. Intercalation generally causes the RRR to increase, consistent with the large increase in carrier concentration and smaller decrease in carrier mobility. The larger RRR values for HOPG–based GICs relative to fibers intercalated with the same intercalant to the

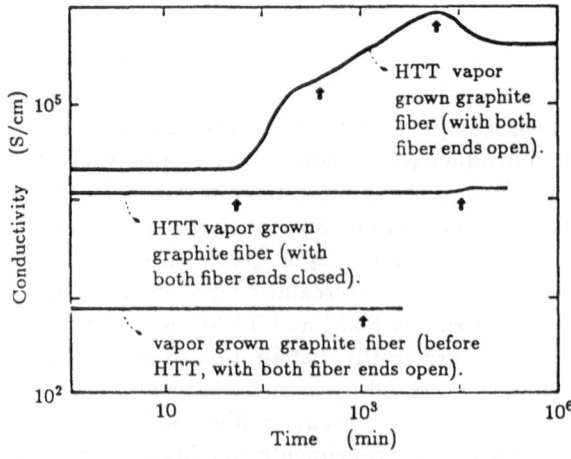

Figure 8: In–situ electrical conductivity *vs.* time for vapor–grown fibers during AsF_5 intercalation under various conditions (see text): ($\uparrow$ indicates AsF_5 recharge).[14]

same stage are consistent with the higher defect density of the fibers.

Of particular interest to the present discussion are several examples of new physics which has been learned about GICs because of the availability of intercalated graphite fibers. The first example pertains to the low temperature electrical conductivity of GICs. In general, 3D metals at low temperature follow the Bloch–Grüneisen law for the temperature dependence of the resistivity

$$\rho_{BG}(T) = A_{BG} + B_{BG}T^n \tag{1}$$

where $n \sim 5$ when only low energy phonons are available for scattering; at high temperatures where all phonons can scatter, a linear temperature dependence for $\rho(T)$ is found.[27] Now for GICs, the functional form of the temperature dependence of the resistivity has been found to be[23]

$$\rho(T) = A + BT + CT^2. \tag{2}$$

This functional form has been verified experimentally for many intercalants for both intercalated graphite fibers and for intercalated HOPG. The room temperature resistivity is predominantly due to carrier–phonon scattering, while defect scattering plays a significant role at low temperatures.

Recently, high resolution resistivity measurements, made possible by the favorable geometry of intercalated graphite fibers have been carried out at low temperatures to investigate whether a Bloch–Grüneisen effect exists for GICs at low temperatures. High resolution resistivity measurements on $CuCl_2$ intercalated vapor phase fibers down to 1.5 K, confirm the validity of Eq. (2), showing that the usual low temperature 3D Bloch–Grüneisen T^n behavior with $n \sim 5$ is not applicable to GICs.[28] It should be noted that the out–of–plane phonons for GICs are almost dispersionless and have very low energies, thus giving rise to unusual low temperature transport phenomena.

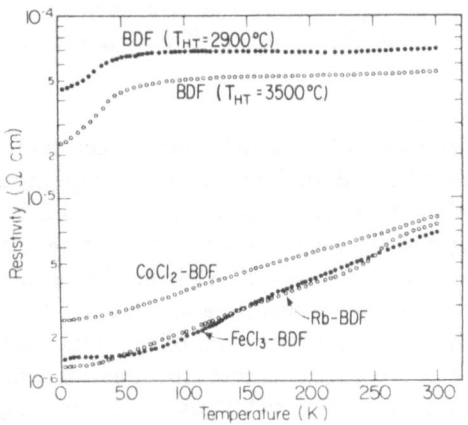

Figure 9: Temperature dependence of resistivity for pristine benzene–derived graphite fibers (BDF) heat treated to 2900 and 3500°C and for fibers ($T_{HT} = 2900°C$) intercalated with donor (desorbed Rb) and air stable acceptors (FeCl$_3$, CoCl$_2$). In all cases the resistivity decreases with decreasing temperature, and the residual resistivity ratio is significantly larger for the intercalated fibers relative to the pristine fibers.[7]

The functional form of Eq. (2) for the intercalated fibers is very different from that for the pristine fibers (see Fig. 9), which is also different from that of pristine HOPG.[2] Both the magnitude of the room temperature resistivity for fiber GICs and the temperature dependence of $\rho(T)$ depend strongly on T_{HT} for a given precursor material.[25] Size dependent resistivity measurements show that the residual resistivity increases as the fiber diameter decreases,[29] consistent with a higher fraction of the sample that is disordered with decreasing diameter.

Vapor grown fibers are of practical interest as thermal conductors, and when heat treated to 3000°C they exhibit the highest thermal conductivity (1300 W/mK) in a fibrous material.[23] Intercalation typically decreases the room temperature thermal conductivity κ by an order of magnitude while increasing the magnitude of κ at low T (electronic contribution).[30]

The availability of graphite hosts in the fibrous form has also revealed new physics regarding thermal conductivity in GICs. Because of the very high anisotropy ratio of the electrical conductivity in acceptor GICs, it is not feasible, using bulk graphite hosts, to make thermal and electrical conductivity measurements on the same physical sample.[23] For this reason the separation of the total thermal conductivity $\kappa = \kappa_e + \kappa_L$ into a lattice contribution κ_L and an electronic contribution κ_e is difficult. Such a separation was carried out approximately using high magnetic fields to exploit the high magnetoresistance of graphite, thereby greatly suppressing κ_e.[31] However, with the vapor grown fibers, it is possible to achieve sufficiently large L/A (fiber length/area) values so that 4–probe measurements can be carried out quantitatively. Indeed, the electrical and thermal conductivities σ and κ have been measured in the same acceptor–intercalated fiber, and using the Wiedemann–Franz law $\kappa_e = L_0 \sigma T$ (where the Lorenz number $L_0 = 2.44 \times 10^{-8} V^2/K^2$), it has been possible to separate κ_L and κ_e directly, yielding the results shown in Fig. 10. The high sensitivity of the measurement technique has made it possible to identify anomalously high κ_L values at low T. The physical origin of the excess low temperature κ_L is presently under investigation.[23,30] The thermopower of a single intercalated vapor grown fiber has also been measured and in this case the temperature dependence of the thermopower (TEP) is very similar to that for the corresponding bulk GICs.[30]

For practical applications of intercalated fibers as conductors, thermal and mechanical stability is of importance. Significant intercalation–induced decreases in tensile strength along the fiber length were found for intercalated PAN and mesophase pitch fibers. The tensile strength is the tensile force that must be applied to rupture the fiber. A decrease in tensile strength by a factor of ~ 2 from the value for the pristine fibers was reported for fibers intercalated with alkali metals to stage 2, and by a factor between 3 and 9 for stage 1.[32] With regard to thermal stability, some intercalated fibers are stable in an argon atmosphere above room temperature, as for example up to ~ 600 K for FeCl$_3$–intercalated vapor grown fibers.[2,13] Thus, both the magnitude of the room temperature electrical conductivity and the thermal stability of intercalated graphite fibers offer promise for utilization as lightweight practical conductors. In this connection, it is interesting to note that intercalated graphite fibers have been successfully applied to improve the performance of thermocells based on nitric acid and bromine intercalated electrodes.[33] For such applications, study of the kinetics of intercalation[21] into graphite fibers is of special interest. Other applications of continuous and chopped intercalated graphite fibers to the microelectronics industry can also be expected, both as separated fibers or in various types of composites.

Magnetoresistance measurements on intercalated graphite fibers are of particular interest because of the information they provide on the temperature dependence of the carrier mobility and because of their favorable geometry for magnetoresistance measurements.[34] Since the magnetoresistance of graphites and graphite fibers yields values for an average carrier mobility μ through the relation $\Delta\rho/\rho(0) = (\mu H)^2$, measurements of $\Delta\rho/\rho(0) = (\rho(H) - \rho(0))/\rho(0)$ have been used in conjunction with zero field resistivity $\rho(0)$ measurements to estimate the temperature dependence of the carrier density for the intercalated fibers. Quantitative interpretations of the measurements are not possible because of the absence of band models for the electron and phonon states. The magnetoresistance of the intercalated fibers shows a quadratic field dependence at very low fields[2] and a linear field dependence at intermediate field values.[7] The significant decrease in carrier mobility associated with intercalation is reflected in the order of magnitude decrease in the magnetoresistance of the intercalated graphite fibers.[7] Intercalation reduces the carrier mobility by increasing the dimensions of the Fermi surface and hence increasing the carrier effective masses, while at the same time introducing additional intercalation–induced scattering centers.

Intercalation of well ordered graphite fibers exhibit sufficiently high carrier mobilities to satisfy the condition for the observation of quantum oscillatory effects such as the Shubnikov–de Haas effect $\omega_c \tau \gg 1$, so that the carriers at liquid helium temperatures complete a cyclotron orbit prior to scattering.[7,32] Because of the fiber geometry, there is no easy way to apply a magnetic field along a single crystallographic direction, except along the fiber axis (the a–axis). Because of the cylindrical nature of the π–electron

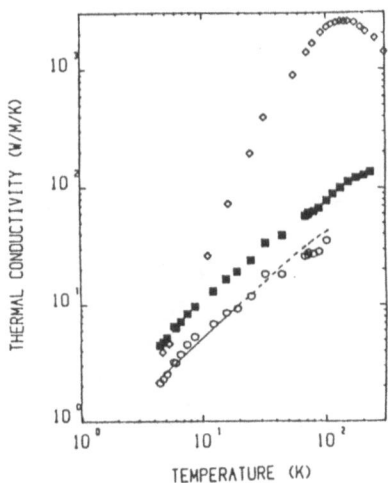

Figure 10: Temperature dependence of the thermal conductivity of $CuCl_2$–intercalated vapor grown fibers compared to that of the pristine fibers (◊) heat–treated at the same temperature (3000°C). The total measured thermal conductivity (■) of the intercalated fiber is separated into its electronic (dashed curve) and lattice (○) contributions (see text).[30]

Fermi surfaces, the most interesting field direction is along the c–axis, which unfortunately is not along a single spatial direction in a fiber. Therefore, the fiber geometry is unfavorable for quantitative interpretation of the observed Shubnikov–de Haas periodicities in terms of band structure models.

Graphite fiber hosts also allow study of 2D weak localization and electron–electron interaction effects in low stage acceptor GICs. Intercalation with acceptors increases the electrical anisotropy by 2 or 3 orders of magnitude, making the transport behavior in acceptor GICs more 2D than in graphite itself. By choosing different types of fiber hosts and different T_{HT}, the amount of disorder in the fibers can be varied so that the amount of localization can be controlled. Control of the Fermi level can be achieved by choice of the intercalant and stage. Such localization and electron–electron interaction effects have recently been observed using high resolution temperature dependent electrical resistivity measurements on low stage acceptor GICs in partly disordered fiber host materials. Both localization and interaction effects appear as a logarithmic increase in the resistivity with decreasing T at very low temperature (see Fig. 11). The magnitude of the localization effect increases with the residual resistivity of the intercalated fiber and is directly correlated with the magnitude of the negative magnetoresistance of the acceptor GIC fibers.[23,35] Magnetoresistance measurements are expected to distinguish between weak localization and electron–electron interaction effects. Negative magnetoresistance phenomena are also observed in disordered non–intercalated graphite fibers, where the phenomena were ascribed to a field–induced (linear) increase in carrier density for a 2D system in the low quantum number limit.

Study of the properties of intercalated graphite fibers is currently a very active research field, both to explore new physical phenomena and for new applications opportunities.

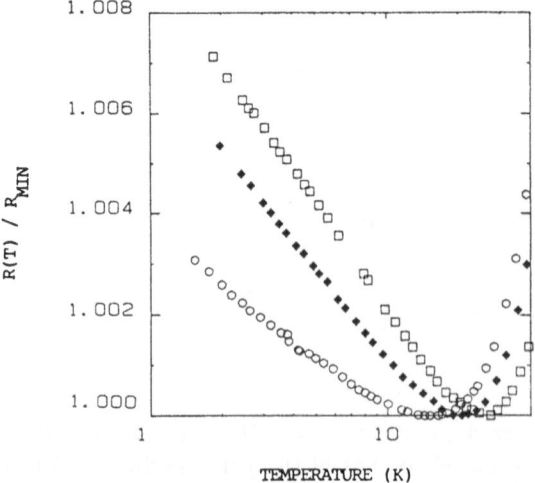

Figure 11: Logarithmic temperature dependence of the resistance at low temperature for three pitch based graphite fiber samples intercalated with CuCl$_2$ and CoCl$_2$, showing 2D localization and electron–electron interaction phenomena.[23,35]

ACKNOWLEDGMENTS

The author gratefully acknowledges years of collaborative efforts on this work with Professor M. Endo at Shinshu University, Japan, with Professor J–P. Issi at the University of Louvain–la Neuve, Belgium, and with members of the research group, Drs. K. Sugihara, T.C. Chieu, G. Dresselhaus, P. Kwizera, L. McNeil and L. Salamanca–Riba. Support by the Advanced Research Projects Agency of the Department of Defense monitored by AFOSR under contract #F49620–85–C–0147 is gratefully acknowledged.

REFERENCES

1. A.W. Moore, *Chemistry and Physics of Carbon*, **11**, ed. P.L. Walker and P.A. Thrower (New York, Dekker), p. 69.

2. M.S. Dresselhaus, *J. de Chim. Phys.* **81**, 739 (1984).

3. I.L. Spain, K.J. Volin, H.A. Goldberg, and I. Kalnin, *J. Phys. Chem. Solids* **44**, 839 (1983).

4. M. Endo, T. Koyama, and Y. Hishiyama, *J. Appl. Phys. Jpn.* **11**, 2073 (1976).

5. M. Endo and M. Shikata, *Ohyobutsuri* **54**, 507 (1985); M. Endo and H. Ueno, *Extended Abstract on Graphite Intercalation Compounds*, Materials Research Society, edited by P.C. Eklund, M.S. Dresselhaus, and G. Dresselhaus, (1984), p. 177.

6. J.A. Floro, S.M. Rossnagel, and R.S. Robinson, *J. Vac. Sci. Technol.* **A1**, 1398 (1983).

7. T.C. Chieu, G. Timp, M.S. Dresselhaus, M. Endo, and A.W. Moore, *Phys. Rev.* **B27**, 3686 (1983).

8. P.C. Eklund, (this volume) p. 323.

9. S.A. Solin, (this volume) p. 173.

10. M.S. Dresselhaus and G. Dresselhaus, Light–Scattering in Solids III, Vol. 51 of *Topics in Applied Physics*, edited by M. Cardona and G. Güntherodt (Springer, Berlin, 1982), p. 3.

11. P.C. Eklund, (this volume) p. 163.

12. G. Hooley, in "Preparation and Crystal Growth of Materials with Layered Structures", edited by R.M.A. Leith (Dordrecht: Reidel) p. 1 (1977).

13. M. Endo, T.C. Chieu, G. Timp, M.S. Dresselhaus, and B.S. Elman, *Phys. Rev.* **B28**, 6982 (1983).

14. J. Shioya, H. Matsubara, and S. Murakami, *Synthetic Metals* **14**, 113 (1986).

15. J.R. Gaier, NASA Technical Memorandum 87275.

16. M.S. Dresselhaus, (this volume) p. 213.

17. S.A. Solin, (this volume) p. 313.

18. L. McNeil, J. Steinbeck, L. Salamanca–Riba, and G. Dresselhaus, *Carbon* **24**, 73 (1986).

19. X.W. Qian, S.A. Solin, and J.R. Gaier, (this volume) p. 477.

20. M. Endo, T.C. Chieu, G. Timp, M.S. Dresselhaus, and B.S. Elman, *Synthetic Metals* **8**, 251 (1983).

21. G. Timp and M.S. Dresselhaus, *J. Phys. C* **17**, 2641 (1984).

22. L. Salamanca–Riba, G. Braunstein, M.S. Dresselhaus, J.M. Gibson and M. Endo, *Nucl. Instr. Meth. Phys. Res.* **B7/8**, 487 (1985); M. Endo, L. Salamanca–Riba, G. Dresselhaus, and J.M. Gibson *Chimie Physique* **8**, 803 (1984).

23. J.–P. Issi, (this volume) p. 347.

24. I.L. Kalnin and H.A. Goldberg, *Synthetic Metals* **8**, 277 (1983).

25. C. Manini, J.–F. Marêché, and E. McRae, *Synthetic Metals* **8**, 261 (1983).

26. D.A. Jaworske and J.D. Miller, NASA Technical Memorandum 87217.

27. C. Kittel, *Introduction to Solid State Physics*, Sixth Edition (J. Wiley & Sons, New York) (1985).

28. L. Piraux, J.–P. Issi, L. Salamanca–Riba, and M.S. Dresselhaus, *Synthetic Metals* (in press).

29. M.Z. Tahar, M.S. Dresselhaus, and M. Endo, *Carbon* **24**, 67 (1986).

30. L. Piraux, B. Nysten, J.-P. Issi, L. Salamanca-Riba, and M.S. Dresselhaus, *Solid State Commun.* **58**, 265 (1986).

31. J. Heremans, M. Shayegan, M.S. Dresselhaus, and J.–P. Issi, *Phys. Rev.* **B26**, 3338 (1982); J.–P. Issi, J. Heremans, and M.S. Dresselhaus, *Phys. Rev.* **B27**, 1333 (1983).

32. V. Natarajan, J.A. Woollam, and A. Yavrouian, *Synthetic Metals* **8**, 291 (1983).

33. M. Endo, Y. Yamagishi, and M. Inagaki, *Synthetic Metals* **7**, 203 (1983); Y. Maeda, H. Kitamura, E. Itoh, and M. Inagaki, *Synthetic Metals* **7**, 211 (1983).

34. L.D. Woolf, J. Chin, Y.R. Lin–Liu, and H. Ikezi, *Phys. Rev.* **B30**, 861 (1984).

35. L. Piraux, V. Bayot, J.-P. Michenaud, J.-P. Issi, J.F. Marêché, and E. McRae, *Solid State Commun.* (in press); L. Piraux, (this volume) p. 375.

A FIELD EMISSION STEM STUDY OF THE Br DISTRIBUTION IN BROMINATED GRAPHITE FIBERS

X.W. Qian* and S.A. Solin

Department of Physics and Astronomy
Michigan State University
East Lansing, Michigan 48824-1116

J.R. Gaier

NASA Lewis Research Center
21000 Brook Park Road
Cleveland, OH   44135

Brominated pitch based graphite fibers exhibit certain novel properties and potential applications [1,2]. For example, it has been observed that an 18 fold reduction of their resistivities of the pitch-based graphite fibers can be achieved if the pristine fibers are allowed to fully interact with bromine [3] and even the residual fiber compounds that are formed by pumping off the bromine from the fully brominated fibers (saturated fibers), exhibit a five-fold reduction in the resistivities relative to that of the pristine fibers [3]. To understand this property, it is essential to know the Br distribution in such fiber compounds. Here we focus on the residual fibers that have a density of 2.2 gm/cm^3 corresponding to a composition of BrC_{45}.

A FESTEM (Field Emission Scanning Transmission Electron Microscope) was used to study the Br distribution in the residual form of the fiber. The fiber under examination was bombarded by a well focused 100 KeV electron beam (beam diameter ~50Å) in such way that the direction of the beam was perpendicular to the longitudinal axis of the fiber. The high energy electrons penetrate into the material, excite the Br atoms in their pathway which in turn give off the characteristic K_α X-rays. A EDX (Energy Dispersive X-ray) detector collects the X-rays. When the electron beam is digitally controlled to scan across the fiber or along the longitudinal axis of the fiber, the cross-sectional or longitudinal X-ray intensity profile can be recorded. These X-ray intensity profiles together with a knowledge of the electron path inside the fiber enabled us to deduce the Br distributions in the fiber.

The details of the incident electron path are established by computer simulation. As the electrons penetrate into the material, they strongly interact with the atoms in the material. They lose energy and are scattered away from the original direction. Consequently, the electron beam diameter and the energy vary with the penetration depth [4]. A Monte Carlo simulation [5] shows that over the range of fiber size (~10 μm), the beam diameter is related to the depth through a power law with an exponent

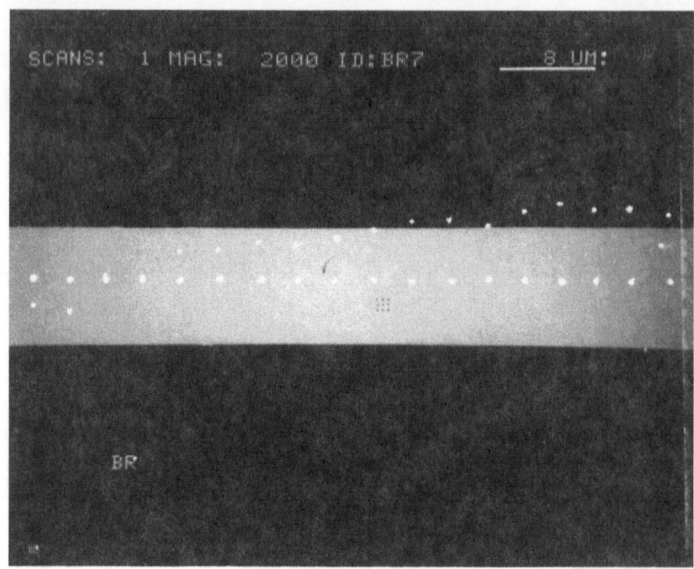

Fig. 1. The Br K$_\alpha$ X-ray intensity variation (curved line of dots) along a PB fiber superposed on the bright field image of the fiber the axis of which is perpendicular to the incident electron beam. The straight dotted line denotes the position of the electron beam as it is digitally scanned along the longitudinal axis if the fiber.

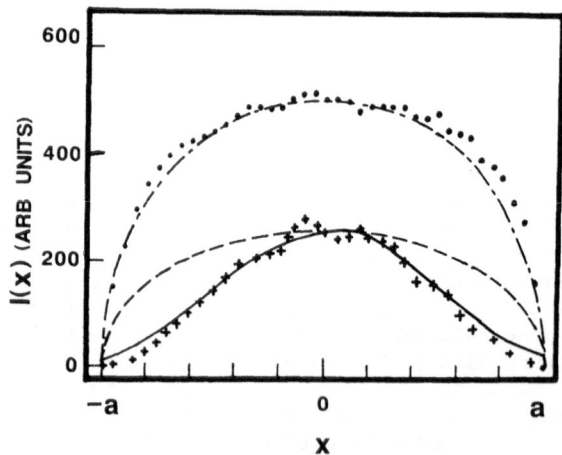

Fig. 2. A plot of Br K$_\alpha$ X-ray intensity vs. the lateral position (x) of the incident electron beam. The dots (·) denote the experimental data for the Br-rich region, and the dashed line is the calculated intensity from a radially uniform Br distribution. The crosses (+) represent the experimental data points in the Br-poor region, and the solid line is the calculated intensity from a Gaussian Br distribution, $\rho_{Br}(r)=\rho_0 e^{-\dfrac{r^2}{0.45a^2}}$, where a is the radius of the fiber and r is the distance from the axis of the fiber. A scaled intensity calculated from a uniform Br-distribution in the Br-poor region is also shown as a broken line.

of 1.5, and the energy of the electrons varies linearly with depth with a rate -0.6 KeV/μm. At 10 μm depth the beam is broadened to 3 μm and the energy is reduced to 94 KeV. Once the electron paths and energies inside the fiber are known, the X-ray intensity profile generated from an assumed bromine distribution can be calculated directly. By comparing the calculated profile to the experimental profile we can deduce the Br distribution in the fiber.

Experimentally we find an inhomogeneous Br distribution along the longitudinal axis of the fiber (see Fig. 1) with two distinct cross-sectional distributions. i.e. a uniform distribution in the Br-rich region and a Gaussian distribution in the Br-poor region (See fig. 2). It is worth noting that a depletion layer of ~1 μm is found on the surface of the Br-poor regions. This is attributed to the porous like structure of the fiber near the surface.

ACKNOWLEDGEMENTS

We thank D.M. Hwang, J.R. Gaier, and D.A. Joworski for their useful discussions. This work is supported by the National Aeronautics and Space Administration under grant NAG-3-595.

*Exxon Fellow

REFERENCES

1. J.G. Hooley and V.R. Diets, Carbon 16, 251(1978)
2. J.R. Gaier and D. Marino, NASA TM-87016,1985.
3. J.R. Gaier, NASA TM-87275,1986.
4. J.I. Goldstein in Introduction to Analytical Electron Microscopy, edited by J.J. Hern, J.I. Goldstein and D.C. Joy (Plenum Press, New York and London, 1979).
5. X.W. Qian, S.A. Solin, and J.R. Gaier, Phys. Rev., submitted.

Contributors

Beguin, F.   457
Behrens, P.   229
Brec, R.   75, 93, 125

Chaiken, A.   387
Chow, P.   303

Davidov, D.   433
Delmas, C.   155
Doll, G.L.   257, 311
Dresselhaus, G.   273, 387, 407, 425, 429
Dresselhaus, M.S.   1, 213, 233, 407, 461

Eklund, P.C.   163, 257, 311, 325, 337

Fan, Y.B.   307

Gaier, J.R.   477
Gao, G,Y.   143
Ghorayeb, A.M.   135

Hao, X.   233
Hatzikraniotis, E.   161
Heremans, J.   381
Hérold, A.   377
Huang, Y.Y.   381

Issi, J–P.   347

Julien, C.   159

Laguës, M.   273
Legrand, A.P.   457
Liang, W.Y.   31, 135, 139, 143

Marêché, J.-F.   377
McRae, E.   377
Metz, W.   229
Moret, R.   185

Neumann, D.   307
Nicholls, J.T.   425, 429

Ohana, I.   345

Palchan, I.   385
Paraskevopoulos, K.M.   161
Piraux, L.   375

Qian, X.W.   381, 477
Quinton, M.F.   457

Rigaux, C.   235
Rouxel, J.   75
Rush, J.J.   307

Selig, H.   433
Shen, T.H.   139
Solin, S.A.   145, 173, 293, 307, 315,
          381, 477
Speck, J.S.   213, 233
Sugihara, K.   429

Talianker, M.   385
Tibbets, G.G.   381

Yacoby, Y.   345
Yang, M.H.   257
Yeh, N.C.   429
Yoffe, A.D.   135, 139, 143

Zabel, H.   303, 307

Subject Index

Acceptor compounds, 25
  graphite intercalation compounds
    structure (see also Structure), 203–
      208, 217, 218
    intercalants (see Intercalation)
    synthesis, 171, 229
    metal ditellurides, 33
    $PtSe_2$, 39, 143, 144
Activation energy, 222
Alkali metals
  distillation, 168
  in dichalcogenides, 32, 54, 79, 155
  in graphite intercalation compounds
    (see Intercalation)
Ammonia, 62, 180–182, 294, 295, 305–
    307, 381–383
Angle resolved photoemission spectroscopy
    (ARPES) (see also Photoemis-
    sion), 144, 279–282
Angular correlation, 88
Anisotropy (see also Electrical conduc-
    tivity, Magnetism, Phonons, Su-
    perconductivity, Transport prop-
    erties), 27, 31, 40, 42,45
Annealing, 88, 461–463
Antiferromagnetic order (see Magnetism)
Auger spectroscopy (see also Electron spec-
    troscopy), 271, 276–279
  Lander model, 277–279
  line shape, 277

Band filling (see also Electronic struc-
    ture), 32, 36
Band gap tuning, 17
Band models (see Electronic structure)
Band offsets, 9
Band structure (see Electronic structure)
Basal spacing (see also Structure), 151,
    296

Batteries (see also Electrochemical Inter-
    calation), 54, 113, 119, 121,
    158, 161, 167, 338
  discharge current, 119
  electrochemical yield, 119
  Li/TiS_2 cell, 54
  overcharging, 340
  oxidation number, 340
BCS theory of superconductivity (see also
    Superconductivity), 45, 390, 391,
    395, 397
Bi–intercalation compounds (see Inter-
    calation)
Biphased systems, 84
Bond length (see also Structure)
  dichalcogenides, 98, 101
  graphite intercalation compound ac-
    ceptors, 188, 189, 340, 341
  graphite intercalation compound donors,
    188, 189, 414
Bound states, 9–12
Bragg law (see X–ray diffraction)

c/a ratio (see also Structure), 56, 62, 88,
    127
Carbon (see also Graphite)
  diamond, 364
Carrier density (see also Charge trans-
    fer, Conductivity, Transport prop-
    erties), 41, 246–248, 260
Cations
  mobile, 75, 99, 125
Charge density waves (CDW), 42, 43, 59,
    64, 68, 135, 398
  commensurate (CCDW), 46
  incommensurate (ICDW), 46
  sliding, 46
  transition temperatures, 52